ALSO BY DANIEL J. KEVLES

The Physicists:
The History of a Scientific Community
in Modern America (1978)

IN THE NAME
OF EUGENICS

IN THE NAME OF EUGENICS

Genetics and the Uses of Human Heredity

by Daniel J. Kevles

UNIVERSITY OF CALIFORNIA PRESS
BERKELEY AND LOS ANGELES

University of California Press
Berkeley and Los Angeles

Copyright © 1985 by Daniel J. Kevles

First California Paperback 1986
Published by arrangement with Alfred A. Knopf, Inc.

Portions of this work originally appeared in *The New Yorker.*

Library of Congress Cataloging in Publication Data

Kevles, Daniel J.
In the name of eugenics: genetics and the uses of human heredity.

Includes index.
1. Eugenics—History. 2. Heredity, human.
3. Race discrimination. 4. Genetic engineering.
I. Title.
HQ751.K48   1985   304.5   84-47810
ISBN 0-520-05763-5

Manufactured in the United States of America

3   4   5   6   7   8   9   10

For my mother and father,
with thanks

CONTENTS

PREFACE

THE WORD "eugenics" was coined in 1883 by the English scientist Francis Galton, a cousin of Charles Darwin. Galton, who pioneered the mathematical treatment of heredity, took the word from a Greek root meaning "good in birth" or "noble in heredity." He intended it to denote the "science" of improving human stock by giving "the more suitable races or strains of blood a better chance of prevailing speedily over the less suitable."[1] Since Galton's day, "eugenics" has become a word of ugly connotations—and deservedly. In the first half of the twentieth century, eugenic aims merged with misinterpretations of the new science of genetics to help produce cruelly oppressive and, in the era of the Nazis, barbarous social results. Nonetheless, in recent years, Galtonian premises have continued to figure in social discourse—notably in the claims of those arguing for a racial basis of intelligence, in certain tenets of human sociobiology, and in some proposals for human genetic engineering.

I was led to write this history of eugenics partly by the recognition that the subject casts a shadow over all contemporary discourse concerning human genetic manipulation. The history of modern physics (a field in which I have previously worked) reveals how unprepared we were to deal with the momentous issues that the release of nuclear energy—a feat requiring only a few years of concentrated effort—suddenly compelled us to confront in 1945. In 1963 the great British biologist J. B. S. Haldane declared that the genetic modification of man was likely to be still millennia away, but he added: "I remember that in 1935 I regarded nuclear energy as an improbable source of power."[2] Acquisition of the knowledge and techniques for human genetic intervention would pose challenges which, while different in kind from those of the nuclear revolution, may be comparable in magnitude, and it is none too soon to examine them in historical context.

I was also convinced that eugenics held a rich variety of opportunities for historical investigation as such. There have been a number of important studies of the subject, but most have dealt with it in only one country or another, tended to view it through the lens of the Holocaust, and halted the story in the early 1930s. I have made this book a comparative history of eugenics in the United States and Britain from the late nineteenth century to the present day, giving attention to its expressions elsewhere, especially in Germany, insofar as they affected Anglo-American developments. The comparative approach has helped to explain certain important features of this history—for example, why a eugenic legislative program succeeded at least partially in the United States but not at all in Britain—that would otherwise have remained puzzling. I have also attempted a critical assessment of Anglo-American eugenicists as they diversely recognized themselves before the Nazis came to power; and the assessment has led me to depart from prevailing interpretations to advance the view instead that eugenics involved not only scientific rationalizations of class and race prejudice but a good deal more, including disputes over how men and, especially, women of the modern era were to accommodate to changing standards of sexual and reproductive behavior.

So much was said and done in the name of eugenics that this book of necessity merges history of science with social, cultural, and political history. It explores the interplay between, on the one hand, the social assertions made by eugenicists and, on the other, advances in pertinent sciences, particularly genetics in relation to man. Since about 1930, that interplay has been strongly affected by research in human genetics. I have here ventured the first historical account of the development of that field through the early sixties, and I have also sketched its remarkable progress since then, not to provide a comprehensive handbook of its specialties—the contemporary state of gene therapy, say—but to deal with such topics in a way that is indicative of emerging problems and possibilities.

This book is thus not an up-to-the-minute technical guide, and it is certainly not a tract for the times. I am under no delusion that a history of eugenics will provide any detailed moral or political map to follow in the uncharted territory of human genetic engineering. What I do expect from such an exploration is at least some assistance in disentangling the benefits we might aim for from the pitfalls we might legitimately fear. I hope that this historical journey will suggest to the reader—as it has to me—how one might think about the human genetic future, and how one might thread a path into it of good sense, reason, and decency.

D.J.K.
Pasadena, California
December 1984

IN THE NAME
OF EUGENICS

soon elected him to its fellowship.[15] Some thirty years later, Galton recorded in an autobiographical fragment that Arnaud's admonition to cross the desert to Khartoum had marked "a division of the ways in my subsequent life."[16]

Not long after his return from southern Africa, Galton met and married an intellectually able young woman named Louisa Butler, a daughter of the longtime headmaster of Harrow and then Dean of Peterborough Cathedral. In 1857, the Galtons settled into a handsome Georgian house in Rutland Gate, off Hyde Park. "Certainly we led a life that many in our social rank might envy," Galton remembered. "Among our friends were not a few notable persons, a full half of whom were first known to me through the connections of my wife." The friendship of many of the others —including Herbert Spencer and Thomas Henry Huxley—reflected Galton's increasingly eminent scientific position. Because of his geographical exploits, he had already been, to his special pleasure, taken into the Athenaeum Club, to which members were ordinarily admitted only after many years of waiting.[17]

Like other Victorian scientists, Galton gave lectures and wrote books for the general public—"Take great pains to describe the subject in tersely forcible language," he once advised a young scientist—and achieved a wide audience with a book on his adventures in southern Africa; another, *The Art of Travel*, rapidly went through five editions.[18] The writings, both popular and scientific, reveal a keen, sometimes eccentric curiosity and sharp powers of observation. At Epsom on Derby Day, Galton scrutinized through an opera glass the "sheet of faces" in the stands opposite, thinking "what a capital idea it afforded of the average tint of the complexion of the British upper classes." He reported to *Nature* that after the horses thundered past, the sheet of faces was "uniformly suffused with a strong pink tint, just as though a sun-set glow had fallen upon it." From Africa, he had informed his oldest brother, Darwin, with obvious relish, that Hottentot women were "endowed with that shape which European milliners so vainly attempt to imitate," adding that "I have seen figures that would drive the females of our native land desperate—figures that could afford to scoff at Crinoline." Unwilling to ask the women for permission to measure what bountiful nature had supplied, Galton sat at a distance with his sextant and "as the ladies turned themselves about, as women always do, to be admired, I surveyed them in every way and subsequently measured the distance of the spot where they stood—worked out and tabulated the results at my leisure."[19]

Galton often said, "Whenever you can, count." The kind of observation he liked best was numerical. Phrenological measurements fascinated him. Although he came to disbelieve the phrenological claim that bumps in the head expressed individual character, he always marveled over the large skulls of many men whom he admired—they included the physicists Lord Rayleigh and Lord Kelvin as well as the mathematician James J. Sylvester—and was puzzled that ability could not be shown mathematically to correlate with head size. Galton rarely took a walk or attended a meeting without counting something, even if it was merely the frequency of fidgets among the audience—which he found inversely related to the degree of audience attentiveness.[20] He had derived particular pleasure from his determinations of latitude and longitude in southern Africa, and at a meeting of the Royal Geographical Society he attacked the explorer Henry M. Stanley for regaling the audience with stories of his adventures on a trek to Lake Tanganyika instead of supplying hard facts. A colleague in the Geographical Society once described Galton's mind as "mathematical and statistical with little or no imagination," and characterized him as "a doctrinaire not endowed with much sympathy."[21] Nevertheless, Galton displayed rich imagination in the adaptation of numerical techniques to scientific subjects, among them the newly developing science of meteorology. In the eighteen-sixties, he published what were probably the first British weather maps. Later in the century, he attempted numerical analysis of fingerprint configurations, became a pioneer in the cataloguing of fingerprints, and campaigned to make them part of the British system of criminal identification.[22] But his propensity for counting and tabulation worked to greatest scientific advantage in his studies of inheritance.

WHY GALTON TURNED to the eugenic analysis of heredity is not at all clear. In *Memories of My Life*, he remarked that the publication of the *Origin of Species*, in 1859, had helped stimulate his thinking along these lines, and so had certain ethnological investigations he had undertaken.[23] But the theory of evolution by natural selection hardly leads directly to research in the heredity of mental characteristics, and Galton was at best vague about the ethnological inquiries. Indeed, though his African travels had confirmed his standard views of "inferior races," racial differences occupied only a minuscule fraction of his writings on human heredity.[24] More influential, perhaps, was an unspoken desire to assert, against lingering self-doubt, the validity of his own success by discovering the origins of success in lines of descent. Moreover, now that he had arrived he may have felt an impulse to social meliorism not atypical among the scions of wealthy, onetime religiously dissenting families.

Like many social improvers a generation or more removed from the manufacturing source of their incomes, Galton had no particular respect for barons of industry; his analyses of ability omitted achievement in commerce or business. He also thought the hereditary aristocracy a "disastrous institution" for "our valuable races"; the younger sons of the peerage, unable to afford a family and simultaneously maintain their position, inclined either not to marry at all or to wed heiresses, who were likely to come from families that were not notably prolific. Hardly a liberal, he did not believe in natural equality; he held that people deserved equal protection but not equal political rights, and he considered mass man the prey of the demagogue.[25] Through emigration, England happily lost "turbulent radicals and the like." (No wonder, Galton wrote, that Americans were "enterprising, defiant, and touchy; impatient of authority; furious politicians; very tolerant of fraud and violence; possessing much high and generous spirit, and some true religious feeling, but strongly addicted to cant.")[26] For Galton the scientist, the professional classes were the prime repository of ability and civic virtue, and his eugenics made them the keystone of a biological program designed to lead to the creation of a conservative meritocracy.

Another factor in Galton's turn to eugenics and heredity may have been the increasingly probable infertility of his marriage. Certainly he took pains to assert the manhood (as Victorians understood the term) of intellectuals. Galton himself was physically powerful and endowed with remarkable endurance; he argued that intellectual capacity was ordinarily associated not with men of "puny frames and small physical strength" but with "vigorous animals . . . exuberant powers." (Had not Queen Elizabeth cast "an eye to the calves . . . of those she selected for bishops?") He insisted that there was "no reason to suppose that, in breeding for the highest order of intellect, we should produce a sterile or a feeble race." He attacked Malthus's preaching of restraint in procreation, on the ground that it would lead to a "pernicious" decline in the numbers of the prudent, abler classes. Galton may well have diverted frustration over his own lack of children into an obsession with the eugenic propagation of Galton-like offspring.[27]

Emotionally, Galton seems never to have been entirely at peace. He was continually plagued by varying degrees of nervous breakdown, including giddiness, dizziness, and palpitations, though he displayed no such symptoms of anxiety in the face of physical danger. On the contrary, during his African travels, his confrontation with the Namaquan chief, and a steamer accident on the Thames that carried him downstream underwater for some two hundred yards, his behavior indicated cool presence of mind.[28] The initial breakdown, at Cambridge, was brought on by his failure to score a first—he ranked a high second—in the intense mathematical competition. (It ought to be discontinued, Galton told his father, because "the satisfaction enjoyed by the gainers is very far from counterbalancing the pain it produces among others.") The later breakdowns were caused by intense absorption in the hard work of learning, in which his family had expected him to excel, and in which—as his friend, acolyte, and biographer Karl Pearson once observed—he was strongly apt to feel himself inferior.[29]

The division in Galton's life marked by the crossing of the desert to Khartoum thus takes on a metaphoric meaning. As he had chosen to explore an uncharted region, so he selected arenas of science without competitors. Although Galton resembled the typical scientific amateur of the nineteenth century in that he was untrained in the research he eventually pursued, he was atypically drawn throughout his scientific career to largely un-populated fields, which in his day included both statistics and studies in human heredity. If at times he embarked on a subject to which others had contributed, he did not begin his research by analyzing the existing body of scientific literature; his library contained hardly two dozen volumes acquired to forward his various inquiries. He learned from the work of others but did not approach it systematically. He came upon useful treatises by chance or sought them out as he happened to need them.[30] Save for a brief debate with Darwin regarding evolutionary mechanisms, he took no part in the late-nineteenth-century disputes on issues related to the theory of evolution. He was a rough-cut genius, a pioneer who moved from one new field to the next, applying methods developed in one to problems in another, often without rigor yet usually with striking effectiveness. Gal-ton's innovativeness in science was intimately bound to his relative intellec-tual solitude—a propensity that arose from a measure of doubt in his abili-ties combined with a compulsion to excel.

Galton once remarked, in a study of English scientists, that "men who leave their mark on the world are very often those who, being gifted and full of nervous power, are at the same time haunted and driven by a dominant idea, and are therefore within a measurable distance of insan-ity."[31] Yet what Galton perceived about others he declined to confront in himself. Neither in his autobiography nor anywhere else did he attempt to puzzle out why immersion in work should have caused him break-downs. Nervous breakdowns were by no means uncommon among nine-teenth-century intellectuals. John Stuart Mill's led him to the introspec-tive conclusion that "the habit of analysis has a tendency to wear away the feelings," and he resolved to give proper place to "the internal culture of the individual." Galton merely reported that a period of rest would cure the affliction, and he diminished its significance by proposing a strong similarity between "a sprained brain and a sprained joint." In gen-eral, Galton seems not to have been given to self-analysis. He remained forever reticent about the details of his personal life.[32] An account of domestic matters would interest no one, he noted in his autobiography, shrouding his wife's mixture of genuine illness and hypochondria and also her discontent with his deep absorption in scientific matters.[33]

Galton also neglected to reveal what contributed to his ideas about religion, a subject that preoccupied many mid-Victorians. His religious attitudes ranged from skepticism to hostility. While he tolerated Louisa's practice of religion in the home, he rarely missed an opportunity to gibe at the clerical outlook. He once tested the efficacy of prayer by investigating whether or not groups for whom people prayed a good deal—for example, members of the royal family—outlived others, and he embarrassed his family by publishing the conclusion that since they did not, prayer must be inefficacious. He indicted the Roman Church for its insistence upon celibacy for clerics and the Anglican Church for its strictures against marriage for Oxbridge dons, because these measures diminished the propagation of the intellectually able.[34] In part, Galton's religious dissent exemplified the pro-scientific rebellion of the day against religious dogmatism, which, in Galton's words, "crushed the inquiring spirit, the love of observation, the pursuit of inductive studies, the habit of independent thought." In part, his beliefs had been shaped by his travels, particularly in the Middle East, where he developed a deep respect for Islam.[35] But what seemed especially to bother Galton about orthodox Christianity was its emphasis on original sin, an emphasis that he seems to have felt with special force.

Galton was troubled during the aimless years after his return from the Middle East. In Syria, established with his Sudan monkeys and his mongoose, he had led what he later called a "very oriental life." His family, who kept everything else he wrote, seems to have kept none of his letters from this period.[36] Enough clues remain, however, to form a plausible interpretation of his later disturbance. One of the few surviving items in Galton's correspondence of the time is a letter from Montagu Boulton, a fellow Englishman also traveling in the Middle East. Boulton reported that he was negotiating for a pretty Abyssinian slave, and added, "The Han Houris are looking lovelier than ever, the divorced one has been critically examined and pronounced a virgin." No doubt such practices and attitudes were common among young Englishmen sowing their oats in the region, probably including Galton. The report of the London Phrenological Institution found men of his head type not only suitable for colonization but also likely to "spend the earlier years of manhood in the enjoyment of what are called the lower pleasures, and particularly of those which the followers of Mahomet believe to form the chief reward of virtue in the realms above"; such reports must have been based on independent knowledge of the subject.[37] One need not assume that Galton's Middle Eastern sojourn was exotically carnal to argue that he occasionally indulged in sybaritic pleasures and was plagued by some degree of guilt on his return. Although he apparently overcame the guilt for a time, it may well have returned to nag him in the eighteen-sixties, as it became more and more likely that his marriage would prove barren. While in the Middle East, Galton seems to have suffered a bout of venereal disease. (Boulton's letter commiserated: "What an unfortunate fellow you are, to get laid up in such a serious manner for, as you say, a few moments' enjoyment.")[38] The effects of such disease were little understood at the time. Galton, with his partial medical training, may have wondered whether venereal afflictions rendered men sterile, and so have blamed himself for the lack of children in his marriage.

Galton's propensity for counting was no doubt reinforced by his inner turmoil. To plumb intangible human depths was to risk self-perception. To enumerate human characteristics required no penetration beneath the phenomenological surface and established a wall of numerical objectivity between the observer and the forces of the heart. Thus Galton reduced the Hottentot women to measurement with a sextant. Thus, a few decades later, he constructed a "beauty map" of Britain by noting the frequency with which he saw attractive women in various towns. His marriage seems to have been built on social and intellectual companionship rather than on passion. (His great-grandnephew Hesketh Pearson reported, "Galton's marriage, as far as I can make out, was not a particularly happy one. . . . I have been told that any comfort which might have given pleasure to his leisure hours was often denied him by [his wife].")[39] Yet at times Galton let slip the veil of enumeration and Victorian propriety. In "Kantsaywhere," an unpublished novel of a eugenic utopia, the women, unlike Louisa, were Rubensian figures—"thoroughly . . . mammalian," Galton called them—and bore their husbands many noble children.[40]

Galton never coped emotionally with his cluster of devils except by breakdown or fantasy. But intellectually, at least, he was able to deal with them after he read Darwin's *Origin*. He rejoiced to his cousin, "Your book drove away the constraint of my old superstition, as if it had been a nightmare." To Galton's mind, the scientific doctrine of evolution destroyed the religious doctrine of the fall from grace. He appropriated Darwin to argue that man, instead of falling from a high estate, was "rapidly rising from a low one."[41] Eugenics would accelerate the process, would breed out the vestigial barbarism of the human race and manipulate evolution to bring the biological reality of man into consonance with his advanced moral ideals. According to Galton, "what Nature does blindly, slowly, and ruthlessly, man may do providently, quickly, and kindly." He found in eugenics a scientific substitute for church orthodoxies, a secular faith, a defensible religious obligation.[42]

Galton eventually gave up on race improvement through the state regulation of marriage, but he continued to hope that the new religion would foster voluntary eugenic marriage practices. After all, religious marriage customs clearly varied across cultures and served particular social purposes. Might not people pursue a procreatively eugenic life, Galton wondered, once eugenics carried the full, authoritative weight of a secular religion?[43] But in the wake of *Hereditary Genius*, Galton came to realize that, whatever the future held, so little was reliably known about heredity that even a Spartan given dictatorial powers over marriage might well produce race degradation rather than improvement. Intent on making a true science of eugenics possible, Galton began trying to ferret out the laws of inheritance.[44]

HE APPROACHED the problem through the infant science of statistics. At the time, no biologist dealt with any part of his subject mathematically; Galton's remarkable methodological departure was of considerable long-term significance for the discipline. It originated, however, not in a conviction on his part that biology needed mathematics but, rather, in something that came naturally to him—counting, and pondering the resultant numbers. The word "statistics" denoted, in Galton's time, "state" numbers—indices of population, trade, manufacture, and the like—the gathering of which aided the state in the shaping of sound public policy. In mid-Victorian Britain, the practice of statistics consisted mainly of the accumulation of socially useful numerical data, with neither theoretical underpinning nor mathematical analysis. But in the late eighteen-sixties, as a result of his meteorological interests, Galton came upon a quite different approach to statistics—the formulation now called the normal, or Gaussian, distribution.[45]

Known at the time as "the law of error," the formulation derived from the analysis by the German mathematician Carl Friedrich Gauss of errors made in the measurement of "true" physical quantities—for example, planetary positions in astronomy. Portrayed graphically, the Gaussian distribution formed the now familiar bell curve; a vertical line bisecting the bell in the center represented the mean of the measurements—which was taken to be the true value of the quantity—and the curve itself expressed the fact that the greater the deviation from the mean in a measurement, the lower the frequency with which such a measurement would occur. Galton's interest, however, was not in the mean but in the distribution of deviations from it. Though he drew upon the few existing authorities in mathematical statistics, he came independently to view the Gaussian distribution not primarily as a way of differentiating true values from false ones but as a tool for analyzing populations in terms of their members' variations from a mean—the kind of variations inevitably manifest in, for example, the heights or weights of a large, randomly selected group of people.[46] Eventually, he concluded that there was "scarcely anything so apt to impress the imagination as the wonderful form of cosmic order expressed by the 'Law of Frequency of Error,' " and added, "The law would have been personified by the Greeks and deified, if they had known of it."[47]

In *Hereditary Genius*, Galton assumed that talent was normally distributed—that deviations in either direction from the mean talent of the population would follow the Gaussian distribution. He used the law to try to estimate the number of men of genius—and of exceptional stupidity—among the British population of 1860.[48] But he made no further use of the law, not least because he lacked data concerning the distribution in human populations of even simple physical characteristics, let alone intelligence. "The work of a statistician is that of the Israelites in Egypt," he later remarked. "They must not only make bricks but find the materials."[49]

In the early eighteen-seventies, Galton began his search for materials by collecting information concerning physical characteristics of schoolboys. For hereditary data, he compared the seeds from a parental generation of *Lathyrus odoratus*, the sweet-pea plant, with those from its progeny. "It was anthropological evidence that I desired, caring only for the seeds as means of throwing light on heredity in Man," he later reported. To obtain human hereditary data, Galton hit upon the brilliant idea of establishing an Anthropometric Laboratory at the International Health Exhibition, which opened at the South Kensington Science Museum in 1884. Within a few months, some nine thousand people, including many parents and their grown children, were measured for height, weight, arm span, breathing power, and the like. At the same time, he published the *Record of Family Faculties*, a questionnaire on heredity, and offered prizes of up to five hundred pounds for the most detailed sets of family data.[50]

Galton scored his first advance in 1876, with the sweet-pea data. He had selected as the parental generation seven groups of seeds, each group containing the same number of seeds of a particular weight. The seven weights were the mean weight of all the seeds and the weights found at three statistical intervals on either side of the mean. He placed ten seeds from each group in separate packets and mailed sets of the seven packets, with detailed instructions for planting, to various friends (one was Darwin), in different parts of England. The sweet pea, a self-fertilizing plant, produces a large number of new seeds in pods. The friends were to harvest the daughter seeds and return them to Galton, placing them in the packets in which the original groups of parental seeds had arrived. When Galton received the complete produce, he was then able to weigh all the daughter seeds individually and analyze the statistical distribution of their weights.[51]

Galton did not discuss how many daughter seeds the complete produce contained, nor at the time did he provide any other concrete numerical details of the outcome. Rather, he dwelt on the general statistical features of the results, which he found astonishing. Each group of parental seeds of the same weight produced a family of daughter seeds in which the weights were distributed around a mean in Gaussian fashion. What astonished Galton was that no matter what the weight of the parent seeds, heavy or light, all the distributions had the same statistical variability; that is, the same proportion of seeds could be found on the bell curve within a given distance from the family mean. He soon realized—"I forgot everything else for a moment in my great delight"—that the laws governing heredity, whether of sweet peas or of men, could be treated mathematically, in terms of units of statistical deviation.[52]

Galton took as the unit of deviation the so-called "probable error" of nineteenth-century scientists, which was arbitrarily defined as the distance along the horizontal axis, or baseline, of the bell curve where a vertical line would divide the area to one side of the bell's center into two equal parts. Twice this distance thus equaled two units of deviation; three times the distance, three units. Taking the sweet-pea data, Galton measured by how many units the mean weights of each parental seed group and its daughter family of seeds, respectively, deviated from the mean weight of the total seed population—the "race." He calculated the ratio for each pair of daughter-parent deviations and discovered that all the ratios were about the same. That striking result complemented another feature of the data: the mean weight of every daughter family fell closer to the mean of the total population than did that of its parent group. Galton interpreted this to suggest that the characteristics of offspring were products not only of the immediate parent but also of numerous forebears. He argued that the effect of ancestry caused the progeny of one generation to revert toward the center of the population, and he dubbed the measure of that tendency, expressed in the common ratio of the daughter-parent deviations, the "coefficient of reversion."[53]

Once Galton had the data from the Anthropometric Laboratory and the *Record of Family Faculties*, he was able to ruminate over the statistics of human heredity. He constructed a table marked with grades of parental height on the left-hand side and of the height of their grown children on the top. For parental height, he used an average of the maternal and paternal heights, which he called the height of the "midparent." An imaginary horizontal line drawn from a given midparental height on the left would intersect an imaginary line dropped vertically from a given child's height. At each point of intersection Galton entered a number denoting the frequency with which, according to his data, a midparent of the height marked on the left produced a child of the height designated at the top. Read across from left to right, the resulting array of numbers expressed the observations that midparents of, say, seventy-one inches in height produced four children with heights of sixty-seven inches, five children of sixty-eight inches, five of sixty-nine inches, and four of seventy inches; or that midparents of sixty-six inches produced four offspring of sixty-five inches, six of sixty-six inches, and four of sixty-seven inches.[54]

In his meteorological work, Galton had liked to connect points of equal temperature or pressure on a weather map. While puzzling over his table of height data, Galton noticed that points of equal frequency—for example, every point labeled "4"—formed a series of concentric ellipses. Equally arresting, a straight line that connected the horizontal tangent points had a slope—the ratio of the line's vertical to its horizontal rate of progress—equal to the coefficient of reversion of the children on the parents; one connecting the vertical tangent points had a slope equal to the coefficient of reversion of the parents on the children. Galton suspected, with considerable insight, that one could construct these ellipses knowing only three things: the probable errors of the parental and filial generations and the reversion coefficient of the latter on the former. Galton was rusty in analytic geometry and unable himself to prove his insight. Disguising the problem as one in abstract mechanics, so as not to prejudice the outcome, he set the task for J. D. Hamilton Dickson, a mathematician at St. Peter's College, Cambridge. Dickson derived Galton's ellipses and their interrelationships, using only analytic geometry and the laws of probability. The outcome was freighted with an implication that delighted Galton. Dickson's result held as a general relationship between any two appropriate variables, not only those linked by heredity. The coefficient of reversion was thus independent of heredity; it was purely a property of statistical manipulation itself. Galton, the onetime aspiring mathematician, had willy-nilly forged a contribution to mathematical statistics. To rid the reversion term of its hereditary flavor, he renamed it the "coefficient of regression."[55]

Not long afterward, Galton became interested in the Frenchman Alphonse Bertillon's system for the identification of criminals, which relied on taking their physical measurements—for example, head and limb size. Galton thought that Bertillon's system suffered from redundancy; it treated different dimensions of the same person as if they were independent, and many of them were not. A tall man, for example, was much more likely than a short one to have a long finger, arm, or foot. To find out whether such characters were in fact independent, Galton tabulated against each other such characteristics as height and arm length. In short order, he noticed that the results fell into a pattern similar to what he had previously found for the heights of parents and children. The tabulation could even be made to produce a similar set of concentric ellipses and mathematical relationships. In consequence, Galton realized that the relationship between measures of two different entities such as height and arm length could be expressed mathematically—just as in regression—by a coefficient of correlation. In fact, Galton concluded, regression was simply a special case of correlational analysis.[56]

The coefficient of correlation, expressed as a number ranging from minus one to plus one, provided a measure of the degree, positive or negative, to which one variable might depend upon another. Statistical correlation could be of particularly powerful assistance in cases—legion in the disciplines of biology and sociology—involving two or more independent variables each of which might be only partly responsible for an observed outcome. Statistical correlation might suggest, for example, that academic performance was negatively correlated with class size—the smaller the class, the better the performance—and, at the same time, positively correlated with the teacher's years of experience. "Some people hate the very name of statistics, but I find them full of beauty and interest," Galton declared shortly after the work on correlation. "Their power of dealing with complicated phenomena is extraordinary. They are the only tools by which an opening can be cut through the formidable thicket of difficulties that bars the path of those who pursue the Science of man."[57]

Galton made a good case for that claim in 1889, when he brought together most of the results of his investigations in heredity and statistics in the scientifically influential *Natural Inheritance*. For all its merits, the book, like much of Galton's mathematical work, lacks rigor and is in places wrong. It is the sort of study to be expected from a pass-degree Cambridge graduate who was neither a formal mathematician nor an intellectually disciplined scientist. Galton proceeded by counting, pondering numerical arrays, constructing mechanical analogues, and relying on geometry and intuition. When he required rigorous mathematical proofs, he had to turn to others. Nevertheless, the core of his work in statistics constituted a sharp and irreversible departure from the mere data gathering that had characterized the science in midcentury. Galton insisted that statistics had to incorporate the theory and methods of mathematical probability. By doing precisely that, he produced, with regression and correlation, a seminally important innovation. His biographer Karl Pearson wrote in 1930, "Thousands of correlation coefficients are now calculated annually, the memoirs and textbooks on psychology abound in them; they form . . . the basis of investigations in medical statistics, in sociology and anthropology. . . . Formerly the quantitative scientist could think only in terms of causation, now he can think also in terms of correlation. This has not only enormously widened the field to which quantitative and therefore mathematical methods can be applied, but it has at the same time modified our philosophy of science and even of life itself."[58]

By "life itself" Pearson meant mainly heredity. *Natural Inheritance* contains numerous obiter dicta—most of them unsupported or erroneous —on aspects of the subject, including the heritability of disease, of the "artistic faculty," and of alcoholism.[59] Galton's mathematical analyses of ancestral or familial hereditary relationships were faulty. And he was in fact unable to shed any real light on the heritability of talent or intelligence— a problem he never solved. But he did contribute crucially to the study of heredity. While scientists before him, including Darwin himself, had spoken vaguely of some force of inheritance, of reversion and variation, or of like begetting like, Galton gave heredity a sharp—albeit, of course, a non-Mendelian—definition: the quantitative, hence measurable, relationship between generations for given characters.[60]

But Galton's heredity studies raised serious problems for his eugenics program. It was clear from his work that in any population the distribution of a given character remained the same from generation to generation; the bell curve for, say, height was the same for children as for parents.[61] More important, even if only members of the population at the extremes of the bell curve—for example, heavier sweet-pea seeds—were chosen for reproduction, Galton's results declared that their progeny, if they were left to reproduce without constraint, would ultimately regress toward the mean of the initial population. It seemed that only by selection of the weightier seeds in every generation could a line of heavy seeds be kept heavy. It was "in consequence impossible that the natural qualities of a race may be permanently changed through the action of selection upon mere variations," Galton believed. "The selection of the most serviceable variations" —presumably he included high ability—"cannot even produce any great degree of artificial and temporary improvement."[62]

If the evolution of new forms did not come about by the selection of small variations, however serviceable they might be, how did it come about? Theorists of evolution had debated the problem of evolutionary mechanisms long before Galton's statistical work. As the debate proceeded, Darwin had cited an early theory of his, called pangenesis, which stated that the environment induced advantageous organic modifications, and that these were transmitted, by particles he called gemmules, via the circulation of bodily fluids to the sexual organs and ultimately to succeeding generations. But in the eighteen-seventies, in an experimental challenge to the theory of pangenesis, Galton had found that gray rabbits whose blood— and, presumably, gemmules—had been mixed with that from whites nevertheless bore not mongrel rabbits but more grays. Heredity, Galton supposed, must be governed by some sort of "stirp" (he took the word from the Latin for "root")—a latent element responsible for the transmission of characters from one generation to the next.[63] In the eighteen-eighties, the German biologist August Weismann independently advanced a similar, though physiologically more substantial, hypothesis with his theory of the continuity of the "germ plasm." Weismann's work reinforced Galton's long-standing belief that race improvement could occur only when nature provided a distinct and heritable organic change—biologists of the day termed it a "sport"—upon which selection, natural or eugenic, could act.[64]

The inability to resolve the controversy over how evolution proceeded cast a certain doubt on Darwin's theory and raised obstacles to Galton's eugenics. In the preface of the 1892 edition of *Hereditary Genius*, Galton acknowledged that "the great problem of the future betterment of the human race is confessedly, at the present time, hardly advanced beyond the state of academic interest." Nevertheless, he insisted that human beings could at least hope to achieve eugenic improvement indirectly. "We may not be able to originate, but we can guide. The processes of evolution are in constant and spontaneous activity, some towards the bad, some towards the good. Our part is to watch for opportunities to intervene by checking the former and giving free play to the latter."[65]

Chapter II

KARL PEARSON
FOR SAINT BIOMETRIKA

G ALTON'S eugenic ideas gradually won a degree of commendation from the scientific community both in Britain and in the United States. "You have made a convert of an opponent in one sense," Darwin told his cousin, "for I have always maintained that excepting fools, men did not differ much in intellect, only in zeal and hard work." In *The Descent of Man*, Darwin canonized Galton: "We now know, through the admirable labours of Mr. Galton, that genius . . . tends to be inherited."[1] But in the Anglo-American world of genteel lay discourse some critics disputed Galton's claim of the heritability of intelligence, and others warned that eugenics would interfere with the freedom and sanctity of marriage. Defenders of the faith rejected his anticlerical views as such and his biological theories because they implied that mental capacity was not implanted by God in every newborn individual. Moral progress, the reasoning went, could not be reduced to biology, because man was predominantly a spiritual rather than a biological creature.[2]

To be sure, in the late nineteenth century a growing body of commentators who introduced Darwinian analogies into social argument thought otherwise. Many social Darwinists insisted that biology was destiny, at least for the unfit, and that a broad spectrum of socially deleterious traits, ranging from "pauperism" to mental illness, resulted from heredity. Such reasoning suggested that the procreation of the fit ought to be encouraged and that of the unfit limited, but most hereditarians of the day on the whole ignored eugenics.[3] Voluntary eugenic measures seemed rather premature; man was as yet insufficiently altruistic to permit eugenic concern for the community to govern his desire for self-reproduction. If not voluntarism, then perhaps coercion. But coercion would violate the dominant doctrine of laissez-faire by requiring state interference with individual liberty, and one of the most private areas of liberty at that. Besides, to the social-Darwinist mind coercion was unnecessary, since in the merciless struggle for survival the unfit were doomed anyway and the fit destined to prevail.[4]

Galton's eugenic doctrines ran into no such obstacles among groups for whom social Darwinism was of little or no consequence. From the eighteen-sixties onward, various sexual radicals raised the banner of better breeding in order to advance the cause of liberty in human couplings. In the United States, Victoria Woodhull repeatedly invoked before lecture audiences "the scientific propagation of the human race" as reason for sexual education and the emancipation of women.[5] John Humphrey Noyes, the founder and patriarch of the Perfectionist Oneida Community, in upstate New York, disapproved of monogamy, which he thought "discriminates against the best and in favor of the worst; for while the good man will be limited by his conscience to what the law allows, the bad man, free from moral check, will distribute his seed beyond the legal limits." In 1865, Noyes's Oneida newspaper, the *Circular*, proclaimed in an editorial that "Human Breeding should be one of the foremost questions of the age, transcending in its sublime interest all present political and scientific questions." Noyes had already established at Oneida a system of "complex marriage," which declared all members of the community wedded to each other, and regulated the permissible yet various sexual bondings. In 1869, inspired by Galton to the further pursuit of perfection, he launched volunteers at Oneida on an experimental program of selective human breeding.[6]

In England, sexual radicalism often combined with Fabian-socialist leanings to produce, typically, the eugenic ideas of George Bernard Shaw and Havelock Ellis: since barriers of class and wealth kept people from eugenically optimal marriages, remove class distinctions and many more biologically desirable unions would be assured. Galton may not have found the Shavians to his taste—he declared that he had never intended to condone the mating of men and women "as we please, like cocks and hens" —but the fact of the matter was that eugenic enthusiasm was highest among social radicals.[7] Indeed, Galton drew his principal successor in eugenics, Karl Pearson, from an offbeat sector of British socialism.

PEARSON WAS THE product of a middle-class Quaker family who had come down from Yorkshire to London so that the father might fulfill his intense ambition in the law. William Pearson, who eventually became a Queen's Counsel, dominated, perhaps to the point of tyranny, his two sons and his wife, an ineffectual woman who retreated into a self-indulgent distractedness. Karl remembered his father as "an iron man" who rose before dawn to prepare his briefs, rushed to the office after a standing breakfast at nine, returned in the evening to hurry taciturnly through dinner, then promptly retired. If Karl entered his father's study, he would be directed to a chair and left to sit for hours entirely ignored. On vacations, he was made to follow along on fly-fishing tramps but instructed not to cast if fish were about. Karl's older brother, Arthur, was sent away to Rugby. Karl was educated in London and thus remained at home, where, unlike Galton, he had no older sister to turn to for warmth. He found his parental affection in his mother, who needed her son's emotional sustenance as much as he needed hers.[8]

Arthur Pearson was made rich by one of his father's clients in exchange for taking his surname; he nominally practiced law, but he spent a great deal of his life on holiday.[9] The senior Pearson's ambitions came to center on Karl, who was also expected to enter the law. For reasons of health, Karl was withdrawn from his London school at the age of sixteen and sent away to a private tutor at Hitchin. He resented the other students there, because they talked indecently, played the banjo and sang while he tried to work, and hardly ever spoke to him. ("I can bear the leaving home," he confided to his mother, "but never speaking to anyone is very hard.")[10] In 1875, he went to King's College, Cambridge, on a mathematics scholarship. He deplored the university, because, as he put it in a bit of doggerel, young men came "to gain social stamp, but not to learn / While teachers only teach to earn." Pearson refused to attend the required divinity lectures and chapel. Though the rebellion was directed more against the requirement than against religion as such, Pearson, like so many Victorian undergraduates, was beset by an agony of religious doubt. Spiritually discontented and more enamored of mathematics than of the law, he went to Germany for postgraduate study, dividing his time among law at Berlin, philosophy at Heidelberg, and mathematics at both.[11]

Pearson declared the Prussians "barbaric" and disliked the Germany of the Kaiser; he found solace for his religious doubts in the Germany of Goethe, giving way to a cow-eyed romanticism.[12] To a Cambridge friend he announced that he would not want any son of his to be a man of the world but would prefer that he make art "his goddess." In 1880, on his return to England, he dutifully entered Lincoln's Inn to prepare for the bar. His heart elsewhere, he changed the spelling of his name from Carl to Karl and dreamed of marrying a German woman.[13] He wrote *The New Werther*, a turgid novel that celebrated lonely idealism; published a Passion play that attacked orthodox Christianity; and resolved his religious doubts in favor of agnosticism and a devotion to Spinoza.[14] Yet Pearson's mixture of idealism and romanticism was accompanied by a certain awareness of socioeconomic reality in late-Victorian Britain. Like many sons of professional fathers, he was hostile to the aristocracies of landed wealth and industrial capital. Above all else, he was a rationalist; history, philosophy, and science, particularly the German variety, had considerably more influence on him than poetry.

Pearson was attracted to the German school's blend of idealism and economic historicism—notably Johann Fichte's insistence that the over-arching good of the people was best expressed in the state. Politically, he drew upon the views of the German socialist left, including, while abroad, its anti-imperialism. (The impoverished millions of Ireland and London, he had predicted with postgraduate confidence, would "make themselves heard in the next twenty years . . . , and woe to those who then have their thoughts in Africa or Asia!")[15] While in Berlin, he had spoken contemptuously of the students who attended DuBois Reymond's celebrated lectures on Darwinism for thinking that "some solution of their social difficulties is to be obtained from the theories of evolution."[16] In Pearson's view, Darwinism buttressed Herbert Spencer's doctrine of individualism and provided a justification of laissez-faire capitalism. But in the England of the eighteen-eighties, reformers were forging Darwinism into a weapon against laissez-faire. Pearson soon followed suit, by sub-stituting competition between national groups for individual struggle. In the era of the pro-imperial Primrose League and of mounting con-cern about the economic rivalry of France, Germany, and the United States, it was a short step from there to social imperialism, to advocating that the nation should be kept internally strong for the sake of the exter-nal struggle.[17]

Yet, unlike other Darwinian socialists, Pearson paid virtually no atten-tion to the intricacies of industrial power. Nor did he show any interest in the individual details of working-class life, either in or out of the factory. Pearson the rationalist and lonely romantic tended to love people more easily in the abstract group than in the particular flesh and blood. Having abandoned religion, he sought a secular creed, and he found one appropri-ate to his personality in a socialism—iron-handed, if necessary—based on the Fichtian imperative of subordinating the mass of citizens to the welfare of the nation-state. Pearson came to equate morality with the advancement of social evolution, the outcome of the Darwinian struggle with the ascend-ancy of the fittest nation, and the achievement of fitness with a nationalist socialism.

Professional self-interest pervaded Pearson's ideas. He insisted upon bringing about the socialist state gradually, through the "enthusiasm of the study," rather than at the barricades. Once the socialist state was achieved, material goods would be divided as equitably as possible among all classes, but the further direction of social progress would fall mainly to workers of the head rather than of the hand. In his scale of values, thinking was as noble a form of labor as stoking a furnace—and more valuable to society. Sharing none of the Fabian enthusiasm for the extension of political democracy, he opposed the vesting of power in the uneducated laboring classes; they were all too easily moved by demagoguery.[18] George Bernard Shaw chided him: "As to an uneducated democracy being worse than a prejudiced aristocracy. For such a view I have an enormous contempt; and so ought you." What Shaw failed to recognize was that Pearson was concerned less with the shape of the new society than with where the Karl Pearsons would fit into it. Pearson called for something akin to a socialist meritocracy, declaring in 1881 that "power intellectual shall determine whether the life-calling of a man is to scavenge the streets or to guide the nation."[19]

While Pearson pondered the future role of power intellectual, his father nagged him to bear down in the present on the study of criminal law. Desperately eager to escape it, Pearson tried for mathematics posts at four universities, and finally, in 1884, landed a professorship at University College, London, then as now on Gower Street a few blocks north of the British Museum.[20] He found much of his intellectual life outside mathematics and the college—in books and public lectures, among his few Cambridge friends, and on the fringes of radical London, where he made the acquaintance of Karl Marx's daughter, Eleanor, George Bernard Shaw, Sidney and Beatrice Webb, Havelock Ellis, and Ellis's inamorata, Olive Schreiner, the South African novelist and passionate feminist.

Pearson soon came to believe that next to socialism the most important issue of the day was "the woman question." Ideas on the question formed part of the coin of socialist London, but Pearson took a special interest in it because of his mother's misery. He felt that she was imprisoned in her marriage—that she lacked the independent economic means to escape. ("There is always a demoralising influence in the power of one individual over another," he once noted, reflecting on his parents' relationship, "and this to a great extent must accompany the power of the purse.") To achieve genuine freedom, women, Pearson was sure, required the economic independence that only socialism could bring. Yet Pearson knew little about women. To teach himself, he founded, in 1885, the Men and Women's Club, for frank discussion of male-female relations.[21]

The club consisted of about fifteen members—mainly Pearson's Cambridge and law friends together with a roughly equal number of female acquaintances. Most of the members were single. Olive Schreiner participated for a time; among the guest lecturers was the birth-control advocate and freethinker Annie Besant. Club discussions ranged over prostitution, venereal disease, contraception—"preventive checks," in the euphemism of the day—marriage, sexuality, and women's economic opportunities and intellectual capacity. It was remarkably daring fare, given the largely middle-class composition of the club and the contemporary limits on what was considered proper in mixed conversation.

Or, rather, daring to a point. Although a faction in the club embraced the free-love theories of the late James Hinton, whose son's recent trial for bigamy was the talk of London, a majority of the club, Pearson included, stood far to the right of that type of sexual radicalism. The club denied admission to Havelock Ellis, and Pearson dutifully played dumb when one day George Bernard Shaw inquired about joining. The club may have agreed upon the desirability of easier divorce and flirted with the idea of sexual experimentation, but most members were in favor of monogamy, preferably in marriage.[22] Even Olive Schreiner, who had her affairs, held that infidelities on the part of either sex were "utterly opposed to the deepest laws of human nature, and are productive of nothing but evil to the individual, the offspring, and society."[23] Before the club disbanded, in 1889, it laid the foundation for at least three marriages among its members, including Pearson.

For Pearson, the club was a means not only of learning about women but of reaching out to them. Approaching thirty, he had never known a youthful passion, or even a serious flirtation. His contacts with the opposite sex—save, it seems, for prostitutes—had for the most part been cerebral. To him, younger women were shallow playthings, whom, he once confessed, he liked to waken out of complacency, with "half cynical intent," by saying "bitter things."[24] He preferred the friendship of older, experienced women, and for a time early in the history of the club he had a liaison of sorts with Olive Schreiner. Pearson suggested that the relationship should exclude sex. Schreiner, who was frightened by her own sexual drives, agreed, but, momentarily estranged from Havelock Ellis, she nurtured an unrequited passion for Pearson that helped drive her to a breakdown. Schreiner, hysterical, denied caring for Pearson with "sex love." Later, preening herself in the post-breakdown calm, she asserted to Ellis that the kind of love Pearson "makes women feel for him is like that of Dante for Beatrice." Schreiner suspected that Pearson's affections had been alienated by one of the club members, Mrs. Henry P. Cobb, the wife of a radical M.P., but the real object of Pearson's interest was Mrs. Cobb's sister Maria Sharpe.[25]

Maria Sharpe was four years older than Pearson, intellectual yet conventional in her experience and attitudes. Having become interested in the woman question as a schoolgirl, she had always supposed that the best way to maintain her independence was to avoid men. She was in her early thirties when she entered the club and thought at the time that prostitution lay at the base of every branch of the woman question ("the region," she put it, "where women are possibly only bodies to men casting a dark shade across all their own relations to the other sex").[26] She wished to be valued for her mind rather than for her body. She sympathized with women who wanted nothing of sexual relations and married only for "mental . . . intercourse with the masculine mind." She considered preventive checks repulsive, since by divorcing the sex act from its procreative consequences they permitted men to use women as mere physical instruments and led to excessive indulgence. Her ideas changed in response to the studies of the club—and the tutelage of Pearson, who courted her by abstract discourse and correspondence on socialism, women, and sex.[27]

Pearson made it clear to Sharpe that he respected her mind and wanted to encourage its development. Although he sometimes referred to middle-class women as "shopping dolls," maintaining that sheer laziness kept them from employing their frequently considerable leisure in literary or scientific pursuits, he conceded that women's comparative lack of intellectual achievement was a result, in some significant degree, of inadequate opportunity. Socialism would provide the necessary opportunities and, by eliminating the economic enslavement of women and rendering their relations with men voluntary, would make possible their free enjoyment of sexual ties. Indeed, contrary to his fellow socialist theoreticians, Pearson held that the right of all to decent labor required the limitation of population, which meant that the purpose of sexual relations would be only secondarily to conceive children and primarily to express "the closest form of friendship between man and woman." Pearson helped convince Sharpe that women possessed a significant sexual drive and might properly seek in union with a man, wedded or not, mutual sexual fulfillment for its own sake.[28] Yet to persuade the mind is not necessarily to reconstruct the psyche, and when, in 1889, Pearson finally proposed marriage, Sharpe had a nervous breakdown. Apparently, she suffered from a deep fear of losing her independence, entering into an actual physical relationship, and living up to the expectations of Karl Pearson. Melancholia enveloped her when she was away from Pearson and even more when she was with him; he described her condition as hysteria. It took six months for Sharpe to regain her composure, and in 1890 she and Pearson were married.[29]

Two years earlier, Pearson had published The Ethic of Freethought, a collection of iconoclastic essays on subjects ranging from religion to socialism and the woman question. "What a very brave thing a man in Pearson's [academic] position has done in printing that book at all," Olive Schreiner marveled to Havelock Ellis, and Shaw wanted Pearson to join the Fabians, where Shaw found himself "absolutely alone . . . on the free-thought question."[30] But Pearson's radical garlands had done nothing for him as a professional mathematician. At University College, where research was regarded as an indulgent luxury, his time was given over primarily to teaching—a good sixteen hours of lectures weekly. Pearson had accomplished little in mathematics. He expected to make his mark with works of history and philosophy, notably *The Grammar of Science*, which was published in 1892.[31] But his mathematical career took a vital turn that year, when he embarked on a collaboration in research with Walter F. R. Weldon, who had been recently appointed to the Jodrell Professorship of Zoology at University College.

WELDON HAD BEEN a prize student at Cambridge of the brilliant biologist Francis Maitland Balfour. Like Balfour, Weldon had studied the morphology of invertebrates and vertebrates in order to illuminate their evolutionary development.[32] Much of late-nineteenth-century life science—paleontology, comparative anatomy, embryology, and botany—was occupied with establishing species "types" and illuminating their evolutionary lines of descent. The results tended to be descriptive and decidedly speculative. One man's guess was as good as another's concerning the functional value for evolution of an organic adaptation, and variation in the structure of organisms made for considerable dispute over what constituted the archetype of a given species.[33] Typically, it bothered Weldon that although changes in larval forms always accompanied evolutionary development no clear functional relationship was evident between the new larval and the new adult characters. Weldon was becoming increasingly dissatisfied with morphological methods when, in 1889, he read Francis Galton's *Natural Inheritance*. Immediately, he wondered whether one might sidestep the inconclusiveness of morphology by subjecting the study of species and their evolution to statistical treatment.[34]

In short order, Weldon measured certain physical characteristics of several large samples of the common wild shrimp. All the samples were drawn from the same species but, having been taken from different sites, represented different "races." In the course of the work he met Galton, who happily helped him perfect the statistical analysis. Weldon found that for each sample group the size of a given organ was distributed normally about a mean—it was the first demonstration that a wild population displayed the normal distribution—but that the probable error in organ size varied from one sample to another. Weldon also applied Galton's method of statistical correlation to a pair of each shrimp's physical features—the lengths of the carapace and of the post-spinous bodily portion. He found that the degree of correlation was high and was approximately equal—the correlation coefficient came to about 0.8 out of a possible 1.0—for all the samples under study.

Weldon then scrutinized two samples of crabs, one group consisting of a thousand taken from Plymouth Bay and the other of a thousand from the Bay of Naples. He measured eleven characters of each sample—a total of twenty-two characters. Although twenty-one were normally distributed, the twenty-second set of measurements—the frontal breadths of the sample from the Bay of Naples—formed itself not into the familiar symmetric bell curve but into an asymmetric curve with a double hump. Weldon wondered whether the double-humped representation meant that this sample might in fact be made up of two different species of crab, each with its own normally distributed frontal breadth. General mathematical training, let alone statistics, was not a standard part of the late-nineteenth-century biologist's professional equipage. Stimulated by Galton, Weldon had begun educating himself in the theory of statistics, but the question presented by the double-humped curve was beyond his competence. It was for this reason that he turned for help to Pearson, who analyzed how to decide whether such a curve represented two normally distributed quantities.[35]

To Weldon, the outcome of the shrimp and crab investigations promised a striking advance for evolutionary studies. Instead of speculative definitions of alleged archetypes, species or races might be defined in terms of the quantitatively certain distribution of a given character around a mean and by the statistical correlation of character pairs. More important, given that natural selection presumably killed off the unfit young before they could reproduce, one could determine the fitness or unfitness of a given organism simply by measuring whether its deviation from the mean for a given character was associated with a greater or lesser death rate; thus, all speculations concerning the adaptive significance of variations would be rendered unnecessary. In Weldon's vision, the entire question of evolutionary process could be formulated and pursued as a concrete problem in statistics.[36]

Such ideas were not wholly new to Pearson. In a comment upon *Natural Inheritance* in 1889, he had pointed to the "considerable danger in applying the methods of exact science to problems . . . of heredity."[37] But in Weldon's crisp reading of it, Galton's approach seemed eminently defensible. Moreover, Weldon's program resonated with the philosophical ideas that Pearson had advanced in *The Grammar of Science*. A remarkably influential work ("The fall or rise of half-a-dozen empires," Henry Adams recalled, "interested a student of history less than the rise of the 'Grammar of Science' "), its intent was to impress upon the English-speaking world the epistemological ideas of Immanuel Kant and, particularly, of the Austrian philosopher and physicist Ernst Mach, whose writings would be so important to Einstein. In it, Pearson held that knowledge of the natural world consisted only of sequences of sense impressions. Man summarized those impressions in such constructs as "atom," "force," and "matter," but these were merely convenient verbal descriptions—shorthand. In this way, man imposed order upon chaos. Rather than "objective reality," they were the products of man's mind.[38]

To Pearson, biology was rife with speculative concepts—"species," "germ plasm," and a variety of life "forces"—that purported to explain vital phenomena yet were beyond operational test. He found Weldon's program appealing because of its positivist determination to deal only with directly observable quantities, to give measurable operational meaning to evolutionary change, to avoid speculative theorizing about unprovable evolutionary mechanisms. Then, too, Pearson was predisposed to be interested in evolution and heredity by virtue of his absorption in social Darwinism and his proto-eugenic leanings; by aiding Weldon he could take "the enthusiasm of the study" beyond historical investigations and lay a solid scientific foundation for gradual social change.[39]

Pearson's intellectual attraction to Weldon's program was reinforced by the Weldon persona. A friend remarked of Pearson that he was inclined to "cut himself off from communion with his fellows by treating any emotional pleasure as a weakness." He was cold, remote, driven; he never really escaped the paternal model. Olive Schreiner said that she learned from him to be ruthless in refusing the demands made upon her for help or advice. Pearson reminded her of "a lump of ice." Characteristically, he argued that the aesthetic value of art depended upon the degree to which, like a law of science, the work confirmed the beholder's experience. (Shaw once chided Pearson for suffering the "worst of training in mathematics," explaining, "You are never exercised on the human factor; and you come at last to be always looking for explanations under the furniture and up the chimney instead of within yourself.")[40] Weldon was more easygoing. During his undergraduate days at Cambridge, he had relished college social life and had kept in his rooms an owl named Pharaoh, which, to the pleasure of his friends, he brought out of its cupboard at night.[41] Weldon took a keen pleasure in literature and painting, particularly French and Italian works. His pleasure was aesthetic; he disliked didacticism in art. He journeyed annually to the Continent, combining biological research with an indulgence of his taste for art and opera. His students found him intensely human and held him in deep affection. So did Pearson. Weldon, three years his junior, was an alter ego, who, he once remarked, "always helped me to feel young."[42]

Pearson and Weldon were first drawn together in a common attempt to reform London University, which was then mainly an examining body for a number of independent colleges confederated under it. The two men campaigned in 1892—unsuccessfully—for a genuinely metropolitan institution under professorial control with uniform academic standards. For the next fourteen years—until Weldon's untimely death, in 1906, at the age of forty-six—they collaborated closely on the statistical study of heredity and evolution. They lunched together daily at the college before their one o'clock lectures, tossing ideas about, jotting notes on menus, and conducting experiments in probability with pellets of bread, then often carried on the discussion in letters posted from Pearson's house, in Hampstead, and from Weldon's, near the college, on Wimpole Street. In their collaboration, Weldon did the biological research, collecting samples and measuring characters. He relied heavily on the aid of his wife, also a former Cambridge student, for the laborious calculations, which had to be done by hand.[43] Pearson developed the necessary statistical theory and pursued its implications for evolution and heredity. The mathematician and the biologist exposed each other to the presumptions of their respective fields. (Weldon once complained to Francis Galton that Pearson cavalierly assumed a particular mathematical ratio to be "the real measure of the importance of variation"—that "he does not see that this is a matter for experiment, and not for *a priori* reasoning at all.") They provided mutual guidance in the framing of problems and did their best, at times with friendly abuse, to keep each other away from embarrassing pitfalls. Weldon was one of the few people who could fault Pearson on a scientific point without inviting demolishing fire in return.[44] So was Galton, whom both regarded as godfather and arbiter of the enterprise.

Weldon and Pearson rarely published jointly, and Weldon put comparatively little into print on his own; much of his work, Pearson said, was done to satisfy himself. Pearson, however, published more than a hundred papers during their collaboration.[45] One of the more important addressed the difficulty that Galton had raised for Darwinian and eugenic theory: that evolution could not proceed by the selection of small variations, because succeeding generations always regressed to the mean of the ancestral population—toward what Galton called the "racial center." Galton had embodied that contention in a mathematical formula, which Pearson later christened "the law of ancestral heredity." On first reading *Natural Inheritance*, Pearson had thought—contrary to Galton—that with proper selection human evolution could in fact be guided. Now he pointed out that Galton was refuted by the data of experience: the facial profiles of human populations, for example, did not regress to those of anthropoid apes. Perhaps, Pearson suggested, the focus of regression was not some ancestral generation but the immediately prior generation of parents. In that case, selective breeding might well change the center of regression from one generation to the next. In short, the mean of the population for a given character might be deliberately moved in an evolutionary line of eugenic advance.[46]

Pearson supported his theory with elaborate statistical analysis and rigorously reworked Galton's law of ancestral heredity. The result was a modified version of the law, predicting that after only a few generations of selection a population would breed true for the selected character. He presented the paper on the revised law to Galton as a New Year's greeting for 1898, announcing that the law was likely to stand as "one of the most brilliant of Mr. Galton's discoveries," for "if Darwinian evolution be natural selection combined with heredity, then the single statement which embraces the whole field of heredity must prove almost as epoch-making to the biologist as the law of gravitation to the astronomer."[47]

"We shall make something of Heredity at last," Galton exulted.[48] Increasingly fastened on that aim, Pearson employed his developing statistical tools to test Galton's original contention that heredity determined mental ability. In the absence of an objective measure of intelligence—the seemingly objective I.Q. tests had not yet appeared in Britain—Pearson obtained teachers' estimates of the abilities and temperaments of schoolchildren. Ideally, Pearson would have correlated the mental capacity of the children with that of their parents, but similar data for the parents was, of course, unobtainable. So Pearson chose to calculate the correlations between siblings. He realized that the correlations might well lump together qualities of nurture as well as nature, but he intended to overcome that problem by comparing the correlations for intelligence with those for physical characters assumed to be entirely uninfluenced by environment—notably eye and hair color.[49]

This procedure raised a technical problem. The existing calculus of correlations measured the relationship between variables—for example, height and arm span—distributed across a continuous range. Pearson was dealing with so-called nominal variables—that is, with discontinuously distributed data. Eye color did not occur at all points along a continuous spectrum; in a given family, eyes might be either, say, blue or brown. The same problem held for Pearson's data on mental capacity. Instead of rating the siblings on a single scale of ability, the teachers' estimates placed them in discrete categories: "very dull," "slow," "quick intelligent," and so forth. To measure relationships between his nominal variables, Pearson invented a new theory of correlation.

The theory proceeded from the assumption that the variables were points on an underlying (and, of course, unobservable) normal distribution curve. He defended the assumption—rather weakly, and in contradiction to his insistence that science must deal only with what can be directly observed—by contending that under all psychic states lay physical states that were presumably normally distributed in the manner of, say, height.[50] Analyzing data, obtained from some two hundred schools on nearly four thousand pairs of siblings, Pearson found that the correlation coefficients for physical characters all equaled about 0.5, and so did those for intelligence. "We are forced, I think literally forced, to the general conclusion that the physical and psychical characters in man are inherited within broad lines in the same manner, and with the same intensity," he announced in the 1903 Huxley Lecture to the Anthropological Institute. "We inherit our parents' tempers, our parents' conscientiousness, shyness and ability, even as we inherit their stature, forearm and span."[51]

"Bravis-is-imo *re* like inheritance of physical and mental!" Galton exclaimed to Pearson. "We have made a firm foothold here"—meaning, as Pearson doubtless understood, a firm foothold in eugenics.[52] The revised law of ancestral heredity suggested that human populations could be permanently improved by biological manipulation. More important, the force of heredity appeared to be so powerful for features like intelligence as to dictate selective breeding as the only means of achieving greater social strength. Weldon, who was not a eugenicist, at times cautioned Pearson against overlooking the role of nurture even in purely biological, let alone social, development, but in these matters Pearson paid less attention to Weldon than to Galton. He spent many hours with Galton at Rutland Gate, discoursing in the white-enameled drawing room, surrounded by relics of the Darwin and Galton families, including Erasmus Darwin's writing table. The relationship with Galton, which had begun somewhat formally in the early eighteen-nineties and had broadened into frequent exchanges on statistics, heredity, and evolution, had ripened by the turn of the century into a warm personal bond, like that between proud father and dutiful son, with the filial Pearson loyal to the paternal eugenic creed.[53]

Yet Pearson hardly came at eugenics from the same angle as Galton —certainly not with the same attitudes toward women and sexuality. Unlike Galton, Pearson, the mainstay of the Men and Women's Club, confronted sexuality head on, with no evident nagging guilt, religious or otherwise. More important, Pearson had no need to fantasize about the eugenic breeding of Pearson-like progeny, for Maria, settling down to a matronly life in Hampstead, had borne him three children within a few years of their marriage.[54] The eugenics of Karl Pearson, husband and father, was charged less with psychosexual energy than with his commitment to social imperialism—the ideological system where, in fact, his eugenic convictions had originated.

In Pearson's view, the imperial nation required more than an economic framework designed to give its citizens a material stake in its power; it also demanded the "high pitch of internal efficiency" won by "insuring that its numbers are substantially recruited from the better stocks." Like Francis Galton, Pearson—a man of lean, athletic build who could tramp or cycle for hours—equated fitness with physique and mental ability, and assumed that it was centered in the middle, and particularly the professional, class.

Unlike Galton, he declared that fitness extended down to the "better" sort of English workingman marked by "a clean body, a sound if slow mind, a vigorous and healthy stock, and a numerous progeny." But, Pearson warned in his Huxley Lecture, Britain was "ceasing as a nation to breed intelligence."[55] Was it not a drawing-room commonplace that Britain suffered a "dearth of national ability"? How else explain that no Englishman had invented the automobile or the airplane? To Pearson, the demographic trend was dangerous. Generalizing mainly from Danish statistical studies, he argued that half of each generation was the product of one-quarter of its married predecessor. That prolific quarter represented only from one-sixth to one-eighth of the adult population and was drawn disproportionately from the "unfit," which in Pearson's lexicon meant "the habitual criminal, the professional tramp, the tuberculous, the insane, the mentally defective, the alcoholic, the diseased from birth or from excess."[56]

Britain, Pearson insisted, was in a state of national deterioration, and he located the trouble in the economic incentives for procreation. He noted that children had never been an economic asset for the "cultured classes"; rather, they were "a luxury which we know we must pay for, and expect to pay for, until after college and professional training, and in the case of unmarried daughters, often until long after our own lives are concluded."[57] No doubt in the interests of economy, the "cultured classes" increasingly indulged in "neo-Malthusianism," as the practice of birth control was also often called, but by limiting family size they failed in their imperial reproductive duty; they deprived the nation of brains. ("With our modern views as to parental responsibility, neither Charles Darwin nor Francis Galton would have been born!" Pearson exclaimed.) Children had been an economic asset for the responsible working class until the passage of such measures as the Factory Acts, Pearson argued. The prohibitions against child labor transformed children into economic liabilities, and the better class of workers quickly reduced their birth rate, leaving the principal task of procreation to the socially worst.[58]

In the eighteen-eighties, Pearson had pinned the excessive reproduction of the socially unfit upon capitalism, which, with its demand for cheap labor, encouraged the immigration of workers below a desirable standard.[59] In the early twentieth century, he found his target in liberal reformism. "We have placed our money on Environment," he quipped, "when Heredity wins in a canter." Thus, assuming that everyone in Britain who could benefit from an education was getting one, Pearson saw no point in expanding schools. "No training or education can *create* [intelligence]," he declared in the Huxley Lecture. "You must breed it." Indeed, he privately asserted to Galton that charities for the children of the "incapables" were "a national curse and not a blessing." In his opinion, such measures as the minimum wage, the eight-hour day, free medical advice, and reductions in infant mortality encouraged an increase in unemployables, degenerates, and physical and mental weaklings. Natural selection, he believed, had been suspended, and replaced by "reproductive selection," which gave the battle "to the most fertile, not the most fit."[60]

Professor and Mrs. Pearson may have done their part to offset any disproportionate contribution by the unfit to the next generation; beyond that, Pearson's response to the fertility crisis was remarkably vague. In his *Ethic of Freethought*, he had declared that the social-imperialist state might well have to intervene in reproductive matters, at least in the families of "anti-social propagators of unnecessary human beings."[61] He advanced no concrete methods of intervention in the early twentieth century. He disavowed even the repeal of liberal reforms, conceding a certain value to environmental improvements. ("Although a keen razor can never be made of bad steel, a good steel requires setting and tempering before it can fitly perform.") Pearson proposed only that Britain deal with the disadvantageous impact of the liberal reforms on national fitness by making sure that national insurance, child allowances, and the like favored the eugenically desirable.[62]

Pearson, the enthusiast of the study, claimed that he had neither the responsibility nor sufficient knowledge to advance legislative programs. He declared that his principal purpose was to explore scientifically the theories—particularly those regarding the relative weights of nature and nurture—on which a sound eugenic policy should be built. To recognize why nations rose and fell, everything that contributed to human character had to be studied "not by verbal argument, but . . . under the statistical microscope," he wrote. "The study of Eugenics centres round the actuarial treatment of human society in all its phases, healthy and morbid."[63] Pearson's purpose, however, was no more disinterested than his eugenics was unprejudiced. His "enthusiasm of the study" was stronger now that he and Weldon had invented a new discipline—"biometry," as they dubbed the statistical study of evolution and heredity—and now that he saw a rich vein of research in the linked subjects of biometry, statistics, and eugenics.

He could hardly afford to go at that research in the manner of an amateur scientist of independent means like Francis Galton. His salary as a professor, his lecture and writing fees, and the income from Maria's capital amounted to more than eight hundred pounds a year—enough to provide for a comfortable three-story brick house on Well Road in Hampton, two domestic servants, the children's education, and rental of a summer cottage. But Pearson worried about money—at least until 1907, when he came into an inheritance upon his father's death.[64] In English science, the day of the Galtons was passing. Succeeding it was the era of the professional—of ambitious men like Pearson, for whom the pursuit of special knowledge required the means of others, the institutionalization of research programs, the establishment of a school.

In 1895, PEARSON HAD begun offering courses in statistics at University College; a few postgraduate students came to work with him in what he soon called the Biometric Laboratory, in a new college building, bright with electric lights, on Gower Street. Weldon and Pearson had both been elected to the Royal Society, which also honored Pearson in 1898 by the award of its Darwin Medal. In 1899, Weldon was appointed to the Linacre Professorship of Comparative Anatomy at Oxford University, and he proceeded to foster biometric studies there—although he found his Oxford colleagues "rank morphologists who prefer speculating . . . to any other more serious inquiry."[65] In 1902, Pearson, Weldon, and Galton founded the journal *Biometrika*. ("We intend to appeal in the first place to biologists," Pearson noted in a letter to a colleague, "and while we shall deal with Statistical Theory at large, we shall clothe it with a biological terminology.")[66] But neither recognition nor a journal made automatically for the establishment of a research school. Pearson remained burdened by a heavy lecture load and students indifferent to his biometric passion. He had acquired a mechanical Brunsviga calculator to ease the laborious arithmetic of statistics, but he was without the computing staff necessary to deal with masses of data. He applied for at least four mathematical professorships elsewhere, each time unsuccessfully, and concluded that a new post was blocked by his outspoken socialism. But his devotion to biometry was no less a handicap. "I fear . . . you are the only part of the scientific public which takes the least interest in my work," he lamented to Galton. "The mathematicians look askance at anyone who goes off the regular track, and the biologists think I have no business meddling with such things."[67]

Though some biologists applauded the mathematization of evolution and heredity, on the whole the Anglo-American biological community responded to Pearson's biometrics with indifference or hostility. Biologists of the day knew little of statistics, and most biologists also inclined as a matter of taste against the mathematical analysis of life forms. Royal Society biologists dealt with Pearson's biometric papers in a manner he considered wholly prejudiced; it was after one such incident that he resolved to found *Biometrika*. The journal won few British subscribers; half the early subscribers were American and many of the rest were on the Continent. Pearson warned prospective students in biometry that there were "no teaching posts, no demonstratorships or fellowships to aid a young man on his way." He predicted that "There will be no doubt one day a demand for statistically trained medical men for registrars, officers of health. . . . But they may have to wait a weary while."[68]

The objections to biometry did not, however, stem entirely from mathematical ignorance or methodological prejudice. At a meeting of the Royal Society, a prominent zoologist attacked Pearson's study of the heritability of intelligence for the eminently sensible reason that teachers' estimates said nothing reliable about students' mental capacity. The study's correlational analysis was indicted not least because it depended heavily upon the assumption of an underlying normal distribution of mental acuity, when there was actually no good reason to suppose that such a distribution, if it existed, was statistically normal.[69] The new Mendelian genetics suggested that Pearson's view of heredity needed a good deal of modification. The results of breeding experiments with plants and animals empirically undermined Galton's law of ancestral heredity, even in its Pearsonian form. And the theory of evolution by the selection of small variations was sharply challenged by the work of the Danish biologist Wilhelm Johannsen, who found that a pure line of beans could not be bred beyond a maximum limit for a given character, no matter what the degree of selection.[70]

Pearson rejected scientifically knowledgeable objections to biometric theories. As even one of his biometrical admirers had to remark, he tended "to take a rather absurd position sometimes in regard to biologically obvious things." He scoffed at Mendelism. When two members of Biometrika's editorial board published a report upholding Johannsen's pure-line work, he summarily removed them from the board.[71] Not only did he often display a relentless closed-mindedness but he frequently took a club to his scientific enemies and slashingly abused even those of his methodological friends who queried his biometry or his eugenics. Yet Pearson rarely, if ever, displayed a mean temper in personal matters; the firestorms erupted over intellectual differences. "It is Saint Biometrika contra mundum!" he once exclaimed.[72] Pearson's search for a secular creed had been distilled from his social imperialism into his science, particularly eugenics. His laboratory walls were adorned with quotations from mathematicians, scientists, and philosophers, including Plato. ("But the best part of the Soul is that which trusts to Measure and Calculation? Certainly.")[73] The calculus of correlation conformed to the icy distance of his character, reinforcing his propensity for dealing with man in the impersonal group. If Pearson responded to criticism with polemics, it was because the dissent struck at his secular church, his methodological character, his intellectual paternity, his self-esteem. When it came to biometry, eugenics, and statistics, he was the besieged defender of an emotionally charged faith.

The emotional charge kept Pearson from recognizing that his positivist pursuit of science as relations among measurable observables was not free of speculative theorizing at all but masked a dependence upon prejudiced hypotheses—particularly in the work on heredity. Johannsen's experimental results demanded serious consideration. And Pearson's argument that equal correlation coefficients for intelligence and physique meant equal force of heredity for both was specious. Nevertheless, many would agree with what the geneticist J. B. S. Haldane once said of Pearson's work: "His theory of heredity was incorrect in some fundamental respects. So was Columbus' theory of geography. He set out for China, and discovered America."[74] Pearson's New World was statistics. Advancing with rigor and generalization beyond Galton's pioneering efforts, he made statistics into something more than a tool for the analysis of elementary variations. He devised the product-moment formula for the regular coefficient of correlation; established the theory of multiple correlation and regression; developed a general theory of probable errors; and introduced the chi-squared test, a measure for "goodness-of-fit"—that is, for how well a theoretical curve conformed to a given set of experimental data. Working in aid of Weldon, eugenics, and biology—a number of his most fundamental papers appeared under the title "Mathematical Contributions to the Theory of Evolution"—Pearson laid the foundations of modern statistical methods.

His achievements made Pearson feel all the more frustrated professionally. At times, he thought about emigrating to the United States, where at least people subscribed to *Biometrika*.[75] A break in the University College situation came in 1903, when the Worshipful Company of Drapers, one of the ancient City of London guilds, granted the Biometric Laboratory a thousand pounds. The Drapers were more interested in general good works than in any particular line of research, but Francis Galton, to whom Pearson frequently confided his troubles, was concerned about the long-term prospects of eugenic studies, especially now that he was in his eighties and deciding how to dispose of his ample estate, which would be valued at a hundred and fifteen thousand pounds.[76]

In 1904, Galton committed five hundred pounds a year to University College for a Research Fellowship in National Eugenics. (By "national eugenics" Galton meant "the study of agencies under social control that may improve or impair the racial qualities of future generations either physically or mentally.") Galton expected the first Eugenics Fellow—he was Edgar Schuster, a student of Weldon's, son of a prominent barrister and a young man, in Pearson's estimate, of "manners, wealth, and some experience"—to establish a register of "able families," so as to ascertain the hereditary ingredients of ability. Schuster opened a Eugenics Record Office on Gower Street and busied himself with the collection of pedigrees— mostly of Fellows of the Royal Society. But, being more zoologist than sociologist, he soon wished he were back experimenting with animals.

"The young Oxford man does not know how to work hard," Pearson complained, and Galton, disappointed, decided after two years to terminate the fellowship. He transformed the Record Office into the Galton Laboratory for National Eugenics, under Pearson's directorship, and provided in his will that the bulk of his estate go to University College for the support of studies in eugenics.[77]

Upon Galton's death, in January 1911, University College received forty-five thousand pounds—enough to provide some fifteen hundred pounds a year for the establishment of a Galton Eugenics Professorship. In accordance with Galton's express wish, the authorities gave the post to Pearson, and he was also made head of a new Department of Applied Statistics, which included the Galton and Biometric Laboratories. The new professorship freed Pearson forever from his burdensome introductory teaching. (In June 1911, he had two hundred examination papers to grade. "I can hardly realise that it may be for the last time," he wrote to a friend.) The Galton money and the Drapers' grant, which was regularly renewed —Pearson's work, the Company judged, was "likely to be of great public service"—supplied funds for staff and publications.[78] The joint laboratories expanded from two rooms to four, but Pearson still complained of inadequate space. Setting the pace for the new breed of empire builder in science, he launched a public subscription drive for larger facilities, which by the start of the First World War brought in pledges of sixteen thousand pounds —the bulk of it from Sir Herbert Bartlett, a wealthy contractor who desired an architectural monument but was indifferent to what the monument contained.[79]

Pearson's new department, which was increasingly the center of the English school of statistics, both pure and applied, drew research workers from England, Scotland, the Continent, the United States, India, and Japan. He had his pick of students, ranging from established professionals who wished to master the statistical gospel to fresh young Cambridge graduates. He lectured regularly on basic statistical theory.[80] Usually generous with advice and consultation, he simultaneously managed between ten and twenty different student projects. According to a review he drew up in 1918, his department had to that point trained more than sixty people, who had come to study under Pearson from disciplines that included mathematics, medicine, biology, anthropometry, criminology, psychology, economics, and agriculture.[81]

A significant part of the department's efforts went into eugenic studies. Pearson relied heavily on numerous volunteers, both on his staff and elsewhere in England. Some were medical men, others social workers. From hospitals, schools, and ordinary homes, they gathered material bearing on the "inheritance" of scientific, commercial, and legal ability, but also of hermaphroditism, hemophilia, cleft palate, harelip, tuberculosis, diabetes, deaf-mutism, polydactyly (more than five fingers) or brachydactyly (stub fingers), insanity, and mental deficiency. Pearson published the raw data, including charts and illustrations, in an occasional compendium called *The Treasury of Human Inheritance*, a publication he considered "a pressing necessity of the time." Whatever its bearing upon social questions, the *Treasury*, though in parts flawed by Pearson's assumptions as to what was hereditary, was one of the first orderly aggregations of data on human heredity, and as such was in fact a scientific treasure.[82]

Roughly speaking, the statistical techniques for dealing with the data were developed in the Biometric Laboratory, and the analysis was carried out in its Galton counterpart, but the symbiosis was so close as to make the distinction meaningless. The work, which involved the gathering and the manipulation of large volumes of quantitative data, required numerous people, and Pearson stretched his research funds by giving about one-third of his regular staff positions—for example, five out of fourteen in 1908—to women. In science, women could be hired comparatively cheaply; Pearson in fact paid the women in the laboratory less on the average than the men, even though some of them, including a few Cambridge graduates, had taken higher academic degrees than their male colleagues. Still, vestigially sensitive to the woman question, Pearson deemed the work of the women "equal at the very least to that of the men," and he treated them as professional equals in rank, publication credit, and position in the staff hierarchy. A few took doctorates. Most seemed to be single, devoted their lives to the laboratory—some, trying to emulate Pearson's pace, suffered breakdowns from overwork—and utterly absorbed his views of the world. The female mainstay of the staff was Ethel M. Elderton, who got her first training in statistics as a personal assistant to Galton, became a clerk to Schuster at the Eugenics Record Office, then worked through Pearson's laboratory from Galton Scholar to Reader—the standard British tenure rank—in Social Statistics at University College. "The calculus of correlations," Elderton once asserted, "is the sole rational and effective method available for attacking . . . what makes for, and what mars national fitness."[83]

Pearson's people calculated the variability of human populations and the correlations among relatives for different diseases, disorders, and traits. Studies emanating from the laboratories typically explored the relationship of physique to intelligence; the resemblance of first cousins; the effect of parental occupation upon children's welfare or the birthrate; and the role of heredity in alcoholism, tuberculosis, and defective sight. It was tedious labor, but between 1903 and 1918 Pearson and his staff published some three hundred works—including a series that Pearson chose to call "Studies in

National Deterioration"—not to mention various government reports and popular expositions of eugenics.[84]

"We of the Galton Laboratory have no axes to grind," Pearson declaimed. "We gain nothing, we lose nothing, by the establishment of the truth."[85] The "truth" that Pearson and his co-workers revealed was often advanced with due genuflection to the necessity for methodological caution and insistence upon the implacable objectivity of correlation coefficients, yet the research program amounted to the convictions of Karl Pearson writ large. Pearson chose and assigned the research problems, guided their execution, and edited the results. Intellectually, he was as domineering in the laboratory as outside it. If staff members or students had private reservations about the validity of the work, it required rare courage for them to make their doubts known. More than two-thirds of the research papers appeared in organs that Pearson controlled—notably *Biometrika*.[86]

Ethel M. Elderton summarized the attitude that suffused the Galton Laboratory's key eugenic endeavors: "Improvement in social conditions will not compensate for a bad hereditary influence. . . . The only way to keep a nation strong mentally and physically is to see to it that each new generation is derived chiefly from the fitter members of the generation before."[87] What Pearson's department produced was a mixture of sound statistical science with usually biased explorations in human heredity. But in the early years of the twentieth century it was the sole British establishment for eugenic research, the principal source of authoritative eugenic science, the scientific benchmark of all eugenic discussion in England.

Chapter III

CHARLES DAVENPORT
AND THE WORSHIP
OF GREAT CONCEPTS

A CONSIDERABLE sector of eugenic "science," especially in the United States, owed less to Karl Pearson than to Gregor Mendel. The son of Austrian peasants (neither Pearson nor Galton would on first principles have taken the Mendel family as eugenically promising), Mendel studied physics, chemistry, mathematics, botany, and zoology at the universities of Olmütz and Vienna. He entered the Augustinian monastery at Brünn in 1843, less because he was interested in taking priestly orders than because he wanted to pursue his scientific interests. The abbot, an enthusiast of agricultural improvement, had an experimental garden, and there Mendel began a program of hybridization research with different varieties of peas. In ten years, he bred some thirty thousand plants, analyzing the distribution from generation to generation of alternative characters, such as tallness or shortness of the plants and wrinkledness or smoothness of the seeds. He studied seven of these character pairs, and the data formed the support for his striking theory, announced in 1865 to the Natural Sciences Society of Brünn, that the characters were determined by hereditarily transmitted "elements."[1]

The process of transmission was described by what later came to be known as Mendel's laws of segregation and of independent assortment. Mendel posited that, in his pea plants, there were two elements for every character—e.g., height. According to the segregation law, they were separated from each other in the formation of gametes, i.e., sperm or eggs. According to the law of independent assortment, the elements for one character recombined independently of those for another. The recombination of the various elements which was made possible by the sexual union of the sperm and egg cells was thus determined by the laws of combinatorial probability. The frequency of occurrence of the hereditary elements among the offspring of a hybridized group of plants could, in fact, be predicted in the same manner as the distribution of marbles of different colors drawn randomly from two bags, each containing a known proportion of each color. For example, if two cross-fertilizing plants each contained one element for tallness and another for shortness, the frequency of possible combinations among their offspring would be: a tall element with a tall; a short with a short; and two tall-with-shorts. The tall-tall combination would yield a tall plant; the short-short, a short one. But though a tall-short union might have been expected to produce a plant of intermediate height Mendel observed that it regularly yielded a tall plant. To account for this phenomenon, Mendel theorized that the element for tallness must always overwhelm that for shortness. He called such elements "dominants" and gave the name "recessives" to elements that did not express themselves except when combined with each other.

In 1866, Mendel published his results in the *Proceedings* of the Brünn Natural Sciences Society, only to have the significance of his work go unrecognized for the rest of the century. It was not that the *Proceedings* of the Brünn Society were unknown—they were distributed to some hundred and thirty-four institutions in various countries, and Mendel sent reprints of his paper to other scientists.[2] The likeliest reason for the lack of recognition of Mendel's epoch-making work was that biologists were fastened on the problem of Darwinian evolution in a way that made them unripe for the advent of Mendelian genetics. Evolutionists of the day focused on the adaptation of species—on change. Mendel's theory accounted for the ongoing transmission of characters—for stability. The work of most biologists was descriptive and speculative; Mendel's was experimental, analytic, and quantitative. And while most biologists dealt with the holistically functioning organism, Mendel resembled the physicists and chemists he had studied under, who saw complicated substances as combinations of elementary particles; he reduced the organism to a set of deterministic, hereditary elements. By the late nineteenth century, however, the same dissatisfaction with the prevailing mode of evolutionary studies which led Weldon to turn to Galton had begun to change the biologists' methodological outlook in a direction more favorable to Mendelism.

Many younger biologists were becoming disenchanted with the speculative and descriptive mode of evolutionary research, and particularly with the traditional approach of studying evolution only in the paleontological raw, which prevented direct observation of evolutionary change. By the eighteen-nineties, these biologists were embarking upon programs of research centered not on raw but on controlled nature—not in the wild but on the experimental farm. The new generation was determined to be analytic rather than descriptive, concrete rather than speculative, even quantitative rather than qualitative. They insisted on asking questions answerable only by experimental test. Exemplifying the trend, biologists on both sides of the Atlantic began calling for the establishment of stations for the experimental study of evolution.[3]

Such research rested primarily on hybridization—the crossing of closely related varieties of plants or animals. The idea was to observe the offspring, look for changes in varietal characters, and test whether new species could be evolved by subsequent crosses. The process seemed likely to reveal which, if any, variations in character were heritable, and whether heritable "sports" of evolutionary significance did in fact occur. Once biologists began to contemplate such an experimental program, they edged into the intellectual frame of Mendel's sort of science. Once they actually made the move to scientifically measured hybridization, they were perhaps destined to stumble upon results similar to those of the isolated Austrian monk. It is thus no surprise that, in 1900, Mendel was rediscovered simultaneously—by Carl Correns, in Germany; Erich Tschermak, in Austria; and Hugo de Vries, in Holland—all working independently of each other on different problems involving hybridization, and in de Vries's case with the aim of changing the way evolution was studied.[4]

In the United States and England, Mendelism was immediately embraced by a number of students of evolution, among them the British biologist William Bateson, and by agricultural breeders like William J. Spillman, a plant scientist at Washington State College, who in 1902, in the course of developing a variety of true winter wheat, discovered that the results of his crosses displayed an astonishing regularity explicable by Mendel's theory. Yet the theory also ran into a good deal of skepticism. What was true for peas or wheat was not necessarily true for the rest of the plant and animal kingdom. The mathematics of Mendelian inheritance seemed to conflict with the one-to-one male-female ratio of sexually reproducing species. Particularly disturbing was that many characters expressed themselves not as alternatives—e.g., tall or short—but in a blended fashion, intermediate between the characters of the parents.

"If only one could know whether the whole thing is not a damned lie!" Weldon exclaimed to Pearson. To Weldon, blending rather than alternative inheritance seemed characteristic of the color of even Mendel's type of pea seeds. To Pearson, Mendelism's reliance upon intrinsic hereditary "elements" violated the epistemological rule of dealing only with measurable, observable phenomena. Not incidentally, both Pearson and Weldon harbored an intense dislike for Mendel's British champion, William Bateson, who was no friend of biometry. Their disagreement with Bateson over the comparative merits of the Mendelian and the biometric approaches to heredity studies exploded early in the century into one of the most vitriolic disputes in the history of science.[5] But the various objections only spurred the Mendelians further into the new discipline that Bateson christened "genetics," and in the years prior to the First World War, especially in the United States, overwhelming evidence accumulated in favor of Mendel's theory.

At Columbia University, in 1902, Walter Sutton, a student in the laboratory of the great cytologist Edmund B. Wilson, showed that in cell division the chromosomes behaved in a way consistent with Mendel's laws of segregation and independent assortment. Three years later, working independently of each other, Wilson and Professor Nettie M. Stevens of Bryn Mawr concluded that the determination of sex, including the one-to-one male-female ratio, was caused in Mendelian fashion by the segregation and reunion of the X and Y chromosomes.[6] Other geneticists extended the boundaries of Mendelian experimentation to incorporate an ever-widening sector of the plant and animal kingdom. While confirming the theory in its essentials, they also modified it with the finding that many traits, including those of an apparent blending nature, were determined by combinations of "genes," as the Mendelian elements came to be known. Guiding a team of brilliant young students, Thomas Hunt Morgan, Wilson's colleague at Columbia, scrutinized the offspring of innumerable generations of the fruit fly *Drosophila melanogaster*; the team identified the chromosome as the seat of the gene, worked out the intricate mechanics of chromosomal determination of heredity, and joined cytological to breeding genetics with a triumphant force that eventually won Morgan the 1933 Nobel Prize in physiology or medicine.[7]

Populations that bred rapidly, such as poultry, rodents, and fruit flies, made the most advantageous subjects for genetic research, but early in the century, scientists began testing Mendel's theory for man, despite the slowness of his breeding—because, of course, he was man. In 1902, the British physician Archibald Garrod, who had been pointed in the right analytical direction by Bateson, convincingly showed that certain "inborn errors of metabolism"—notably alcaptonuria, a disease signaled by a darkening of the urine shortly after birth—were caused in a Mendelian manner by recessive genes. In 1907, the Mendelian inheritance of human eye color was demonstrated in Britain by C. C. Hurst, one of Bateson's allies in the war against the biometricians, and in the United States by the accomplished biologist Charles B. Davenport, who extended the analysis to hair and skin color, and soon launched his career as America's leading eugenicist.[8]

A MEMBER OF THE NEW, anti-speculative generation of biologists, Charles Davenport had studied engineering in preparatory school, acquiring mathematical skills rare for a biologist. While an instructor in zoology at Harvard in the eighteen-nineties, he read Karl Pearson's papers on the mathematical theory of evolution, lectured and published on variation and inheritance, and, in an influential book on morphology, pleaded for the infusion into biology of the exact methods of the physical sciences. He was soon recognized as an important pioneer in biometry, and in 1899 he left Harvard for an assistant professorship at the recently founded University of Chicago. Early in the new century, on a trip to England, he visited Galton, Weldon, and Pearson—the high priests of biometry—and, after a dinner at Rutland Gate, returned home with "renewed courage for the fight for the quantitative study of Evolution."[9] Davenport, who early showed signs of being an energetic organizer, successfully persuaded the munificent new Carnegie Institution of Washington—its ten-million-dollar endowment from Andrew Carnegie then exceeded the total endowment for research in American universities—to establish a station for the experimental study of evolution.[10]

The station was set up in 1904, under Davenport's directorship, at Cold Spring Harbor, some thirty miles from New York City on Long Island's North Shore. Davenport was already the head of the summer Biological Laboratory of the Brooklyn Institute of Arts and Sciences, also at Cold Spring Harbor. The two laboratories bordered the sound on twelve acres of woodland, field, and marsh, with abundant fauna and a freshwater stream. The new station was well funded; its budget for 1906 was twenty-one thousand dollars—more than twice Pearson's at University College. William Bateson regarded the new station with "wonder and admiration," telling Davenport, "How any decent competition is to be kept up on our side I scarcely know!"[11] Davenport recruited a small staff, in part from able students who had passed through the Biological Laboratory, and set it to work on research projects in variation, hybridization, and natural selection. On the whole, the staff contributed respectably to the developing fields of biometry and genetics, and so did Davenport. His work with poultry and canaries played an important role in the early Mendelian analysis of inheritance in animals. After the work on eye, hair, and skin color, he was eager to explore the force of heredity across a broad range of human traits.[12]

Unable, of course, to experiment with human breeding, Davenport had to find his inheritance data by collecting extended family pedigrees. Galton and Pearson had gathered data only for parents and children, because they were concerned with what geneticists came to call the "phenotype"—the organism's set of observable characters. Davenport was interested in the "genotype"—the individual's genetic makeup. Not directly observable, the genotype had to be inferred from scrutiny of as many related phenotypes as possible, in and beyond the immediate family. A lot of rough, unsystematic data of that sort were scattered through medical journals, where over the years doctors had recorded the familial incidence of various diseases. (A member of Davenport's staff once remarked that many works on eugenics reeked like "a medical museum of morbid anatomy with its charnel odors and gruesome sights.")[13] To gather data on normal as well as abnormal characters, Davenport drew up a "Family Records" form and distributed hundreds of copies to medical, mental, and educational institutions; to numerous individuals, especially scientists; and, through the aid of his sister Frances, to the Association of Collegiate Alumnae. Hundreds were returned, filled out for at least three generations. They formed the basis of a widely noted book he published in 1911, *Heredity in Relation to Eugenics.*[14]

Wherever the family pedigrees seemed to show a high incidence of a given character, Davenport concluded that the trait must be heritable and attempted to fit the heritability into a Mendelian frame. He observed that single Mendelian elements—he called them "unit characters"—might well account for such abnormalities as brachydactyly, polydactyly, and albinism, and for such diseases as hemophilia, otosclerosis, and Huntington's chorea. Although he noted that single elements did not seem to determine important mental and behavioral characteristics, he did argue that patterns of heritability were evident in insanity, epilepsy, alcoholism, "pauperism," criminality, and, above all, "feeblemindedness"—a catchall term of the day, used indiscriminately for what was actually a wide range of mental deficiencies.[15] Like many scientists of his time, Davenport held that physiological and anatomical mechanisms made some people alcoholics, others manic-depressives, still others "feebleminded." Such people had often inherited "a general nervous weakness—a neuropathic taint—showing itself now in one form of psychosis and now in another."[16] Davenport similarly reduced pauperism to "relative inefficiency [which] in turn usually means mental inferiority." Of course, he conceded, human breeding was complicated, and human progeny were the products of both "conditions *and* blood." But attention to environment was not to obscure the crucial role of protoplasm in human fate. Heredity determined the characteristics both of Negroes—Davenport's views on black Americans conformed for the most part to the standard racism of the day—and of the immigrants then flooding into the United States.[17]

Like many of his colleagues, Davenport equated national and "racial" identity, and assumed as well that race determined behavior. He held that the Poles, the Irish, the Italians, and other national groups were all biologically different races; so, in his lexicon, were the "Hebrews." Davenport found the Poles "independent and self-reliant though clannish"; the Italians tending to "crimes of personal violence"; and the Hebrews "intermediate between the slovenly Servians and Greeks and the tidy Swedes, Germans, and Bohemians" and given to "thieving" though rarely to "personal violence." He conceded that "the great influx of blood from Southeastern Europe" was less prone than the native variety to burglary, drunkenness, and vagrancy, and "more attached to music and art." Some of the best professors of science with whom Davenport was acquainted came from a Hungarian family. Yet on the whole Davenport expected that the new blood would rapidly make the American population "darker in pigmentation, smaller in stature, more mercurial . . . more given to crimes of larceny, kidnapping, assault, murder, rape, and sex-immorality."[18]

Like Galton and Pearson, Davenport identified good human stock with the middle class—especially "intellectuals," artists and musicians, and scientists. In his American context, he also gave high marks to the native white Protestant majority. With the aim of improving the national protoplasm, Davenport to a degree embraced the eugenics of Galton, with its stress on the procreation of the good stock. He looked forward to the day when a woman would no more accept a man "without knowing his biologico-genealogical history" than a stockbreeder would take "a sire for his colts or calves . . . without pedigree."[19] Yet his concern with fostering the increase of the good stock was decidedly outweighed by his emphasis on what came to be called "negative eugenics"—preventing proliferation of the bad.

Anxious that the nation's protoplasm was threatened from without, Davenport favored a selective immigration policy. In his biologically considered view, "no race *per se*, whether Slovak, Ruthenian, Turk or Chinese, is dangerous and none undesirable." He thus took sound immigration policy to mean not the wholesale exclusion of national groups but the denial of entry to individuals and families with poor hereditary history. "The idea of a 'melting pot' belongs to a pre-Mendelian age," he wrote to a fellow exclusionist. "Now we recognize that characters are inherited as units and do not readily break up." Defective germ plasm from abroad would therefore not be obliterated by mixture with the healthy variety; it would persist. If the family history of all prospective immigrants could be investigated, people with hereditarily "imbecile, epileptic, insane, criminalistic, alcoholic, and sexually immoral tendencies" could be detected and kept out.[20]

To counter the threat from within, negative eugenics called for preventing the reproduction of the genetically defective, possibly by state-enforced sterilization. If the state could take a person's life, Davenport judged, surely it could deny the lesser right of reproduction. (In 1911, six states already had sterilization laws on the books.) Yet scientifically it was not clear who should be sterilized; "feeblemindedness" was hardly as sharply defined as, for example, polydactylism. Besides, Mendelism taught that the union of a so-called feebleminded person with a normal person could produce normal offspring. Why sterilize people unnecessarily? Davenport preferred to eliminate the nation's defective protoplasm by the sexual segregation of defectives while they were capable of reproducing. The state would eventually be repaid the cost of care, because in the long run the policy would drastically reduce the need for state institutions.[21]

Davenport could occasionally give eugenics a flavor of humane good sense with his warnings that the victims of Huntington's chorea—"this dire disease"—or the sisters of hemophiliacs should not have children; or with the logical declaration that there was no point in imprisoning the insane, the feebleminded, or the criminal whose antisocial behavior was genetically determined—their fate ought to be decided by physicians and eugenicists rather than by judges, and they belonged in homes and hospitals rather than in prisons. Yet Davenport was prepared to curtail other people's rights in order to promote the race—to ensure the common protoplasmic good. He remarked to a prospective patron that "the most progressive revolution in history" could be achieved if somehow "human matings could be placed upon the same high plane as that of horse breeding."[22] His protoplasmic vision was on the whole offensive, in part cruel. Equally indefensible, although it was advanced with the authority and prestige attendant on one of America's most powerful biology directorships, it proceeded from science that, even by the standards of his own day, was usually dubious and often plain wrong.

Davenport's Mendelism was generally up to date, and the research program he set up at Cold Spring Harbor addressed fundamental genetic issues. Imaginative in its extension of Mendelism to human heredity, his work on the inheritance of traits—color blindness, for instance—that lent themselves to a pedigree approach contributed usefully to the early study of human genetics. But Cold Spring Harbor—richly budgeted and equipped, the envy not only of Bateson but of Pearson, a warm-weather watering hole for many able biologists—amounted scientifically to much less than it might have. "The success of these things always lies in the individual who dominates the whole," Pearson remarked to Francis Galton, "and our friend Davenport is not a clear strong thinker."[23]

Galton agreed, and so did even some of Davenport's pro-Mendelian American colleagues. He combined Mendelian theory with incautious speculation. He knew that certain traits expressed combinations of elements —that is, were polygenic in origin—and had advanced the notion in his own research on skin color, yet his analysis of mental and behavioral traits usually neglected polygenic complexities. Davenport thought in terms of single Mendelian characters, grossly oversimplified matters, and ignored the force of environment. Sometimes he was just ludicrous, particularly in various post-1911 studies on the inheritance of "nomadism," "shiftlessness," and "thalassophilia"—the love of the sea he discerned in naval officers and concluded must be a sex-linked recessive trait because, like color blindness, it was almost always expressed in males.[24] His eugenic analyses rested on pedigrees gathered without rigorous rules of evidence concerning the traits they purported to show. His analytical concepts drew uncritically on vague, unproved notions—notably the neuropathic basis of mental illness.

Davenport's friend Smith Ely Jelliffe, a New York psychiatrist who was a pioneer of Freudianism in America and an authority on mental illness, chided him for lumping its various expressions under the word "insanity." The term might be warranted in legal practice, but it was "nonsensical" to employ it in medical matters for what was caused in one man by a head injury, in another by too much alcohol, in a third by typhoid infection, in a fourth by uremia, in a fifth by ongoing emotional disturbance. "Is it logical to take such an enormous complex of conditions as all the psychoses and try to make them all fit in one artificial box?" Jelliffe asked. "It is the same way with the epilepsies. . . . There is no *one* epilepsy." He pointed out that convulsions could arise from a hard blow to the head, a motor-area thrombus provoked by infection, or poisoning by santonin, and asked, "Is there any heredity here—or chance of it?" If eugenics was to be "correctly started," Jelliffe noted, "we must sharpen up our conceptions, and that very markedly."[25]

Davenport tossed aside Jelliffe's sensible caution and continued to claim that mental disease seemed for the most part to be heritable. From the point of view of hereditary transmission, differentiation into classes was useless, and he saw no reason for such diseases not to be lumped together in the formulating of eugenical advice.[26] Greatly given to oversimplification and little to self-critical reflection, Davenport possessed neither Galton's idiosyncratic imagination nor Pearson's formidable intellectual power. Like Pearson, he was blinded by eugenic prejudice. But, unlike Pearson, Davenport did not base his eugenics on any political world view; he had none. His eugenics arose from the combination of professional circumstance and personal background which shaped his life.

DAVENPORT'S CHILDHOOD HOME—on Garden Place, in Brooklyn Heights—was dominated by a quick-tempered, puritanical paterfamilias. Amzi Davenport, the father of eleven children by two wives, was a former teacher who had become a successful real estate and insurance broker and an ardent temperance advocate. Before the Civil War, he had been an abolitionist; he was a founder, deacon, and ruling elder of Henry Ward Beecher's Plymouth Congregational Church. In the Davenport house, the day began with an early prayer meeting. It continued, for Charles, with a full morning of factotum's duties in his father's office, an afternoon of solitary study there, and an evening of paternal instruction and quizzes—on his religious and academic lessons; he was sent to bed immediately if he failed on any count. Charles spent the summers in a similar regimen at the family farm, on Davenport Ridge, near Stamford, Connecticut. His boyhood diary lists chores accomplished and sermons listened to; he hardly mentioned jokes, pranks, friends, or pleasures. He found a kind of outlet in writing, particularly the *Twinkling Star*, a small monthly that he published for two years and filled with some humor and a lot of family, including paternal, news.[27]

"O! I want to go to school," to escape the office "prison house," Charles confided to his diary. There was no discouragement from his mother. She was Amzi Davenport's second wife—the granddaughter of a wealthy judge and the daughter of a prominent Brooklyn builder. Self-confident, easy in her piety to the point of religious skepticism, she was openly affectionate and, besides bearing nine of her husband's children, pursued serious interests in French, gardening, and natural history. Though Charles's father felt himself to be an adequate tutor, she wanted her children educated right through college. In 1879, at the age of thirteen, Charles was permitted to enter the Brooklyn Collegiate and Polytechnic Institute. Although his office duties continued after school, he began to collect insects, became a recorder of data for the United States Weather Service, and celebrated in a student theme "the privilege of adding to human knowledge by studying the stars, by investigating the lives of animals and plants, by revealing the secrets."[28]

Near the end of his Polytechnic career, Charles proposed to devote one-quarter of his time during the coming summer on Davenport Ridge to his father, "to compensate directly for my indebtedness to you for my support," and three-quarters to a variegated program of scientific research, including agriculture, meteorology, and surveying. Charles hoped that his father would not regard the plan merely as "a selfish scheme to get rid of work," since by now he had found his vocation in science. It was almost two months before his father replied to the entreaty, and then he announced that Charles had "failed somewhat in meeting my views on the *practical* parts of the subject." The surveying would be fine. "As to spending so much time in looking after the geological character of the place, the nature of the soil, the adaptations of manures and chemical appliances to the improvement of the land, etc., I think . . . that you are too theoretical." Charles graduated first in his class at the Polytechnic and dutifully catered to his father's inclinations by becoming a surveyor. He stuck it out for nine months, and then fled to Harvard and his mother's cherished subject, natural history. In 1891, he received an instructorship, and the following year a Ph.D. and entrance to the world of professional biology.[29]

During his early Cambridge Sundays, Davenport would attend church in the morning, read in the afternoon, and write to his mother in the evening. The rest of the week was taken up by biology and, after a while, by Gertrude Crotty, the daughter of a Kansas rancher and a graduate student in zoology at the Society for Collegiate Instruction of Women, as Radcliffe College was then known. She married Davenport in 1894 and became his closest friend and collaborator, eventually helping to write the work at Cold Spring Harbor on the heritability of human eye, skin, and hair color. She also spurred his ambition. Strong-willed and increasingly money-conscious after the birth of their two children, Gertrude wanted Davenport to move beyond the Harvard instructorship; she is said to have scanned the death notices in *Science* for the likelihood of academic vacancies.[30] In 1904, the year he became director of the experimental station at Cold Spring Harbor, with an annual salary of thirty-five hundred dollars and the promise of a raise the next year to four thousand dollars—the equal then of the very best-paid professorships in the United States—the Davenports bought six acres and a house on the shore at Cold Spring Harbor to rent to laboratory staff. They soon added a nearby nineteen-acre farm, bought in Gertrude's name, to their holdings. "Quite an empire for us, isn't it?" Davenport proudly exulted to his wife.[31]

Cold Spring Harbor was then a semi-rural area of large, wealthy estates. One side of the laboratory property was bounded by the country seat of the Tiffany family. The laboratory's well-paid, amply propertied director easily took on the general political coloration of early-twentieth-century Nassau County. Davenport organized taxpayers' associations, railed against the spending of public funds for more social workers, and called for more police control in the Cold Spring Harbor neighborhood to drive home the idea of law and order to laborers brought out from the city, and "especially to the young recent immigrants to this country who . . . mistake liberty for license." He may have argued against barring the entry of particular national groups, but he believed that the European nations sent over disproportionately large numbers of their worst human stock, that immigrants rapidly outbred the native population, and that they supplied an excess of public charges. Davenport deplored the fact that the government had to support tens of thousands of insane, mentally deficient, epileptic, and otherwise handicapped wards, not to mention prisoners and paupers, at a cost he estimated to be about a hundred million dollars a year.[32] In part, his negative eugenics simply expressed in biological language the native white Protestant's hostility to immigrants and the conservative's bile over taxes and welfare.

But only in part. Davenport had rejected his father's piety, but he replaced it with a Babbitt-like religiosity, a worship of great concepts: Science, Humanity, the Improvement of Mankind, Eugenics. The birth-control crusader Margaret Sanger recalled that Davenport, in expressing his worry about the impact of contraception on the better stocks, "used to lift his eyes reverently and, with his hands upraised as though in supplication, quiver emotionally as he breathed, 'Protoplasm. We want more protoplasm.' "[33] Davenport may have embraced his mother's beloved science, but he could never exorcise the paternal insistence upon practicality and success, or the paternal implication that he was somehow inadequate to achieve either. Thus, in his scientific work, he went from topic to topic, from biometry to Mendelism to poultry to people, exploring each with shallow carelessness. Thus he plunged into eugenics, with its mixture of science and social utility. Constantly craving approval, he joined numerous editorial boards, took out memberships in sixty-four organizations, accepted ten executive posts. He found his sense of identity in his work. When his scientific papers were attacked, he lapsed into depression, confusion, and petulant bitterness.[34]

Davenport, having been virtually a stranger to pleasure in his boyhood, was a driven man, uncomfortable with enjoyment to the point of guilt. Revealingly, he described his daughter Jane as "methodical and self-controlled . . . a fine girl." He found his daughter Millia—nicknamed Billy—something of a trial. Divorced after a hasty marriage, she established herself in Greenwich Village in the nineteen-twenties as a breathless flapper, attached herself to an avant-garde arts magazine, designed costumes for the Provincetown Players, ran up large bills at Wanamaker's, and irritatingly challenged her father's eugenic convictions. ("The world," she observed, "is not made up of college professors' children, at least, not the dearest part I've found.")[35] Billy's father cast a pall over life at Cold Spring Harbor, where the remoteness from any town of consequence made the social life of single researchers none too happy to begin with. Demanding, suspicious, quick to charge discontented staff with disloyalty, he strongly objected when two young women invited a male colleague to their room for a late-night cup of soup.[36]

Davenport bridled at the merest hint of sexual indulgence. Sexually continent before his marriage, he remarked with seeming knowledge after it that "a man who has never been sexually active can more readily be continent than one who has had such experiences," adding, "Instincts develop and are strengthened by exercise."[37] Yet the piety-bound Victorian childhood he knew had been pervaded with disapproval of most kinds of physical gratification. Davenport deplored birth control not only because of its dysgenic effects among the families of intellectuals but also because of its "aid to luxury and convenience." He regularly grouped sexual immorality with such eugenically adverse traits as feeblemindedness and criminality. He was anti-Semitic partly because of his conviction that Jews showed "the greatest proportion of offenses against chastity and in connection with prostitution, the lowest of crimes."[38] Probably Davenport was emotionally ambivalent toward sexuality. Certainly in his eugenic theories he gave particular attention to sexual abandon in others, and even advocated drastic measures to suppress it.

In a study of "wayward" girls, Davenport concluded that the cause of prostitution was not economic circumstance but an "innate eroticism," determined by a dominant Mendelian element. He believed that the brain contained a center for eroticism similar to that for, say, speech. In normal people, the erotic center would be of moderate strength and inhibited by a genetically determined governor; in abnormal people, the erotic center would be excessively energetic and would lack the inhibiting mechanism. The release of so much erotic energy, Davenport claimed, resulted not only in sexual licentiousness but also in violent outbreaks of temper and derivative crimes. People thus afflicted fell into a class that Davenport named, in analogy with the feebleminded, the "feebly inhibited."[39] While he preferred segregation to sterilization as a means of preventing the reproduction of the unfit, he argued that any sterilization of the unfit should be accomplished by castration instead of vasectomy. Vasectomy, he knew, prevented paternity but not lust, and he believed that physiologically divorcing the sex act from responsibility for its procreative consequences might well encourage rapists. Davenport maintained that castration, unlike vasectomy, "cuts off the hormones and makes the patient docile, tractable and without sex desire."[40]

Davenport paid no attention to Freud. He was as wrong in his neurological theories as in many of his genetic ones, but, perhaps self-protectively, he acknowledged that eugenicists were far from possessing the knowledge required to advise people on what constituted fit marriages— how to "fall in love intelligently," as he put it—or to decide who, exactly, ought to be prevented from propagating. He thought it imperative that the eugenicist avoid reproach for marching beyond clear, certain knowledge into the thickets of hereditary policy. "Our greatest danger," he once warned the farmers and biologists of the American Breeders' Association, "is from some impetuous temperament, who, planting a banner of eugenics, rallies a volunteer army of utopians, free lovers, and muddy thinkers to start a holy war for the new religion." To Davenport the professional scientist, the watchword of eugenics for the time being had to be "investigation." He dreamed of gathering enormous quantities of human hereditary data, recording them in a central bureau of study, and ultimately throwing light on "the great strains of human protoplasm . . . coursing through the country."[41]

FINANCING SUCH AN ENTERPRISE was a challenge, but not one beyond Davenport's reach, given the nature of the era. In the early twentieth century, a swelling chorus of commentators held that solutions to the complex issues of modern society required knowledge that could come only from research and from consultation with experts. The endowment of research was joining the agenda of philanthropic works; Andrew Carnegie exemplified the trend. State university scientists were successfully persuading their legislatures to appropriate money for research, and the federal scientific establishment was steadily growing in response to the need for data essential to commerce, trade, and regulation. Institution building was the order of the scientific day, and Davenport, with his protoplasmic social purposefulness, was, more than anything else, an entrepreneur of the knowledge business. In 1909, his eugenic ambitions in mind, he approached Mary Harriman in the hope of stimulating the philanthropic interest of her mother, Mrs. E. H. Harriman, who had recently taken over the management of her late husband's immense railroad fortune.[42]

Mary Harriman had spent part of the summer of 1905 at the Cold Spring Harbor Biological Laboratory while an undergraduate at Barnard College. A founder of the New York Junior League for the Promotion of Settlement Movements, she was a social activist with a liberal bent. (She would eventually break with her family's Republicanism to join her brother Averell in supporting the presidential candidacy of Al Smith, become a friend of Eleanor Roosevelt and Frances Perkins, and head the Consumers' Advisory Board of Franklin Roosevelt's National Recovery Administration.) Eugenics struck her as a means of social improvement, and she brought Davenport together with Mrs. Harriman, who later said that both her husband's and her father's interest in breeding racehorses had suggested to her that the laws of heredity might also be used for the amelioration of man. Over luncheon, in February 1910, Mrs. Harriman agreed to support Davenport's ambitions for eugenic research on a grand scale. "*A Red Letter Day* for humanity!" Davenport wrote in his diary.[43]

Later that year, Mrs. Harriman funded the establishment of a Eugenics Record Office, on seventy-five acres of land she bought for the purpose up the hill from Davenport's Cold Spring Harbor experimental station. Eugenic research, Davenport held, was best conducted by scientifically trained personnel and in proximity to studies on the heredity of other organisms. Drawing on his prior experience, Davenport intended to ferret out human hereditary data by making house-to-house surveys and by scrutinizing the records of the nation's numerous prisons, hospitals, almshouses, and institutions for the mentally deficient, the deaf, the blind, and the insane. Information about the relatives of a given "defective," he thought, could be "obtained with a high degree of precision by tactful field workers." People might object to such surveys on the ground that heritable traits were "private and personal matters," but surely theirs was "a narrow and false view." Besides, Davenport proposed to keep all records confidential, and to employ them only statistically.[44]

The site came with a house for use by the staff, including a fireproof addition for the storage of the expected pedigrees. Mrs. Harriman also provided funds—amounting initially to some twenty thousand dollars a year—for operating expenses. Evidently pleased with the work, she continued to contribute handsome sums to the Record Office until 1918, when she turned the entire installation over to the Carnegie Institution of Washington, which soon incorporated it with Davenport's original station as its Department of Genetics. Mrs. Harriman's gift to the Carnegie Institution came with an endowment of three hundred thousand dollars, bringing the total of her eugenic patronage between 1910 and 1918 to more than half a million dollars.[45]

Part of Mrs. Harriman's money paid for the field workers that Davenport wanted, and so did an additional twenty-two thousand dollars over four years from John D. Rockefeller, Jr. With the money, the Eugenics Record Office provided scholarships for young men and women to come to Cold Spring Harbor in the summer for training in human heredity and field-research techniques. After the summer training course, the trainees, at a salary of seventy-five dollars a month, began a year's work in the field. Davenport had expected most of the field workers to be women's college graduates with some training in biology, and many did indeed come from Radcliffe, Vassar, and Wellesley, joining graduates of Harvard, Cornell, Oberlin, Johns Hopkins, and other reputable schools. Once trained, they were armed with a "Trait Book" for guidance and sent to study albinos in Massachusetts; the insane at the New Jersey State Hospital in Matawan; the feebleminded at the Skillman School, in Skillman, New Jersey; the Amish in Pennsylvania; the pedigrees of disease in the Academy of Medicine records in New York City; and juvenile delinquents at the Juvenile Psychopathic Institute of Chicago. The only cost to the institutions was the workers' expenses. The institutions got hereditary information concerning their charges—which they used for any number of purposes, including reports to legislative committees—and the Eugenics Record Office got the data for cataloguing and eventual analysis. By 1913, these amounted to thousands of items—"a sort of inventory of the blood of the community," *Scientific American* noted.[46]

Davenport assembled the active field workers for a conference each year to keep them in touch with the latest research techniques and eugenic theories. Other participants in the conferences included directors of epileptic colonies and hospitals for the insane or the feebleminded, and, on one occasion, the medical examiner from Ellis Island. In 1916, the Record Office began publishing—also for the edification of the field workers—the *Eugenical News*, which looked in format like a slightly more substantial version of Davenport's old *Twinkling Star*. ("Serious we have to be, Working for posterity," they sang one summer at Cold Spring Harbor.) Field work was hard work. Only from twenty to thirty percent of the field workers did it well, Davenport estimated. The rest suffered from timidity or from ignorance of what they were supposed to be studying. Able or not, more than two hundred and fifty field workers were sent out by the Eugenics Record Office between 1911 and 1924, when the training program ended.[47] Their efforts centered on subjects like the "feebleminded," since, as wards of the state, they had family histories that were comparatively simple to get. The data from the field—analyzed, indexed, and entered on about three-quarters of a million cards at Cold Spring Harbor—served as the source of bulletins, memoirs, and books, on such topics as sterilization, the exclusion from the United States of inferior germ plasm, and the inheritance of pellagra, multiple sclerosis, tuberculosis, goiter, nomadism, athletic ability, and temperament. Davenport consulted his cards to respond to numerous inquiries about the eugenic fitness of proposed marriages. The publications, he proudly reported in 1920 to the Carnegie Institution, had enjoyed a "marked educational influence."[48] So had the field workers, many of whom became teachers of genetics and eugenics or members of state commissions and other institutions dedicated to the reduction of hereditary degeneration and defect. Like Karl Pearson's research program, their work supplied ample "authoritative" material to the Anglo-American eugenics movement, which gathered increasing popular force after the turn of the century, with no small impact upon education and immigration policy and such sectors of social distress as the so-called feebleminded. "What a fire you have kindled!" Davenport wrote to Mrs. Harriman shortly after the founding of the Record Office. "It is going to be a purifying *conflagration* some day!"[49]

Chapter IV

THE GOSPEL
BECOMES POPULAR

ONE DAY IN JANUARY 1901, Karl Pearson took a moment from his biometrical labors at University College, in London, to write to his friend Francis Galton on a subject of "the greatest national importance—the breeding from the fitter stocks." He told Galton that Britain needed "some word in season, something that will bring home to thinking men the urgency of the fertility question in this country." Certainly no one, he added, would be listened to on the matter more than Galton himself, who had after all "set the whole scientific treatment of heredity going."[1]

Galton, seventy-eight years old, ailing, yet still a spirited enthusiast of eugenics, bestirred himself during the next few years to take to the podium for the science of human improvement. What he had to say about eugenics in the new century differed little from his pronouncements in the heyday of Victoria's reign, but the response was hardly the same. In 1904, a large audience—including medical men and scientists, not to mention H. G. Wells—turned out to hear him at the Sociological Society, in London, and his address was reprinted on both sides of the Atlantic. In 1909 Galton was knighted, and the next year was awarded the Copley Medal—the Royal Society's highest honor.[2] In the last few years of his life, among the thinking classes of the Anglo-American community, Francis Galton and his eugenics were suddenly very much in season.

"You would be amused to hear how general is now the use of your word *Eugenics!*" Pearson exulted to Galton in 1907. "I hear most respectable middle-class matrons saying if children are weakly, 'Ah, that was not a eugenic marriage!'" Pearson's writings had also helped stimulate popular eugenic discussion—most of it serious (few took his dire warnings about the future of British society anything but seriously), some of it droll (his studies on tuberculosis suggested that firstborn children tended to be weaker than others, which implied, members of the press were quick to point out, that the House of Lords, populated as it was by firstborn sons, must be degenerating). A London woman, pregnant and enterprisingly Lamarckian, betook herself to plays and concerts, conversed with H. G. Wells among other writers, and in 1913 gave birth to "Eugenette Bolce," who was widely hailed as England's first eugenic baby. A Brighton physician lamented that the word "eugenics" had become "a mere catch phrase which covers any rubbish which any crank chooses to inflict upon the world." Eugenics meetings would bring out "all the neo-Malthusians, antivaccinationists, antivivisectionists, Christian Scientists, Theosophists, Mullerites (who have strange ways of having a bath and of breathing deep breaths), vegetarians, and the rest! Poor Sir Francis Galton."[3]

In America, thousands of people filled out their "Record of Family Traits" and mailed the forms to Charles B. Davenport's Eugenics Record Office, at Cold Spring Harbor, Long Island. The undergraduate F. Scott Fitzgerald wrote the song "Love or Eugenics" for the 1914 Princeton Triangle Show ("Men, which would you like to come and pour your tea. / Kisses that set your heart aflame, / Or love from a prophylactic dame"). In both countries, demands for lecturers on eugenics came from ethical, debating, health, and philosophical societies; school and university campuses; women's clubs; medical and nursing associations; and Y.M.C.A.s. In London, the Ladies Emily Lutyens and Ottoline Morrell opened their Bloomsbury drawing rooms to eugenics speakers and students thronged to hear lectures on the topic at the Bedford College for Women.[4] British and American newspapers frequently published articles on eugenics and a steadily increasing number of such articles appeared in popular magazines. Hardly a year went by without a spate of books on eugenics—from scientists like Davenport as well as from enthusiastic laymen. Virtually all the literature paid homage to Francis Galton, and many rehearsed the data, theories, and opinions of Davenport, Pearson, and their collaborators.[5]

The outbreak of the war thrust eugenics into the background of public discourse, although eugenics theorists in Britain and the United States did worry in print about the impact of the war upon the quality of their national protoplasm. To some, war and militarism were clearly dysgenic: whether armies were formed by voluntary or selective service, war took the best and the bravest and exposed them to death, probably before they had managed to procreate; it also left the biologically less fit at home to father the next generation. Even theorists who disputed this analysis agreed that battlefield losses meant a reduction in the ratio of marriageable men to marriageable women. Some predicted a eugenic result—the remaining men would choose only the ablest and most beautiful women; others expected a dysgenic outcome—women would be reduced to scrambling after even unworthy men.[6]

Whatever the outcome, public attention to eugenics was renewed after the Armistice with a force that made Galton's religion as much a part of the secular pieties of the nineteen-twenties as the Einstein craze. One of the leading popularizers of the creed was Albert E. Wiggam, the journalist, author, and Chautauqua lecturer. Wiggam had begun to educate himself in eugenics just before the war, with visits to geneticists and to Davenport's Eugenics Record Office. Adding eugenics to his lecture repertoire, he distributed family-record blanks to his Chautauqua audiences for completion and mailing to Cold Spring Harbor. During the twenties, Wiggam promoted eugenics in articles and in three widely read books, including the 1923 best-seller *The New Decalogue of Science*. While many writers reported soberly upon standard eugenic doctrines, Wiggam stood out for the way he melded eugenic science with statesmanship, morality, and religion. Eugenics was "simply the projection of the Golden Rule down the stream of protoplasm." Indeed, had Jesus returned in the nineteen-twenties, he would have given the world a new commandment: "the biological Golden Rule, the completed Golden Rule of science. *Do unto both the born and the unborn as you would have both the born and the unborn do unto you.*" Biologists tended to find him inaccurate and breezy. But Wiggam was pro-science, pro-biology, pro-evolution. In the era of the Scopes trial, scientists no doubt forgave him his errors because of the banner he carried.[7]

The vogue for eugenics derived energy from the organizational efforts of its advocates. In 1907, inspired by Galton, a national Eugenics Education Society was founded in Britain. "Its purpose," a charter member explained to Galton, "is to stir up interest . . . and is, on the whole, frankly propagandist." Galton hesitated to join but accepted membership in 1908, and was thereupon elected honorary president. Branches of the society sprang up in Birmingham, Cambridge, Manchester, Southampton, Liverpool, Glasgow, and Sydney, Australia.[8] Local eugenics groups sprouted across the United States, including the Galton Society, which met regularly at the American Museum of Natural History, in New York; the Race Betterment Foundation, in Battle Creek, Michigan; and eugenics education societies in Chicago, St. Louis, Wisconsin, Minnesota, Utah, and California. Eugenic themes diffused into groups devoted to sex education and sex hygiene, and were evident in the baby-health competitions that spread to some forty states before the war. Various efforts—the promoters included Davenport, Alexander Graham Bell, and Luther Burbank—were mounted to organize eugenics on a national basis, along the lines of the British society; they culminated in the formation in 1923 of the American Eugenics society, which rapidly spawned twenty-eight state committees and a southern California branch.[9]

Nominal membership in the British society never exceeded seventeen hundred, and in the American probably no more than two-thirds of this, but what the organizations lacked in size they made up for by what an early British member predicted would be "the advantage of excellent patronage." Local British and American groups listed leading townspeople among their members, and the national councils included distinguished scientists and social scientists, prominent lawyers, clerics, physicians, schoolmasters, intellectuals, and—in Britain—several knights of the realm. In 1911, the Oxford University Union moved approval of the principles of eugenics by a vote of almost two to one, and meetings of a eugenics society at Cambridge University before the war drew hundreds of people, including high college officials, Nobel laureate scientists, powerful senior professors, and the young John Maynard Keynes.[10] The prime mover in the American Eugenics Society was the well-known Yale economist and public health advocate Irving Fisher. The president (from 1911 to 1928) of the British society bore a name to conjure with in matters of descent—he was Major Leonard Darwin, a son of Charles Darwin.

Like Galton before them, Anglo-American eugenicists reckoned that, before a eugenics revolution could occur, the public would have to be taught to be "eugenic-minded." Dues and endowments gave the eugenic societies of both countries funds for the sponsorship of lectures and meetings. By the late nineteen-twenties, the British society's budget amounted to a modest thirty-five hundred pounds a year. The American Society's annual budget, a few thousand dollars at first, was supplemented by gifts from John D. Rockefeller, Jr., George Eastman, and Fisher himself and rose to forty thousand dollars by the end of the decade.[11] A large fraction of the British society's budget went to establish a quarterly journal, the *Eugenics Review*, which Galton thought "rather feeble" at first, while conceding that it might mend. In fact, it rapidly became self-supporting and, according to the society's secretary, was to be found in many public and scientific libraries in the United States, Europe, India, and Japan.[12] The American Eugenics Society left the publication of a journal to its sister American Genetics Association, which put out the *Journal of Heredity*, an organ of research devoted in part to heredity in human beings.

Both eugenics societies supplied speakers, who gave dozens of lectures yearly. The British group produced a film on eugenics and showed it free of charge in small-town cinemas throughout England, Wales, and Scotland. Both societies distributed pamphlets and study materials to clubs, libraries, and schools; one of these texts, put out by the American society, explained that, since the ultimate fruits of eugenics would naturally require many generations, the eugenics movement, unlike the usual short-lived political or social movement, "is, rather, like the founding and development of Christianity, something to be handed on from age to age."[13]

In 1926, the American society published *A Eugenics Catechism*, which assured readers that eugenics was not a plan for making supermen or for breeding human beings as if they were animals. The catechism did promise that eugenics would "increase the number of geniuses," foster "more selective love-making," and produce more love in marriage. It continued:

Q: Does eugenics contradict the Bible?

A: The Bible has much to say for eugenics. It tells us that men do not gather grapes from thorns and figs from thistles. . . .

Q: Does eugenics mean less sympathy for the unfortunate?

A: It means a much better understanding of them, and a more concerted attempt to alleviate their suffering, by seeing to it that everything possible is done to have fewer hereditary defectives. . . .

Q: What is the most precious thing in the world?

A: The human germ plasm.[14]

That same year, the society was moved to launch a eugenics sermon contest, whose judges included Charles Davenport and the Yale literary critic William Lyon Phelps, who was also a deacon in the Calvary Baptist Church of New Haven. An estimated three hundred sermons were inspired by the competition, and some sixty were submitted in the judging for the prizes—of five hundred, three hundred, and two hundred dollars.[15] In Kansas City, Missouri, Rabbi Harry H. Mayer chose a special Mother's Day service convoked by the Council of Jewish Women and the Temple Sisterhood to declare, "May we do nothing to permit our blood to be adulterated by infusion of blood of inferior grade." If the Protestant sermons were to be believed, the Bible was indeed a eugenic book and Christ was born into a family representing "a long process of religious and moral selection."[16] The Reverend Dr. Kenneth C. MacArthur of the Federated Church in Sterling, Massachusetts, sermonized upon the heritability of intelligence and speculated that moral and spiritual qualities were similarly determined, submitting in evidence the biblical words of Paul to Timothy which, in his paraphrase, celebrated "the unfeigned faith which dwelt first in thy grandmother Lois, and thy mother Eunice; and in thee also." The Reverend Dr. MacArthur, whose sermon won the second prize, later became a member of the society's Massachusetts branch, and informed the society's president that he had been deeply interested in eugenics for years, was concerned with problems of genetics as a breeder of purebred cattle, and was the proud winner of a silver cup in the Fitter Families Contest at the Eastern States Exposition of 1924.[17]

The Fitter Families contests had started in Topeka, in 1920, at the Kansas Free Fair. Under the aegis of the American Eugenics Society, they were soon being featured—together with eugenic exhibits—at seven to ten state fairs yearly; by the end of the decade, requests for help with such contests were coming to the society from more than forty eager sponsors a year. Local publications gave front-page attention to the competitions and their winners. At the state fairs, the Fitter Families competitions were held in the "human stock" sections. ("The time has come," a contest brochure explained, "when the science of human husbandry must be developed, based on the principles now followed by scientific agriculture, if the better elements of our civilization are to dominate or even survive.")[18] Any healthy family could enter. Contestants had only to provide an examiner with the family's eugenic history. All family members had to submit to a medical examination—including a Wassermann test and a psychiatric assessment—and take an intelligence test. At the 1924 Kansas Free Fair, winning families in three categories—small, average, and large—were awarded a Governor's Fitter Family Trophy, presented by Governor Jonathan Davis. "Grade A Individuals" won a Capper Medal, named for United States Senator Arthur Capper and portraying two diaphanously garbed parents, their arms outstretched toward their (presumably) eugenically meritorious infant. A fair brochure noted that "this trophy and medal are worth more than livestock sweepstakes or a Kansas oil well. For health is wealth and a sound mind in a sound body is the most priceless of human possessions."[19]

In both Britain and America, exhibits at various fairs and expositions often included a depiction of the laws of Mendelian inheritance—usually an array of stuffed black and white guinea pigs arranged on a vertical board so as to express the inheritance of coat color from generation to generation. At the Kansas Free Fair in 1929, the exhibits included charts illustrating "laws" of Mendelian inheritance in human beings: Cross a "pure" with a "pure" parent, and the children would be "normal." Cross an "abnormal" with an "abnormal," and the children would be "abnormal." Cross a "pure" with an "abnormal," and the children would be "normal but tainted; some grandchildren abnormal." Cross "tainted" with "tainted," and of every four offspring, one would be "abnormal," one "pure normal," and two "tainted." Another chart declared: "Unfit human traits such as feeblemindedness, epilepsy, criminality, insanity, alcoholism, pauperism and many others run in families and are inherited in exactly the same way as color in guinea pigs."[20] At the Sesquicentennial Exposition in Philadelphia, the American Eugenics Society exhibit included a board which, like the population counters of a later day, revealed with flashing lights that every fifteen seconds a hundred dollars of your money went for the care of persons with bad heredity, that every forty-eight seconds a mentally deficient person was born in the United States, and that only every seven and a half minutes did the United States enjoy the birth of "a high grade person . . . who will have ability to do creative work and be fit for leadership." An exhibit placard asked, "How long are we Americans to be so careful for the pedigree of our pigs and chickens and cattle—and then leave the *ancestry of our children* to chance or to 'blind' sentiment?"[21]

To a later generation, it may all seem like material out of Sinclair Lewis, but Anglo-American eugenicists approached hereditary matters with utmost seriousness, aware that they were part of a worldwide movement. After the turn of the century, eugenic efforts—often called "race hygiene"—had also developed in Sweden, Norway, Russia, Switzerland, Germany, Poland, France, and Italy; in the nineteen-twenties, the movement spread to Japan and Latin America. In 1912, some seven hundred and fifty people from Britain, Europe, and the United States attended the first International Eugenics Congress, in London, where the Right Honourable Arthur Balfour delivered the inaugural address, receiving hearty applause when he mentioned the "dignity of motherhood." Participants in the congress delivered some thirty papers, and its sponsoring vice-presidents included the Lord Chief Justice of Britain, the Right Honourable Winston Churchill, the Right Reverend the Lord Bishop of Ripon, Alexander Graham Bell, and Charles William Eliot, the former president of Harvard University.[22] British and American eugenicists also maintained particularly close links, through transatlantic publication of their books and articles, the election of each other to their respective societies, and personal contact.

Two years before the eugenics congress, the essay on "Civilization" in the new eleventh edition of the *Encyclopaedia Britannica* confidently stated that the lines of future progress were sure to include "the organic betterment of the race through wise application of the laws of heredity."[23] So sanctified, Francis Galton's scientific program seemed at long last to have been launched, virtually as a planetary revolution.

It is in the nature of social movements that they often command the support of disparate groups who share few ideas in common other than those of the movement itself. In 1908, the American geneticist Raymond Pearl noted that eugenics was " 'catching on' to an extraordinary degree with radical and conservative alike, as something for which the time is quite right."[24] In Britain, eugenics united such social radicals as Havelock Ellis, Ottoline Morrell, George Bernard Shaw, Harold Laski, and Beatrice and Sidney Webb with such establishmentarians as Leonard Darwin, who after twenty years in Her Majesty's Royal Engineers had retired to good causes and the country gentry, and Dean William Inge of St. Paul's Cathedral— the Gloomy Dean, as he was known—who relished the Duke of Wellington's alleged remark that the Battle of Waterloo had been won on the playing fields of Eton. (Dean Inge told his neighbor Francis Galton that "we are living in a 'stiff-necked and perverse generation,' who will listen to any guides except those who tell them the truth. The democracy seem quite unteachable.")[25] In the United States, the eugenics movement brought together conservatives like Davenport with progressives like Gifford Pinchot, Charles R. Van Hise, Charles W. Eliot, and David Starr Jordan and radicals like Emma Goldman and Hermann J. Muller, a future Nobel laureate for his work in genetics, who was a Marxian socialist and (for a time) an admirer of the Soviet Union. For all their political differences, eugenicists shared a concern for a set of issues they considered pertinent to heredity—some actually were—and also, generally, the same social milieu.

Eugenics enthusiasts in the United States and Britain were largely middle to upper middle class, white, Anglo-Saxon, predominantly Protestant, and educated. The movement's leaders tended to be well-to-do rather than rich, and many were professionals—physicians, social workers, clerics, writers, and numerous professors, notably in the biological and social sciences.[26] Leaders and followers alike had the time and inclination to attend lectures and debates, interested themselves in public affairs, and thought it necessary to keep abreast of science and to set their social compasses by the new discoveries. Fully half the membership of the British eugenics society consisted of women, and so did about a quarter of its officers. In the United States, women played an insignificant role in the national society but a prominent one in local groups. In both countries, women constituted a large part of the eugenics audience.[27] Eugenics, concerned ipso facto with the health and quality of offspring, focused on issues that, by virtue of biology and prevailing middle-class standards, were naturally women's own.

It was a commonplace among eugenicists that men and women alike would be better equipped for race regeneration the more they knew about family and maternal health. Eugenic writings warned, for instance, that "tobacco decreases in a marked degree the sexual power, the organs becoming relaxed and shrivelling in proportion to the amount of tobacco used," and that "the system of the wife becomes saturated with the nicotine and her reproductive cells also are poisoned. Surely strong, healthy offspring cannot come from such sources."[28] Like the campaigns against alcoholism, prostitution, and pornography, eugenics brought women into the domain of public affairs and provided them with a respectable avenue of social activism. It also brought women, as social activists if not as researchers, into direct involvement with the world of science, from which they were otherwise largely barred.

In a sense, the eugenics movement was Karl Pearson's Men and Women's Club—with its determination to explore the relations between the sexes—enlarged to encompass the transatlantic educated community.

And like the members of the club, eugenicists divided on pertinent issues, particularly those rooted in sexuality. Eugenics complemented—and perhaps in part grew out of—the late-nineteenth-century social-purity movement. That movement had proposed to work a moral reform of a society given to prostitution and the like. Mixing standard medical texts with moral prescriptions, it had tended to deny that women's sexual energy matched that of men, and it had insisted upon the reduction of male sexual expression to the female level and the replacement of male lustfulness by female tenderness, spirituality, and moral concern. It had thus encouraged women to take greater control over their marital sex and, in consequence, over the frequency with which they would bear children. Honoring motherhood, the movement aimed to make motherhood voluntary, an achievement that it claimed would not only benefit women but would promote the eugenic interest of the race.[29]

Social-purity attitudes found their way into the eugenics literature. One text declared that too much sexual activity led to a "squandering of the life principle." The author invoked the results of an unnamed scientist to show that after a period of sexual indulgence the sperm were "languid" and after continence stronger, larger, and more vigorous, and concluded, "Therefore may we not believe that children born of depleted parents will probably be physically feeble, literally 'born tired?' " In the contention of another text, no one should need to be told nature's plain law that women ought to avoid sexual relations during pregnancy. For by indulging their sexual appetites while their wives were pregnant, men implanted "in the coming life the seeds of sensuality, besides greatly increasing the suffering of the mother before and during the child's birth."[30] Such attitudes persisted even into the late nineteen-twenties. According to Paul Popenoe, the founder of the southern California branch of the American Eugenics Society and the head of the Institute of Family Relations in Los Angeles, "divorcees represent a type that is eugenically less desirable than the average. They have a higher frequency of mental diseases, shorter expectations of life, and a high degree of sterility, even in cases where the divorce occurred after many years of marriage."[31]

Quite different views, of course, were held in the social-radical wing of eugenics, a good deal of which carried forward the late-nineteenth-century utopian impulse. Havelock Ellis effectively advanced the view that women were just as capable of sexual pleasure as men. For Ellis, who incorporated remarks on eugenics in the sixth volume of his *Studies in the Psychology of Sex*, eugenic improvement required women's sexual liberation from the shroud of repressive Victorian attitudes. Social radicals pronounced the restrictions against divorce dysgenic, because they encouraged the production of children by mismated parents. Some agreed with the radical feminist Victoria Woodhull that the entire marital system was an "obstacle to the regeneration of the race."[32]

Prince Morrow, a professor of medicine at the University of the City of New York and president of the American Society of Sanitary and Moral Prophylaxis, told a 1910 child welfare conference that "the sex problem lies at the root of eugenics." Social radicals and social purists alike agreed with Morrow that sex instruction ought to be given to "the rising generation, the future fathers and mothers of the race." In Morrow's view, "the most serious obstacle in the way of instructing young people in the laws and hygiene of sex is the traditional sentiment which has invested everything relating to the sexual life with an atmosphere of shame and secrecy, and has decreed that this 'holy silence' must not be broken."[33] Defying that sentiment, the magazine *Arena*, influential among middle-class Americans, had begun, in the eighteen-nineties, to publish articles on sex education, free love, heredity, and marriage. In 1911, the Eugenics Education Society heard Edith Ellis, Havelock's wife, lecture on "sexual inversion," the term then used for homosexuality. Many eugenicists considered it essential to instruct adolescents about the physiology of sex, if only to prevent venereal disease, and partly through the aid of eugenics groups, parent-and-teachers' clubs made available to schoolchildren "social hygiene" lectures on sex, heredity, and marriage. Withal, in the name of preserving or improving the qualities of the race, eugenics took up subjects that had formerly been outside the bounds of respectable discussion, thus helping to bring about a transformation in public discourse which moved one writer, by 1914, to remark upon the "obsession of sex which has set us all a-babbling about matters once excluded from the amenities of conversation."[34]

Eugenics also helped to cast the light of science upon superstitions concerning conception, pregnancy, and childbirth, notably the law of maternal impressions—a commonplace assumption, rooted in folk belief and Lamarckian theory, that the characteristics of offspring were shaped by the experiences of the pregnant mother. In 1887, adumbrating Mrs. Bolce's experiment with her fetus Eugenette, Alice Stockham, an advocate of marital hygiene and birth control, had typically suggested that pregnant women might study natural history or botany so as to produce another Agassiz, Humboldt, or Audubon. The University of Wisconsin zoologist and eugenicist Michael Guyer listed some of the long-standing theories: "The mother sees a mouse with the result that a mouse-shaped birthmark occurs on the child . . . or she produces beauty in the child by long contemplation of a picture of a beautiful child. . . . The favorite is usually the production of a red birthmark or marks on the child's body by strong desire on the part of the mother for strawberries, tomatoes, etc.—the fruit must be red since the mark is red—or by fright from seeing a fire."[35] Some eugenic literature continued to advance such wrongheaded notions, but authoritative writers like Guyer dismissed them and instead tried to introduce modern medical and biological sense into questions of childbearing. "While parents can do nothing toward modifying favorably such qualities as are predetermined in their germ-plasm," Guyer advised, "nevertheless they must come to realize that bad environment can wreck good germ-plasm. . . . Their one sacred obligation to the immortal germ-plasm of which they are the trustees is to see that they hand it on with its maximal possibilities undimmed by innutrition, poisons, or vice."[36]

Of course, poisons, vice, and the like made no difference to certain human debilities; a common topic of eugenic writings was the heritability of disease. Medical opinion on the subject was divided. Karl Pearson studied biometrically the inheritance of alcoholism and tuberculosis. He outraged both physicians and temperance reformers (the two groups overlapped a good deal) by his outspoken insistence that a tendency to contract tuberculosis was heritable—which made a mockery of public health measures to combat it—and that a tendency to alcoholism was not. ("People are very savage about your memoir [on alcoholism]," a friend told him, "some on the ground that 'it cannot be true because it is such a wrong bad thing to get drunk,' and others because 'it may be true but it is calculated to encourage people to drink.' ")[37] Physicians disagreed, too, over whether disease or simply a predisposition to it was hereditary. But the research of Charles Davenport and others did make clear that numerous afflictions—for example, Huntington's chorea—were indeed inherited.[38]

Davenport's Eugenics Record Office received perhaps hundreds of queries regarding the heritability of diseases in the writer's own or prospective spousal family. Barren women sometimes wrote to Pearson begging to know how they could achieve fecundity. An American journalist, writing in the July 1913 issue of *Cosmopolitan*, celebrated "the inspiring, the wonderful, message of the new heredity" particularly when set against the sorrow of bearing offspring that were "diseased or crippled or depraved," and told his readers that "the one simple, all-encompassing rule is this: do not marry into a family that carries a defect of a kind that is carried also in your own family strain."[39] That such issues were openly raised bespoke the accuracy of the observation made early in the century by Charles Reed, chairman of the American Medical Association section on obstetrics and the diseases of women: "The subject of marriage, especially in its relation to the great problem of heredity, may now, upon proper occasions, be discussed in the drawing room without violence to 'good form.' The family newspapers and the magazines discuss the questions without reserve. The school teacher and the minister of the gospel are within the pale of propriety, when they consider it in their respective stations. Clubs are formed, books are printed and lectures are delivered on this subject, all with not only the approval but the patronage of good society."[40]

Spousal choice and parental practice among the middle classes had long been shaped by family tradition in tandem with religious authority. Now the latitude, mobility, and diversity of urban life were diminishing familial constraints, and religious authority had of course long since eroded in the storms of scientific skepticism. Even clerics felt compelled to align themselves with the modernist doctrine of harmonizing religion and morals with the methods of science and the known laws of nature. The Reverend Harry Emerson Fosdick of the Riverside Church in New York City announced that "few matters are more pressingly important than the application to our social problems of such well-established information in the realm of eugenics as we actually possess," and Dean Inge carried the eugenic banner to the British public, telling an audience at the Bedford College for Women that some knowledge of eugenics would "in many cases prevent falling in love with the wrong people."[41]

Like Francis Galton, literate Americans and Englishmen, conservative as well as reformist, had undergone their religious crisis, cast off biblical religion and—some with enthusiasm, others by default or in despair—had embraced a religion of science. Galton had expected eugenics to provide a secular substitute for traditional religion, and in the opening decades of the twentieth century, amid the turbulence of Anglo-American urban industrial life, it was said to have accomplished just that. In *The New Decalogue of Science*, Albert Wiggam intoned: "God is still doing the same thing. However, in our day, instead of using tables of stone, burning bushes, prophecies and dreams to reveal His will, He has given men the microscope, the spectroscope, the telescope, the chemist's test tube and the statistician's curve in order to enable men to make their own revelations. These instruments of divine revelation have not only added an enormous range of new commandments—an entirely new Decalogue—to man's moral codes, but they have supplied him with the technique for putting the old ones into effect."[42]

So it seemed, given the material benefits—electric lights, trolleys, and machinery; phonographs, cinema, and radio; dyestuffs, fertilizers, and gasoline; anesthesia, medicines, and diagnostic X rays—that science had conferred upon the Anglo-American world since the late nineteenth century. Charles Van Hise, president of the University of Wisconsin and a distinguished geologist, declared: "We know enough about agriculture so that the agricultural production of the country could be doubled if the knowledge were applied; we know enough about disease so that if the knowledge were utilized, infectious and contagious diseases would be substantially destroyed in the United States within a score of years; we know enough about eugenics so that if the knowledge were applied, the defective classes would disappear within a generation."[43]

With the modern miracles went a modern priesthood: the scientists— no small number of them geneticists. In America, the eugenic priesthood included much of the early leadership responsible for the extension of Mendelism—besides Davenport, there were Raymond Pearl and Herbert S. Jennings, both of the Johns Hopkins University; Clarence C. Little, the president of the University of Michigan and later the founder of the Jackson Laboratory, in Maine; and the Harvard professors Edward M. East and William E. Castle. In Britain, eugenicists could count on the aid not only of Pearson but of the horticulturalist Charles C. Hurst; F. A. E. Crew, the Scottish animal geneticist; the brilliant statistician Ronald A. Fisher, who would succeed Pearson in the Galton Eugenics Chair; J. B. S. Haldane, a groundbreaker in population genetics and an outspoken social radical; and the evolutionary biologist Julian Huxley, a grandson of Darwin's great defender. Some of these scientists lent themselves to the work of eugenic organizations. Others—notably the leading British Mendelian William Bateson, and his American counterpart Thomas Hunt Morgan, of Columbia—awarded eugenics tacit support for some years either by declining to criticize it publicly or, more important, by providing forums for it at scientific meetings and in scientific journals. Many geneticists wrote books that favored eugenics. Among them was Castle's *Genetics and Eugenics*, the most widely used college text in its field, going through four editions in the fifteen years after its first publication, in 1916. The large majority of American colleges and universities—including Harvard, Columbia, Cornell, Brown, Wisconsin, Northwestern, and Berkeley—offered well-attended courses in eugenics, or genetics courses that incorporated eugenic material.[44]

Geneticists warmed easily to their priestly role. The new industrial order had elevated practitioners of the physical sciences to positions of power and public service. Physicists and chemists found themselves in demand by innovative firms like Western Electric, Du Pont, and Standard Oil of New Jersey, which were opening research laboratories; and the requirements of public policy formation in such areas as food and drugs, communications, and aeronautics were bringing physical scientists into the orbit of government. Geneticists experienced no comparable demand. For them, the science of human biological improvement provided an avenue to public standing and usefulness. Herbert Jennings, in his 1930 book *The Biological Basis of Human Nature*, remarked on the new "eagerness to apply biological science to human affairs," and observed with evident satisfaction: "Gone are the days when the biologist . . . used to be pictured in the public prints as an absurd creature, his pockets bulging with snakes and newts. . . . The world . . . is to be operated on scientific principles. The conduct of life and society are to be based, as they should be, on sound biological maxims! . . . Biology has become popular!"[45]

Chapter V

DETERIORATION AND DEFICIENCY

THE ENTHUSIASTS OF eugenics were unquestionably stimulated by the advent of Mendelian genetics in 1900 and its application to human heredity. Yet among the audience for the creed, a climate of receptivity to eugenic ideas had already been forming, in both the United States and Britain. Social Darwinism, with its evocation of natural selection to explain diverse social phenomena, had brought about a flow of proto-eugenic writings that foreshadowed the salient concerns of the post-1900 movement, particularly the notion that "artificial selection"—state or philanthropic intervention in the battle for social survival—was replacing natural selection in human evolution. Some regarded the possibilities of artificial selection as an opportunity, others worried that it was leading to the degradation of the race. Alfred Russel Wallace reported in 1890: "In one of my last conversations with Darwin he expressed himself very gloomily on the future of humanity, on the ground that in our modern civilization natural selection had no play, and the fittest did not survive. Those who succeed in the race for wealth are by no means the best or the most intelligent, and it is notorious that our population is more largely renewed in each generation from the lower than from the middle and upper classes."[1]

Wallace, humane and generous, preferred to think that environmental improvement, rather than the elimination of "inferiors," would produce social advance. But he was compelled to admit that "grave doubts" had been cast upon this view by the work of Galton and August Weismann. If Galton's statistical studies of heredity strongly suggested the constancy of populations for a given character, Weismann had seemingly provided a mechanical underpinning for the result in his germ-plasm theory that the force of heredity resided in a substance impermeable to environmental influence. Henry Fairfield Osborn, the paleontologist and director of the

American Museum of Natural History, declared in 1891 that Weismann's theory, if true, "profoundly affects our views and conduct of life."[2] The theory may have been more a contention than a proved scientific fact; advanced to the English-speaking world in the eighteen-nineties through translations of Weismann's books, it helped bolster an emerging pre-Mendelian hereditarianism which held that environmental reforms, however well intended, could work little if any social improvement over the long run because people's germ plasm remained the same—because nature defied nurture.

Important to the eugenics movement was the increasingly widespread notion that heredity determined not simply physical characteristics but temperament and behavior. In the late nineteenth century a growing body of social-Darwinist writings had commonly held that paupers spawned paupers and criminals bred criminals. The research of the Italian criminologist Cesare Lombroso convinced a generation of social analysts that there existed a criminal "type," defined not only by behavioral but by physical characteristics. The biology of criminality had it that, since the crime-producing features of the physical organism must be hereditary, so must be criminality, especially since criminals tended to mate with each other.[3]

Perhaps no single work suggesting the hereditary nature of social pathology was better known than Richard Dugdale's famous study of the Jukes family, published in 1877. Dugdale, who traced the ancestry of a large group of criminals, prostitutes, and social misfits back through seven generations to a single set of forebears in upstate New York, actually attributed the Jukes's misfortunes in significant part to the degradation of their environment. The misinterpretation of his work simply reflected the mounting hereditarian propensity of the day. Arthur Estabrook, a field worker for the Eugenics Record Office, would later confirm Dugdale's gloomy results in a follow-up study, *The Jukes in 1915*. The study reported the latter-day descendants to be as unredeemed—in Charles Davenport's summary, as beset with "feeblemindedness, indolence, licentiousness, and dishonesty"— as their predecessors had been when Dugdale brought them to national attention.[4]

In Britain, there was no Dugdale, but there was Charles Booth's extensive survey of the London poor in the eighteen-eighties and eighteen-nineties, which was taken to show that an irreducible fraction were doomed to remain impoverished. And after 1900 there was also a good deal of exploitation of Mendelism to account for behavior. One study proposed that the excitable religious temperament revealed two characters, religious feeling and instability. These might be transmitted separately, the study warned, with the result that "one son may possess religious feeling of a steady normal type, while another, inheriting instability unchecked by religion, and finding of necessity the home environment uncongenial, may go to support the common idea that the sons of extremely religious parents are apt to run to excess in riotous living."[5]

Yet neither the literature of eugenics nor the preexisting intellectual climate of social Darwinism in which it came to flourish were enough to create a eugenics movement. Essential to that were the social changes straining both Britain and the United States after the turn of the century: industrialization, the growth of big business, the sprawl of cities and slums, the massive migrations from the countryside and (in the United States especially) from abroad. Urban Anglo-America may have always known prostitution, crime, alcoholism, and disease, but neither society had ever before possessed the weight of statistical information, expanding yearly by volumes, that numerically detailed the magnitude of its problems. Statistics revealed, with seeming mathematical exactitude, that afflictions such as "mental defectiveness" and criminality were worsening every year.[6] Both societies had long absorbed the foreign-born, but the United States experienced an especially large immigration of Eastern and Southern Europeans, who, beginning in the late eighteen-eighties, came by the millions across the Atlantic and settled in the major cities of America. In 1891, the economist Francis Amasa Walker, who had directed the 1870 and 1880 United States Censuses, advanced a striking statistical case that immigrants were breeding at a much higher rate than native-born Americans. Britain, too, knew its immigration; Irish Catholics settled in Liverpool and Birmingham or huddled with Polish and Russian Jews in the East End of London. The stresses of immigration alone, Irving Fisher wrote to Davenport in 1912, provided "a golden opportunity to get people in general to talk eugenics."[7]

Why this new "cult" of eugenics? a contributor to the *Yale Review* asked in 1913. In part because of the rediscovery of Mendel's laws, he noted, but also because of the growing demands on the taxpayer. "Statistics have shown a rapid and steady increase in the ratio of pauperism, insanity, and crime to the whole population," he pointed out, "proving that the support of these defectives has become a veritable burden upon the taxpaying community, and that, although there might be individual improvement in those thus cared for, these very persons 'breed back,' so to speak, to their degenerate ancestors, their very betterment but affording the opportunity for them to propagate their unfit kind." In England, it was said that "the number and kind of people born into a nation . . . are points of vital importance to every sane person, and are brought home to him in a practical manner every time the rate collector calls at the door [or] the Income Tax Commissioners deliver their demand." In the era of Al Capone, the American Eugenics Society announced that crime cost the average family about five hundred dollars per year. "It must be remembered," the Society pointed out, "that the majority of criminals have either defective intelligence, defective emotions or a combination of both defects."[8]

Yet eugenics expressed more the social than the economic anxieties of the white Protestants who were its chief supporters. They were old stock in America, older still in Britain—though eugenic ranks included hardly any members of the hereditary aristocracy. (British eugenicists tended to denigrate the hereditary nobility; some proposed to reconstitute the House of Lords in accordance with eugenic principles.) No doubt the aristocracy did not suffer from the social insecurity that led members of the British and American middle and upper middle classes to celebrate the qualities of the Nordic or Anglo-Saxon "race" (the terms were often used interchangeably) and to disparage those who seemed—by virtue of their hereditary endowments or lack of them—to threaten their respective nations' "racial" strength.[9]

Confidence in such strength meant a good deal in imperial Britain, where the German naval challenge was provoking apprehension over British hegemony on the seas and the protractedness of the Boer War had kindled widespread questioning of John Bull's mettle. Signs of physical degeneration had cropped up during the Boer conflict when the Inspector General of recruiting reported that eight out of eleven volunteers in Manchester had to be rejected as physically unfit. In 1903 Parliament was stirred to establish a commission on "national deterioration." To many British, the general fiber of their nation—its overall moral character, intelligence, energy, ambition, and capacity to compete in the world—was declining.[10]

The English physicist W. C. D. Whetham and his wife addressed the issue of Britain's racial strength in 1909, in their widely noted book *The Family and the Nation*. The Whethams, themselves the parents of six, were decidedly distressed by the restriction of births among the abler classes. They called the desire to limit the number of children to those who could be well provided for a "mistaken kindness, . . . an imminent danger to the country, and high treason to the human race." It was all an old story in the history of nations, the Whethams concluded; such practice had been "the prelude to the ruin of States and the decline and fall of Empires." The German birthrate, they warned darkly, had fallen far less than the British.[11]

The American Eugenics Society sponsored a contest in 1928—first prize, a thousand dollars—for essays on the causes of decline in "Nordic" fertility. The psychologist G. Stanley Hall, president of Clark University, raised the specter of "the yellow and Oriental peril," asserting that "the future belongs to those people who bear the most and best children and bring them to fullest maturity. They will in the end wield all the accumulated resources of civilization, and infertile races will fade before them."[12] And Theodore Roosevelt, the bully imperialist, outdoorsman, and rollicking paterfamilias, whose beloved first wife had died in the aftermath of childbirth, scolded the middle and upper middle classes for committing "race suicide" by restricting their births. The progressive reformers prominently identified with eugenics tended, like Roosevelt, to take as supreme the national as opposed to the merely local interest, to put the welfare of the group over and above that of the individual, to celebrate America's new imperial power. It was disturbing to such eugenicists that late-nineteenth-century Harvard graduating classes had, twenty to twenty-five years later, accounted for male progeny equal only to half to two-thirds their original number.[13]

Perhaps no datum was more frequently cited in the Anglo-American literature of eugenics than Karl Pearson's on the differential birthrate. A 1906 demographic study of a number of London districts, carried out by David Heron of the Galton Laboratory, substantiated his warning—that half of each succeeding generation was produced by no more than a quarter of its married predecessor, and that the prolific quarter was disproportionately located among the dregs of society. The Whethams maintained that social reforms and advances in medical skills extended life "for the members of weak and unsound stock" and—what was more significant—reduced their children's mortality rate.[14] The prospect of "national deterioration" prompted the socialist Sidney Webb, in a Fabian tract, to enlarge upon Pearson's conclusion that the lower classes were outreproducing everyone else. Webb pointed out that poorer districts characterized by prolific breeding were heavily populated by Irish Catholics and Jews, who tended to be fruitful and multiply for religious reasons. "In Great Britain at this moment," Webb wrote, "when half, or perhaps two-thirds of all the married people are regulating their families, children are being freely born to the Irish Roman Catholics and the Polish, Russian, and German Jews, on the one hand, and the thriftless and irresponsible—largely the casual laborers and the other denizens of the one-roomed tenements of our great cities— on the other. . . . This can hardly result in anything but national deterioration; or, as an alternative, in this country gradually falling to the Irish and the Jews. Finally, there are signs that even these races are becoming influenced. The ultimate future of these islands may be to the Chinese!"[15]

Racism—in that era racial differences were identified with variations not only in skin color but in ethnic identity—was a feature of both British and American eugenics. Eugenicists solemnly discussed the racially hereditary features of non-white Protestant groups. Pearson praised Galton's attempt to depict the Jewish type by composite photography ("we all know the Jewish boy," Pearson said); in the mid-nineteen-twenties, Pearson reported that Jewish children in the East End of London, while no less intelligent than Gentiles, tended to be physically inferior and somewhat dirtier.[16] Charles Davenport informed a high and interested officer in the American Telephone & Telegraph Company that if a Jew and a Gentile mated, ninety percent of the offspring would resemble the Gentile parent: "In general, the Jewish features are recessive to the non-Jewish." Whatever good qualities Jews and other aliens might possess, the Whethams asserted, "they are not those typical of the Anglo-Saxon; and these immigrants cannot be regarded as a satisfactory equivalent to the native population."[17]

Anglo-American eugenicists embraced the standard views of the day concerning the hereditarily biological inferiority of blacks. Some eugenicists expected that, just as with vigorous hybrids, miscegenation might yield racially beneficial results: Samuel J. Holmes, a biology professor at the University of California at Berkeley, told a meeting of the Commonwealth Club in San Francisco that, because of white males impregnating Negro females, the Negro race was gradually being "bleached" and the white race "nowhere nearly so appreciably tanned," adding that "from the white point of view, this is a fortunate type of race assimilation." The weight of eugenic opinion, however, lay with Michael Guyer, who observed that "many students of heredity feel that there is great hazard in the mongrelizing of distinctly unrelated races no matter how superior the original strains may be." So Davenport believed he and Morris Steggerda, a young zoologist, had demonstrated in a 1929 study, *Race Crossing in Jamaica*, which examined the characteristics of three groups of a hundred adults each: "full blooded Negroes (Blacks), Europeans (Whites), and hybrids (Browns)." The characteristics included those traits of temperament that, as Davenport had explained to Steggerda, "bear upon our main problem: the relative capacity of negroes, mulattoes, and whites to carry on a white man's civilization." The authors concluded not only that blacks were inferior in mental capacity to whites but that a larger proportion of browns than of either pure group were "muddled and wuzzle-headed."[18]

Especially in the United States, assumptions of genetic differences between white Protestants of Northern European stock—"Wasps," in the term of a later day—and the country's substantial numbers of blacks and Jewish and Catholic immigrants figured significantly in the eugenics movement. The influential New York City circle, grouped around the Galton Society and the Eugenics Record Office, included the Park Avenue socialite and eugenicist Madison Grant, who wrote *The Passing of the Great Race*, a book, first published in 1916, that enjoyed considerable vogue in the nineteen-twenties, and who insisted that the intermarriage of Nordics— which Grant alleged to be the highest-order group in the white race—and the lesser Alpines or, worse, Mediterraneans inevitably led to debilitating "mongrelization."[19]

Racism figured much less markedly in British eugenics. Francis Galton, the founding father, had been no less a racist than most Victorians, but such considerations entered very little into his eugenic theorizing. Although Karl Pearson disparaged Jews—and blacks, for that matter—he took a certain pleasure from outraging a Newcastle cleric by telling an audience there that, since Neanderthal man was undoubtedly dark-skinned, the original Adam must have been "negroid."[20] British society was ethnically more-or-less homogeneous; most Jews and Irish Catholics were concentrated in a small number of cities, and the United Kingdom had not yet experienced the significant non-white immigration of later decades. Indeed, some Jews, like the physician Solomon Herbert, were prominent in the British eugenics movement.[21] While British eugenicists talked of the threat of immigrants from Ireland and the Continent, they fretted a good deal more about the threat to the national fiber arising from the differential birth rate and the consequent weakening of their imperial competitive abilities in relation to France and Germany. British eugenics was marked by a hostility decidedly more of class than of race.[22]

An unabashed distrust, even contempt, for democracy characterized a part of eugenic thinking in both Britain and America. Henry Fairfield Osborn, the president of the American Museum of Natural History, welcomed his fellow eugenicists to the second International Eugenics Congress with the declaration that "the true spirit of American democracy that all men are born with equal rights and duties has been confused with the political sophistry that all men are born with equal character and ability to govern themselves and others, and with the educational sophistry that education and environment will offset the handicap of heredity."[23] But if Anglo-American eugenicists resented challengers from the social bottom, they displayed no great admiration for the economic top of modern society. Business talent was generally not recognized in the pantheon of eugenically desirable traits, and hardly any businessmen were to be found among the leadership of organized eugenics in either country. The eugenics movement enabled middle- and upper-middle-class British and Americans to carve out a locus of power for themselves between the captains of industry on one side and lower-income groups—both native and foreign-born—on the other. Socialist, progressive, liberal, and conservative eugenicists may have disagreed about the kind of society they wished to achieve, but they were united in a belief that the biological expertise they commanded should determine the essential human issues of the new urban, industrial order.

LIKE FRANCIS GALTON, whom they took as their patron saint, eugenicists identified human worth with the qualities they presumed themselves to possess—the sort that facilitated passage through schools, universities, and professional training. They tended to equate merit with intelligence, particularly of the academic sort. And like Galton, they were predisposed to think that intelligence was inherited. Karl Pearson had sought to test that heritability by relying upon teachers' estimates of mental ability; earlier, Galton had relied on social or professional place as an inferential proxy of it. In the eighteen-eighties, Galton had helped pioneer a quantitative approach to the psychology of individual differences by measuring reaction times and the like. Inspired by his innovation, psychologists on the Continent and in the United States attempted to establish a relationship between mental ability and physical characteristics. By the turn of the century, it was clear that no such connection existed, but the idea of systematically measuring intelligence had captured the attention of the French psychologist Alfred Binet, an acolyte of Galton's quantifying aims, if not of his particular methods.[24]

In 1904, the French government, expanding its educational system, asked Binet for ways to detect mentally deficient children. Binet drew up a series of tests consisting of numerous short problems designed to probe such qualities as memory, ratiocination, and verbal facility. In collaboration with a colleague, Théodore Simon, he also devised a scheme for classifying each test taker according to his "mental age." A child's mental age was defined as that of the chronologically uniform group of children whose average test score he matched. Thus, if a six-year-old's test score matched the average score of ten-year-olds, the six-year-old's mental age would be ten; similarly, if a ten-year-old scored the same as the average of six-year-olds, his mental age would be six.[25]

The American psychologist Henry H. Goddard brought the Binet-Simon tests from Europe to the United States in 1908. At the time, American psychology was breaking away from its traditional association with philosophy and, under the leadership of innovators like G. Stanley Hall, was moving in an independent, experimentally oriented direction. Goddard, a student of Hall, was an exemplar of the trend, and he was naturally impressed by the tests, not least because they at long last seemed to provide a direct, quantitative measurement of intelligence. He employed the Binet-Simon examinations at the Vineland, New Jersey, Training School for Feeble-Minded Boys and Girls, where he had recently been appointed director of a new laboratory for the study of mental deficiency—one of the first established in this country.

The tests did seem to classify the Vineland pupils in a way consistent with his staff's direct experience of them; the "boys and girls" of the Vineland School ranged in age up to fifty, yet none scored on the tests at a mental age greater than twelve. By 1911 Goddard had extended his Binet-Simon testing program to many more subjects, including some two thousand children. The tests, Goddard confidently believed, were "amazingly accurate and would be very easily applied by any field-worker without anybody realizing that they were being tested."[26]

Goddard noted with particular interest that the test results revealed wide variations in degree of "feeblemindedness"—a term then used to denote a wide range of mental deficiencies and, as well, of tendencies toward socially deviant behavior. The results also provided a way to distinguish among the differences. Turning numbers into categories, Goddard eventually classified as "idiots" those among the feebleminded whose mental age was one or two, and as "imbeciles" those whose mental age ranged from three to seven. Those who scored between eight and twelve he dubbed "morons," a word he took from the Greek for "dull," or "stupid."[27]

Some of Goddard's earlier field studies had revealed several families with a high incidence of mental deficiency—in one case nearly three hundred members of a family of six hundred people. Like many scientists of his day, he strongly suspected that "feeblemindedness" was inherited. With regard to the genetics of the disability, he confessed to Charles Davenport, he had "much more zeal than knowledge." Davenport, who started consulting with Goddard on the matter in 1909, made the heritability of feeblemindedness a subject of increasing importance at the Eugenics Record Office and provided field workers to help Goddard carry out a systematic study of the mental characteristics of the Vineland students and their relatives in the local population.[28]

Using such data, Goddard, in 1912, published *The Kallikak Family: A Study in the Heredity of Feeblemindedness*, which examined a pseudonymous family—the name was constructed from the Greek words *kalós* (good) and *kakós* (bad)—in the Pine Barrens to the north of the Vineland Training School. He followed that two years later with *Feeble-mindedness: Its Causes and Consequences*, in which he speculated that the feebleminded were a form of undeveloped humanity: "a vigorous animal organism of low intellect but strong physique—the wild man of today." In Goddard's view, it was essential to distinguish between the moron and the insane person: the latter's mind was diseased; the former's was, functionally, "a dwarf brain." He stressed that, unlike idiots or imbeciles, morons might appear normal but in fact were not.[29]

Further surveys of intelligence, including administration of the Binet-Simon tests, revealed a high incidence of mental deficiency among the inmates of prisons, reformatories, and homes for wayward girls. The feebleminded, Goddard argued, lacked "one or the other of the factors essential to a moral life—an understanding of right and wrong, and the power of control." Children thus afflicted became truants because they could not succeed in school. They grew up to become criminals because they lacked the power to "do the right and flee the wrong"; paupers, because they found the burdens of making a living too heavy; and prostitutes, because they were weak-minded and unintelligent.[30] Goddard was unsure whether mental deficiency resulted from the presence in the brain of something that inhibited normal development or from the absence of something that stimulated it. But whatever the cause, of one thing he had become virtually certain: it behaved like a Mendelian character. Feeblemindedness was "a condition of mind or brain which is transmitted as regularly and surely as color of hair or eyes."[31]

In both the United States and England, Goddard's research impressed the corps of people who concerned themselves professionally with social deviants. It was increasingly believed that the root of antisocial behavior lay in the mental rather than in the physical type, and that, in the words of Michael Guyer, "a considerable amount of crime, gross immorality and degeneracy is due at bottom to feeblemindedness." The so-called feebleminded in America were variously estimated at one to three percent of the population, and were commonly said to constitute a "menace." On both sides of the Atlantic, workers with the mentally handicapped began to examine the family histories of their charges. While some cases of mental deficiency were recognized as the result of disease or accident, the common opinion concerning the principal cause was summarized by Havelock Ellis in 1912: "Feeble-mindedness is largely handed on by heredity."[32]

Goddard's tests stimulated other psychologists to experiment with different schemes for the quantitative assessment of mental capacity. Various new testing systems were devised, for normal as well as mentally deficient children. Among the most prominent was the revision of the Binet-Simon tests, published in 1916 at Stanford University by the psychologist Lewis Terman—another of G. Stanley Hall's students—who had come to mental testing and to a hereditary view of intelligence through research with precocious children, including his own. It was Terman who introduced the term "I.Q." to the language. I.Q., of course, stood for "intelligence quotient," a concept invented by the German psychologist William Stern, in 1912; it was expressed as the ratio of a child's mental age to his chronological age, times one hundred: if the ratio was 1, the child's I.Q. would be 100; if nine-tenths, 90; if eleven-tenths, 110; and so on. (Terman was pleased to note that his own boy and girl tested consistently between 125 and 140.)[33]

Before the First World War, there was a good deal of resistance to intelligence testing. Tests had to be administered individually, usually (many psychologists claimed) by a trained psychologist. Because of the expense, for the most part they were used only for the identification and classification of mentally handicapped schoolchildren. Perhaps more important, since they were associated with the measurement of mental deficiency, many people assumed that testing a child amounted to questioning his or her intelligence.[34] In 1916, the case of Esther Meyer came before the New York State Supreme Court. Meyer had been recommended for confinement in a custodial institution because of seeming low intelligence, and her parents had protested. Justice John W. Goff refused to admit the Binet-Simon test results as evidence of Meyer's alleged mental deficiency. "Standardizing the mind is as futile as standardizing electricity," Judge Goff declared, warning that the "votaries of science or pseudo-science" could too easily make prejudiced testimony of the tests. (The New York *Times* decried Justice Goff's opinion: "The Binet-Simon tests, intelligently applied, are as trustworthy as the multiplication table.")[35] The Justice's misgivings were soon forgotten. During the First World War, extensive testing was used to sort out the hundreds of thousands of draftees who flooded into the United States Army.

The chief wartime tester was the comparative psychologist Robert M. Yerkes. Yerkes's scientific attitudes had been partly shaped by Francis Galton, to whose works Charles Davenport had introduced him, in 1898, when Yerkes was a Harvard graduate student and Davenport one of his instructors. As a young Harvard faculty member, Yerkes helped pioneer the separation of psychology from philosophy, insisting that the study of psychological phenomena must be based on fact rather than on speculation and must be tied to an experimental, preferably quantitative, methodology.[36] Fascinated by the study of mental capacity, Yerkes began experimenting with mental tests, in 1913, at the Boston Psychopathic Hospital, working in conjunction with Professor Ernest E. Southard of the Harvard Medical School, an ally of Goddard's, an adviser to Davenport, and a confirmed eugenicist. With James W. Bridges, a graduate student in psychology at Harvard and an intern at the hospital, Yerkes developed the Yerkes-Bridges scale, a rival to Terman's Binet-Simon system for measuring mental ability. In 1916, the same year that Yerkes was elected to the presidency of the American Psychological Association, Harvard declined to award him academic tenure—largely, it seems, because the administration considered his field unworthy. Eugenically inclined and ambitious for his science ("theoretically," he once declared, "man is just as measurable as is a bar of steel"), Yerkes had special reasons to demonstrate its utility during the war emergency.[37]

The National Academy of Sciences had meanwhile established a National Research Council to mobilize scientists for defense. In May 1917, under the auspices of the Council, a group of psychologists headed by Yerkes and including Terman and Goddard set out to design an Army testing program, "not primarily for the exclusion of intellectual defectives,"

Yerkes noted, "but rather for the classification of men in order that they may be properly placed in the military service." To the end of introducing a scientific system of classification, the committee devised two sets of examinations: the alpha tests for literates in English, the beta tests for everyone else. Unlike most intelligence tests of the day, these examinations could be administered to adults en masse. "If the Army machine is to work smoothly and efficiently," Terman remarked, "it is as important to fit the job to the man as to fit the ammunition to the gun."[38]

The alpha tests consisted of the sort of questions—number sequences, word analogies, arithmetic problems, synonym-antonym puzzles, and commonsense queries—that would become familiar to generations of students; the beta tests consisted largely of pictorial problems involving the comparison of forms and the completion of partial drawings. The regular military had its doubts about both the purpose and the practical utility of the tests, some officers suspecting Yerkes and his crew of making the camps into laboratories for their own purposes. Then, too, most seasoned officers considered themselves quite capable of determining without any tests who would or would not make a good soldier. At Fort Dix, a draftee with a very low test score was, according to his commander, "a model of loyalty, reliability, cheerfulness, and the spirit of serene and general helpfulness." "What do we care about his intelligence?" the commander wondered.[39]

The Army critics penetrated to a difficulty that would continue to plague mental testers. The tests were biased in favor of scholastic skills, and the outcome was dependent upon the educational and cultural background of the person tested. Yerkes and others claimed that the tests were almost entirely independent of the environmental history of the examinees, and that they measured "native intelligence." But certainly one of the questions on the alpha test—"The Knight engine is used in the: Packard/Stearns/Lozier/Pierce Arrow"—demanded a knowledgeability that could hardly be supplied by native intelligence. Examinees were also bound to fare better with the word analogies and the arithmetic problems of the alpha test if they had had extensive schooling. Illiterates and non-English-speaking recruits had to cope in the beta test with the vagueness and uncertainty of orally communicated directions. Many of the beta examinees had never taken a written test before. "It was touching," one examiner recalled, "to see the intense effort . . . put into answering the questions, often by men who never before had held a pencil in their hands."[40] Still, the Army did have to sort out the immense numbers of draftees, and the tests did provide some indication of mental ability. The testing program went forward. By the Armistice, some one million seven hundred thousand recruits had been tested. Younger career officers, at least, had come to value the tests for personnel placement. During the war, the contents of the tests were classified as military secrets.[41]

The wartime trial of the tests worked a dramatic transformation in the public's attitude toward intelligence testing. After the war, Yerkes was inundated with hundreds of requests for the now declassified alpha and beta examinations. In 1919, with a grant from the Rockefeller Foundation, Yerkes and his co-workers drew up a standard National Intelligence Test, which sold more than half a million copies in less than a year. Intelligence testers examined ever more paupers, drunkards, delinquents, and prostitutes. Business firms incorporated mental tests in their personnel procedures. Intelligence tests were administered annually to a few million primary and secondary school students, and a number of colleges and universities began to use intelligence-test results in the admissions process.[42]

The postwar testing vogue generated much data concerning the "intelligence" of the American public, yet the volume of information was insignificant compared with that from the wartime test program. The National Academy of Sciences summarized that experiment in 1921, in a hefty volume entitled *Psychological Examining in the United States Army*.[43] Drawn up by Yerkes, Terman, and their colleagues, the report presented the test procedures and broke down a large sample of the test results by geographical region and ethnic or racial background. Two inches thick, five pounds in weight, and containing more than a half a million words, the volume was hardly a best-seller, but it formed the basis of numerous popular books and articles about intelligence tests and their social import. Almost four hundred thousand draftees—close to one-quarter of the draft army—were unable to read a newspaper or to write letters home. Particularly striking, the average white draftee—and, by implication, the average white American—had the mental age of a thirteen-year-old.[44]

The psychologist Carl Brigham, one of the wartime Army testers, extended the analysis of the Army data in 1923, in his book *A Study of American Intelligence*. The Army data, Brigham said, constituted "the first really significant contribution to the study of race differences in mental traits." In the early stages of analyzing the data, he had privately confided to Charles Davenport that "we are all on the right track in our contention that the germ plasm coming into the country does not carry the possibilities of that arriving earlier." In 1917, Henry Goddard had reported—on the basis of the results of the Binet-Simon test given four years earlier to a small group of "average" immigrants at Ellis Island—that two out of five of those who arrived in steerage were "feebleminded." Now Carl Brigham found that according to their performance on the Army tests the Alpine and Mediterranean "races" were "intellectually inferior to the representatives of the Nordic race," and he declared, in what became a commonplace of the popular literature on the subject, that the average intelligence of immigrants was declining.[45]

The average intelligence of black Americans, apparently, was just as low as most white Americans had long liked to think it. Anyone doubting the claim could turn to Brigham's analysis of the Army test data, and various test surveys disclosed that blacks accounted for a disproportionately large fraction of the "feebleminded." The Army tests also appeared to indicate that the average black person in the United States had the mental age of a ten-year-old.[46]

Clearly a variety of causes, including the cultural bias of the Army tests themselves and the poor education of many of the test takers, might have accounted for the results. Yet the supposedly objective test data further convinced eugenically minded Americans not only that mental deficiency was genetically determined but that so was intelligence. White college students scored very well on the alpha tests, and so did high school students from Anglo-Saxon or white-collar homes. This was taken to mean that gifted students came from homes that, in the words of one educator, "rank high racially, economically, intellectually, and socially." Terman and other psychologists were quick to point out that opening up avenues of opportunity to the children of lower socioeconomic groups probably made no sense; they did not have the I.Q. points to compete. President George B. Cutten of Colgate University took the Army test results as a starting point to attack the democratization of higher education and wondered aloud in his inaugural address whether democracy itself was possible in a country where the population had an average mental age of thirteen.[47]

British eugenicists had no similar array of data to sustain their convictions regarding the hereditary nature of intelligence, but they did have the psychologist Cyril Burt. The son of a country doctor, Burt was inspired intellectually in his youth by the aged Francis Galton, who was one of his father's patients. While a student at Oxford early in the century, and later as an instructor of physiology and psychology at the University of Liverpool, Burt imbibed Galton's hereditarian doctrines, Karl Pearson's statistical techniques, and the Mendelian theory of inheritance. He became an early member of the Eugenics Education Society. Between 1909 and 1911, he tested boys from a preparatory school and a higher elementary school in Oxford, and a school in the Liverpool slums. Most of the children in the first group were the sons of university and college academics, Fellows of the Royal Society, or bishops; the second, of small tradesmen; the third, of laborers. The preparatory-school boys did better than the elementary school students, whose performance was considerably superior to that of the boys from the Liverpool slums. "Among individuals," Burt declared, announcing the position he would popularize with increasing tenacity, "mental capacities are inherited. Of this the evidence is conclusive."[48]

Burt read with admiration, in the *Eugenics Review*, a Brigham-like account by Robert Yerkes of the United States Army test results. "Your work in the American Army has given psychology an immense impetus in this country," he wrote to Yerkes. Still, like British eugenic thought in general, Burt's work showed little if any of the racial themes characteristic of the American school. In fact, he concluded after a review of British and American research on racial characteristics that while for the individual the influence of heredity was "large and indisputable," for the race it was "small and controversial." A onetime settlement house worker, Burt also recognized that the causes of crime and delinquency could hardly be pinned entirely upon mere feeblemindedness. But his writings did give a good deal of support in the interwar years to the belief that intelligence was not only heritable but highly correlated with socioeconomic position—that is, with the hallmark of British eugenic concern, class.[49]

Whatever their prejudices, American and British eugenicists were alike distressed over the trend in their respective nations' intelligence. Before the First World War, eugenicists like Karl Pearson and Charles Davenport had warned that excessive breeding of the lower classes was giving the edge to the less fit. The growth of I.Q. testing after the war gave a quantitative authority to the eugenic notion of fitness. For the vogue of mental testing did more than encourage fears regarding the "menace of the feeble-minded." It also identified the principal source of heedless fecundity with low-I.Q. groups, and it equated national deterioration with a decline in national intelligence.

The majority of mental testers and their audience, their views shaped in considerable part by the racial or class prejudice that pervaded eugenics, found the biological theory of intelligence, advanced in the seemingly neutral language of science, persuasive. In the Vanuxem Lectures at Princeton University in 1919, Henry Goddard himself stated their thesis: "the chief determiner of human conduct is the unitary mental process which we call intelligence. . . . This process is conditioned by a nervous mechanism . . . and the consequent grade of intelligence or mental level for each individual is determined by the kind of chromosomes that come together with the union of the germ cells . . . [and] is but little affected by any later influence except such serious accidents as may destroy part of the mechanism. As a consequence, any attempt at social adjustment which fails to take into account the determining character of the intelligence and its unalterable grade in each individual is illogical and inefficient."[50]

Chapter VI

MEASURES
OF REGENERATION

IN 1891, IN A BOOK entitled *The Rapid Multiplication of the Unfit*, Victoria Woodhull had observed: "The best minds of today have accepted the fact that if superior people are desired, they must be bred; and if imbeciles, criminals, paupers, and [the] otherwise unfit are undesirable citizens they must not be bred." After the turn of the century, Anglo-American eugenicists talked increasingly about how to accomplish those aims. The proposals were as diverse as the social convictions of the movement's members. Nevertheless, the courses of action could be divided into two at times overlapping approaches: "positive eugenics," which aimed to foster more prolific breeding among the socially meritorious, and "negative eugenics," which intended to encourage the socially disadvantaged to breed less—or, better yet, not at all.[1]

Francis Galton had been principally a positive eugenicist, and his heirs included visionaries, many of them conservative like himself. Their ranks included Alexander Graham Bell, who advocated marriage between the deaf and people from families with no deafness, in the expectation that the deficiency would eventually be weeded out; the Bishop of Ripon, who urged procreation among the fit in the imperial interest, so that the colonies might be populated with able members of the British "race"; and Theodore Roosevelt, who noted to Charles Davenport that "someday we will realize that the prime duty, the inescapable duty, of the *good* citizen of the right type is to leave his or her blood behind him in the world."[2] Yet the most vigorous advocates of positive eugenics in the United States and Britain after the turn of the century tended to be social radicals, many of them inclined to utopian visions.

"We generate the race; we alone can regenerate the race," Havelock Ellis declared in 1911. Ellis hoped to exploit the new knowledge of heredity to increase the numbers of the fit. So did George Bernard Shaw, even though he did not spare the eugenics movement his unpredictable mockery. (Murderers, he once declared, shocking the British press, ought not necessarily to be punished, since they might remove eugenically undesirable people from society; people prone to homicidal mania could perhaps be taught to kill only those whom society could eugenically do without.) But though Shaw acted the outrageous buffoon at times, he took his eugenics seriously enough to subscribe to Karl Pearson's *Biometrika,* to stay in touch with Pearson, and to make himself a figure in the British eugenics movement.[3]

Shaw invested *Man and Superman* with eugenic doctrines. "Being cowards, we defeat natural selection under cover of philanthropy: being sluggards, we neglect artificial selection under cover of delicacy and morality," he declaimed in the Preface. To Shaw, mere environmental reforms would by no means usher in a eugenically golden age. He scoffed at negative eugenics, which society seemed ever ready to carry out "with considerable zest, both on the scaffold and on the battlefield," and insisted that considerably more attention be given to the biological amelioration of so deplorable a piece of work as man. "We have never deliberately called a human being into existence for the sake of civilization, but we have wiped out millions," he reasoned. "We kill a Thibetan regardless of the expense, and in defiance of our religion, to clear the way to Lhasa for the Englishman; but we take no really scientific steps to secure that the Englishman, when he gets there, will be able to live up to our assumption of his superiority."[4] Shaw's positive eugenics distilled Galton's chiliastic goal—the elimination of original sin by getting rid biologically of the original sinner—to a socially imperative essence. For without the Superman, without the enlargement of man's moral and ethical capacity, social progress would amount only to an illusion.

Social-radical eugenicists considered environmental reforms eugenically essential. Typical in outlook was the young socialist Harold Laski. Prior to entering Oxford University, Laski spent six months working in Pearson's laboratory and while there published an article on eugenics that caught the eye of Galton himself, who invited him to tea. ("Simply a *beautiful* youth of the Jewish type," he told Pearson, adding to his diary that Laski would make his mark if he stuck to eugenics.) Laski, who at Oxford formed a Galton Club for eugenic discussion, thought that the time was surely coming when society would regard "the production of a weakling as a crime against itself." But he parted company from Pearson on the question of such social legislation as the Factory and Education Acts, which, by keeping children out of the work force, Pearson said, had perhaps discouraged sturdy working families from bearing more offspring. Laski considered the aims of the acts "worthy of all praise" and their dysgenic effects best dealt with through the payment of a minimum wage high enough to permit parents of desirable children to afford them.[5]

In the view of social-radical eugenicists, the most important environmental reform was to ease—or, better, to abolish—class distinctions. The presumption was, as Ellis put it, that the "best stocks" were not "necessarily the stocks of high social class" but were spread through all social classes, with those of the lower classes being "probably the most resistant to adverse conditions." Poverty, the argument went, resulted from indiscriminate breeding among men and women prevented from choosing genetically optimal partners. Once class distinctions were destroyed, human and social evolution could proceed, not haphazardly but by a conscious act of the collective will. In his Preface to *Man and Superman* Shaw observed: "To cut humanity up into small cliques, and effectively limit the selection of the individual to his own clique, is to postpone the Superman for eons, if not forever. Not only should every person be nourished and trained as a possible parent, but there should be no possibility of such an obstacle to natural selection as the objection of a countess to a navvy or of a duke to a charwoman."[6]

In 1910, Shaw roused a Caxton Hall audience to cheers with the suggestion that, for the eugenic good, women should be permitted to become respectable mothers without having to live with the fathers of their children.[7] Obviously Shaw agreed with Ellis's claim that "the question of Eugenics is to a great extent one with the woman question." Liberals on that question considered it a eugenic necessity for a woman to control her own life—and not only its physical side. Without independent careers, women were forced to marry, too often taking as husbands diseased or dissolute men. Careers would enable them to avoid eugenically disadvantageous marriages—though not, it was hoped, marriage or parenthood altogether. Havelock Ellis, impotent and childless though (or perhaps because) he was, avowed that "the realization of eugenics . . . can only be attained with the realization of the woman movement in its latest and completest phase as an enlightened culture of motherhood."[8]

Yet many social-radical eugenicists doubted that people with hereditary deficiencies would refrain from marriage for the good of the race. That, Ellis noted, had been the utopian fantasy of Francis Galton—with the result, he added, that eugenics was "constantly misunderstood, ridiculed, regarded as a fad." But now that the mechanical control of reproduction was ever more reliably at hand, Ellis argued, eugenics no longer needed to be impractical, ridiculous, or contrary to natural human desires. Sexual satisfaction ought to be separated from procreation, specifically through birth control. This would render sexual practice purely a matter of private pleasure and invest the act of procreation with responsibility to the race. Ellis added that the limitation of births would ease the financial burdens upon lower-income families, safeguard the health of the mother, and permit better care for the children who were produced.[9]

These arguments, familiar now, were controversial in Ellis's day. In America, Comstockery had long suppressed even the discussion, let alone the distribution, of birth-control methods. Even people who sympathized with the cause of contraception at times assisted in the suppression of the subject by a prudish unwillingness to discuss it publicly.[10] Before the First World War, the cause of birth control was strongly opposed on both sides of the Atlantic by the numerous eugenicists who adhered to the dominant attitudes of the movement—"mainline eugenicists," we may call them. Many of these were of a conservative bent, and their views on such issues as the "woman question" were markedly different from those of Ellis and Shaw.

Mainline eugenicists held that contraceptive methods ("preventive checks") would permit the separation of passion from the responsibilities of procreation, and thus foster licentiousness. As late as 1932, Henry Fairfield Osborn complained that birth control led in "fundamentally unnatural" directions, and noted that the country employing birth control in its "most radical form" was the Soviet Union, "where it is connected with a great deal of sexual promiscuity." Leonard Darwin kept the subject of birth control out of the deliberations of the Eugenics Education Society and the pages of the *Eugenics Review*. It was not simply that so many members of the society found the subject distasteful but that they considered birth control—in Darwin's words—"racially" devastating.[11] Although contraceptive methods might in principle help halt the proliferation of lower-income, less educated groups, they tended in practice to be ignored in those sectors of the population; instead, they were used disproportionately by the upper classes—precisely those groups whose declining fecundity alarmed so many eugenicists.

Within both English and American mainline eugenics, it was a morally injunctive commonplace that middle- and upper-class women should remain at home, hearth, and cradle—that it was their duty, as Dean Inge intoned and Theodore Roosevelt trumpeted, to marry and bear children (four per marriage was the number thought necessary to maintain a given stock). Edwin Grant Conklin, professor of embryology at Princeton University and one of the prominent biologists of his day, declared in 1915 that the feminist movement was "a benefit to the race" insofar as it brought women greater intellectual and political freedom, but insofar as it demanded "freedom from marriage and reproduction it is suicidal." In *The Family and the Nation*, W. C. D. and C. D. Whetham—by themselves the

Eugenics Education Society's most anti-feminist wing—cried, "Woe to the nation whose best women refuse their natural and most glorious burden!"[12]

Mainline eugenicists were alarmed by the higher education of women. Education, so the reasoning went, diverted women's biological energy from the task of reproduction to the burdens of intellectual or worldly activities. Early in the century, studies showed that women college graduates tended not to marry, and that those who did bore—on the statistical average—fewer than two children, less than half the number necessary to keep up their social stock. (Actually, the principal of Newnham College, Cambridge, had demonstrated statistically in 1890 that college-educated women were just as healthy, just as likely to be married, and just as fertile as their less educated female relatives; the degree of spinsterhood and childlessness was a mark of their social class, not of their higher learning.) An American (male) analyst feared that the large proportion of female professors in women's colleges encouraged susceptible college girls to yearn for careers. Osborn refused to endorse the women's suffrage amendment on the ground that it would interfere with human evolution.[13]

Nevertheless, hardly able to stem the tide of women's suffrage, higher education, or sexual revolution, mainline eugenicists joined social radicals to make education, at least in eugenics, a cardinal point of their program. The American Eugenics Society was pleased to note in the nineteen-twenties that courses dealing with eugenics were then offered in some three hundred and fifty colleges and universities. Young men and women were to be sensitized to their procreational responsibilities to the race (the marriage brought about merely for the desire of happiness, Karl Pearson lectured in the nineteen-twenties, was "born in selfishness, and is antisocial"). More than that, for the sake of the overall racial welfare, even mainline eugenicists endorsed teaching the young not only the laws of heredity but the facts of sex hygiene, venereal disease, pregnancy, and child care—so that they would know the consequences to offspring ultimately inherent in the act of love, so that, in Havelock Ellis's phrase, "the new St. Valentine will be a saint of science rather than of folk-lore."[14]

While the educational effort was intended to foster positive eugenics, it aimed at least as much at encouraging a negative eugenic sensibility: matings between healthy and "tainted" or diseased individuals were to be avoided. The British biologist J. Arthur Thomson suggested that eugenics education would arouse a "wholesome prejudice against the marriage and especially the intermarriage of subjects in whom there is a strong hereditary bias to certain diseases—such as epilepsy and diabetes," and asked, "Is it Utopian to hope . . . that the ethical conscience of the average man will come more and more to include in its varied content 'a feeling of responsibility for the healthfulness of succeeding generations'?"[15] To eugenicists,

that healthfulness depended a great deal upon reducing the social differential in the birthrate, an issue which increasingly spotlighted the merits of
contraception. Even mainliners could recognize a certain validity in the
assertion of Emma Goldman's monthly, *Mother Earth*, that those who
denied access to birth-control methods would "legally encourage the increase of paupers, syphilitics, epileptics, dipsomaniacs, cripples, criminals,
and degenerates."[16]

By the nineteen-twenties, the Freudian invasion of middle-class mores
was well along. Women were said to expect sexual fulfillment in marriage
without fear of pregnancy. Birth control had come to stay, and so, it
seemed, had a steady decline in the birthrate of the upper classes. As
Margaret Sanger put it, the sensible eugenic response to the differential
birthrate was to make available to lower-income and less educated groups
the contraceptive knowledge and opportunities enjoyed by others. Before
the war, Sanger had linked birth control with feminism. Now, like her
British counterpart Marie Stopes, she tied contraception increasingly to the
eugenic cause. In 1919, she wrote: "More children from the fit, less from the
unfit—that is the chief issue of birth control."[17] Even Leonard Darwin was
eventually persuaded to lend his name to the birth-control movement.
Stripped of its assertive feminism, contraception became acceptable to conservative eugenicists, for there was a natural harmony between their social
predilections and the pro–birth-control rationale advanced by Havelock
Ellis: "The superficially sympathetic man flings a coin to the beggar; the
more deeply sympathetic man builds an almshouse for him so that he need
no longer beg; but perhaps the most radically sympathetic of all is the man
who arranges that the beggar shall not be born."[18]

MANY EUGENICISTS EXPECTED their program of race improvement, whether
positive or negative, to rest on voluntarism—thus the stress on education,
moral injunction, and the need for contraception. Control over those most
private areas of life—marriage, sex, and childbearing—were to be left in
private hands as matters of private choice. Radicals or mainliners, many
eugenicists were moral reformers, who held with Havelock Ellis that "the
only compulsion we can apply in eugenics is the compulsion that comes
from within." Voluntarism was also deemed essential because, in the view
of various genetic authorities, little was yet known about the laws of heredity in human beings.[19] Moreover, to invoke the state against the proliferation of degenerates might, in the remark of an English sociologist made
long before the Nazis took power, "renew, in the name of science, tyrannies
that it took long ages of social evolution to emerge from." Soon after its
founding in 1907, the Eugenics Education Society had announced that its

policy was not to advocate eugenic "interference by the state." Yet a number of British and American eugenicists came to the view suggested by Galton himself at the end of his life: that perhaps with regard to certain critical problems—notably the proliferation of degenerates—the situation was so clear-cut, and so dire, as to warrant state intervention of a coercive nature in human reproduction.[20]

Given the energy that eugenics drew from social Darwinism, it may seem puzzling that eugenicists, particularly the conservatives, were so ready to resort to governmental action. For in the standard view of that creed, competition led automatically to survival of the fittest, and the intervention of government would corrupt a process best left to the independent operation of Darwinian law—the natural selection of the fit. But in Darwin's theory, fitness meant the ability to survive and multiply in a given environment, and the fit in that sense included precisely the lower-order types responsible, as Karl Pearson demonstrated, for the high side of the differential birthrate. Edgar Schuster, the first Eugenics Fellow at University College London, remarked that fitness meant something else to eugenicists: "In good condition or of good quality, physical and mental . . . a sort of biological ideal of what a man should be."[21] If natural selection yielded the Darwinian fit, only artificial selection—by governmental means, where appropriate—could multiply the eugenically fit.

This reasoning moved conservatives in particular (reformers and radicals of the day needed no special impulsion) to depart from laissez-faire with a program of positive-eugenic measures. The program included government involvement in the procreation of the better sort through a variety of financial incentives. "If a woman can, by careful selection of a father, and nourishment of herself, produce a citizen with efficient senses, sound organs, and a good digestion, she should clearly be secured a sufficient reward for that natural service to make her willing to undertake and repeat it," Shaw wrote in the Preface to *Man and Superman.* Proposed incentives included tax rebates to help cover the costs of maternity and child-rearing, especially for meritorious families (an idea which seemed inappropriate to Chancellor of the Exchequer Winston Churchill, who, though he considered encouraging the fertility of the professional classes praiseworthy, declared that this aim had no connection with the budget, whose only preoccupations were to finance the government). The recommendations also included educational allowances and grants to make up for lost salary to women from the industrious laboring classes who had to leave employment during pregnancy, birth, and early infant care.[22]

Appalled at the battlefield loss of so many of their "best" young men in the First World War, British eugenicists asked the military to issue "eugenic stripes" to the meritorious wounded; worn on the uniform sleeve, they were presumably to offset the injuries that might make such men less attractive to women. Eugenicists also petitioned the government to award special bonuses to such veterans who married.[23] The logic of positive eugenics impelled even conservatives to recommend that the wages of respectable working-class men be raised so that their wives could remain at home to bear and care for children instead of joining the work force. The Whethams thought that the state might well selectively endow parenthood, giving honors and rewards to those in all ranks of life who produced strong, healthy, and able offspring.[24]

The willingness to depart from laissez-faire was more forceful still on the side of negative eugenics. Socialist principle led H. G. Wells to claim that "the children people bring into the world can be no more their private concern entirely, than the disease germs they disseminate or the noises a man makes in a thin-floored flat." Social workers, psychiatrists, reformatory superintendents, and the like, convinced of the hereditary origin of social deficiency, felt compelled to endorse governmental intervention as perhaps the only way to reduce to manageable size the magnitude of their task. Havelock Ellis called it Sisyphean for a society to attempt social improvement while conceding "entry into life . . . more freely to the weak, the incompetent, and the defective than to the strong, the efficient, and the sane."[25]

Nowhere did that logic seem more evident than in the response to the "menace of the feebleminded." Eugenicists of every stripe found common ground in the righteous idea of wiping out social defect by preventing the procreation of the eugenically undesirable.[26] Suggestions to accomplish that end ran the gamut from the cruel (putting degenerates painlessly to death or permitting mothers to smother children possessing inherited deformities) to the mocking (the abolition of alcoholism by letting the intemperate drink themselves to death, or the punishment of a murderer by hanging his grandfather). Eugenicists generally refused to consider abortion to halt the birth of the unfit—the American Eugenics Society regarded it as murder, unless performed on strict medical grounds.[27] Instead, the eugenics community fastened most seriously and persistently upon marriage restriction, sexual segregation, sterilization, and—in the United States especially—immigration restriction.

The marriage of the feebleminded, the insane, and the diseased, particularly the venereally diseased, was of special concern to eugenicists. Some suggested that the grounds for divorce might be enlarged to include epilepsy, mental deficiency, criminality, and drunkenness. Others went so far as to recommend that all prospective bridegrooms be compelled to obtain a physician's certificate testifying to their freedom from venereal or mental disease.[28] However, Henry Goddard, for one, pointed out that marriage-

restriction laws were no panacea, particularly in the case of the feebleminded; since many of them lacked an inherent sense of morality, barriers to marriage hardly constituted an obstacle to procreation. Many eugenicists on both sides of the Atlantic thought that the problem of the mentally
handicapped demanded their institutionalized sexual segregation. "By legislative reform," said the Whethams, "we may segregate the worst types of
the feeble-minded, the habitual criminal, and the hopeless pauper, and thus
weed out of our race the contaminating strains of worthless blood."[29]

Segregation and control were simply the other side of education and
care; the two approaches easily complemented each other in eugenic thinking. The American Eugenics Society explained, in its catechismal pamphlet, the decided benefits to be expected from segregation:

Q. How much does segregation cost?

A. It has been estimated that to have segregated the original "Jukes"
for life would have cost the State of New York about $25,000.

Q. Is that a real saving?

A. Yes. It has been estimated that the State of New York, up to 1916,
spent over $2,000,000 on the descendants of these people.

Q. How much would it have cost to sterilize the original Jukes pair?

A. Less than $150.[30]

A few officials at American institutions for the mentally incapable had
begun to advocate a policy of sterilization as early as the eighteen-eighties,
and in 1889, at the Pennsylvania Training School for Feebleminded Children at Elwyn, Superintendent Dr. Isaac Newton Kerlin had, with parental
permission, castrated some of his charges. No state had legally authorized
sterilization, but such experiments upon the "unfit" were continued, notably by Dr. Harry C. Sharp, physician to the Indiana State Reformatory at
Jeffersonville, who in 1899 pioneered the sterilization of criminals by vasectomy. By 1907, Sharp had performed vasectomies on four hundred and
sixty-five males, more than a third of whom were said to have requested the
operation. "Vasectomy consists of ligating and resecting a small portion of
the *vas deferens*," Sharp reported. "The operation is indeed very simple and
easy to perform; I do it without administering an anaesthetic, either general
or local. It requires about three minutes' time to perform the operation and
the subject returns to his work immediately, suffers no inconvenience, and
is in no way impaired for his pursuit of life, liberty and happiness, but is
effectively sterilized."[31] While men were dealt with by vasectomy, women
were sterilized by the more hazardous and painful salpingectomy, or tubal
ligation.

Pro-sterilization eugenicists were found all across the political spectrum; all took as higher the good of society over the rights of individuals.
"It is the acme of stupidity . . .," declaimed Dr. William J. Robinson, a New

York urologist and a sex radical, "to talk in such cases of individual liberty, of the rights of the individual. Such individuals have no rights. They have no right in the first instance to be born, but having been born, they have no right to propagate their kind." In the United States, a strong consensus in favor of sterilization—supporters ranged from Margaret Sanger to Theodore Roosevelt—grew among eugenicists. In Britain, no such consensus existed. Comparatively few British eugenicists were convinced of the necessity for the procedure, although among those who considered it were Francis Galton ("except by sterilization I cannot yet see any way of checking the produce of the unfit who are allowed their liberty and are below the reach of moral control") and H. G. Wells, who pondered improving the human stock by "the sterilization of failures."[32]

Many American eugenicists worried that marriage-restriction laws and sterilization programs would be useless if the threat from abroad to the nation's biological strength were allowed to continue. In the first fifteen years of the twentieth century, immigration accounted for roughly half the increase in population. Michael Guyer warned in 1916 that this trend was particularly alarming in view of the fact that, "once in our country, the alien far outbreeds the native stock." (In his prize-winning eugenics sermon, the Reverend Kenneth C. MacArthur warned, "At the present rate one thousand Harvard graduates of today will have only fifty descendants two hundred years hence, by which time one thousand Roumanian immigrants will have increased to one hundred thousand.") High scientific authority —geneticists, psychologists, anthropologists—drew upon expert "evidence," notably Henry Goddard's I.Q. tests of immigrants and Carl Brigham's analysis of the Army intelligence test results, to proclaim that a large proportion of immigrants bordered on or fell into the "feebleminded" category and that their continued entrance into the country made, in Robert Yerkes's phrase, for the "menace of race deterioration." Whatever the symbolic meaning of the Statue of Liberty, American eugenicists like Charles Davenport stressed that the nation's future had to be taken into account. Davenport wrote to his brother William—a minister who was devoting his life to settlement-house work with Italian immigrants in Brooklyn, New York—"Just imagine what sort of country it will be . . . [in two hundred years] if the gates have, in the meantime, been wide open and population encouraged to come hither by the transportation companies and by the employers of the cheapest labor? . . . We don't want to make a State of Mississippi or worse out of New York City and Long Island."[33]

Charles Davenport argued that the selection should be on an individual basis, that no national group could be classified as undesirable. But by the early nineteen-twenties the eugenic principle of selection on the basis of individual biological and mental quality had been submerged in a principle of racial- or ethnic-group selection. The shift no doubt bespoke the weight of the national clamor for immigration restriction; it also expressed the patent racial prejudices of many eugenicists, prejudices which took the form of biologically celebrating Wasps and denigrating non-Wasps. A cardinal point of the American eugenics program had come to be the restriction of immigration from Eastern and Southern Europe. Eventually, the program was enlarged to permit the immigration only of pure Caucasians; to require a minimum grade of "C"—the presumed average grade of the American population—on the Army intelligence-test scale; and to require certification, based on an assessment of near kin, that the prospective immigrant would become a biological asset to the United States. "Immigration," said the American Eugenics Society, "should be first of all considered a long-time investment in family stocks."[34]

Chapter VII

EUGENIC ENACTMENTS

IN BRITAIN, the Eugenics Education Society had hardly been founded before it formed a watchdog committee to monitor all parliamentary bills of eugenic interest. British eugenicists sent their opinions to the *Times* and deputations to Westminster on matters of the poor laws, divorce, education, venereal disease, and the feebleminded.[1] In the United States, such matters for the most part did not fall within the purview of the national government, and the political efforts of eugenicists tended to take place at the state level. But immigration was, of course, a matter of federal rather than of state policy, and on the issue of immigration from Europe American eugenicists entered the national political arena to lobby for restrictive laws.

They were not alone. Since the late nineteenth century, various interest groups had been pushing for immigration laws that would stem the flow of new arrivals. The immigration of Asians had already been severely restricted, notably by the Chinese Exclusion Acts of 1882 and 1902 and by the so-called Gentleman's Agreement between Japan and the United States in 1907–08. The groups opposed to immigration included organized labor, worried that the influx would adversely affect wages; staunch nativists, convinced that foreign influences adulterated the American character; social workers, eager to reduce the flow so as to deal better with the disadvantaged already in the country; and assorted businessmen, who feared immigrants as infectious carriers of radicalism. Though racism figured in the arguments against unrestricted immigration, economic factors, including fear of radicalism, tended to dominate the debate through 1921, when Congress—in the wake of the Red Scare and postwar unemployment—passed an emergency restriction act by which immigration from any European country was limited annually to three percent of the foreign-born of that nationality listed in the 1910 U.S. census.[2] The more zealous restrictionists felt that this feature did not discriminate enough against the most recent wave of immigrants—those from Eastern and Southern Europe—and, joined by eugenicists, they campaigned for a still stricter, permanent immigration law. However, unable to make any headway toward a measure more to their liking, they had to content themselves with a two-year renewal of the emergency law after its expiration, in 1922.

Eugenicists did a good deal to make racial differences of an alleged genetic sort a prominent feature of the immigration debate. In 1922, Robert Yerkes urged Carl Brigham's publisher to bring out the latter's *Study of American Intelligence* in time for consideration at the next round of immigration-restriction hearings; he also called the attention of the immigration committee chairmen in both houses of Congress to the seeming mental inadequacies of Eastern and Southern Europeans as revealed by the Army intelligence tests. In 1923, the House Committee on Immigration and Naturalization began holding hearings on a permanent bill. Many witnesses argued that "biology" demanded the exclusion of most members of the Eastern and Southern European "races." Soon after the conclusion of the House hearings, Congressman Samuel Dickstein, Democrat of New York, who was a Jew and one of the two committee members to vote against sending the measure to the floor, remarked: "If you had been a member of that committee you could not help but understand that they did not want anybody else in this country except the Nordics." The committee was dominated by members from the South and West, and both the House and the Senate were controlled by conservative Republicans. On both sides of Capitol Hill biological and racial arguments figured prominently in the floor debate on the bill. Congressman Robert Allen, Democrat of West Virginia, declared: "The primary reason for the restriction of the alien stream ... is the necessity for purifying and keeping pure the blood of America."[3]

In April 1924, the Immigration Act was passed by overwhelming majorities in the House and Senate and quickly signed into law by President Calvin Coolidge (who as Vice-President had publicly declared: "America must be kept American. Biological laws show . . . that Nordics deteriorate when mixed with other races"). The act limited the influx to the United States from any European country, through 1927, to a small percentage of the foreign-born of the same national origin recorded in the census of 1890. The shift of the reference point back by two decades, to a date when fewer Eastern and Southern European immigrants were in the country, made immigration policy more discriminatory against newcomers from those areas. A permanent provision of the law, which took effect on July 1, 1927, was based on the 1920 census, but it had the same consequences, because the quotas were now to be apportioned in accordance with the distribution of national ancestries in the total population. The new law was widely acclaimed by eugenicists for what they considered its biological wisdom.[4]

Comparatively few British eugenicists—aside from Karl Pearson, who was lukewarm about the issue anyway—agitated for immigration restriction in this period. Immigration from Eastern Europe, consisting largely of Jews, had been heavy since the eighteen-eighties, but it had already been stemmed by the passage of the Aliens Act in 1905, which empowered the government to bar the entrance of steerage passengers who were diseased, or were criminals or potential paupers.[5] Non-white immigration to Britain from the Empire was negligible. Among the variety of other issues with which British eugenicists concerned themselves, one above all—the control of the mentally deficient—engaged their political energies.

Legislative attention to the problem of mental deficiency in Britain was to no small degree stimulated by Mrs. Hume Pinsent. Née Ellen Parker, she was a minister's daughter and the sister of the eminent lawyer and chancery judge Robert Parker, one of Karl Pearson's closest college chums and lifelong friends; with her brother and her future husband, Ellen Parker was an active member of Pearson's Men and Women's Club. When her engagement to Pinsent became known, Pearson remarked: "I suppose Miss Parker will now devote herself to housekeeping and possibly the piano. She might have done excellent work, if she had had the ordeal of getting her own living by some profession for a few years, instead of passing from home to home." Actually, Ellen Parker Pinsent, whose husband became a successful lawyer in Birmingham, got herself elected to the city council, became a formidably effective activist in the school program for mentally handicapped children, wrote a book on mental health policy—not to mention four novels—and eventually was made a Dame of the British Empire. A member of the Royal Commission on Care and Control of the Feebleminded from 1904 to 1908, she emerged a firm advocate of preventing the proliferation of the mentally deficient by compulsory institutionalization on a sexually segregated basis.[6]

Through Ellen Pinsent, the Eugenics Education Society joined forces with the National Association for the Care of the Feebleminded, which she had helped found, to demand of every candidate for Parliament in the 1910 election: "Would you undertake to support measures . . . that tend to discourage parenthood on the part of the feebleminded and other degenerate types?" Winston Churchill, Home Secretary in the Asquith government, told a delegation from Mrs. Pinsent's association that, while the thousands of feebleminded in Britain deserved "all that could be done for them by a Christian and scientific civilization now that they were in the world," they should, if possible, be "segregated under proper conditions [so] that their curse died with them and was not transmitted to future generations." (To Asquith, Churchill privately described the proliferation of the mentally deficient, combined with the "restriction of progeny among all the thrifty, energetic, and superior stocks," as a "very terrible danger to the race.")[7]

In May 1912, the government introduced a mental-deficiency bill; by the end of that year, the Home Office had received hundreds of resolutions urging passage of some such measure from public bodies—county and borough councils, education committees, and boards of guardians— throughout Britain. The government's bill passed, with only three dissenting votes, in July 1913. The opposition included the radical libertarian M.P. Josiah Wedgwood, who attacked the measure as the suggestion of "eugenic cranks" and mocked Ellen Pinsent for her "wonderful ability, such as only ladies seem to possess these days." Wedgwood's substantive objections—they concerned state interference with individual liberty— were sufficiently felt in the liberal wing of press and parliament to render the law something less than a eugenicist's dream. It recognized that the varieties of mental deficiency ranged from cretinism or mongolism to inability to benefit from education. It made the test of such deficiency, and of the need for care, not heredity but social incapacity—an inability to look after oneself. It also provided the possibility for many victims of mental deficiency, even while under care and control, to live in the outside community rather than in institutions. The law, in short, did not impose mandatory segregation of all mentally handicapped people to prevent their reproducing themselves, and there was no mention of sterilization.[8]

Yet the Eugenics Education Society took the Mental Deficiency Act as a victory for the eugenics movement. The law did, after all, grant a central authority compulsory powers to detain and segregate certain of the "feebleminded," a feature which would result in some curbs upon the multiplication of the unfit. And "defectives" subject to the Act were defined to include not only paupers and habitual drunkards but women on poor relief at the time of giving birth to, or being found pregnant with, an illegitimate child. The *Eugenics Review* celebrated the act as the "only piece of English social law extant, in which the influence of heredity has been treated as a practical factor in determining its provisions."[9]

Britain passed no sweeping law preventing the marriage of the mentally deficient, but in America, by 1914, some thirty states had enacted new marriage laws or amended old ones. Three-quarters of the statutes declared voidable the marriages of idiots and of the insane, and the rest restricted marriage among the unfit of various types, including the feebleminded and persons afflicted with venereal disease. The ostensible ground of most of the laws was that such partners were incapable of making contracts, marital or otherwise, but in some of them, the restrictions were justified on eugenic grounds. The first of this type, passed in Connecticut in 1896, prohibited marriage (as well as extramarital relations) to the eugenically unfit if the woman was under forty-five, and set the minimum penalty for violation at three years' imprisonment. The Connecticut statute was extolled as a model for other states, but the marriage law that captured the most attention was Indiana's—a three-part measure passed in 1905, which forbade the marriage of the mentally deficient, persons having a "transmissible disease," and habitual drunkards; required a health certificate of all persons released from institutions; and declared void all marriages contracted in another state in an effort to avoid the Indiana law.[10]

By the nineteen-twenties, many states had enacted measures forcing a delay between license application and the actual wedding, a policy that eugenicists advocated in the interest of discouraging hasty and ill-considered unions. Eugenic arguments also figured in the anti-miscegenation statutes of the day.[11] Clearly, eugenicists did not single-handedly cause the passage of the large variety of restrictive marriage laws enacted in the first quarter of the century; they were part of a coalition that put the laws on the books, and they provided prior (or, at times, post hoc) biological rationalizations for what other interest groups wanted. But American eugenicists played a dominant role in bringing about the passage of state sterilization laws.

The first state sterilization law was passed in 1907, in Indiana, where Dr. Sharp of the State Reformatory had mounted a campaign for the measure. ("Indiana is working much on sterilization," a Johns Hopkins physician remarked in 1910. "Practice hurries ahead of inquiry there.") Between 1907 and 1917, sterilization laws were enacted by fifteen more states, representing every region of the country except the South. Virtually all of the prewar statutes gave the states the power to compel the sterilization of habitual or confirmed criminals, or persons guilty of some particular offense, like rape. Also included within the scope of most of the statutes were epileptics, the insane, and idiots in state institutions. Most wide-ranging was a law passed in Iowa in 1911. It made eligible for sterilization inmates in public institutions who had been incarcerated for a variety of reasons, including drug addiction, sexual offenses, and epilepsy. The Iowa statute compelled the sterilization of twice-convicted sexual offenders, of thrice-convicted other felons, and of anyone convicted just once of involvement in white slavery.[12]

BRITISH EUGENICISTS MARVELED at the extent to which their American counterparts managed to write such a comprehensive negative-eugenics program onto the statute books. Perhaps contributing to the divergent legislative outcomes, at least with regard to marriage and sterilization, was the jurisdictional difference—in Britain such matters fell to a national body,

Parliament, while in America they were the province of state legislatures, whose level of deliberation even today leaves a good deal to be desired. Yet the Parliament at its worst has often been inferior to a number of American state legislatures at their best, and American standards of civil liberties have often been higher than the British. To account for the legislative differences, it also bears keeping in mind that the eugenics movement depended upon the authority of science, and that it was a coalition, united by a belief in the significance of heredity in human affairs yet, particularly in Britain, divided along a cluster of social fault lines.

In the United States during the opening decades of the century, it came to be a hallmark of good reform government to shape public policy with the aid of scientific experts. In many states the practice was modeled after the "Wisconsin Idea," advanced by the progressive governor Robert La Follette, of drawing upon experts in the state university for advice in complicated policy areas like taxes, agriculture, regulation, and public health. Eugenics experts aplenty were to be found in the biology, psychology, and sociology departments of universities or colleges, and among superintendents of state mental institutions. The fount of expertise was Charles Davenport's Eugenics Record Office, with its numerous publications and field workers. After their stint with Davenport, the field workers fanned out to various states, where they took jobs on the staffs of institutions for the mentally incompetent and, often with the aid of state appropriations for the purpose, conducted investigations among the local population regarding the heritability of mental and social defect.[13]

The field workers, the professors, and the institution superintendents not only could provide expert advice on eugenic issues to state legislative committees and commissions; together they might form a small yet influential lobby for eugenic legislation, particularly under reformist state administrations, and usually in the absence of equally expert opposition.[14] In the state of Wisconsin itself, the prime mover in the sterilization movement was Albert W. Wilmarth, the superintendent of the Home for the Feebleminded and a firm believer in the existence of hereditary "moral imbeciles" unable to resist the "animal emotions" that led to the promiscuous production of endless criminals, prostitutes, paupers, and tramps. In the campaign for a sterilization law, Wilmarth enlisted the State Medical Society and found additional allies at the university, including the well-known progressive sociologist Edward A. Ross and the biologists Michael F. Guyer and Samuel J. Holmes, whose teachings helped create a pro-eugenic climate of opinion.[15]

Davenport himself was somewhat ambivalent about the employment of Record Office information for legal or legislative purposes. Field worker data ought not to be given to governmental authorities without his permission, Davenport insisted, even to save a man from the gallows.[16] Yet, not surprisingly, his objection to governmental involvement was selective. Indeed, he indulgently tolerated the rather influential and significant foray into policymaking of Harry H. Laughlin, his appointee, in 1910, as superintendent of the Eugenics Record Office, his right-hand man at Cold Spring Harbor, and an advocate of views that accorded with Davenport's own social prejudices.

Laughlin was the product of small midwestern towns and a family that teetered impecuniously between the religious and the academic life. His father, a preacher in the Christian Church, wandered through the unstable world of sectarian colleges, from presidencies to pastorates to professorships; he finally landed in Kirksville, Missouri, where, in 1892, he became chairman of the English Department at the Kirksville Normal School. Laughlin did not take the fundamentalist doctrines of his father's church too fundamentally. As an undergraduate in history at the Kirksville School, he reconciled science and religion by identifying God with some sort of force, "a universal ether—perfectly elastic, granular and uniform." After 1896, while holding school posts in rural Iowa, he developed an interest in agriculture. He took several terms of work in the subject at the state college, and in 1907 returned to the Kirksville Normal School to head a one-man Department of Agriculture, Botany, and Nature Study.[17] That year he wrote to Charles Davenport for advice about certain breeding experiments with chickens. He soon came to Cold Spring Harbor for a summer course —"the most profitable six weeks that I ever spent," he later told Davenport —and proceeded to turn himself into a professional biologist, specializing in heredity. He published workmanlike papers in genetics and achieved a certain degree of professional recognition, including a doctorate of science in biology from Princeton University. He also became a convert to eugenics, just as soon as Davenport introduced him to the creed.[18]

Like Davenport, Laughlin was a workhorse—humorless, intolerant of criticism, and continually afire with dogmatic secular zeal. The stern force in Davenport's background had been the father; in Laughlin's it was the mother, a determined, energetic woman who was a women's suffragist, an activist in women's missionary societies, and a contributor to the religious press. A member of the Women's Christian Temperance Union, she prevailed upon Harry to sign a temperance pledge, which he stuck to through life. "If I can't be great," Harry wrote his mother during his Iowa years, "I certainly can do much good. And I intend to do it."[19]

The "good" he intended to do centered increasingly on eugenic research, particularly on "feeblemindedness" and on the characteristics of immigrants. Proud that his family could be traced back to the American Revolution, he thought the new immigrants from Eastern and Southern

Europe afflicted, from generation to generation, by a high degree of insanity, mental deficiency, and criminality. A disproportionately large number, he maintained, were to be found in institutions for the insane and the feebleminded.[20] In the spring of 1920, Laughlin went down to Washington, D.C., to present a sheaf of eugenic petitions to the House Committee on Immigration and Naturalization, which was then working on the original emergency restriction act. The majority chairman of the committee was Congressman Albert Johnson, Republican of Washington, a rough-hewn, heavy-drinking politico with a hatred of radicals, Japanese, and open-door immigration policies. In short order, Johnson appointed Laughlin "Expert Eugenical Agent" of the committee. Given the franking privilege and letterheads, Laughlin was to study alien inmates and inmates of recent foreign extraction in a number of state institutions.[21]

In November 1922, Laughlin reported his results to the committee, plastering the walls of the hearing room with numerous charts and tables, including a gallery of Ellis Island photographs labeled "Carriers of the Germ Plasm of the Future American Population." Laughlin assured the congressmen that he had been wholly objective throughout, interested only in the truth of the situation. Actually, he twisted the facts (often he had found proportionately more native- than foreign-born in asylums) and indulged in blatant prejudice (recent immigrants, he said, might themselves be healthy, but they carried bad recessive genes, which would sooner or later out). In Laughlin's view, the "evidence"—mainly the results of a survey that he had conducted of the comparative incidence of mental deficiency among the foreign-born—led implacably to the conclusion that the recent immigrants were biologically inferior and that they jeopardized the blood of the nation.[22]

Quickly, Laughlin became known in Washington as an indispensable authority on the "biological" side of the immigration issue. Without much difficulty, he won over influential members of the committee, including not only Albert Johnson but also the minority leader, John C. Box, to the eugenic point of view. In 1923, Johnson joined the Committee on Selective Immigration of the newly founded American Eugenics Society; the committee issued a compendious report at the end of 1923 which added up to an endorsement of the permanent immigration restriction bill. After its passage, Johnson and his colleagues called upon Laughlin from time to time for new studies, which were done and duly published. Laughlin's writings and testimonies were much cited in the restrictionist literature of the decade. In 1929, John Box told the head of the Carnegie Institution of Washington, the parent agency of the Eugenics Record Office, that he knew of no one else "so thoroughly competent and so free from factional or political bias as is Dr. Laughlin." Box added, "Personally, I shall always regard his work in connection with this [immigration] question as monumental."[23]

IN THE UNITED STATES, eugenic politics and eugenic research were symbiotically linked; if so much negative eugenics was written into law in America, it was probably not least because American eugenics activists could draw upon the publications or allies of the Eugenics Record Office—the principal scientific and authoritative institution in its field. In Britain, the counterpart institution was the Galton Eugenics Laboratory, headed by Karl Pearson, at University College, London. But Pearson steadfastly refused to join the Eugenics Education Society, to participate in political activity, or to make available his institutional resources and expertise for the support of legislative measures. In the days of their mutual absorption in socialism, George Bernard Shaw had admonished Pearson: "Your aim is never to give yourself away, never to make a fool of yourself. . . . You are full of reasons for doing nothing, all excellent reasons—reasons for not making speeches in Trafalgar Square, for not writing plays, for not printing them, reasons for not living, not loving, not working, . . . an infinity of nots."[24] For all the bastinado, Shaw did penetrate to an aspect of Pearson's personality: by temperament, he was simply not a joiner, let alone a political activist. Intent upon establishing eugenics as an academic discipline, he liked to pursue research and publish primarily in scholarly journals. He might lecture widely upon his research results and their social implications, might even send copies of his eugenic writings to Members of Parliament, including Arthur Balfour and Winston Churchill; he preferred to leave the messy business of politics to others.

At the outset, Pearson considered the Eugenics Education Society a mixture of wisdom and folly, yet more than might have been anticipated, he noted to Francis Galton, and possibly a valuable aid in spreading ideas. The rosy hopefulness disappeared once Pearson saw the shape of the society, particularly the presence of people like Havelock Ellis among its leaders. In 1909, he bluntly told Galton that if the Eugenics Laboratory were mixed up in any way with Ellis and his ilk, "we should kill all chance of founding eugenics as an academic discipline." Physicians were supplying considerable data for Pearson's studies of human inheritance; Pearson worried that Ellis and his allies were "red rags to the medical bull, and if it were thought we were linked up with them, we should be left severely alone."[25]

Setting himself a high standard of scientific rigor, Pearson was contemptuous of eugenic work that he considered careless or slipshod. To Pearson, the Eugenics Education Society indulged itself too little in science and too much in rank propaganda.[26] A prominent member of the society was Caleb W. Saleeby, a physician, medical researcher, and temperance advocate, who preached that the evils of alcohol not only afflicted those who drank but devolved upon their offspring as well. Even before the storm that followed Pearson's announcement, in 1910, that parental alcoholism produced no adverse effects upon children's intelligence, physique, or tendency to disease, Saleeby had attacked the Eugenics Laboratory as a waste of Galton's money. Pearson fumed to Galton that "we are the only people who have really endeavoured to measure the relation of alcoholism in parents to the mental and physical condition of the children, and that only in this laboratory is the relation of alcoholism to crime and insanity actually known." Pearson sneered at Saleeby's "rhetoric and fustian," and his opinion of the Eugenics Education Society as a group of mush-minded propagandists deepened.[27] Yet the scientific fissure between Pearson and the eugenic activists was more fundamental. While Pearson was a zealous biometrician, organized eugenics tended to be dominated—in Britain as in America—by Mendelians.

The dispute between Mendelians and biometricians which so infected British genetics early in the century plagued British eugenics with almost equal virulence. Pearson's charges of unscientific propagandizing against the Eugenics Education Society were as much an assault on its Mendelian claims as on its lack of rigor. Particularly offensive to Pearson was the society's embrace of the Mendelian heritability of mental defect. Pearson had not strayed from his own conviction that mental deficiency was inherited, but his belief rested on biometrical studies, which he thought reliable compared, for example, to the American studies popularized by Charles Davenport. In 1913, in a critique of them published from the Galton Laboratory, David Heron concluded of Davenport's data that it had been "collected in an unsatisfactory manner," and "tabled in a most slipshod fashion," and that "the Mendelian conclusions drawn have no justification whatever."[28]

Amid the acrimonious cockfight that followed, thoughtful observers could hardly be blamed for concluding that not a good deal was known for certain about the heritability of mental defect. The dispute, pitting Pearson, the principal eugenic researcher in the land, against his colleagues, undermined the scientific authority with which British eugenicists could speak. On behalf of the society, Leonard Darwin kept trying to reach some sort of rapprochement with Pearson; repeatedly failing, he deplored the sharp rift in the camp of British eugenics.[29] Yet if that rift became perhaps the chief weakness of the British movement, British eugenic activism was also hampered by a rift within its own ranks on the appropriateness of pressing for eugenic legislation.

Some of the opposition came from scientifically knowledgeable mainliners like Edgar Schuster, who recognized that knowledge of human heredity was primitive and thought many of the eugenic laws in America "hasty and ill-considered." William Bateson, sympathetic to eugenics but thoroughly opposed to enforcing it by act of Parliament, declared, in the Herbert Spencer Lecture at Oxford in 1912: "It is not the tyrannical and capricious interference of a half-informed majority which can safely mould or purify a population." But the principal opposition to legislative action came from the social-radical wing of British eugenics. Havelock Ellis salted his eugenic writings with warnings against it: "Public opinion is the only lever at present, and legislative action must be impossible—and futile—for a very long time to come."[30]

Ellis was particularly reluctant to resort to compulsory sterilization. Not offended by sterilization as such, he thought it possibly a useful and effective method of preventing the procreation of the unfit, particularly the mentally deficient. He was convinced, however, that in almost all cases sterilization must be voluntary rather than coercive. Voluntary submission to the procedure was to be accomplished by educating the subjects to their civic duty, their responsibility to the race. Compulsion was to be applied only to that tiny, "irreducible nucleus of the incapable group," and then only after using such social inducements to foster voluntary avoidance of procreation as the group might be amenable to. The opposition of Ellis and other social radicals probably tempered whatever inclination the Eugenics Education Society may have had for compulsory sterilization; though it had flirted with the idea in the period before passage of the Mental Deficiency Act of 1913, it refrained from advancing a mandatory program in the years that followed.[31]

There was no social-radical counterpart in the American eugenics movement to the group identified with Havelock Ellis. Many American social radicals—Max Eastman, John Reed, Lincoln Steffens, Mabel Dodge Luhan, and the like—seem not to have been drawn to eugenics, undoubtedly because the mainline posture was so anti-feminist, anti-birth control, and, above all, anti-immigrant. American radicals who did dabble in eugenics, like Emma Goldman, were outside the movement's organized leadership. American eugenics of course included numerous progressive reformers, but many seem to have been drawn from that wing of progressivism which tended to an anti-immigrant racism. To the reformers as well as to the conservatives in American eugenics, sterilization went hand in hand with immigration restriction and was just as defensible.

Yet it was hardly as effective in diminishing the population of the "unfit." From 1907 to 1928, fewer than nine thousand people had been eugenically sterilized in the United States, as against a "feebleminded menace" of—in Henry Goddard's estimate—three hundred thousand to four hundred thousand people.³² Nevertheless, American eugenicists seemed to have few misgivings about sterilization. Indeed, they were confident, even enthusiastic about the policy—enthusiastic enough to make one speculate about the psychodynamics of their attitudes.

EUGENICISTS GAVE A GOOD DEAL of attention to the sexual behavior of the "feebleminded," some authorities discerning excessive sexuality among the males, others claiming that mentally deficient males were actually under-sexed. Whatever the disagreement about males, there had long been no dispute about females; they were reputed to be sources of debauchery, licentiousness, and illegitimacy. In the eighteen-eighties, the trustees of the New York State Custodial Asylum for Feeble-minded Women had argued, typically, that retarded women required special care because they were "easily yielding to lust." Henry Goddard, although he suspected that the feebleminded were not as sexually promiscuous as was generally believed, attributed an overdevelopment in the sexual instinct to a lack of inhibition. Mary Dendy, one of Britain's leading workers with the mentally deficient in the decade before the First World War, remarked: "the weaker the *Intellect* . . . the greater appears to be the strength of the reproductive faculties. It is as though where the higher faculties have dwindled the lower, or merely animal, take command."³³

For all the scientific theorizing, there was a good deal of circularity to the analysis. Immoral behavior was taken ipso facto as evidence of fee-blemindedness, which in turn was claimed to produce immoral behavior. The circularity arose from the tendency of eugenicists to identify as depravity most sexual expression that fell outside the bounds of prevailing middle-class standards. William J. Robinson proposed that "some of these pseudo-eugenicists would, if they had the power, castrate or sterilize every man or woman who is not strictly moral according to *their* standard of morality, who smokes, drinks a glass of beer, indulges in illicit sexual relations, or dares to doubt the literal veracity of the Bible."³⁴ Why such distress at social deviancy? An entire sociopolitical movement can hardly be put on the analyst's couch, but the attention given eroticism, the denunciation of feminism, and the genital attack implicit in sterilization all suggest the possibility that mainline eugenics was driven in part by the psychic energy of a repressed discomfort with sexuality.

Lionel Penrose, a British physician and a world authority on mental deficiency, knew of little evidence that the retarded male or female had abnormally strong sexual drives. Just why the respectable classes should think they did, and should want to sterilize them for it, stimulated the young Penrose to advance in the early thirties a Freudian speculation: "It is a well-known psychological mechanism that hatred, which is repressed under normal circumstances, may become manifest in the presence of an object which is already discredited in some way. . . . An excuse for viewing mentally defective individuals with abhorrence is the idea that those at large enjoy themselves sexually in ways which are forbidden or difficult to accomplish in the higher strata of society. The association between the idea of the supposed fecundity of the feebleminded and the need for their sterilization is apparently rational, but it may be emphasized by an unconscious desire to forbid these supposed sexual excesses. It has been pointed out that the advocates of sterilization never desire it to be applied to their own class, but always to someone else."[35]

Dr. Harry Sharp first experimented with sterilization in Indiana partly for eugenic ends but also to reduce sexual overexcitation in delinquent boys. Dr. Charles Carrington, surgeon to the Virginia Penitentiary, sterilized a dozen men by vasectomy around the turn of the century and reported, "in every instance but two, the subjects were insane, persistent masturbators, and in every case masturbation has ceased, patients have invariably improved mentally and physically. . . ."[36] In reality, vasectomy did not diminish male sexual desire. Neither did tubal ligation diminish female sexual drives. Indeed, some of those opposed to sterilization emphasized the point that it would foster rather than diminish sexual license because pregnancy could not follow indulgence. The problem made some knowledgeable eugenicists like Charles Davenport ambivalent towards sterilization, but not Harry Laughlin, who thought "it ought to be a eugenic crime to turn a possible parent of defectives loose upon the population." There was no more passionately outspoken advocate of sterilization in America than Laughlin, who made himself an authority on the subject, including its legal as well as biological intricacies.[37]

The prevailing popular tendency seems to have been to confuse sterilization with castration and to assume that sterilization reduced sexual energy. According to a 1932 study of sterilization laws in the United States by Jacob H. Landman—a lawyer who had earned a doctorate for his investigation of the subject—sexual offenses or moral degeneracy figured explicitly in the grounds for sterilization found in almost half the state statutes then on the books. In the rest of the sterilization statutes, sexual license was implicitly covered in the provisions concerning "feeblemindedness." A review in 1938 of sterilizations at the Virginia Colony for Epileptics and Feebleminded noted that two-thirds of the inmates sterilized had been in trouble with the law, with sexual infractions ranking third among the offenses committed by the males and first among those committed by the females. In California, three out of four of the sterilized women had been judged sexually delinquent prior to their institutional commitment.[38]

In many states, sterilization measures ran afoul of the courts, of legislative opposition, of executive refusal to enforce, and of gubernatorial vetoes. In 1905, Governor Samuel W. Pennypacker rejected the sterilization act of the Pennsylvania legislature with the ringing broadside: "It is plain that the safest and most effective method of preventing procreation would be to cut the heads off of the inmates." (Not long afterward, Pennypacker wisecracked down a raucous political audience: "Gentlemen, gentlemen! You forget you owe me a vote of thanks. Didn't I veto the bill for the castration of idiots?") Many of the laws were couched in punitive rather than eugenic terms. Most did not provide elementary procedural protection to those singled out for possible sterilization. Most also confined eligibility for sterilization to people in state institutions. Thus, the objections centered on violations of the constitutional safeguards against cruel and unusual punishment, due process of law, and equal protection of the laws.[39]

One of the most biting critiques of the sterilization statutes was published in 1913 in the *Journal of the American Institute of Criminal Law and Criminology* by the prominent New York lawyer Charles A. Boston. Boston's numerous distinctions included the vice-presidencies of the New York Council of the American Bar Association and the Society of Medical Jurisprudence, whose Committee on the Law of Insanity he chaired. To the principal constitutional objections to the statutes, Boston added the prohibition against bills of attainder and against double jeopardy. Holding up to scrutiny the Indiana law, which began: "Whereas heredity plays a most important part in the transmission of crime, idiocy and imbecility . . . ," Boston charged that the legislature had "accepted as established fact, the finest shading in the laws of heredity, which are not yet established as a fact in their very broadest outlines." Boston guessed that the number of convictions for rape annually in Indiana must be smaller than the number of persons killed by automobiles. "If criminal tendencies were hereditary, then there would be more substantial reason for sterilizing reckless chauffeurs than 'rapists.' "[40] He indicted the law as without practical effect, having calculated, on the basis of the number of sterilizations carried out in Indiana, that the number of children prevented from birth in that state during a half century would amount to only one half of one percent of the population. And that at considerable threat to civil liberty. "If a legislature can constitutionally sterilize a criminal or an insane person . . . it could sterilize multi-millionaires," he wrote, ". . . for it might declare in a preamble that the sons of these tend to become a menace to the community, as an idle and licentious class; similarly it could sterilize clergymen, pursuant to a preamble that their sons are frequently charged with being, on the average, worse than other men's sons." In all, Boston deemed that sterilization laws belonged to that class of legislation better "left behind in the cast-off junk of ignorant efforts, with which the past is filled."[41]

By the outbreak of the First World War, sterilization laws were in such dispute as to have been de facto suspended in their operation in a number of states. The courts had also declared unconstitutional not only the stringent Iowa statute but less sweeping measures in six other states. Advocates of eugenic sterilization, frustrated at the legal impasse, wanted to take the issue to the Supreme Court. In Virginia, eugenicists helped draw up a sterilization statute, passed by the legislature in March 1924, that was designed to meet the constitutional objections. The opportunity to press a test case arose that June, when a seventeen-year-old girl named Carrie Buck, who seemed definable as a "moral imbecile," was committed to the Virginia Colony for Epileptics and Feebleminded, in Lynchburg.[42]

Carrie's mother, Emma, had lived at the Colony since 1920 and was also certified to be feebleminded. Carrie herself had conceived a child out of wedlock, and shortly before her commitment, she gave birth to a daughter, Vivian. Carrie was given the Stanford revision of the Binet-Simon I.Q. test and was found to have a mental age of nine years, well within Henry Goddard's definition of "moron." Carrie's mother was found to have a mental age of slightly under eight years. Thus, according to these results, there was mental deficiency in two successive generations. If Vivian could be shown to be feebleminded too, Carrie would be a perfect subject for a test of the Virginia sterilization statute. In September 1924, the Colony's board of directors ordered Carrie Buck sterilized, and a court-appointed guardian initiated legal proceedings by appealing the order in a suit on Carrie's behalf against the superintendent of the Colony, Albert S. Priddy.[43]

In preparing their case, Virginia officials consulted Harry Laughlin at the Eugenics Record Office. Laughlin examined the pedigrees of Carrie, her mother, and her daughter, and information about them given him by Colony officials, and—without ever having seen them in person—provided an expert deposition that Carrie's alleged feeblemindedness was primarily hereditary. Carrie and her forebears, Laughlin submitted, "belong to the shiftless, ignorant, and worthless class of anti-social whites of the South." At the time of Laughlin's deposition, however, there was no evidence at all that Vivian was mentally deficient. To clarify the matter, Caroline E. Wilhelm, a Red Cross worker who had placed Vivian in a foster home, was prevailed upon to examine her there. At the initial hearing, in the Circuit Court of Amherst County, she testified that there was "a look" about Vivian (who at the time of the visit was seven months old) which was "not quite normal." Evidence also came from Arthur Estabrook of the Eugenics Record Office, who had subjected Vivian to a mental test for an infant and concluded that she was below average for a child her age. In the court proceeding, Estabrook testified that the feeblemindedness in the Buck line conformed to the Mendelian laws of inheritance, and the judge upheld the sterilization order.[44]

The case—now known as *Buck* v. *Bell*, because Priddy had in the meantime died and been replaced as the defendant by the Colony's new superintendent, John H. Bell—was carried to the Virginia Supreme Court of Appeals in 1925, and the sterilization order was again upheld. In April 1927 it was argued before the United States Supreme Court. Carrie's defense counsel, I. P. Whitehead, a onetime member of the board of directors of the Colony, attacked the sterilization statute, warning that under this type of law a "reign of doctors will be inaugurated and in the name of science new classes will be added, even races may be brought within the scope of such a regulation and the worst forms of tyranny practiced." Nevertheless, the Court was persuaded not only that Carrie Buck and her mother were "feebleminded" but also—because Vivian was, too (or so all the experts said)—that the feeblemindedness was heritable. The Court, whose membership ranged in political conviction from William Howard Taft to Louis D. Brandeis, upheld the Virginia statute by a vote of eight to one. The sole dissenter was Justice Pierce Butler, a conservative, and he kept his minority opinion to himself. The decision declared that sterilization on eugenic grounds was within the police power of the state, that it provided due process of law, and that it did not constitute cruel or unusual punishment.[45]

The Court's opinion was written by Justice Oliver Wendell Holmes, an enthusiast of science as a guide to social action, who managed to find a link between eugenics and patriotism: "We have seen more than once that the public welfare may call upon the best citizens for their lives. It would be strange if it could not call upon those who already sap the strength of the State for these lesser sacrifices . . . in order to prevent our being swamped with incompetence. . . . The principle that sustains compulsory vaccination is broad enough to cover cutting the Fallopian tubes." With deliberate punch Holmes asserted: "Three generations of imbeciles are enough."[46]

Eugenicists naturally rejoiced at *Buck* v. *Bell*. For some years prior to the decision, the American Eugenics Society had promoted what it thought might be a constitutional revision of the faulty sterilization statutes. Apart from procedural and technical changes, the revisions centered on making the laws eugenic rather than punitive in intent. After *Buck* v. *Bell*, what was constitutional was clear. By the end of the nineteen-twenties, sterilization laws were on the books of twenty-four states, with the South no longer a regional exception. (Though now severely restricted by federal regulation, they are still on the books of twenty-two states today.) The laws were not uniformly enforced, but Carrie Buck was sterilized soon after the Court's decision, and officials at the Virginia Colony subjected other inmates to the procedure—a total of about a thousand in the next ten years. By the mid-thirties, some twenty thousand sterilizations had been legally performed in the United States.[47]

Buck v. *Bell* generally stimulated either favorable, cautious, or—most commonly—no editorial comment. Few if any newspapers took notice of the impact of the decision on civil liberties in the United States. The I.Q. tests used in the Buck case have long since been discredited as indicators purely of general intelligence. With regard to the allegedly hereditary nature of mental defect in the Buck line, it is of interest that Carrie's daughter Vivian went through the second grade before she died of an intestinal disorder in 1932. Her teachers reportedly considered her very bright.[48]

Chapter VIII

A COALITION OF CRITICS

In 1930 the eugenic publicist Albert Wiggam told a gathering at the American Museum of Natural History: "Civilization is making the world safe for stupidity." Mainline eugenicists may have long worried about differential birthrates or declining national intelligences, but their apprehension deepened considerably in the era of the Great Depression.[1]

In the thirties, eugenicists marshaled statistical evidence that America's mental institutions would soon house more than half a million people, one for about every two hundred and fifty persons in the country. It was said that twice as many families sent a child to an institution for the feebleminded as those who sent one to college. But no American statistics matched in authority the evidence provided, after a five-year survey, by the 1929 report of the British government's Joint Committee on Mental Deficiency. According to what the committee considered a conservative estimate, there were at least three hundred thousand mental defectives in England and Wales, which meant that since 1908, when the last survey was completed, the incidence of deficiency had doubled.[2] Three-quarters of the mentally deficient tended to come from families persistently below the average in income and social character. These families included, in the words of the Joint Committee Report, a much larger proportion of "insane persons, epileptics, paupers, criminals (especially recidivists), unemployables, habitual slum dwellers, prostitutes, inebriates, and other social inefficients" than did families with no mentally deficient members. In the nomenclature of the report, they constituted a "social problem group" comprising about a tenth of the English and Welsh population.[3]

The committee's report was hedged with cautions: the increase in the number of mental defectives did not necessarily result from proportionately more unfortunates being born; it no doubt expressed, among other things, that more were surviving into adulthood because of improved public-health services. But the central conclusions of the report, not the cautions, commanded the headlines. The future Prime Minister Neville Chamberlain solemnly told the House of Commons that the doubling in the estimated incidence of deficiency "must give serious anxiety and apprehension among all who care for the future physical and mental condition of our people."[4]

Mainline eugenicists attributed the economic condition to biological deterioration, a consequence of the differential birthrate. They insisted that the unemployed were by reason of biological destiny mentally incompetent, improvident, irresponsible, and thriftless. In the United States, in 1931, *Cosmopolitan* magazine reported that in the opinion of President Herbert Hoover all deficient children suffered from malnutrition. The White House was moved to issue a correction: the statement did not represent the President's views, and it of course contravened "all scientific knowledge of heredity."[5] British eugenicists called special attention to *Heredity and the Social Problem Group*, a study of the poor in London's East End, published in 1933, after a quarter of a century of research, by E. J. Lidbetter, a longtime relief worker in the area. The study, essentially a vast compendium of genealogical data, tentatively concluded that the poor constituted a biological class of their own, which was marked by a considerable degree of defectiveness and which they tended to perpetuate by marrying each other. "The best in civilization is the best biologically," Lidbetter averred. "What is therefore necessary today is attention to the problems of reproduction and its control."[6]

The Depression added a strong fillip of interest to the pro-sterilization arguments of mainline eugenics. In Britain, the Eugenics Society—as the Eugenics Education Society had been renamed in 1926—printed ten thousand copies of a pamphlet explaining the advantages of sterilization; demand was so brisk that it had to print ten thousand more. The editor of the respected scientific journal *Nature* devoted space to sterilization matters, including the view of one British biologist who proposed "compulsory sterilization as a *punishment* for parents who have to resort to public assistance in order to support their children." In the United States, people talked of similar measures for potential parents on relief beyond a certain length of time. A 1937 *Fortune* magazine poll revealed that sixty-three percent of Americans endorsed the compulsory sterilization of habitual criminals and that sixty-six percent were in favor of sterilizing mental defectives. The country, said E. A. Hooton, professor of physical anthropology at Harvard University, had to do some "biological housecleaning."[7]

The housecleaners on both sides of the Atlantic insisted that sterilization was humane as well as practical, and in proof of the point they cited *Sterilization for Human Betterment*, a report published in 1929 by the American eugenicists Ezra Gosney and Paul Popenoe on the history of the procedure in California. Since early in the century, California had led the nation in sterilizations and by 1929 had 6,255 operations to its credit—almost twice as many as those of all other states combined. According to Gosney and Popenoe, sterilization was prescribed with kind judiciousness by a group of doctors, social workers, and mental-health professionals in consultation with the family, and the operations—vasectomy or tubal ligation—were carried out in a scrupulously professional fashion, with a very low rate of infection and a minuscule number of fatalities. Many of the patients—or at least their families—were reported to have welcomed the procedure. Particularly grateful were numerous women, "many of them pathetic in their expression of gratitude and their wish that other women who faced the combination of pregnancy and psychosis might have the same protection."[8]

Aided and abetted by the Depression, sterilization drew diverse support in the United States and Britain which went far beyond eugenicists. Its advocates ranged from college professors to elementary school principals, from clubwomen to mental-health workers, from the British Conservative Women's Reform Association to the New Jersey League of Women Voters, from private congresses to the 1930 White House Conference on Children and Health, from Anglican bishops to the Newark Methodist Conference, from Lord Horder, physician of King George VI and the Prince of Wales, to H. L. Mencken, who suggested that the federal government pay a thousand dollars to every "adult American" who volunteered to be sterilized.[9] Governments in Sweden, Denmark, Finland, and even a canton of Switzerland also enacted eugenic sterilization measures. By 1933, Paul Popenoe proudly estimated that sterilization laws were in effect in jurisdictions comprising some hundred and fifty million people.[10]

In Britain, however, sterilization on any ground was assumed to be repugnant to the law. No statute forbade it directly, but various laws were held to imply that the eugenic sterilization of the mentally deficient would be illegal. Most frequently cited was the Offences Against the Person Act of 1861, which made it a crime for one person to cause grievous bodily harm to another. Although sterilization would probably not constitute such an injury, British doctors were reluctant to perform the operation even on volunteers, let alone on anyone who, by virtue of mental incompetence, could not in either legal or commonsensical terms make a voluntary choice.[11]

In the United States, despite *Buck* v. *Bell,* the large majority of the mentally afflicted were safe from the surgical knife. Mencken talked of voluntary sterilization because he rightly recognized that "the sharecropper, though he may appear to the scientist to be hardly human, is yet as much under the protection of the Bill of Rights as the president of Harvard," adding, "He may not be jailed unless he has perpetrated some overt act forbidden by law, and he may not be gelded unless his continuance at stud is plainly and undoubtedly dangerous to society."[12] The import of the Court's decision was to sanction compulsory sterilization only of the inmates of state mental institutions whose disabilities were judged to be hereditary. Yet if the road was narrow, the constitutional traffic light remained green. By 1935, four more states had passed sterilization laws, and bills for the same purpose were pending in the legislatures of another seven. Through the nineteen-twenties, the national sterilization rate had annually run between two and four per hundred thousand. In the mid-thirties the rate shot up to fifteen and climbed to twenty by the end of the decade; the national sterilization total would reach almost thirty-six thousand by 1941. Moreover, from 1932 to 1941, sterilization was actually practiced—as distinct from merely legislated—in a greater number of states than before: California's share, while still the largest, was about a third of the national sterilization total. Second in rank, with a seventh of the national total, was Virginia, where *Buck* v. *Bell* had originated.[13]

Howard Hale, a former member of the Montgomery County, Virginia, Board of Supervisors, recently recalled that the state sterilization authorities raided whole families of "misfit" mountaineers in the thirties. At the time, Hale was the proprietor of a small candy store that catered to those families. "Everybody who was drawing welfare then was scared they were going to have it done on them," he remembered. "They were hiding all through these mountains, and the sheriff and his men had to go up after them. . . . They really got them up on Brush Mountain. The sheriff went up there and loaded all of them in a couple of cars and ran them down to Staunton [Western State Hospital] so they could sterilize them." Hale added that "people as a whole were very much in favor of what was going on. They couldn't see more people coming into the world to get on the welfare."[14]

Sterilization of males at the state colony at Lynchburg was carried out regularly on Tuesdays; females were sterilized on Thursdays. Still, Dr. Joseph S. DeJarnette, long a powerful voice in Virginia eugenics and a major influence in its sterilization program, thought that the state was sterilizing too few people. In 1934, he urged the legislature to broaden the sterilization law. "The Germans are beating us at our own game," he said.[15]

Adolf Hitler's cabinet had promulgated a Eugenic Sterilization Law in 1933. Going far beyond American statutes, the German law was compulsory with respect to all people, institutionalized or not, who suffered from allegedly hereditary disabilities, including feeblemindedness, schizophrenia, epilepsy, blindness, severe drug or alcohol addiction, and physical deformities that seriously interfered with locomotion or were grossly offensive. The counselor of the Reich Ministry of the Interior, who had drawn up the law, called it an exceptionally important public-health initiative. "We want to prevent . . . poisoning the entire bloodstream of the race," he told a group of foreign correspondents in Berlin. "We go beyond neighborly love; we extend it to future generations. Therein lies the high ethical value and justification of the law."[16] After January 1, 1934, when the law went into effect, physicians were to report all "unfit" persons to hundreds of Hereditary Health Courts established to adjudicate the German procreational future. Each court consisted of a jurist and two physicians, including at least one specialist in heredity. Decisions could be appealed to a higher eugenic court, whose rulings were final and could be carried out by force if necessary. Within three years, German authorities had sterilized some two hundred and twenty-five thousand people, almost ten times the number so treated in the previous thirty years in America. About half were reported to be "feebleminded."[17]

Sterilization was, of course, only the beginning of the Nazi eugenic program. In the interest of improving the German "race," the government provided loans to biologically sound couples whose fecundity would likely be a credit to the *Volk*; the birth of a baby would reduce the loan indebtedness by twenty-five percent. A number of German cities established special subsidies for third and fourth children born to the fitter families. To foster the breeding of an Aryan elite, Heinrich Himmler urged members of the S.S. to father numerous children with racially preferred women, and in 1936 he instituted the *Lebensborn*—spa-like homes where S.S. mothers, married and unmarried, might receive the best medical care during their confinements.[18]

For a time, the Nazi sterilization program ran independently of the regime's anti-Semitic policies. Anti-Semitism had not markedly characterized the pre-1933 German eugenic leadership; in fact, before 1933, the leading German eugenic journal had assumed that the Jews of Germany were virtually members of the Aryan race. Jacob H. Landman, who was a Jew and a critic of sterilization, concluded in a 1936 issue of *Survey Graphic* that the German program was "not intended to exterminate non-Aryans but to improve the German national stock." He continued: "It does not include in its scope the sterilization of Semites or other non-Aryan groups. There is no evidence that the law has been violated so as to cause the sterilization of patients exclusively because they were non-Aryans."[19]

But as Hitler turned ever more overtly against the Jews, Nazi racial and eugenic policies merged. The regime promulgated eugenic marriage laws prohibiting the espousal of persons with mental disorders, certain infectious diseases, or different "racial" backgrounds. Exceptions to the marital ban on the mentally disordered were permissible if the prospective partners had been sterilized, but after the Nuremberg Laws of 1935 no exceptions were to be made in the case of unions between "Jews" and "Germans." In 1939, the Third Reich moved beyond sterilization to inaugurate euthanasia upon certain classes of the mentally diseased or disabled in German asylums. Among the classes were all Jews, no matter what the state of their mental health. Some seventy thousand patients were eventually designated for euthanasia. The first groups were simply shot. Later victims were herded into rooms disguised as showers, where they were gassed.[20]

In the early years of the Nazi regime, most mainline eugenicists in the United States and Britain could not know—and likely did not want to imagine—that a river of blood would eventually run from the sterilization law of 1933 to Auschwitz and Buchenwald. Shortly after the law was passed, an officer in the American Eugenics Society advised several newspaper editors that Hitler's sterilization policy showed great courage and statesmanship. Other observers, including Havelock Ellis, echoed Landman's report that the Nazi sterilization program was without nefarious racial content. German eugenicists, flattering to their American counterparts, said that they owed a great debt to American precedent, including the report of Gosney and Popenoe on the California program. In 1936, the University of Heidelberg voted an honorary doctorate of medicine to Harry Laughlin, still a sterilization enthusiast and in charge of the Eugenics Record Office, at Cold Spring Harbor, Long Island. Laughlin, who accepted the degree at the German consulate in downtown Manhattan, wrote to the Heidelberg authorities that he took the award not only as a personal honor but also as "evidence of a common understanding of German and American scientists of the nature of eugenics."[21]

THE BARBAROUSNESS OF NAZI policies eventually provoked a powerful anti-eugenic reaction, but the reaction, perhaps because of its pervasive power, obscured a deeper historical reality: many thoughtful members of the British and American public had already recognized that a good deal was wrong with mainline eugenics. Indeed, long before the Nazis came to power a growing, influential coalition had turned against the mainline movement. The opposition came from diverse sources both secular and religious. Prominent among them were Liberals and Labourites in England and civil libertarians in the United States, social workers, and social scientists, among them members of minority groups who were entering the academic world. The critics included eugenicists who had never been part of the mainline —usually social radicals and feminists—and mainliners who had become apostates. Also among them were Protestants of various denominations, Jews, and especially Catholics.[22]

The Catholic dissent rested intellectually on the Church's doctrine that in the scheme of God's creation man's bodily attributes are secondary, his spirit paramount. What to the eugenicist were biologically unfit people were, to the Church, children of God, blessed with immortal souls and entitled to the respect due every human being. In 1912, in a study for the Catholic Social Guild, Father Thomas J. Gerrard dubbed radical eugenic doctrines "a complete return to the life of the beast" and criticized the more commonplace versions for holding that man was primarily and essentially animal in nature and that "his betterment is chiefly if not entirely a matter of germ plasm, milk, fresh air, sentimental art, and illuminated certificates [of eugenic worth]." The Church stressed the role of love and religious ethics, rather than parental perfection of physique and intelligence, in producing offspring with eugenic qualities. "The Church declares the root cause of degeneracy to be sin," Gerrard said, "and the root cause of betterment to be virtue."[23] If parents were in danger of producing hereditarily disabled offspring, the Church insisted upon abstinence rather than contraception; the latter not only allowed, in Gerrard's words, for the "perversion of the appetite within the marriage state" but also made for race suicide. Catholic authorities linked eugenics with the modern permissiveness that threatened the integrity of the family, the obedience of wife to husband, the subordination of erotic passion to moral will. Pope Pius XI revealed the congeries of Catholic fears in his encyclical *Casti Connubii,* of December 31, 1930, in which he condemned eugenics along with divorce, birth control, companionate marriage, and the celebration of animal passion in films, the press, and the theater.[24]

Secular critics of eugenics hardly agreed with all the Church's sweeping denunciations, but there was a good deal they could applaud in the Catholic attack—particularly the assault on the biological reductionism of the mainline creed. Distressed by the implications of such reductionism for women, the British liberal theorist L. T. Hobhouse allowed that the mainline eugenicist was "within his rights in calling attention to the dwindling of the family among the more educated classes," but declared him wrong "if he insists on quantitative reproduction at the expense of qualitative life, if he returns to the conception of woman as limited in her function to the bearing and rearing of children."[25] James Joyce's Stephen Daedalus, ruminating upon notions of beauty, disliked to think that "every physical quality admired by men in women is in direct connection with the manifold functions of women for the propagation of the species." That led "to eugenics rather than to esthetic"—and to professorial lectures "that you admired the great flanks of Venus because you felt that she would bear you burly offspring and admired her great breasts because you felt that she would give good milk to her children and yours."[26]

Critics of a humanist bent identified the offensiveness of eugenics with their more general resentment toward the ever-mounting authority of science. G. K. Chesterton, in the years before the war, had fired salvos of biting essays against mainline pretensions, indicting its advocates for having discovered "how to combine the hardening of the heart with a sympathetic softening of the head," and for presuming to turn what common decency held to be commendable deeds—marriage to an invalid, for example—into "social crimes."[27] The essays, collected in the early nineteen-twenties— about the time of Chesterton's conversion to Catholicism—into his *Eugenics and Other Evils*, became a staple of the anti-eugenic arsenal on both sides of the Atlantic. Chesterton linked eugenics to Prussianism, to the "same stuffy science, the same bullying bureaucracy and the same terrorism by tenth-rate professors that have led the German Empire to its recent conspicuous triumphs." In his view, science had long aimed to tyrannize. Through eugenics, it proposed to extend its tyranny "to reach the secret and sacred places of personal freedom, where no sane man ever dreamed of seeing it; and especially the sanctuary of sex." He predicted that eugenics would mean forcible marriage by the police.[28]

When various American states enacted eugenic marriage laws, practical analysts scoffed that, like such sumptuary legislation as prohibition, the measures would be largely unenforceable—people could avoid marriage laws in one state by wedding in another. *The Nation* predicted "evasion . . . false swearing . . . maladministration," and ultimate "immorality."[29] However shrill a Chesterton or practical the legislative analysts, it was commonly understood that eugenic interference with marriage and, more fundamentally, with procreation was an unwarranted and dangerous invasion of civil liberty. Various critics pointed to the mainline eugenic movement's distrust of democracy, to its claims that men were not created equal even in political rights, to its threat to establish some sort of caste system of government. Bertrand Russell speculated that eventually opposition to a given government would be taken to "prove imbecility, so that rebels of all kinds will be sterilized." Writing in *The American Mercury*, in 1926, Clarence Darrow warned that if the state was invested with eugenic authority, "those in power would inevitably direct human breeding in their own interests," and continued: "At the present time it would mean that big business would create a race in its own image. At any time, it would mean with men, as it does with animals, that breeding would be controlled for the use and purpose of the powerful and unintelligent."[30]

Principle fired the anti-mainline dissent, but principle was strengthened by the experience of social workers who confronted face to face the human objects of eugenic attack in charitable agencies, settlement houses, and institutions for the mentally deficient. No doubt typical was Charles

Davenport's older brother the Reverend William E. Davenport, who drew upon his experience as founder and head of the Italian Settlement Society of the United Neighborhood Guild in Brooklyn, New York. True, like so many native Americans of his day, he tended to attribute behavioral traits to ethnic groups—for example, tendencies toward violence or intoxication —but his experience suggested that the Italians he knew did not merit the biological animadversions visited upon them by eugenics. The Reverend Davenport told his brother that "over and over again young men . . . rated very low mentally by competent [reformatory] examiners . . . have come out and subsequently evinced excellent capacity in their home relations and social obligations and more frequently still in their capacity to get and hold on to their money. . . . I know personally half a dozen bootleggers whose resistance to the temptation to get easy money has been extremely poor, while their resistance to the temptation to part with it quickly has been undeniably marked."[31]

Principle also drew energy from the eugenic threat to lower-income groups. Catholic theologians denounced eugenics not only because they found it incompatible with the canons of the Church but because so much of the flock were the poor immigrants of Liverpool and London's East End, of New York City, Boston, Philadelphia, and Chicago. If Catholic theologians attacked the eugenic embrace of birth control, it was partly because, with Father John J. Burke, general secretary of the National Catholic Welfare Conference, they thought its advocates were "recommending that the lower classes be less productive on account of economic conditions, holding that infant mortality, arising from want of care and from the prevalence of ignorance and disease, should be reduced not by improving social conditions and curbing those who exploit the poor, but by fitting the habits of these classes to their condition."[32] In the impassioned view of many dissidents, to rank the merits of the national germ plasm of the future ahead of the human needs of the socially disadvantaged in the present seemed morally outrageous. Social reformers argued from hard experience that what needed to be halted was social rather than racial decline; and that what needed to be furthered was not racial but social betterment.[33]

Principle was also reinforced by the newer findings of various social sciences—notably anthropology, sociology, and psychology—which, often without specific regard to eugenics, were trending in anti-eugenic directions. Yet for some years principle had to do without the endorsement of genetics.[34] Many geneticists, both British and American, either were themselves caught up in the mainline creed or were reluctant, in the self-professedly apolitical community of science, to offend their pro-eugenic colleagues. Nevertheless, a growing number worried that mainline eugenics was tarnishing the genetics enterprise. Eugenic writings, with their attention to sexuality, baby health, and family life, smacked of a deplorable pop science. Others found mainline eugenics morally or socially offensive.[35] But important above all for most scientists, much of what passed as eugenic research was slipshod in method, evidence, and reasoning. There was, in fact, a widening disjunction between the chief scientific claims of eugenics and the results of modern genetic science. The more genetics advanced in the first third of the century, the more its practitioners recognized on scientific grounds that they were, in the words of the Harvard geneticist William E. Castle, scarcely able to do more eugenically "than make ourselves ridiculous."[36]

During the First World War, a number of geneticists began to separate themselves from mainline eugenics, declining office in eugenic organizations, objecting to meetings that combined eugenics with genetics, insisting that journals of genetics refrain from publishing eugenic material. Thomas Hunt Morgan of Columbia University resigned from the eugenically connected American Breeders' Association, privately denouncing its journal for "reckless statements" and "unreliability." In the numerous editions of his *Heredity and Environment in the Development of Men*, the Princeton embryologist Edwin Grant Conklin called into question the more extravagant mainline claims. The geneticist Hermann J. Muller, an outspoken socialist, roundly condemned the mainline creed at the Third International Eugenics Congress itself. And Raymond Pearl, professor of biometry and vital statistics at Johns Hopkins and intimate of Mencken's iconoclastic circle, lambasted the "biology of superiority" in the November 1927 issue of *The American Mercury*, asserting that eugenics had "largely become a mingled mess of ill-grounded and uncritical sociology, economics, anthropology, and politics, full of emotional appeals to class and race prejudices, solemnly put forth as science, and unfortunately accepted as such by the general public."[37]

THE LEADING SCIENTISTS IN the anti-mainline assault, those most powerful and sustained in their critique, were the British biologists J. B. S. Haldane, Julian Huxley, and Lancelot Hogben and their American colleague, Herbert S. Jennings. Haldane was professor of genetics, and later of biometry, at University College London; Huxley headed the Zoological Society of London; Hogben was professor of social biology at the London School of Economics; and Jennings was professor of zoology at Johns Hopkins. They were advocates of the new mode in biology—experimentalism, the interpretation of life phenomena in terms of physics and chemistry, and the subjection of biological problems, where appropriate, to mathematical analysis. "An ounce of algebra is worth a ton of verbal argument," Haldane said for all of them.[38] With different emphases within the new mode of practice, all four made salient contributions to the increasingly related fields of genetics and evolutionary biology. Huxley's research accomplishments included theories of the evolution of behavior in birds, particularly the rituals associated with courtship and mating. Jennings demonstrated that the asexual reproduction of paramecia yielded genetically uniform descendants and he used this result to untangle the roles of heredity and environment in their development and behavior. Hogben's work ranged from cytogenetics to the inheritance of intelligence. Haldane, a virtuoso, covered physiology and biochemistry as well as biometry and genetics. Most important, he was one of three scientists—his British colleague Ronald Fisher and the American Sewall Wright were the other two—to deploy mathematics in aid of establishing the theory of evolution on a genetic basis, the overall achievement that Huxley summarized in his classic book of 1942, *Evolution, The Modern Synthesis.*[39]

The British wing was united by personal attachment: Haldane and Huxley had become friends at Eton. Hogben joined with them in 1922, at the University of Edinburgh, to help found the *Journal of Experimental Biology.* Jennings, who had gotten a late professional start, was the old man of the anti-mainline leadership, forty-six at the outbreak of the First World War, when the others were just finishing their university training or getting settled in their first jobs. He knew well the lesser lights in the American anti-mainline attack, especially Raymond Pearl, his colleague at Johns Hopkins. Although separated by the Atlantic Ocean most of the time, the British and American biologists were early tied together by Huxley, who had taught from 1912 to 1916 at Rice Institute, in Houston, with Hermann Muller. Muller had taken a job there after completing the requirements for his Ph.D. with Thomas Hunt Morgan.[40]

What Haldane, Hogben, Huxley, and Jennings knew so intimately about genetics by no means turned them as one man against the chief doctrines of mainline eugenics. To be sure, from early in his career Hogben uncompromisingly opposed the mainline movement, identifying it with "ancestor worship, anti-Semitism, colour prejudice, anti-feminism, snobbery, and obstruction to educational progress." But Haldane, who as a young man joined the Oxford Eugenics Society, sympathized for a time with aspects of the creed, particularly its denigration of the lower classes and eagerness to reduce their rate of reproduction, while Huxley, at the beginning of the Depression, proposed that unemployment relief be made contingent upon the male recipient's agreeing to father no more children. Jennings, who had been a student (and a tenant) of Charles B. Davenport at Harvard, belonged to groups of a mainline character.[41] The rapidly advancing field of genetics helped turn all four men against mainline eugenics, but so did factors of background, temperament, and sociopolitical belief.

Haldane and Huxley were the products of England's intellectual aristocracy. Huxley, who remembered sitting upon his grandfather Thomas Henry's knee, came naturally to science. So did Haldane, who learned Mendelian genetics at home as a boy by breeding guinea pigs. Haldane's father, an Oxford physiologist and frequent consultant to government as well as industry, educated his young son as an assistant. (Once, in a mine, Haldane, at his father's behest, learned about the effect of fire damp by standing up and reciting a Shakespearean speech until, panting, he collapsed on the floor where the air was all right.)[42] Hogben grew up on the southern coast of England, where his father, a minister of the Plymouth Brethren, delivered fire-and-brimstone sermons on the Portsmouth beach, presided over daily family prayers, and proscribed card playing, alcohol, and dangerous—including scientific—thoughts. It was because his mother dreamed of her son's becoming a medical missionary that he was permitted to read books on botany and zoology. Once the family moved to London and he entered a secular school, there was no quelling his autodidactic intellectual appetite. Jennings was raised in Tonica, Illinois, a tiny town with three churches and no saloons. Yet like Haldane and Huxley, he came to science as a birthright. His father, an impecunious physician, founded the local literary society, became an apostate from the strict Protestant faith of his ancestors, and earned a reputation as the town infidel. He middle-named his future biologist son Spencer, his other son Darwin, and taught his children the new religion of evolution.[43]

"I'm an atheist, thank God," Hogben liked to say. So were Haldane and Huxley, while Jennings tended to a general religious indifference.[44] In Tonica, when Jennings was not reading books on natural history, he devoured Shakespeare, and he jubilated upon his arrival at Harvard over the architectural styles that one could see firsthand and over the operas—he loved Wagner—plays, and lectures that one could, and he frequently did, attend. Hogben and Huxley were polymaths of sorts, but Haldane towered over them all. When he had left Eton, Haldane could read Latin, Greek, French, and German, had a fair knowledge of history and contemporary politics, and knew enough chemistry and biology to take part in research. A strapping two-hundred-pounder of indomitable physical courage, he often performed taxing physiological experiments on himself, including the imbibing of hydrochloric acid to test its effect on physical activity, or arduous exercise to measure the change in the pressure of carbon dioxide in the lungs. An awed French geneticist said of him: "*Ce n'est pas un homme, c'est un force de la nature.*"[45]

Hogben, who married the feminist and economist Enid Charles not long after the First World War, made the emancipation of women part of

his formal credo. Haldane and Huxley were both acquainted in Bloomsbury and counted among their friends Ottoline Morrell, Lytton Strachey,
D. H. Lawrence, and Bertrand Russell, not to mention Julian's brother
Aldous. They made a point of declaring that sexual compatibility was
essential to the happy marriage, that women deserved sexual satisfaction as
much as men, that there was nothing wrong or degrading about sexual
pleasure dissociated from procreation. Of course, they endorsed divorce
and birth control. Huxley actively campaigned for contraception, earning
the condemnation of Lord Reith for sullying his BBC ether by discussing
the subject on the airwaves.[46]

Yet Haldane, Huxley, and Hogben were caught between the internal-
ized morality of their Victorian upbringings and their rebellious codes of
reasoned belief. Huxley suffered repeated nervous breakdowns—one occurred after his honeymoon—which he attributed to "my unresolved con-
flicts about sex." The Hogbens tried to arrange their married life so that
the wife, a brilliant mathematician, could pursue her desired career as a
statistician, but four children came along and so did the usual differentiation
of sex roles. (Hogben often flaunted his own familial fecundity against the
barren marriages of many eugenicists, whom he called "childless rentiers
—twentieth century bourbons who have earned nothing and begotten
nothing.")[47] In an unpublished autobiographical fragment, Haldane took
the trouble to note that he did not join in the homosexuality rampant at
Eton and that he was sexually—meaning heterosexually—ill at ease until
much later. Still, his biographer Ronald Clark found in him "a shyness with
women which he never overcame, an inferiority complex which he tried
to disguise by an open bawdiness." Once a student entered Haldane's
college rooms for his first tutorial and was told to be seated while Haldane,
chamberpot in hand, finished his natural duty. Inept, like many other
theoretically inclined scientists, at the manipulation of laboratory equipment, he would tell women students, "I do claim to be an accomplished
exponent of the use of the paternal apparatus." Haldane once harrumphed
at a proper Cambridge dinner party that he never went in "really seriously
for bestial sodomy" and he exclaimed while riding on a Glasgow bus filled
with Sunday churchgoers, "That's the place where I came to fornicate as
a boy."[48]

There was no anti-Victorian sexual bravado to Jennings. As a young
man, he thought it adventurous to read Olive Schreiner's *Dreams* with a
young lady in a cool cellar one sultry afternoon, while eating strawberries
and sugar. He and his fiancée—she was Mary Burridge, whom he had met
while she was a biology student at Michigan—unflinchingly endured an
engagement of some years until familial and financial circumstances permitted them to marry. At Harvard, where he took his Ph.D., Jennings found

irritating a female graduate student who constantly attributed every setback to discriminatory intent on the part of the university authorities. "She really does the cause of women a great deal of harm, for people think if that's the way women will do in science or the university, they [will] want no more of them." Mary Burridge Jennings attended to the business of her children, her husband, and his career, aiding in his research and illustrating some of his books. In one of his major works, Jennings briefly debated freedom in sexual relationships and concluded that, from a biological point of view, the needs of human beings would probably best be met over the long-term evolutionary future by monogamy.[49]

Lancelot Hogben, whose family's fundamentalist religiosity set him apart as a boy, remained a prickly outsider. By his undergraduate years, at Cambridge, the evangelical Christianity had been transmuted into a fervent anti-imperialist socialist radicalism. During the First World War, Hogben served in the Friends' Ambulance Unit and then, declining his medical student's eligibility for exemption from military duty, he refused call-up on conscientious grounds and went to Wormwood Scrubs prison. In postwar London, he gave time and energy to socialist and labor groups, including his friend and neighbor Sylvia Pankhurst's Workers' Federation.[50]

J. B. S. Haldane, by contrast, volunteered for the Scottish Black Watch within days of Sarajevo, went to the trenches, and discovered, to his discomfort, that he actually liked killing. He risked his life above and beyond the call of duty and, twice wounded, was commended by Sir Douglas Haig as "the bravest and dirtiest officer in my Army." Exposure to the common soldier taught Haldane that the lower orders of society might be worth redeeming after all. In 1924, he complained that genetic theory was being used in Britain "to support the political opinions of the extreme right, and in America by some of the most ferocious enemies of human liberty." Like so many intellectuals of his generation, he came away from the war disillusioned by the failure of liberal aims, particularly the ingrained Liberalism of the Haldane family. But though turning to a nominal socialism, he continued to be an imperialist sympathizer, patronizing toward colonials, and a studied inactivist.[51]

Julian Huxley, whose politics had tended to a tepid middle-of-the-roadism, was jolted to the left by the Depression and the Fascist threat. So, even more, was Haldane, who went to Spain as an adviser on civil defense to the Republican forces. Doubting the ability of either a Conservative or a Labour British government to stand up to the Nazi menace, Haldane became a committed Marxist; in 1942 he would follow his wife, the journalist Charlotte Burghes, into the Communist Party. Hogben, uncomfortable with Marxist certainties and Soviet repressiveness, hewed to an independently idiosyncratic radicalism. His open criticism of Soviet Russia provoked disapproval among British radicals, including many of his colleagues at the London School of Economics, but he was commonly identified with Haldane and Huxley as a prominent member of the country's scientific left.[52]

Jennings, who had known firsthand the bitter experience of persistent penury, found his own politico-economic perceptions compellingly mirrored in the English journalist William Stead's *If Christ Came to Chicago*, the influential tract that infuriated conservatives by asking what the Christ of the Sermon on the Mount might have to say about the brawling city's mercenary churchgoing establishment. In the eighteen-nineties, Jennings had listened to a preacher say, in one of the best sermons he had ever heard, that "it was more a man's duty to go to caucuses and elections than to go to church and prayer meetings, and that the Lord had more interest in what the political parties of this country were doing than in what its churches were doing."[53] Jennings never embraced political activism, nor did he ever become self-consciously political in the manner of Haldane, Hogben, or Huxley. Nevertheless, in the eighteen-nineties, he had been a Populist sympathizer, and in later life he strongly tended to the progressive side of the political spectrum.

Their political liberalism-to-radicalism inclined all four to recognize that mainline eugenics expressed race and class prejudice. Hogben's convictions on the point were sealed during a stint in the late twenties as professor at the University of Cape Town in South Africa. Back in London, he persuaded Huxley that his mainline views concerning the unemployed merely aided and abetted Nazism. Through the lens of his socialism, Haldane saw—and was offended by—the sociopolitical presumptions, so many of them increasingly contrary to his own, hidden in mainline doctrine. Little was yet known about human heredity and, as Haldane put it, "many of the deeds done in America in the name of eugenics are about as much justified by science as were the proceedings of the inquisition by the gospels."[54] The three British biologists, too, all outspokenly modernist on issues pertaining to women's rights, were at odds with the sexual repressiveness of mainline eugenics, and Jennings, while conventional in his own attitudes toward women, sex, and marriage, was tolerant of the unconventional in others.

In the interwar years, these four men were among the leading public biologists—writers of books and articles for laymen on the content and social import of advances in the life sciences. Their works were published and read on both sides of the Atlantic.[55] Hogben wrote with uncompromising force, Huxley with supple lucidity, Jennings with vigorous straightforwardness, and Haldane with wit and irreverence. (Haldane called Einstein "the greatest Jew since Jesus," and he ventured that the hemophilia gene in Queen Victoria's pedigree had most likely arisen from a mutation "in the nucleus of a cell in one of the testicles of Edward, Duke of Kent, in the year 1818.") Jennings noted in his 1930 book, *The Biological Basis of Human Nature* —it won the *Parents' Magazine* award for the year's best book on heredity —that "a lot of fallacies" appeared to be circulating "under the guise of biological principles applicable to human affairs. . . . Particularly abundant appear such fallacies in the attempts to apply to human problems, to social reforms, the results of scientific study of heredity." The fallacies included, in Jennings's view, the notion that biology "requires an aristocratic constitution of society" and—Huxley's critique—"the assumption of the eugenic superiority of the more prosperous classes over the artisan and labourer mass." In 1930, Hogben complained in *The Nature of Living Matter*, that eugenicists had encumbered social biology "with a vocabulary of terms which have no place in an ethically neutral science."[56] During the interwar years, Haldane, Huxley, Hogben, and Jennings, together with their fellow public biologists, took it upon themselves to expose the fallacies, to disencumber the vocabulary, to cleanse the use of their science. The knowledge they injected into public discourse combined with the lay dissent to form a corrosive and increasingly effective case against the authority of mainline eugenics.

Chapter IX

FALSE BIOLOGY

M AINLINERS MAY HAVE HELD that the race was degenerating as a result of the differential birthrate, but the figures they cited suggesting, for instance, an alarming rise in criminality were much less alarming when set against the total population. In England, in the fifty years prior to 1911, the crime rate per hundred thousand of population had actually fallen by forty percent, and in the United States, between 1890 and 1904, prisoners per hundred thousand had dropped by about twenty-five percent. In 1904, a British government committee appointed to look into the physical degeneration issue concluded that "the impressions gathered from the great majority of the witnesses examined do not support the belief that there is any general progressive physical deterioration."[1]

By the twenties, a growing body of lay and professional opinion held that there had been no progressive intellectual deterioration either. Walter Lippmann, in a series of *New Republic* articles in 1922, passionately attacked the conclusions drawn from the Army's I.Q. testing program, bluntly declaring: "The statement that the average mental age of Americans is only about fourteen is not inaccurate. It is not incorrect. It is nonsense." Lippmann assailed the more fundamental pretension that the Army tests or any others measured hereditary intelligence. That claim, he said, had "no more scientific foundation than a hundred other fads, vitamins and glands and amateur psychoanalysis and correspondence courses in will power, and it will pass with them into that limbo where phrenology and palmistry and characterology and the other Babu sciences are to be found."[2]

In Lippmann's view, the basic flaw in any hereditarian interpretation of I.Q.-test results lay in the insistence of psychologists like Lewis Terman that there was some concrete, invariant entity called intelligence that could be unambiguously measured. "Intelligence," Lippmann insisted, "is not an abstraction like length and weight; it is an exceedingly complicated notion which nobody has as yet succeeded in defining." In the nineteen-twenties,

American psychologists increasingly came to recognize the notion's complexity, in large part because of the strong links of test performance with social and educational environment. Also troubling was the way that commonsense ideas of intelligence eluded capture by examinations. The Columbia University psychologist Edward L. Thorndike aptly remarked that verbal and mathematical tests said little about "ability to understand and manage things and people as they *exist in concrete reality*." In 1930, the entire subject was reviewed by Professor Carl Brigham of Princeton University, whose 1923 *Study of American Intelligence* had helped so much to promulgate the fear of mental degeneration. The more he studied the data, the more he came to believe that the tests—verbal, mathematical, and behavioral—measured only how well the examinee did on a particular examination. To say that the scores, taken together, indicated something called general intelligence, Brigham concluded, was to indulge in "psychophrenology," to confuse the test name—e.g., "verbal"—with the reality of the trait, and to misidentify the summed traits with intelligence.[3]

What I.Q. tests revealed about innate abilities was much more hotly contested in the United States than it was in Britain. Lancelot Hogben and his social biology group at the London School of Economics were virtually alone in Britain in mounting a research program to untangle the relative weights of nurture and nature in measures of intelligence. (Hogben's laboratory was also virtually alone in arguing, on the basis of considerable data, that, of children aged nine to twelve with I.Q.s greater than 130, a large fraction came from lower-income families and that only about a quarter of the children with such scores who went to state-supported schools would go on to a secondary school.)[4] British psychologists argued about the *technology* of testing—about whether particular tests truly assessed intelligence; they disputed little whether general intelligence could be truly assessed.

In the United States, however, the clash between the white Anglo-Saxon Protestant majority and various minority groups helped make I.Q. testing a volatile issue. The British lacked the polyglot social groupings of America, and they had no test results as comprehensive as those from the American Army to fight about. I.Q. tests may have swept through American primary and secondary schools, but in Britain a good half of local educational boards successfully resisted their introduction. Teachers considered the tests threatening to their authority.[5] Besides, since Britain was not committed to the democratization of higher education, children did not normally have to be sorted into grades of academic potential. No matter that perhaps valuable academic talent might await its chance unnoticed at the bottom of the social scale: the British had their class, if not ethnic, differences, and neither psychologists nor educators were on the whole disposed to query what they believed in their bones to be true—that the lower classes were on the average less intelligent than the upper.

But although the British were comparatively uninterested in the social issues of I.Q. testing, they were far from indifferent to the closely related question of the decline of national intelligence, particularly to the fraction of the fall accounted for by the alleged increase in the rate of mental deficiency. Obviously, if the meaning of "intelligence" was vague and the measures of it faulty, the claims that it was declining were, to say the least, dubious. British social scientists would eventually put the issue of decline to direct test by comparing two studies of the "intelligence" of Scottish schoolchildren—one carried out in 1932 with eighty-seven thousand students, the other in 1947 with seventy-one thousand. The mean mental test score of the later group differed only slightly—and in the higher direction —from that of the earlier one. The improvement was attributed to greater test savvy in 1947, along with better nutrition.[6]

In the early nineteen-thirties, responding to the outcries over the alleged doubling in the incidence of mental deficiency since the early years of the century, Lancelot Hogben pointed out: "This increase is far too great to have resulted from genetic selection in less than a single generation." Social observers argued that if mental deficiency had in fact increased, the rise was far more likely the result of the brutalizing impact of poverty than of any deterioration of stock. Hogben himself doubted the reality of the increase, and he preferred to interpret any evidence of it as indicating that "the criteria of defect and methods of ascertainment have changed." In England, he noted, individuals were not certified as feebleminded unless they appeared before the police court, applied for poor-law relief, or were sent to special institutions for the retarded. There was therefore no means of estimating the prevalence of deficiency "among the prosperous classes, where eccentricity fades into the diplomatic service." (According to J. B. S. Haldane, the proceedings of bankruptcy courts showed that "a considerable number of the nobility are incapable of managing their own affairs. They are not, however, segregated as imbeciles on that ground.") In the United States, analysts could find no basis for the claim that the mentally deficient had increased out of proportion to population growth. If more people were found in facilities for the mentally deficient, it was because such facilities had steadily expanded, making room for more patients.[7]

On both sides of the Atlantic, the statistics of mental deficiency tended to be strongly class-biased. The more well-to-do the family, the more likely that a mentally deficient member would be entrusted to private care and escape the statistical net. The lower the income, the more likely that such a person would be consigned to a public institution and counted. Thus poverty could with ease be attributed to mental deficiency. Thus British mainliners could malign the differential birthrate as a dire threat to British society. And thus their American brethren could identify the supposed degeneration of their society principally with the proliferation of the new immigrants from Southern and Eastern Europe.

Harry Laughlin, of the Eugenics Record Office, had officially reported to the House Immigration and Naturalization Committee in late 1922 that immigrants were disproportionately present in America's mental institutions. An editor of *The Survey* magazine, fearful that the report, if true, would be difficult to combat, asked Herbert Jennings to write a piece on Laughlin's handiwork. Jennings found Laughlin's conclusions so prejudiced, and as such so offensive to his sense of scientific propriety, that he formally contested them not only in *The Survey* but also in a letter to the leading professional journal *Science* and in his own statement to the House committee.[8]

True enough, he admitted, first- and second-generation immigrants showed up more frequently than native Americans in public homes for the mentally deficient, and in prisons and public charity wards as well. But, he suggested, the effects of poverty, ignorance, and difficulty with the English language rather than biology might render immigrants more susceptible than natives to mental, moral, and physical breakdown. Laughlin's own results revealed that only twenty percent of the country's mentally deficient were in the care of public institutions. Jennings asked, "Would not statistics from expensive private institutions in all probability show a reversal in the proportions of native-born and foreign-born?" Particularly infuriating to Jennings was Laughlin's claim that mental deficiency was commoner in Eastern and Southern European immigrants than in other "racial" groups. If Laughlin's own data were to be believed—and Jennings had his doubts—the Irish contributed the most to mental deficiency in America, the Austro-Hungarians the least. Czechs, Poles, and Yugoslavs were thus more desirable than a large class of Northern Europeans. Jennings saw "no warrant" for the claim that recent immigrants to the United States suffered defects and diseases arising from heredity. "It is particularly in connection with racial questions in man," Jennings complained, "that there has been a great throwing about of false biology."[9]

The false biology proceeded from a false anthropology. Biology supplied no evidence for the mainline-eugenic assumption that Italians, Poles, Lithuanians, or other national groups were biologically uniform. Moreover, where biological similarities were sufficient to warrant a racial identification, there was no evidence that genetic differences between groups were at all socially significant. Thomas Hunt Morgan, in the 1925 edition of his *Evolution and Genetics*, added a chapter on human heredity in which he took the trouble to declare: "Least of all should we feel any assurance in deciding genetic superiority or inferiority as applied to whole races, by which is meant not races in a biological sense, but social or political groups bound together by physical conditions, by religious sentiments, or by political organizations." It was virtually impossible to determine the genetic basis of behavior within even homogeneous groups, Morgan noted. How extraordinarily more difficult to attempt such a task between groups that differed in material advantages arising from location, climate, soil, and mineral wealth, as well as in traditions, customs, religion, taboos, conventions, and prejudices. "A little goodwill might seem more fitting in treating those complicated questions than the attitude adopted by some of the modern race-propagandists," Morgan concluded.[10]

The goodwill seemed all the more necessary once the Nazis came to power. In 1935, Julian Huxley and the former Cambridge University anthropologist A. C. Haddon published *We Europeans: A Survey of 'Racial' Problems*, which castigated works like Madison Grant's *The Passing of the Great Race* ("When . . . we read in [Grant's book] that the greatest and most masterful personalities have had blond hair and blue eyes, we can make a shrewd guess at its author's complexion. A flaw in his line of thought is that the same claims are made by brunets!"). Going beyond *ad hominem* ridicule, Huxley and Haddon advanced the genetic and anthropological consensus that the concept of "race" made no biological sense. What seemed like a racial group actually consisted of the intermixture of many biological types, the product of successive migrations and intermarriages.[11] The Nazis might claim that Jews constituted a racial type, but in fact in every country Jews overlapped with Gentiles in every conceivable physical characteristic. Jews of one area differed genetically from those of another; they were biologically no more uniform than any people of Europe—including so-called pure Germans. The Nazis might celebrate a Teutonic type—fair, long-headed, tall, and virile; Huxley and Haddon wondered how close a composite of the black-haired Hitler, the broad-faced Rosenberg, the slight Goebbels, and the rotund Goering would come to the Teutonic ideal. Populations differed from each other, Huxley and Haddon stressed, only in the relative proportions of genes for given characters that they possessed. "For existing populations," they maintained, "the word *race* should be banished, and the descriptive and non-committal term *ethnic groups* should be substituted."[12]

Huxley nevertheless supposed that, though it had not been proved, different human groups must possess "innate genetic differences" with regard to intelligence. So did J. B. S. Haldane, who insisted that simply because no racial differences had been proved it did not follow that "the theory of absolute racial equality" was correct. Still, even Haldane allowed that, in the absence of equal environmental opportunity, one could not easily know the type and degree of innate racial differences. More important, whatever they might be, they were only statistical; that is, they applied to group averages and not to individuals. "It is quite certain," Haldane declared, "that some negroes are intellectually superior to most Englishmen."[13]

BY THE OPENING OF THE nineteen-thirties, psychologists were coming to recognize that within given racial or ethnic groups, I.Q. test results varied widely; a large number in every group scored higher than the middle of the overall highest group. To be sure, whites tended to score higher than blacks, natives outperformed immigrants. But such results began to be seen as indicative of faults in the tests themselves rather than as evidence of innate "racial" differences. In his 1930 review of the field, Carl Brigham concluded that "comparative studies of various national and racial groups may not be made with existing tests," and he courageously added that "one of the most pretentious of these comparative racial studies"—his own—was "without foundation."[14]

Brigham's remarkable mea culpa, based mainly on technical considerations, also suggested a substantive dissatisfaction that was of special importance to assessments of mental differences among racial or ethnic groups. Walter Lippmann had adumbrated the issue in 1923, when in a letter he scolded Robert Yerkes for presuming to think that the test results proved, among other things, that Irish children were inferior to English children. "You are in no position to assess the effects of the history of Ireland upon the Irish intelligence [test] behavior," Lippmann wrote. "You are in no position to disentangle the biological from the traditional causes of the result. You are in no position to disentangle the emotional disturbances of a migration not only across the sea but from a peasant to an industrial environment. You cannot examine the effects of clericalism, or the effects of a disintegration in America of the clerical tradition."[15] Later in the decade, a growing number of American psychologists edged toward Lippmann's position: performance on I.Q. tests was considerably affected not only by education but by social and cultural environment.

In academic circles, the trend was given impetus by a group of social scientists centered on Professor Franz Boas of Columbia University, a German-Jewish immigrant who had become the country's best-known anthropologist. Boas included a chapter highly critical of mainline eugenics in his *Anthropology and Modern Life*, published in 1928. Openly suspicious of I.Q. tests in general, he held that a person passing a test was proficient in what the test tested—the meaning of the score was impossible to get at. Angered by the "Nordic nonsense" advanced by theorists like Madison

Grant, he was certain that there was no proof of hereditary, racially specific mental or behavioral traits in blacks, immigrants, or any other group, and he provided technical consultation on the issue to Congressman Emanuel Celler in the latter's losing battle to beat back the immigration restrictionists. Boas also stimulated a good deal of academic research into questions of race and intelligence. Among the products was the master's thesis of Margaret Mead, who studied the children of Italian immigrants and demon-strated that their performance on I.Q. tests depended on their families social status and length of residence in the United States, and also on the extent to which English was spoken in the home.[16] A decidedly more sustained product—Mead was already deeply into her Samoan studies when she published her results—was the psychologist Otto Klineberg's authoritative body of work on race and I.Q.

In 1979, the American Psychological Association honored Klineberg with its award for psychological work in the public interest, citing his "long series of notable research and publications that shattered the claims for innate racial differences in intelligence, sensory-motor performance, and in other psychological functions."[17] Klineberg—a sparkling octogenarian who recently returned to New York City after twenty years of teaching in Paris—likes to point out that he got into the field of race differences by accident. His professors at McGill University, from which he graduated in 1919, had discouraged his ambition to go into academic psychology. There were very few university jobs available in Canada, and Klineberg, the grandson of Austrian-Jewish immigrants to Quebec, was well aware of the anti-Semitism then prevalent in the academic world. After a year of post-graduate study at Harvard, he returned to McGill for medical school, thinking that he might at least become a psychiatrist, and went on in 1925 to graduate work in psychology at Columbia University.

Out of general curiosity, he took a course during the summer of 1925 titled "Culture and Personality," with the anthropologist Edward Sapir, a specialist in American Indian ethnology and comparative linguistics. Klineberg recently recalled that the course had made him begin to think that "what we were talking about in psychology made no sense if we knew only people in our own culture and our own background." It was "a kind of religious conversion—suddenly feeling that it's ridiculous to talk about human psychology if you knew only one particular group of human beings. The anthropologists knew that, but the psychologists didn't."[18] Klineberg avidly pursued both disciplines, slipped easily into the Boas circle, and regularly crossed the Hudson with the other anthropologists for discussion soirées at Boas's home in Grantwood, New Jersey. Even though Klineberg was technically taking his degree in psychology, Boas became, as he puts it, "*my* Papa Franz, too." Before entering Columbia, Klineberg had tended to accept uncritically the prevailing idea that racial and ethnic groups differed genetically in qualities of mind and character. Exposure to Boas and his disciples inclined him to the opposite view.

The following summer, two of Boas's students asked Klineberg to drive with them in an old Ford to Washington State, where they intended to do field work among the local Indians; perhaps Klineberg could give the Indians some tests. He jumped at the chance—both to see the continent and to study the Indians in a way that combined anthropology and psychology. In Washington, he applied performance tests—another class of tests used to assess mental ability and so-called because they relied on the doing of physical, as distinct from verbal or mathematical, tasks—to children of the Yakima tribe and to white children in the town of Toppenish. He later described the results as "unexpected and exciting," noting that "the Indian children worked much more slowly than the white but, perhaps as a consequence, made fewer errors. They seemed entirely indifferent to the amount of time required to complete the problem, and my exhortation to 'do this as quickly as possible' fell on deaf ears." The outcome excited Klineberg because, though alive to the importance of culture in such matters as family relationships, behavioral motives, and the like, he had never thought of it in connection with such seemingly technical characters as speed of performance. That was the novelty. Suddenly it struck him that culture intruded into even a very simple performance test like taking a piece of wood and putting it in the right place. He hadn't realized how culture could go "deep down into the little movements of the hands."

The Yakima project changed Klineberg's life. He had expected to exploit his medical knowledge by specializing in some field such as psychopathology; now he determined upon work in questions of race. In 1927, he embarked on research into the claim of Carl Brigham—who had not yet repudiated it—that the Nordic, Alpine, and Mediterranean "races" differed in native ability. At the time, most anthropologists, even Boas, believed in the biological reality of such races—though obviously not in innate mental or behavioral differences among them. Shortly before Klineberg finished the project, Brigham recanted, but Klineberg's results added considerable substantive force to Brigham's essentially methodological turnaround. Brigham had originally inferred racial abilities from the I.Q.-test scores of nationality groups in America. Klineberg, who considered Brigham's procedure ridiculous, had gone abroad and given performance tests to the purest Nordic, Alpine, and Mediterranean groups he could find. He was able to report that—at least in the kinds of abilities measured by the performance tests—the three groups displayed no significant differences.[19] Back at Columbia in 1929, Klineberg turned to the subject he had first taken up in his doctoral thesis—black-white differences in the United States. "I suppose my moral attitudes contributed to the shift away from European racial topics," he recalled. "I thought the problem the most important for an American race psychologist to study. Besides, it was tremendously interesting."

In Europe, Klineberg had observed that the performance test scores of city dwellers among the three racial groups tended to be higher than those of rural residents. In the United States, he had also noticed that blacks in the urban North had on the average scored higher on the Army I.Q. tests than certain white groups in the rural South. One of the prevailing explanations of the phenomenon was the theory of "selective migration": People in an urban or sectional region scored better on mental tests because the more intelligent had migrated there from areas where people scored worse. The more intelligent blacks, in short, tended to leave the South for the North. Klineberg was inclined to an alternative explanation: the superior test performance of northern blacks was attributable to their advantageous cultural and educational environment.

Klineberg devoted his research of the early thirties to deciding between the two theories. He examined school records of black children in three southern cities to determine whether the students who had gone North were any more "intelligent" than those who had remained in the South. He also gave intelligence tests to southern-born blacks who had lived in New York City for different lengths of time, reasoning that if selective migration was at work, length of residence in the North should make no difference in the scores, but that if environment counted, then the scores should rise in proportion to time in the North.

During his travels in the South, Klineberg, a warm human being, partied, picnicked, and became friends with many blacks. The experience was an eye-opener for a white man of the time, even someone who, like Klineberg, already suspected that mental ability had little if anything to do with the biology of race. "I could see some of the ways in which culture and race got confused," he recalled. "I remember being once at a football game between two black college teams, and at every time-out the band would begin to play. Whenever the band played, many black mothers would wave the arms of their babies in time to the music. I said to myself, Well, here you see mothers teaching their children rhythm. You very rarely see that when two Ivy League teams play football. I noticed that same sort of thing with dancing. I also ran into a lot of blacks who had no sense of rhythm and who couldn't sing. The personal experience led me to query all the stereotypical ideas about blacks, and that skepticism came to be rather important in the work."

In 1935, Klineberg reported the full results of the study in his pathbreaking *Negro Intelligence and Selective Migration*. In his conclusions:

"The superiority of the northern over the southern Negroes to approximate the scores of the Whites, are due to factors in the environment, and not to selective migration. There is, in fact, no evidence whatever in favor of selective migration. The school records of those who migrated did not demonstrate any superiority over those who remained behind. The intelligence tests showed no superiority of recent arrivals in the North over those of the same age and sex who were still in southern cities. There is, on the other hand, very definite evidence that an improved environment, whether it be the southern city as contrasted with the neighboring rural districts, or the northern city as contrasted with the South as a whole, raises the test scores considerably; this rise in 'intelligence' is roughly proportionate to length of residence in the more favorable environment."[20]

In numerous subsequent publications, Klineberg continued to press the argument against the biological nature of racial differences in measurements of mental ability, often pointing to the superiority in test scores of northern blacks over various southern whites. "I wasn't the first to notice those data," he remembered with a smile. "But I used them. I did use them. My friendly enemies attributed the discovery to me and called it 'the Klineberg twist.'" The enemies rapidly diminished in force. The tide of thinking about innate racial differences was with Klineberg and a growing number of like-minded psychologists, anthropologists, and geneticists. Not all of them believed that there were absolutely no biologically determined mental differences between races, but virtually all held that no such differences had been scientifically demonstrated. By the end of the Second World War, with the aid of the Nazis, that view had replaced the orthodoxies of mainline eugenics.[21]

In 1950, UNESCO issued a strong "Statement on Race." It was the product of an internationally distinguished effort—the drafters and commentators included Otto Klineberg, Hermann Muller, and Julian Huxley —and its principal points summarized the new views on the biology of race: The idea of race was merely a convenient tool of classification. Differences between human groups resulted from various combinations of heredity and environment. Racial groupings did not necessarily coincide with ethnic and cultural differences. The results of intelligence tests depended on some combination of innate mental ability and environmental opportunity, and there was no proof that the groups of mankind differ in their innate mental characteristics, whether in respect to intelligence or temperament.[22]

WALTER LIPPMANN EARLY RECOGNIZED that the stakes in the I.Q. issue went far beyond race. "The whole drift of the propaganda based on intelligence testing is to treat people with low intelligence quotients as congenitally and hopelessly inferior," he wrote in 1922 in the *New Republic*. The prominent testers believe "that they are measuring the capacity of a human being for all time and that this capacity is fatally fixed by the child's heredity." Lewis Terman rose to the defense of hereditarian psychology, charging Lipp- mann with having brought more feeling than thought to the issue, even to the point of denying heredity a role in intelligence. Writing in reply, Lippmann admitted to being emotional about the matter. "I hate the impu- dence of a claim that in fifty minutes you can judge and classify a human being's predestined fitness in life. . . . I hate the abuse of scientific method which it involves. I hate the sense of superiority which it creates, and the sense of inferiority which it imposes." And he had not refused to recognize the hereditary factor in mental ability. He had simply denied Terman's "unproved claim" to have isolated it.[23]

A common task in science is to determine exactly how a given result may depend upon one among several variables—how, for example, the time of travel of a bullet from gun to target depends upon its shape, or the temperature of the air, or the velocity of the wind, etc. It is a long-standing article of scientific method that to determine this dependence experimen- tally, the result, e.g., the time of the bullet's flight, should be measured while holding constant all the variables save one—shape, say—which is allowed to change. Lancelot Hogben stressed that the hereditarian interpretations of intelligence violated this dictum. Following Karl Pearson's example, I.Q. studies had grown mathematically more sophisticated in their use of corre- lational analysis. Hogben pointed out "the danger of concealing assump- tions which have no factual basis behind an impressive façade of flawless algebra," stressing that a particular hazard inhered in using correlation coefficients to measure the degree to which, within a group, the variability of a trait—say, I.Q. scores—depended, on the average, upon hereditary factors. A given coefficient might be predicted on a purely genetic hypothe- sis, but obtaining the same coefficient from the data did not necessarily imply that genetics alone accounted for the variability in question. The identical coefficient could be the result of the combined effects of heredity and environment. Hogben warned that "when used without proper regard for the limitations imposed by the way in which data are collected correla- tion methods yield conclusions which throw more light upon the social prejudices of the investigator than upon the problem of nature and nur- ture."[24]

The prejudices were reflected in the usual I.Q.-test survey, which controlled for hereditary or environmental variables either insufficiently or not at all. Across families, environment obviously varied with social class, occupation, educational background, and income. Environment might vary even within families, not least, Hogben noted, because nature interacted with nurture. Fraternal twins might tend to greater diversity in their relationships than their identical counterparts, while siblings of different ages might experience sharply different childhoods.[25] To distinguish the force of nature from that of nurture in I.Q., the tests would have to be administered to groups similar in heredity whose environments varied, or vice versa. Francis Galton had pointed to a way of accomplishing the first approach: study identical twins, especially those reared apart. The same approach could be taken with fraternal twins or with ordinary siblings brought up in foster homes. Both types of non-identical siblings shared enough genes on the average to qualify for the similarity of heredity necessary to the procedure.

In articles published in British scholarly journals in 1943, 1955, and 1966, Cyril Burt would claim to have found and tested ever more numerous sets of identical twins reared apart—in the last article, the number reached fifty-three—with the results strongly favoring the hereditarian theory of intelligence. This work has recently been exposed as fraudulent. There were virtually no records of the twins he claimed to have studied, and the correlations for different pairs of children were so often the same as to strongly imply that Burt simply fabricated the numbers. In the nineteen-twenties, investigators who wanted to measure the I.Q.s of twins reared apart found precious few cases. They did better in searching out ordinary siblings raised separately or comparing foster and natural children raised in the same environment. One of the first studies of the latter type was carried out in the mid-twenties by Barbara Burks, an associate of Lewis Terman at Stanford. Burks found that only seventeen percent of variability in the I.Q.-test performance of her subjects was attributable to environment. The rest was the result of heredity.[26]

Lancelot Hogben, along with psychologists like Otto Klineberg, judged the Burks study questionable on the ground that many of the environmental influences she chose to take into account—they included "neatness" and "artistic taste"—had no necessary bearing on I.Q.-test performance. In contrast, both Hogben and Klineberg awarded the highest grade to the sophisticated investigation—it soon came to be recognized as a methodological benchmark—carried out in the late twenties by three educational psychologists at the University of Chicago, Frank N. Freeman, Karl J. Holzinger, and Blythe C. Mitchell.[27]

Freeman, Holzinger, and Mitchell studied some four hundred foster children, with the aim of determining changes in measures of I.Q. both for children of similar heredity brought up in different environments and for children with diverse heredities brought up in similar environments. The children were tested before their foster placements, then retested after several years in their foster homes. The I.Q. scores of children placed in superior foster homes tended to improve. And the higher the quality of the foster home and/or the longer the residence there, the greater the degree of improvement. Furthermore, siblings raised together tended to be closer in I.Q. than were those raised in separate foster homes. Indeed, I.Q.s were particularly dissimilar for siblings separated before the age of six. No less significant, unrelated children raised in the same foster home tended to be more alike in I.Q. than siblings raised in different homes.[28]

In the nineteen-thirties, Freeman and Holzinger collaborated with the University of Chicago biologist Horatio H. Newman in a rigorous analysis of twins. At the core of the subject group were nineteen pairs of identical twins raised apart, the total number of such pairs that Newman had been able to locate in over a decade of searching. At the end of the nineteen-twenties, Newman had been a strong hereditarian and eugenicist. After comparing the nineteen pairs with a control group of identical twins reared together, he concluded, with his co-investigators: "If the environment differs greatly as compared with heredity, the share of environment in determining traits which are susceptible to environmental influence is large. If, on the other hand, there is large genetic difference and small environmental difference, the share of heredity is relatively large." More important, the authors found themselves "disillusioned" with regard to their original ambition to isolate definitively the relative contributions of heredity and environment to human characteristics. Their twin study reinforced what the foster-children research had already suggested: that nurture interacted with nature to produce many observable characters, particularly "intelligence." Freeman, Holzinger, and Newman remarked how they rather sympathized with the dictum that what heredity could do, environment could also do.[29]

With a change in environment, not even "feeblemindedness" was stable. In the early nineteen-thirties, the Iowa Child Welfare Research Station inaugurated a series of experiments with feebleminded children. The experiments had been partly suggested by the history of two babies in an orphanage at Davenport, Iowa. The parents of both had been found to be mentally deficient, and both babies tested at feebleminded levels, with I.Q.s of 35 and 45. Sent from the orphanage to a school for the feebleminded, they were assigned by chance to a ward with some "high-grade moron" girls, the brightest in the school, who played with them a lot. About six months later, their I.Q.s were found to have risen substantially; at the end of the second year, one had an I.Q. of 95, the other of 93. Placed in average foster homes, they maintained their mental level. Having monitored the progress of these two babies, the Iowa Research Station set up an unusual experiment.[30] One of the investigators later recalled the results for the eugenically minded journalist Albert E. Wiggam:

In cooperation with . . . the school for feebleminded at Glenwood, Iowa, we selected thirteen babies in the Davenport orphanage with I.Q.s around 65 and sent them as visitors to the school . . . where they were placed, like these previous two babies, with the brighter girls. They remained there as "visitors" for two years, and by that time their I.Q.s [averaged around] 91. Two are of superior intelligence—one 115 and one 117. When we sent them to Glenwood only two had I.Q.s above 85. As a contrast to this group, we have studied twelve children who remained in the Davenport orphanage and whose I.Q.s at the age of eighteen months, when they were brought in, averaged 87—dull-normal, but not feebleminded. After remaining in the orphanage two years, their average I.Q. was 60, definitely feebleminded. That is, the thirteen children who had had the moderate stimulus of toys and of being played with even by high-grade moron girls and their attendants had gone up in their mental scores over twenty-five points, and those who had remained in the deprived environment of the orphanage had lost the same amount.[31]

By the late thirties, the Iowa Research Station had data on some three hundred children from low-income, low-I.Q. groups who had been placed in good adoptive homes and repeatedly tested between the ages of two and eight. Those whose natural mothers had been tested as "dull-normal" to "feebleminded" displayed no different a range of mental development—the average I.Q. score was 115—from that of the children of brighter mothers. Widely reported both in the United States and abroad, the Iowa results were summarized by one of the research team leaders: "Fantastic as it may sound, it is possible to take a group of pre-school-age children of average intelligence and change them into dull-normal children of sluggish intellect or to change them into very superior children." Geneticists would later disagree that intelligence was quite so plastic, and some psychologists had their quarrels with the procedures of the Iowa studies, but, taken together with the investigations of Freeman and Holzinger, first with Mitchell and then with Newman, the results were not only striking but strikingly consistent. They revealed that environment could either accentuate or reduce apparent genetic differences. And they strongly suggested that what I.Q. tests measured was some combination of nurture and nature.[32]

HERBERT JENNINGS KNEW FROM biology itself that, from the fetal stage onward, nurture acted upon nature to shape the organism. The chemical and physical environment could affect germ cells—sperm and ova—prior to fertilization. Afterward, in the womb, developing cells destined to become one part of the organism—for example, skin—would form into another—for example, the spinal cord—if the embryo were suitably disrupted. Jennings declared that "what a cell becomes, what line of development it follows, depends, not merely on what it has within it, but on its relation to the other cells; on its relation to the other parts of the embryo."[33] Jennings's own research showed that genetically identical paramecia would appear phenotypically diverse under differing environmental conditions. Fruit-fly geneticists had discovered that while certain strains of *Drosophila* might have a gene for an irregular abdomen, the gene would not express itself if the flies were raised in a moist rather than a dry atmosphere. Certain types of corn had a gene for red color. The red appeared only if the corn were grown in sunlight; plants grown in the shade would be green. Similarly, a variety of primrose would flower white if grown at hothouse temperatures, red otherwise. And then there was the axolotl, a large salamander beautifully adapted to living in water. If the young axolotl were given great quantities of thyroid material, its gills would disappear, its bodily features would change dramatically, and it would be transformed into a land salamander.[34]

As with salamanders, so with human beings. The 1904 committee studying alleged physical degeneration in the United Kingdom had noted that all the evidence pointed to "active, rapid improvement, bodily and mental, in the worst districts, as soon as [the inhabitants] are exposed to better circumstances." Diet alone counted for a lot. By the mid-thirties, so much more had come to be known about the satisfactory human diet that it became clear that many physical defects were the result of undernourishment. In 1936, Sir John Boyd Orr published his celebrated *Food, Health and Income*, a probing investigation of British dietary patterns, which revealed that one-tenth of the population was forced to depend on foods inadequate in fats, proteins, calories, and vitamins. Manipulate the nutritional environment one way and biological organisms could be made to appear genetically sound. Manipulate it another and they could be made to appear genetically inadequate.[35]

What made for social pathology, hundreds of reformist treatises declared, was the way environment interacted with human potential. Jennings cried, "Nonsense," to the mainline eugenic claim that environmental improvement—public-health measures, social services, better wages and working conditions—fostered the survival of the "unfit." One might as well disparage the Promethean bringing of fire for having preserved the weak or claim that clothing, tools, vaccination, and the like outweighed in their degenerative effects their dividends to the staying power of the race. J. Arthur Thomson of the University of Aberdeen ridiculed the eugenicist's tendency to fault modern hygiene for denying human evolution the selective effect of deadly microbes: "This seems a little like saying that the destruction of venomous snakes in India is eliminating a most valuable selective agency which has helped to evolve the Wisdom of the East. . . . Which microbe? Surely not that of the plague, which strikes indifferently, and is no more discriminately selective than an earthquake. Surely not that of typhus, which used to kill weak and strong alike. Surely not that of typhoid, which may strike anyone, and does not confer more than a passing immunity. And so on through a long list."[36]

H. G. Wells had speculated that many criminals were "the brightest and boldest members of families living under impossible conditions." Any man who had searched his heart, Wells once exclaimed, knew that to call "criminality" a specific human quality was "a stupidity." (Wells added that every man "knows himself to be a criminal, just as most men know themselves to be sexual rogues. No man is born with an instinctive respect for the rights of any property but his own, and few with a passion for monogamy.") In due course, Wells's speculation concerning criminal brightness was reinforced by a library of studies, including an American investigation of one thousand juvenile offenders, which concluded that there was no evidence for the heritability of criminality as such, and Cyril Burt's analysis of juvenile delinquents, which convincingly argued, in line with its American counterparts, that social environment had a good deal to do with delinquency. Burt, who had spent the first nine years of his life in a seedy section of inner London and who could drop his Oxford accent for a pure Cockney, reported what he had apparently learned at first hand—that among young offenders, "paucity of educational attainments and peculiarities of emotional attitude will debase their performances and impoverish their replies to a degree that may be gravely deceptive; and unless duly discounted, may engender an unwarrantable suspicion that the bulk of them are mentally defective." I.Q. surveys in America discerned that on the average criminals scored at least as well as the draft Army. Echoing Wells with data drawn from a comparative assessment of college students and prisoners, the American psychologist and authority on criminality Carl Murchison remarked that quite likely the characteristics which "make for worldly success in business or professional life also make for success in crime."[37]

In 1914, the London *Times* reported on an address by James Crichton-Browne, an authority on mental and public health, noting his contention that slum life favored "the survival of those who could subsist on a relatively small amount of nourishment and light and air; but . . . ruthlessly stamped out those who were strong and sensitive, and who demanded a copious supply of nourishment." The article added that "intellectual gifts, emotional refinement, and moral sentiment had little chance in slumdom against low cunning, blunted feelings, and vicious propensities." Slum children became thieves or prostitutes because they grew up among numerous models to emulate. It was a commonplace of the anti-mainline attack that social pathology was communicated, not inherited, that moral habits were learned, not determined by the germ plasm. A staple of anti-mainline literature was the case of the American colonial Elizabeth Tuttle Edwards and her descendants. Sister to a woman who murdered her son and to a brother who murdered another sister, Edwards was herself divorced by her husband on grounds of adultery and gross immorality. Yet despite this "evil taint," one of Elizabeth Tuttle's grandsons was the philosopher and divine Jonathan Edwards; her later descendants included college professors and presidents, physicians, clergymen, lawyers, authors, Army officers, judges, and congressmen.[38] Anti-mainliners stressed that the lower reaches of the social order had produced enough geniuses—including Shakespeare, Franklin, Pasteur, and Lincoln—to stock a pantheon. If they had not produced more, it was because social conditions, in Julian Huxley's observation, condemned their Darwins and Einsteins, like their Miltons, to be "mute and inglorious." The British educational system, Huxley remarked, left "vast reservoirs of innate intelligence untrained in children from the lower social strata."[39]

Mainline doctrine presumed that like produced like—that superior or inferior parents spawned, respectively, superior or inferior offspring through the transmission of traits by single Mendelian characters—unit-characters as they were known. It was here that the principal disjunction lay between mainline ideas and the advance of genetics. While geneticists knew that many physical characteristics were inherited, and a number of them also thought there might indeed be a biological basis for mental and behavioral traits, they also knew that even in the simplest version of Mendelism like did not necessarily produce like. Among the reasons was that what counted in breeding was the genes of the organism—the genotype, not the expression of them—the phenotype. One could not expect to produce superior progeny simply by breeding together phenotypically superior parents. By the First World War, the unit-character doctrine had generally been pronounced dead, though Herbert Jennings remarked in the twenties that, like the decapitated turtle, it was not yet sensible of its demise.[40] In *Prometheus, or Biology and the Advancement of Man*, a book he published in 1925, Jennings explained what he thought people concerned about eugenics ought to know:

> Neither eye color, nor tallness, nor feeblemindedness, nor any other characteristic, is a unit character. . . . There is, indeed, no such thing as a 'unit character' and it would be a step in advance if that expression should disappear. . . . Into the produc-

tion of any characteristic has gone the activity of hundreds of the genes if not all of them; and many intermediate products occur before the final one is reached. In the fruit fly at least fifty genes are known to work together to produce so simple a feature as the red color of the eye; hundreds are required to produce normal straight wing, and so of all other characteristics.[41]

In the terminology of genetics, inheritance was understood to rest on a polygenic base. Characters that were continuous, like height, as distinct from those that were apparently discrete, like eye color, were obviously the products of multiple genes. Intelligence, which of course occurred in continuous grades, was accordingly assumed to be polygenic, too. Even if the environmental circumstances could be reproduced, no way was known to duplicate the genetic combination that yielded Plato or Newton, Dante or Darwin, Bach or Einstein. In human as in virtually all forms of sexual reproduction, genes from one partner were sorted, then combined with those from the other in an infinite variety of unpredictable ways. And the new combinations for most characteristics were likely—as Francis Galton had discovered in his law of regression to the mean—to be closer to the average of the population. Jennings elaborated the point in *Prometheus*:

> When they are taken apart, the new combinations made are almost certain to be the commoner types, less valuable than their parents. What occurs in such cases is seen when one of the valuable fruits—a fine variety of apple or orange—is allowed to reproduce by seed, forming thus new combinations of genes. Among the offspring are many types, mostly inferior ones, thorny, irregular, weak plants with worthless fruits. Almost never is one produced that equals the parent. This is the sort of thing that occurs regularly in man. . . . The same is true for the poor combinations. They, too, must disintegrate and pass into new groupings; and now the offspring may be better than the parents; certainly they will be diverse. And from the large population of commonplace types appear continually, as the generations pass, a few rare ones —for genius or for inferiority—then after a generation these drop back again into the great reservoir.[42]

Geneticists understood that, in man, the only way for like to produce exactly like was to take a Shakespeare, for example, and multiply him without change of genetic makeup—in short, in the language of a later decade, to clone him. "If this could be done," Jennings averred, "man would have his fate in his own hands. He could multiply the desirable combination until the entire population consisted of that type." Short of that, human traits might just possibly be bred in—or bred out—of the population if people were willing to submit to the rigorously controlled selection procedures of animal husbandry, and animal husbandry boards. Still, unlike animal breeders, eugenicists hardly knew what types to encourage. H. L. Mencken pronounced the "great moral cause" of eugenics "much corrupted by blather," adding, "In none of the books of its master minds is there a clear definition of the superiority they talk about so copiously." Definitions of human perfection were in fact as diverse among eugenicists as among everyone else. If Anglo-American eugenicists commonly attached high importance to superior scientific or professional intelligence, they also made a cult of physical health and moral character. Yet it was sensibly observed that "a man may be a criminal and otherwise a perfect physical creature; a man may be diseased and yet be intellectually and morally a giant."[43]

Many geneticists held that the biological strength of the human race lay in the vast diversity of its genetic makeup. The diversity allowed for variety of types, and such variety was essential, not only for the endlessly different tasks that man asked himself to perform but also for the variation in environments, both present and possibly to come, to which he had to adapt. J. B. S. Haldane held forth on the matter in 1932, from the steps of a building at Cornell University, where he was attending the Third International Congress of Genetics. A society composed of uniformly perfect men, he said, would be highly imperfect. The essence of perfection among plants, animals, and most certainly man was variety. The ideal society had to have room for all sorts of people, each best at some one thing or other.

But would it not be desirable to produce more Leonardo da Vincis? a reporter wondered.

Da Vinci, Haldane remarked, would have been sterilized in some American states because of certain abnormalities.

F. A. E. Crew, of the Institute of Animal Genetics at the University of Edinburgh, came wandering by. "Crew," Haldane said, "what is the perfect man?"

"There isn't any," Crew replied, with an eye to the importance of matching man to his environment. "Define us a heaven and we will tell you what an angel is."[44]

Chapter X

LIONEL PENROSE
AND THE COLCHESTER SURVEY

In 1919, Dr. Walter E. Fernald, the leading American authority on mental deficiency, reflected, at a meeting of the National Committee for Mental Hygiene, on the subject that had figured most prominently in the mainline-eugenic diagnosis of social problems. "A dozen years ago we had practically settled all the problems of feeblemindedness," he told his colleagues. "We had decided that the feebleminded were all of hereditary origin, that they were pretty much all vicious and depraved and immoral, that they were not capable of self-support." Now, Fernald went on to say, there were a number of reasons not to be so sure.[1]

The reasons included the growing conviction among psychologists that the diagnosis of mental deficiency had depended too heavily upon the results of intelligence tests. Mental-health professionals learned from experience what the Iowa Studies eventually demonstrated—that a number of people committed to institutions as feebleminded on the basis of the Binet-Simon tests were capable of leading successful independent lives. Mary Dendy, nationally prominent in England for her work in the Manchester-based Lancashire and Cheshire Society for the Permanent Care of the Feeble-Minded, had early thought it "perfectly normal for some people to be excessively slow and dull at 'lessons,' and . . . [yet] have extremely good common sense and be useful and sensible members of society." By the late nineteen-twenties, Henry H. Goddard himself had, as he said, gone over "to the enemy," conceding that only a small percentage of the people who tested at mental ages of twelve or less were incapable of handling their affairs with ordinary prudence and competence.[2]

Suspicions of the heritability of mental deficiency derived in part from dissatisfaction with the methods of data gathering—particularly the field surveys of relatives—that scientists like Goddard had used. Provided with only brief training, the field workers tended to be amateurs overly willing to diagnose by "rule-of-thumb recipes," as a critic put it, and to confidently enter into their notebooks the mental or behavioral characteristics of people dead three or four generations.[3] Then, too, the data were interpreted by scientists like Charles Davenport in ways that David Heron of the Galton Laboratory had attacked as "Mendelism run mad." Heron's critique may have been energized by his commitment to biometry, but as early as 1911 William Bateson, the leading Mendelian in Britain, had reviewed Goddard's tables and concluded that "feeblemindedness will not do as a dominant." Within a year he had come to doubt that it could qualify as a recessive either.[4]

Questions about the genetic basis of mental "defect" were suggested by the fact that the children of men and women admitted to asylums often did not themselves appear to be similarly afflicted. Some deficiencies were in fact inherited, but matings between mentally deficient people did not necessarily produce deficient offspring in the numbers predicted by the Mendelian unit-character theory. In the speculation of geneticists, the reason was precisely that inheritance was polygenic. Then, too, the mental deficiency suffered by one parent might originate in a different set of genes from that found in another. In that case, Herbert Jennings pointed out, in *Prometheus*, "experimental breeding shows that the two parental stocks may supplement one another, so that the defect will not appear in the offspring. The characteristics that are predictable are extremely few; a new combination is produced with every child."[5] But Jennings's explanation amounted merely to a well-intentioned extrapolation of the polygenetics of plants and animals; however plausible, it was based on little if any direct evidence from human beings. Just what genetic combinations made for mental deficiency were, to say the least, unclear. Mental deficiency was found in many forms. Complex in its expression, it was presumably diverse in its causes. Certain forms of it seemed to be hereditary; many others did not; and many of those that did appeared to flout Mendel's laws, even in their polygenic form.

Few scientists were closer to the confusion than Edmund O. Lewis, a British physician and an expert on mental health, who had conducted the survey on which the British government's Joint Committee on Mental Deficiency had based its influential 1929 report. Lewis was trained in both experimental psychology and medicine, and he had an acute interest in social conditions. In the course of the survey, which sampled six areas of Britain, he had encountered more than five thousand cases of mental deficiency. He was struck by the complexity of the data—particularly the diversity of case types and their relation to social and geographical circumstance. Among the "feebleminded"—in British usage, the term denoted the highest grade of mental deficiency, composed of both "intellectual" and "moral defectives"—the "moral defectives" frequently did not lack intellectual ability, but the "intellectual defectives" were often morally incapable. The two categories were not sharply separated; indeed, many individuals displayed both inadequacies. Then, too, Lewis noticed that "defectives" of the lower grades—idiots and imbeciles—seemed to occur in all social classes, while the merely feebleminded made up the three-quarters of the mentally deficient who tended to be concentrated in the "social problem group." He also observed that some types of mental deficiency seemed to be familial, while others did not.[6]

Lewis was disturbed by the apparent large increase in the rate of mental deficiency since the 1908 survey—especially because the data suggested that the rise had occurred mainly in rural areas. "A prosperous future in agriculture," he declared in his own section of the 1929 report, "is impossible if our rural population has an unduly large proportion of men and women of low mentality. Agriculture is becoming more scientific every year; and this trend makes an increasing demand for a higher level of intelligence among all rural workers." Yet Lewis, sympathetic to the British countryman, cautioned that mental capacity had to be judged against the standards of the subject's own community. There should be "no confusion of mere rusticity with feeblemindedness." Lewis thought it "impetuous" to conclude from his survey that "rural inhabitants as a group are generally inferior in mental endowment to the inhabitants of urban areas." For Lewis, the study for the Mental Deficiency Committee raised many more questions than it answered. Was the increase in rural mental deficiency real? If so, was the rise attributable to the urban migration of the more intelligent, to local inbreeding, or to something else? More generally, just how did one form of deficiency differ from another? And how were the various types to be categorized?[7] Despite the newer views of mental deficiency, the standard practice in Britain was to divide all forms of it into primary or secondary amentia. By definition, primary amentia was the result of heredity; secondary, of environment. But Lewis suspected that the categories of primary and secondary amentia were simplistic and faulty. Further research into the problem was imperative.[8] So he insisted to anyone who would listen, including, in 1929, Ruth Darwin, one of Charles Darwin's granddaughters and a principal of the newly created Darwin Trust, who was very glad to listen indeed.

The object of the Darwin Trust was to foster research into "mental defect, disease, or disorder." It had been formed to administer the income —some two hundred and twenty-five pounds a year—from a property owned by Ruth Darwin's recently deceased father, Sir Horace Darwin; the property in question was rented to the Royal Eastern Counties' Institution, a hospital for the mentally deficient. In 1930, at the instigation of Lewis, who had been made a Darwin officer, the Trust proposed to the Medical Research Council—the British equivalent of the nascent United States National Institutes of Health—an authoritative scheme for research in mental deficiency, to be funded cooperatively by the Council, the Trust, and its institutional tenant. Lewis undoubtedly saw a special opportunity in a research venture involving the Royal Eastern Counties' Institution, a major facility with more than a thousand patients. It was located at Colchester, about fifty miles northeast of London, in Essex. One of the six areas sampled in his survey, the region was heavily rural and, of the six, had the highest incidence of mentally deficient children. Lewis's guiding hand was evident in the stated importance of the scheme: "Of all problems, the causation of mental deficiency is the one in which research is most needed. The classification of causes given in the most modern textbooks can scarcely be regarded as satisfactory." A broad attempt at classification might well reveal that "the underlying conceptions are unsound and misleading from both the biological and clinical standpoints."[9]

In those interwar days, the Medical Research Council, not yet the sprawling bureaucracy governmental research agencies have since become, operated largely out of the office of its chief, Sir Walter Morley Fletcher, who relied upon a small cadre of expert advisory groups, not to mention his own discerning taste and judgment. Within weeks, Fletcher, who considered the research scheme "likely to be one of the soundest pieces of work we are supporting in relation to mental disorder," committed the Council to an annual grant of five hundred and fifty pounds—enough, combined with the income from the Darwin Trust and three hundred and seventy-five pounds from the Royal Eastern Counties' Institution, to provide a workable budget of about twelve hundred pounds a year. The majority of the money was to pay the salary of a medical investigator appointed to the Colchester staff, who was to undertake a complete physical and mental classification of all the patients and attempt "to discover the causes of the mental defect in each case, more especially as to whether the cause is what is now called primary or secondary amentia." By October 1930, the Darwin Trust had found its investigator. He was Lionel S. Penrose, a physician in his early thirties. "He is not an administrator," Ruth Darwin explained to the chief of the Medical Research Council, "but there is no doubt that he is a thinker."[10]

LIONEL PENROSE WAS A product of the type of well-to-do, polymathically capable British family whose fecundity eugenicists liked to celebrate. His father was an accomplished portrait painter and fellow of the Royal Hibernian Academy. His brother Roland Penrose was a prominent surrealist painter and art critic. Penrose himself possessed the sort of crisp, incisive, pellucid mind that makes for exquisite science. With the Colchester appointment, he embarked on a career that rapidly led to pioneering preeminence in the field of mental deficiency and that, by the nineteen-fifties, had inspired J. B. S. Haldane, a man not given to overstatement, to call him "the greatest living authority on human genetics."[11]

The Penrose parental wealth came from the mother's side of the family. She was the daughter of Alexander Peckover, a Quaker and the proprietor of a successful family bank who finally became Baron Peckover of Wisbech, in Cambridgeshire. Peckover made himself into a bibliophile and philanthropist (the type who understood the Anglican hymn line: "the rich man in his castle, the poor man at the gate" to be prescriptive; he would sit of a morning in his handsome Georgian house tossing "begging" letters into the fire). The Baron's wife died young, so his children, including Penrose's mother, were raised by his spinster sisters, especially Priscilla Hannah, who was so frugal that, instead of lighting a candle at night, she would read by the light of the streetlamp.[12]

Growing up, Penrose knew a rather more stern Quakerism than had either Francis Galton or Karl Pearson. Roland recalled that the family was "ruled by remote control" from Wisbech by grandfather Peckover, his two surviving sisters and two unmarried daughters, "all virgins and, in contrast to the old patriarch, all strictly teetotal."[13] In the Penrose household, the physical demonstration of affection was rare, and the expression of feeling was strongly discouraged. Such indulgences as fiction, theater, and music were prohibited, although games like chess were allowed; card games, too, so long as jack, queen, and king were replaced, as though they were biblically proscribed graven images, by 11, 12, and 13. On Sundays, the reading of books was encouraged, including those on natural history and astronomy, since they revealed God's handiwork.[14]

Penrose was sent to Leighton Park, a Quaker school, where he earned a teacher's commendation for declaring that Jesus' message for the Pharisees was "to do away with their traditions and rites and to look at the things which really mattered." To young Penrose, what really mattered were mathematics, science, and chess. The Quakerism counted for a lot, too, especially the pacifism to which the Peckovers were unbendingly committed. Lionel greatly admired his great-aunt Priscilla Hannah, who ran an International Peace Society from Wisbech until she died in her nineties, corresponding with members all over the world in many languages, including Esperanto, which she preferred. During the First World War, Penrose served in a Friends Ambulance Train Unit. One evening in France, during a break in the work, he heard a lecture on Freud's theory of dreams and was, he recalled, "astonished to hear that some fairly reasonable explanation

could be given of the apparently disordered sequence of ideas in the noctur-
nal theatre with an audience of one."[15] By the time he matriculated at
Cambridge, in 1919, the knowledge Penrose cared about had gone beyond
mathematics and science to include an increasingly intense interest in
Freudian psychology.

At Cambridge, with brother Roland, Lionel plunged into the forbidden—the work of the dramatic society, the pleasures of classical music (he
became a lifelong Mozart addict). He also became a loyal member of the
select Cambridge Society of Apostles, that remarkable, informal hothouse
of so many illustrious intellectuals. He came to think during this period that
religion "*stunts* our mental growth," that religious belief ought to take a
back seat to knowledge. Many years later, his daughter remarked, "To him,
God was simply too vague a concept. It was one of those ideas that you
couldn't quantify or test."[16] Academically, Penrose pursued the Moral
Sciences Tripos of psychology, mathematical logic, and philosophy. He did
brilliantly at the mathematical logic, disliked the philosophy (despite his
admiration for the principal Apostle, G. E. Moore), and was utterly disappointed by the limited range of studies in psychology. Where he had hoped
to find a forest of psychological knowledge, he confronted what he remembered as "an intellectual desert"—an emphasis on the semantics, rather than
the substance, of such matters as thought, sensation, feeling, memory, and
perception.[17]

But here and there he found brave Freudian shoots. Among them were
the lectures of W. H. R. Rivers, the anthropologist and neurophysiologist
and the leading Freudian on the faculty. There was also John Rickman, a
physician, whose example helped steer Penrose in the direction of professional work in mental illness. Rickman, a marvelous raconteur, had often
come to Leighton Park, where he was an "old boy," to regale the students
with his tales, and Penrose met him again by chance in the street. Rickman
was working at the nearby Fulbourn Asylum. "The difference between me
and the patients," he explained to Penrose, "is that I have a key and they
haven't." Soon after this meeting, attracted by the new Freudian psychology, Rickman went to study in Vienna. "So it came about," Penrose
recalled in an unpublished memoir, "that, after learning nothing at Cambridge except a little mathematical logic, . . . I set off [in 1922] . . . to Vienna
with the vague idea of following in Rickman's footsteps."[18]

In Vienna, Penrose met Freud, made his way into the circles of Viennese psychiatry, and underwent analysis for about a year. But gradually a
certain skepticism concerning psychoanalysis set in. A friend remembers
his remarking at the time that the aim of psychoanalysis was "the acquisition of a quiet effrontery." The skepticism was evident in a notebook
jotting, a heavy-handed "psychoanalysis of chess" that described the game

as "a sadistic activity" whose object, checkmate, was "strictly the castration
of the opposing party." Penrose remained fascinated by Freudian insight,
but he came to consider psychoanalytic theory too elusive, too slippery for
scientific test. His dissatisfaction with it, like that with God, boiled down
to the fact that you couldn't quantify its terms.[19] Increasingly, his interests
swung toward the abnormal mind, including the biological role in mental
disorder. (Important among the reasons he was drawn in this direction was
a love affair that had ended traumatically because the woman was mentally
disturbed.) He needed a solid grounding in medicine, so in 1925 he returned
to Cambridge and earned a medical degree while spending some of his time
each week as an analyst at the London Clinic of Psycho-Analysis.[20] He took
his doctorate at the Cardiff City Mental Hospital with a thesis on a set of
schizophrenics, among them one of special interest who had been there for
twenty-two years and had established his own complete universe—including a calendar, astronomy, natural history, theology, and social order—in
a series of notebooks.[21]

Penrose may have seen a bit of himself in the patient with the fabricated universe. His cheerless, undemonstrative childhood had made him
self-contained, absorbed in his own thoughts, distant from the lives of those
around him. His original fascination with Freudian psychology had perhaps been stimulated by his own sense of emotional isolation. The isolation
ended somewhat when, in 1928, he married Margaret Leathes, the daughter
of a British physiologist. Though he rarely discussed personal subjects even
with the four children who eventually came along, he bubbled with childlike enthusiasm to them about mathematics, science, Mozart, and chess,
especially the mental version with no board but the players' minds—and
about wonderful toys and puzzles, physical and mental. He kept a small
pedal saw at home with which he constantly fashioned ingenious wooden
games and devices. He once remarked that "those who consider logic and
amusement incompatible terms will perhaps prefer to reverse it. A paradox
is an amusement in logic."[22]

Through the toys, the games, the logical puzzles, the chess that so
occupied him, he gave vent to the emotions that as a child he had been
taught to suppress. The seeming diversions were also how he made human
contact, with children, friends, colleagues, whomever. He held that the best
way to strike up a conversation on a train was to take your watch apart with
a nail file and put the pieces in a matchbox. Sooner or later, strangers would
ask you what you were doing.[23] There was no distinguishing Penrose's
playful inventiveness from his character as a human being or as a scientist.
In the late nineteen-fifties, he pedaled his pedal saw to produce an ingenious
alternative to the Watson-Crick DNA model of genetic reproduction.
Resembling interlocking pieces of a puzzle, the wooden units were capable

of mechanically reproducing themselves. "I wish I had thought of that myself," J. B. S. Haldane announced when he saw them. "An insult to nucleic acid," Francis Crick snapped.[24] Penrose was, of course, aware that such a model's function is to suggest ideas, but for him there was no sharp break between devices for play and those for serious science—what started as one, whether the product of mind or saw, might turn into the other.

Penrose was the quintessential anti-religious scientist, but he continued, as Haldane once said of him, to hold Quaker views in everything save theology. Visitors to the Penrose household in midwinter would find the coal fires out and the family wearing overcoats against the chill. According to various familial explanations, he disliked burning excessive amounts of coal either because it overburdened the miners or because the warmth symbolized the comforts of the rich. His children remember that he always seemed to be writing out checks for one good cause or another, and the house was often filled with guests, many of them political refugees from various parts of the world.[25] A lifetime pacifist, Penrose was generally liberal in his politics—and acidly skeptical toward any sweeping doctrine that pretended to unite theories of biology, medicine, and society.

Penrose early objected to the foundations of mainline-eugenic doctrine, particularly to the assumption that social pathology was genetically determined. He twitted theorists of the Mendelian inheritance of a propensity for crime by pointing out that in Jukes-like families the incidence of criminality was far higher than Mendelian expectations would allow.[26] Mainline eugenics also offended his acute moral and social sensibilities. Eugenicists might claim that the "degenerate mind" was inherited; Penrose found, as he remarked to a lecture audience in Birmingham in 1933, "considerable variation in opinion as to what constitutes a degenerate mind." He noted: "It is customary to use the term to designate the peculiarities of individuals belonging to a social sphere different from that of the user." "In the upper classes, poverty is sometimes regarded as evidence of degeneracy. Similarly, the poor can complain of the degenerate, idle and dissolute behavior of the rich."[27] Penrose thought that a society should be judged by how well it cared for its mentally incompetent. To his mind, the "menace of the feebleminded" was no menace at all. There were not that many. All of them, he insisted in 1933, could be given the institutional care they needed at a public cost totaling no more than five percent of what Britain was then spending on armaments, and he thought that the advocacy of sterilization revealed more about the neuroses of its proponents than about any behavioral tendencies among its objects.[28]

At the end of the Second World War, Penrose would be appointed Galton Eugenics Professor, and head of the Galton Laboratory for National Eugenics, at University College London; there he constantly irritated the

Eugenics Society by, among other things, relentlessly contesting Cyril Burt's theories of intelligence and the renewal of the claim, by Burt and others, that the British national intelligence was declining. The case for the decline hinged on the fact that I.Q.-test surveys showed that in larger families children's test scores were on the average lower than in smaller ones. No matter, Penrose objected. Given the mean test scores from the 1932 and 1947 Scottish surveys, it was obvious that the national intelligence had not fallen. More important, the inverse relation between children's test scores and family size had been observed for decades. Something was keeping the average intelligence level constant; "otherwise," Penrose remarked at a Eugenics Society symposium shortly before the 1947 Scottish survey was completed, "by now there would be nothing but defectives left in the population."[29]

WITH HIS HUMANE Quakerism, precision of mind, and implacably skeptical temperament, Penrose began work at Colchester in 1931 oriented against the simplistic ideas of so many of his predecessors in the field of mental deficiency and endowed with considerable sympathy for the unfortunate human beings he was to investigate.[30] He found a supportive spirit in the superintendent, Dr. Frank Douglas Turner, a kindly man, in Penrose's recollection:

> His manner was direct and could be forcible but he was always benign and his modesty seemed to be emphasized by a slight forward stoop. The domain over which he ruled contained more than one thousand idiots, imbeciles and feebleminded people, for so we used to designate them in those days. . . . The first medical superintendent, Dr. P. M. Duncan, invented the classifications idiot, imbecile, and simpleton for the patients of different mental levels and gave a very early description of a patient now recognizable as a mongol. The next superintendent, a layman, Mr. Millard, was accustomed, on the occasion of each new patient's admission, to kneel in prayer with the parents. Dr. Turner seemed doubtful about the value to the patient of this procedure but agreed that it shared the responsibility if treatment was unsuccessful.[31]

Though it did not accept the insane, the Royal Eastern Counties' Institution housed, at the time of Penrose's arrival, "defectives" of all grades and numerous variety. It was Penrose's task not only to get to know each of the patients but to ascertain everything that might illuminate the nature and causes of their respective deficiencies, especially whether these were primary or secondary in origin.[32] He quickly recognized that, as he put it in 1932, "there are a great number of different types of retarded mental development, many of which have almost nothing in common with one another except the inability to perform those functional acts which society regards as being an index of intelligence." But reliable differentiations among the various types required reliable criteria of difference. Penrose rejected out of hand legal grades of mental deficiency, which hinged on social aptitude, as scientifically worthless. ("They are about as much use from the biological standpoint as a classification of aquatic organisms based upon their suitability for consumption as articles of human diet.") He also recognized that legal standards were even less reliable in action than in principle, since he knew that liability to certification as mentally deficient hinged on social class. Penrose insisted upon approaching the study of mental deficiency as "a branch of human biology." He preferred a set of criteria expressive more of the patient as such than of the patient's interaction with the social order.[33]

Edmund O. Lewis, a veteran at struggling with the problem, had ventured such a scheme. His version divided mental defectives into two types: One, which he called the "subcultural" group, consisted of people who fell on the low side of the intelligence distribution of the general population. The other consisted of people made mentally deficient by disease. Penrose saw a certain guiding utility in Lewis's system; the Colchester patients included people who, though intelligent, were mentally deficient by reason of epilepsy or psychosis. But Penrose found the system inadequate to his rigorous scientific purpose. Intelligence was distributed in continuous grades through the population, from the highly capable to the mentally deficient. Lines drawn at any given point excluded or included numerous borderline cases arbitrarily. Besides, the pathologically afflicted might also belong otherwise to the naturally low-intelligence group, and the naturally dull might also suffer from such physical diseases as syphilis or such mental disorders as severe neurosis. In short, depending upon which symptoms the investigator might emphasize, the categories "subcultural" or "pathological" could often be applied to the same people. Penrose noted that three-quarters of the Colchester patients could be classified into either group.[34]

Penrose's preliminary investigations at Colchester revealed that the origins of his patients' afflictions were indeed confusing. Superintendent Turner thought that mental disorder was too often attributed to heredity, and Penrose was inclined to agree that environment played a major role in the etiology of defect. But Penrose recognized the possibility that hereditary factors might "enter significantly into every case of mental deficiency."

If environmental determinants were involved in nearly every case, too, then any attempt to classify mental deficiency as primary or secondary—genetic or acquired—was, in his opinion, "foredoomed to failure." He proposed to start from a classificatory tabula rasa—to sort out as far as possible all the pure clinical types, then to determine whether any given patient was an example of a pure type or a mixture of more than one.[35]

To identify the types and untangle their causes, Penrose gathered extensive clinical data on all the Colchester patients. He also oversaw the investigation of each patient's social background and family history. The work was laborious, long, and, at times, frustrating—some families, afraid that the Colchester investigators were the harbingers of a sterilization program, refused to provide information. The investigation was meticulously conducted through interviews with patients' relatives, friends, schoolteachers, and ministers. Investigators took note of the family's social class and of such home conditions as number of rooms per person. A psychologist administered intelligence tests—designed at Penrose's instigation so that the results would depend as little as possible upon the extent of the test-taker's education—not only to the patients but to members of their families. Along with Lewis, Penrose knew that "apart from hereditary likenesses, the child's mentality is, in many ways, copied or modeled on that of the parents"—that "the parents' social status and ability determine the physical and mental nutrition of the children."[36]

Penrose was acutely sensitive to the methodological shoddiness that even in the nineteen-thirties continued to plague the field. Although he pursued family medical histories, he understood that queries as to whether the patient seemed to be suffering from a hereditary complaint risked "a large initial probability of mistake or concealment in the answer." He laid emphasis on data concerning stillbirths, infant deaths, and the like, realizing that neglect of such information could well produce too low an estimate of a given condition's familial incidence, with the consequence that a disease that was really hereditary might seem otherwise. Although he used mental tests, he believed that one could not rely solely on such devices to assess intelligence. For Penrose, test results were just one item in a much larger evidentiary context, and the family histories were to be sifted, re-sifted, and, if necessary, gone after again to get at the truth. Penrose made it his overall aim "to understand, as far as possible, the mental outlook of the patients and to relate this to their upbringing, [education], and past emotional experiences."[37]

As the survey proceeded, he accumulated evidence confirming the hereditary nature of certain afflictions. Some, including Huntington's chorea, neurofibromatosis, and epiloia, were genetically dominant; others—for example, congenital diplegia, microcephaly, cerebromacular degeneration, and cretinism—were recessive. Penrose found particularly interesting— because of the way it unambiguously announced itself—the recessive con- dition identified in 1934 by the Norwegian scientist Ivar Asbjörn Fölling.

Fölling had analyzed the urine of four hundred and thirty mentally deficient patients. He detected phenylpyruvic acid in ten of the samples. As soon as Penrose saw Fölling's paper, he analyzed the urine of his Colchester patients. If the acid was present, the urine would turn green upon the addition to it of iron trichloride. After four hundred and fifty-one samples were treated, the urine of a teenage boy revealed the telltale green color; it took five or six hundred more before Penrose found a second. The family histories of both these cases strongly suggested that the condition was caused by a rare recessive gene that, when expressed, caused an inborn error of metabolism. It was soon learned that the error occurred in the liver in infancy and that it affected the development of the brain. Juda H. Quastel, a biochemist and a collaborator in the study, coined the word for the disease: "phenylketonuria," which was ultimately contracted to PKU.[38]

Penrose recalled in his memoirs that at the time, in another English institution, "there was a school of investigators, headed by the eminent anatomist . . . R. J. A. Berry, who believed that mental deficiency could almost always be ascribed to inadequate development of the brain, induced by 'rotten' heredity." He continued:

> Dr. Berry's methods of research included estimating the number of nerve cells in the brains of his patients and relating this to the intellectual capacity. . . . Berry represented a powerful influence, depressing to those who sought to elucidate and to specify exactly the causes of mental retardation. His attitude represented a widely accepted and fatalistic point of view. Against this background of popular belief, it is easy to understand how delighted Dr. Turner was when in 1934 I was able to tell him that a new and quite unsuspected cause had been discovered, by a Scandinavian biochemist, in certain cases of imbecility. . . . The origin of the abnormality seemed probably to be recessive hered- ity but the mental defect arose because the patient had something wrong with his liver, not his brain. I remember how Dr. Turner's eyes lighted up with excitement at this news and we went on to discuss the possibility in the future of rational treatment for such patients by altering their diet at an early age.[39]

The discussion was prophetic, but the dietary treatment for PKU lay many years in the future. At the time, only a comparatively small fraction of the mental diseases that Penrose encountered seemed attributable to so definite a genetic, let alone treatable, origin, either dominant or recessive. Although the prevailing wisdom had it that some eighty percent of mental deficiency could be classified as primary amentia, heredity alone seemed to account for only about a quarter of the Colchester cases.[40] A number of the rest seemed to originate from environmental forces, although just what these were was not clear.

Notable for the confusion regarding its etiology was the disease then termed mongolian imbecility. The first systematic identification of the disease had been made in 1866 by the British physician John Langdon Haydon Down. Down described a syndrome that, along with severe retardation, included an enlarged head and a prolonged, or epicanthic, fold to the eyelid. In Down's time, Western physicians had observed the syndrome only in Caucasians. Down supposed that the disease indicated a biological reversion in its victims to the Mongols of Asia, whom he thought they physically resembled, and who he assumed were a surviving example of an earlier human type. Down interpreted the "fact" that Caucasians could produce Mongols as evidence for "the unity of the human species"—a liberal idea running counter to contemporary theories that "inferior" human races had sprung from separate biological origins. Down believed the disease to be congenital rather than hereditary, and he speculated that the reversion might be caused by parental tuberculosis.[41]

The identification of the imbeciles with the Mongols of Asia—or, at least, with some general primitive type—persisted. In the nineteen-twenties, in the widely noted book *The Mongol in Our Midst,* the British physician F. G. Crookshank furthered this view by arguing that the syndrome might derive from a recessive "unit character," a vestige of man's evolutionary past, and that some Mongol blood no doubt flowed in the veins of many Europeans. "It is the 'Mongolism' rather than the idiocy that it is important to stress," Crookshank claimed, and he added that a portion of the native British population possessed "a kind of physical and psychical makeup that is coarsely and brutally displayed and accentuated in certain idiots and imbeciles."[42]

A third edition of Crookshank's book was published in 1931—by which time Penrose had begun an extensive study of mongol patients. There were only forty-two of them at Colchester; he had to search out others from local and London hospitals and through mental-health organizations, going so far as to track down an afflicted child whom he spotted on the street. He took special care to be certain that each patient he found was an actual victim of Down's syndrome—a not inconsequential problem. Some cases were borderline; the severe retardation aside, one or more characteristics of the syndrome—besides the epicanthic fold and a high cephalic index, they included a fissured tongue and the so-called simian crease, a pronounced transverse palm line—could be found among normal people.[43] Penrose, confident that Crookshank's ideas were utter trash, surveyed the blood types of one hundred and sixty-six mongols and of a control group of two hundred and twenty-five other mental patients. He found that the distribution of blood types in the mongol group was about the same as that in the control group. The results meant, he wrote to a fellow physician, that "mongolian imbeciles are no more racially Mongolian than other imbeciles."[44] To Penrose, the very term "mongolian imbecility" seemed scientifically inappropriate; foreshadowing current practice, he came to prefer the phrase "Down's syndrome."

The outcome of the blood-type study gave Penrose special pleasure. He liked mongolian imbeciles. He liked them for their gentle, childlike quality, for what he called "their secret source of joy." He may have warmed to them, too, because their simple, trusting nature encouraged him to break through his normal reserve. Mongolian imbecility remained a major subject of Penrose's research to the end of his career. In later years, he set aside Saturdays for work with Down's-syndrome children, observing and playing with them in the kindergarten swirl of the Galton Laboratory.[45] Yet from the beginning he judged that Down's syndrome merited special scientific attention, because it seemed so forcefully a product of action on the fetus by the intrauterine environment.

It was noticed early in the century that Down's-syndrome births were related to the age of the mother, occurring much more frequently among women over thirty-five than among younger women. Nevertheless, there was considerable dispute about the role of maternal age in the origins of the syndrome. Some authorities claimed that what counted was not the mother's age but the father's. Others insisted that the critical factor was the place of the Down's offspring in the family birth order: the mongol was often the last in a long line of children, and it was therefore theorized that the syndrome resulted from the mother's "reproductive exhaustion." Then, too, a mother often produced a mongolian imbecile long after the birth of her last previous child, so length of time between births was also advanced as a cause.[46]

Beginning at Colchester, Penrose worked to untangle the truth from among the conflicting theories. To choose among the important factors in the birth of a Down's-syndrome child, he adopted a simple statistical procedure: calculate the expected number of afflicted offspring on the hypothesis that one factor (for example, maternal age) made a difference while others (for example, birth order) did not; then compare the calculated expectation with the observed incidence. If the two figures matched closely enough, the hypothesis would be demonstrated. ("His statistics are definitely 'low brow,' " Haldane once remarked, "but I think effective for the purpose for

which they are designed.")[47] The entire procedure demanded the gathering
of complete and accurate family data. Penrose found that official case records of Down's-syndrome patients were of little value. Richly rewarding
were personal visits to the families (some of whom rebuffed him) to gather
data on the victims' parents, siblings, and other relatives; on numbers of
miscarriages, stillbirths, and infant deaths; on the ages of children, parents,
and grandparents. In due course, he had extensive information concerning
some hundred and fifty families. Analysis of the data revealed that the birth
of a Down's-syndrome child did not depend upon paternal age. It did not
depend upon birth order. It did not depend upon the length of time elapsed
since the birth of the last previous child. In most cases, it depended only
upon the age of the mother, with the probability of occurrence rising
sharply for women over thirty-five.[48]

Just why advancing maternal age raised the probability of a Down's-
syndrome birth, no one, including Penrose, could say. The prevailing
medical speculation included degeneration of the ovum or an inadequate
supply of nutrients to the fetus. Penrose himself wondered whether, at least
in some cases, genetics might be at work. The evidence for a genetic role
in Down's syndrome was slight but real enough. It consisted mainly of the
facts that some mongolian imbeciles were identical twins, and that the
syndrome sometimes manifested itself in more than one child in a family
or occurred with higher than random incidence among the offspring of
cousins. However, there was no way to distinguish between a genetic and
an environmental hypothesis. Down's-syndrome children born to the same
mother gestated in the same intrauterine environment. The explanation for
both random and familial occurrence would have to await the development
of human chromosomal genetics in the late nineteen-fifties.[49] Although
Penrose was unable to clarify the causes of Down's syndrome completely,
his conclusions about its dependence on maternal age and its likely genetic
origins in cases of familial incidence were definitive and rapidly came to be
recognized as such.

In 1938, seven years after he had begun, Penrose published the full
results of his Colchester survey in *A Clinical and Genetic Study of 1280 Cases
of Mental Defect*. Apart from his conclusions concerning Down's syndrome
and phenylketonuria, he reported that, unlike either disorder, most of the
Colchester cases were in origin principally neither environmental, pathological, nor genetic but some combination of the three. In his summary
view, "the aetiology of mental defect is multiple, and a facile classification
of patients . . . into primary or secondary . . . cases would only have led
to a fictitious simplification of the real problems inherent in the data."[50]

Penrose had hoped from the outset that the Colchester survey would
take the field of mental deficiency far beyond the simplicities of mainline

eugenics. E. O. Lewis felt that he had succeeded handsomely and understood that one of the reasons was the rare arsenal of expertise—the combination of genetics, medicine, psychology, and psychiatry—that he had brought to his task. "I know of no other investigator who has made such a thorough genetic and clinical analysis of a large group of mentally defective patients," Lewis wrote to an official of the Medical Research Council. "Unless I am much mistaken this work by Penrose will be the basis of most researches in mental deficiency during the next few decades. His definite findings on a large number of specific problems are valuable scientific contributions, but it seems to me the chief merit of the work is the new orientation it gives to our genetic approach to this complex problem."[51]

Chapter XI

A REFORM EUGENICS

IN 1935, THE AMERICAN GENETICIST and future Nobel laureate Hermann J. Muller was moved to write that eugenics had become "hopelessly perverted" into a pseudoscientific façade for "advocates of race and class prejudice, defenders of vested interests of church and state, Fascists, Hitlerites, and reactionaries generally." By the mid-thirties, mainline eugenics had generally been recognized as a farrago of flawed science. Jacob Landman summarized the failings of the creed: "It is not true that boiler washers, engine hostlers, miners, janitors, and garbage men, who have large families, are necessarily idiots and morons. It is not true that college graduates, people in 'Who's Who,' and some 'successful' people, such as racketeers and bootleggers, are necessarily physically, mentally, and morally superior parents. . . . It is not true that celebrated individuals necessarily beget celebrated offspring . . . [or] that idiotic individuals necessarily beget idiotic children. . . . It is not true that, because the color of guinea pigs is transmissible in accordance with the Mendelian theory, therefore human mental traits must also be. . . . It is not true that, by any known scientific test, there is a Nordic race or that the so-called Nordic race is superior to any other race."[1] It was not true, either, many others would have added, that the unemployed were any more unfit than the employed. And it was not true, most geneticists had come to understand, that eugenic sterilization could rapidly rid society of the eugenically undesirable.

Sterilization might sharply reduce the incidence of dominant hereditary traits, like Huntington's chorea, but its effectiveness with the many recessive genetic diseases was, to say the least, debatable. One could reduce the incidence of such diseases by sterilizing people who were homozygous for the recessive trait—that is, who carried two genes for it—and in whom, consequently, the trait was expressed. But single recessive genes would continue to be transmitted by the more numerous heterozygous members of the population, in whom the trait was not expressed. Mating at random, the heterozygous group would once again produce a certain number of homozygous progeny, who, expressing the disease, would have to be sterilized in turn. To rid the population of harmful recessive traits would thus require sterilizing a certain fraction of the population in each succeeding generation.

In 1917, Reginald C. Punnett, the Balfour Professor of Genetics at Cambridge University, had calculated the number of generations it would require to reduce the incidence of the "feebleminded" by a given fraction if in each generation all of them were sterilized. Assuming that "feeblemindedness" was the product of a unit-recessive character and that mating occurred at random in the population, Punnett concluded that to diminish the frequency from 1 in 100 to 1 in 1,000 would require twenty-two generations, to 1 in 10,000 ninety generations, and to 1 in 1,000,000 more than seven hundred generations—all of which argued that sterilization promised no quick fix to the problem of mental deficiency.[2]

In 1924, Ronald Fisher, the British mathematical geneticist and eugenics advocate, attacked Punnett's approach to the issue as misleading. Fisher preferred to pose the question: "What reduction would the sterilization or segregation of all the 'feebleminded' produce in one generation?" Proceeding from a polygenic model of mental deficiency and aware, as well, that the feebleminded did not tend to mate randomly but assortatively—that is, with each other—Fisher calculated that the segregation or sterilization of the feebleminded of one generation would yield a thirty-six percent reduction of incidence. This was, he asserted, "of a magnitude which no one with a care for his country's future can afford to ignore."[3] Thanks to Herbert Jennings, Fisher's estimates were not ignored, though he may not have liked the way they were noticed. In his *Biological Basis of Human Nature,* Jennings scrupulously reported Fisher's assortative estimate of the reduction in the incidence of mental deficiency. But in his discussion, he also drew upon an alternative estimate of Fisher's which was made on assumptions of random mating and single-gene inheritance—and which led Jennings to declare that only about a tenth of the feebleminded in each generation were born of feebleminded parents. What took hold in the United States was Jennings's inferential observation—Fisher had never made the point explicitly—that approximately nine out of ten children mentally deficient by reason of heredity were the offspring of normal parents. In 1932, the New York *Times,* citing Jennings's rendition of Fisher, editorially summarized the view increasingly common on both sides of the Atlantic: "The evidence is clear that normal persons also carry defective genes which may manifest themselves in an insane progeny. . . . Even if we discovered the carriers of hidden defective genes by applying the methods of the cattle-breeder to humanity, the process would take about a thousand years."[4]

J. B. S. Haldane noted that estimates of the proportion of the mentally deficient who derived from "defective" parents ranged from five percent to fifty percent. Population geneticists now knew that the rate and effectiveness of selection for a character depended in a complicated way upon whether the genetic trait was dominant or recessive, sex-linked or not, polygenic or not. Estimates thus varied as to what the rate of reduction in the incidence of mental deficiency would be under the drastic policy of sterilizing all of the allegedly feebleminded in each generation—Haldane thought it would come to no more than twenty percent. Whatever their disagreement on the numbers, Haldane, Fisher, and most geneticists could support Jennings's warning: To encourage the expectation that the sterilization of defectives will "solve the problem of hereditary defects, close up the asylums for feebleminded and insane, do away with prisons, is only to subject society to deception."[5]

Besides, by the mid-thirties, the weight of authoritative opinion concerning mental deficiency was rapidly shifting to the truths that Lionel Penrose was demonstrating: that the term "feebleminded" was carelessly used to cover a spectrum of mental disabilities, most of them ill-defined; that many of the disorders were caused by deprivation or disease; and that apart from a few deficiencies, little reliable was known about the actual dependence of mental disability upon heredity. Authoritative opinion also had it that the feebleminded were not proliferating at a menacing rate; their fertility was on the whole no greater than that of the general population, and the reproduction rate of the most severely ill was in fact much lower.[6]

In 1934, a special blue-ribbon committee of the British government, appointed two years earlier to look into the sterilization issue, made its report. Headed by Laurence G. Brock, the head of the Joint Committee on Mental Deficiency, the special committee included Fisher, Ruth Darwin, and E. O. Lewis, all of whom were well aware of Penrose's work. The Brock report took substantial note of the considerable ignorance and uncertainty surrounding the biological origins of mental deficiency, and observed that "the more closely individual records are examined the more difficult it becomes to fix on one cause to the exclusion of others, or to say with certainty that the genetic endowment of any individual is such that it must produce a given result." In the United States, in 1936, a committee of the American Neurological Association, headed by the Boston psychiatrist Abraham Myerson, also issued a report on eugenic sterilization. The Myerson group drew on the findings of the Brock committee, came to similar conclusions, and emphasized a special point: "There is at present no sound scientific basis for sterilization on account of immorality or character defect. Human conduct and character are matters of too complex a nature, too interwoven with social conditions . . . to permit any definite conclusions to be drawn concerning the part which heredity plays in their genesis."[7]

Both committees flatly declared that there was no established case for compulsory sterilization, eugenic or otherwise. Both observed that sterilization might be warranted for the few disorders that were demonstrably genetic in origin. Both insisted that any such sterilization should be entirely voluntary.[8]

The two reports commanded widespread attention in informed transatlantic circles. The Myerson report helped arm anti-eugenicists in the United States. But in Britain, where still no law permitting sterilization existed, the Brock report was welcomed by eugenicists for its endorsement of voluntary sterilization in cases of indisputably hereditary disorders. Even if the reduction in the incidence of mental deficiency would require many generations, in the view of British eugenicists it was worth voluntarily starting down the long road. In 1931 and 1932, the Eugenics Society had seen to the introduction into Parliament of two bills to legalize voluntary sterilization; neither had stimulated more than back-bench debate, or even reached formal consideration. Now the society, embracing the position of the Brock report as its own, renewed the campaign for legalization. It was joined by mental-welfare workers who believed that the mentally deficient capable of caring for themselves ought to be permitted to live in the general community. Sterilization was considered advantageous for these people—not only because it would prevent the transmission of heritable disorders but because many of the deficient, though able to care for themselves, were unable to shoulder the responsibilities of parenthood.[9] Voluntary sterilization also won the endorsement of Julian Huxley and Lancelot Hogben, of some Labour groups, and of women's organizations.[10]

Legalizing voluntary sterilization was said to be a matter of social justice and—like birth control then, and abortion later—of a woman's right to control her own reproduction.[11] But the British public was divided on the issue. Voluntary sterilization was denounced on the floor of the House of Commons as anti–working class.

J. B. S. Haldane, though recognizing in principle the utility of voluntary sterilization, was rather more cautious than Hogben and Huxley in promoting it. The more he moved to the left, the more was he ready to allow that "a man who can look after pigs or do any other steady work has a value to society, and . . . we have no right whatever to prevent him from reproducing his like." To Haldane, sterilization smacked of economic class legislation. He noted that mental defect was "often not certified among the rich, although a glance at the press will convince anyone that they include a number of persons who satisfy the legal criterion of imbecility." It was axiomatic to Haldane that "any legislation which does not purport to apply, and is not actually applied (a very different thing), to all social classes alike, will probably be unjustly applied to the poor."[12]

Haldane's axiom was consistent with the sterilization record in the

United States. State sterilization laws applied only to the inmates of public mental institutions, whose residents were disproportionately from lower-income and minority groups. In Virginia, the overwhelming majority of those sterilized were poor; perhaps as many as half of them were black. In California, more than half the insane males sterilized were unskilled or semi-skilled laborers. The foreign-born were more likely to be admitted to state mental institutions and to be sterilized once there. While they accounted for about a fifth of the California population in 1930, they represented at least a third of the group compelled to undergo the sterilization procedure.[13]

A significant fraction of foreign-born patients eligible for sterilization were undoubtedly Catholic. Thomas Gerrard had warned: "Feeblemindedness is so often a cause of poverty, and poverty so often a cause of feeblemindedness, that there is danger of confusing one with the other. Catholics, therefore, need to exercise a strong vigilance lest, under the pretense of eugenic reform, the rights of the poor are infringed." Supreme Court Justice Pierce Butler, who cast the sole dissenting vote in *Buck* v. *Bell*, was a practicing Roman Catholic and the father of eight children. In Britain during the nineteen-thirties, voluntary sterilization was opposed by the Roman Catholic press, and the Secretary of the Labour Party discouraged taking the campaign for legalization to the rank and file on grounds that a number of them were Catholic and the measure would be controversial among them.[14]

The debate over sterilization also called attention to the point that vasectomy or tubal ligation did not diminish sexual energy or capacity. The publicity given that fact of life perhaps helped undermine enthusiasm for sterilization precisely in the mainline constituency where sexual repressiveness was so entwined with eugenic ardor. The Brock committee had cautioned that sterilization might foster promiscuity. Catholic theologians rejected sterilization not least because, in the same manner as contraception, it permitted sexual indulgence without procreational consequences, and Pope Pius XI condemned it in the same encyclical that attacked birth control, divorce, and companionate marriage.[15]

George Bernard Shaw, tossing his insouciant tuppence into the debate, attacked sterilization for the "unfit" on grounds that, had it been practiced a few generations earlier, he would not have been born. More important, with Superintendent Frank Turner at Colchester, many mental-health authorities wondered just what "voluntarism" could mean in the case of the mentally deficient; despite the insistence of its advocates that sterilization should not be a condition of release into the community, might it not become precisely such a condition in practice? It was charged that what was advanced as voluntarism today might turn into compulsion tomorrow—compulsion addressed not simply to the mentally deficient but to everyone whom eugenicists had for so long been deeming "unfit."[16]

The specter of compulsion gradually overshadowed the legalization fight in Britain. It provoked rising opposition from the political left as well as from Catholics and other religious groups; they formed an increasingly potent anti-sterilization coalition on both sides of the Atlantic.

The forceful American Catholic liberal Father John A. Ryan, among others, warned that once the sacredness of the individual was weakened, all human rights were placed in jeopardy. Compulsory sterilization of the mentally deficient might well lead to compulsory sterilization of the "socially inadequate," then to "the killing of all sorts of incurables."[17] From Germany, it was reported that authorities in Saxony were demanding the sterilization of twenty thousand children yearly; that in Kiel a girl who had cheated in school had been sterilized; that zealots in Freiburg were going after "moral defectives" as though they were psychopaths; that sterilization was practiced upon otherwise sound people with webbed fingers or clubbed feet; that some enthusiasts were calling for the sterilization of diabetics in the interest of racial health. It was estimated that sterilization killed between one and two percent of healthy German women who underwent the operation. Twenty-eight thousand women were said to have been sentenced to the procedure in 1934 alone.[18]

During the war, news reports trickling back to the United States indicated that the Nazis were deploying eugenic sterilization on an even broader scale. When the full horrors of the death camps were revealed at the Nuremberg trials just after the war, witnesses testified that Nazi doctors had established centers for experimental sterilization. Men were used to test castration procedures; women, to assess sterilization by X rays, injections, and electrical destruction of their reproductive organs. Marie Claude Valliant-Couturier, a former inmate at Auschwitz, reported: "The Germans said they were looking for the best method of sterilization so they could repopulate all western European countries with Germans within one generation after the war."[19]

Well before Nuremberg, the reports from Germany had joined with the scientific, the political, and the religious opposition to turn the tide against eugenic sterilization. In Britain, the move to legalize voluntary sterilization failed utterly and was dead as a legislative issue by 1939. In the United States, Catholics in particular mobilized to beat back the passage of new state sterilization laws and to block the enforcement of those already on the books. Enforcement of United States sterilization laws plummeted sharply in the early forties and was minuscule by 1950.[20]

THE FORTUNES OF THE mainline-eugenic movement fell with those of the drive for eugenic sterilization. The Third International Congress of Eugenics, held in New York City in 1932, attracted fewer than a hundred people.

In the following years, the number of books and articles published on eugenics steadily declined. Yet if the mainline movement collapsed, the eugenic idea by no means died with it. On the contrary, eugenics continued to compel attention among a small group of enthusiasts—including a number of the principal critics of mainline theories and programs—who were tantalized still by the dream of human biological improvement.[21]

The large majority of these advocates differed considerably from their mainline predecessors. Some—like Ronald A. Fisher, Karl Pearson's successor as the Galton Eugenics Professor and director of the Galton Laboratory for National Eugenics at University College London—were antiracist conservatives; others were social radicals in the tradition of George Bernard Shaw and Havelock Ellis. The prominent biologists among them ranged from the moderate left to the Marxist left—from Julian Huxley and Herbert S. Jennings to Lancelot Hogben, J. B. S. Haldane, and Muller. Whether right or left, they were united in recognition that advances in anthropology, psychology, and genetics had utterly destroyed the "scientific" underpinnings of mainline doctrine, and that any new eugenics had to be consistent with what was known about the laws of heredity.

Similar convictions characterized the new generation of leaders in organized eugenics, particularly Frederick Osborn, in the United States, and C. P. Blacker, in England. Osborn belonged to a New York mercantile and banking family—his father was William Church Osborn, a prominent New York corporate lawyer, and Cleveland H. Dodge was an uncle—in which the crudities of money-making had long since been replaced by a taste for art and a concern for the public welfare. After graduating from Princeton in 1910, Osborn went into railroads and banking, commuting between Wall Street and his ample home in Garrison-on-Hudson, where the walls were hung with pictures by Monet, Gauguin, and Pissarro. A Princeton geology course had aroused his interest in the evolution of man, and the interest remained with him into the Wall Street years; he often discussed it with another uncle—Henry Fairfield Osborn, the paleontologist, eugenicist, and president of the American Museum of Natural History. In the late nineteen-twenties, now an advocate of eugenics and convalescing from an illness, Osborn quit Wall Street to devote himself, along with some of his considerable funds, to a kind of intellectual philanthropy centered on eugenics. In 1929, he installed himself in an office in his uncle's museum, and for the next four years he read widely in demography, differential psychology, and genetics. His reading turned him against the mainline creed, particularly the racist and anti-immigrant claims so central to the American eugenics movement.[22]

C. P. Blacker, a physician, was a graduate of Eton who had served during the First World War in the Coldstream Guards and had gone on to Oxford. A tall, spare man, stern in manner, he married the daughter of a British Army major, fathered three children, and ran five miles before breakfast every morning until the age of sixty-five. Yet there was a good deal more to Blacker than conventional upper-crust English vigor. The "C" in his name stood for Carlos. On his father's side, he was once-removed from a mid-nineteenth-century connection with the Peruvian aristocracy, and the elder Blacker was a latter-day Dickensian figure who called himself "gentleman," married the daughter of a Union Army general from St. Louis, and, dividing his time between England and the Continent, befriended Anatole France and Oscar Wilde. Blacker had lost a younger brother in action during the war and was himself decorated for gallantry. He held that war was dysgenic, both because it killed people who tended to be above the physical average and because the prospect of it deterred sensitive and thoughtful people from parenthood. Blacker himself seems to have been deeply shaken by the experience of the trenches, particularly the death of a friend and fellow officer who was blown up before his eyes during the Battle of the Somme in 1916. Evidently in order to come to grips with things, he began to read deeply in Freudian psychology while at Oxford. His immersion in Freud, to whom he counted his intellectual debt "immense," helped lead him into a career in psychiatric medicine.[23]

Blacker was never himself psychoanalyzed, and he was never an uncritical devotee of Freud. One counterbalance was his Oxford training, under Julian Huxley, in evolutionary biology. Another was his practice at the Maudsley Hospital, in London, where he encountered diverse forms of mental illness. In his judgment, Freudian psychology unduly stressed the universals of mental disorder; more attention needed to be given to the idiosyncrasies of individual patients. Like numerous members of the British school of psychiatry, Blacker preferred to locate mental disorder in concrete biology, and, as a trained zoologist, he tended to think strongly in terms of the biology of evolution. (Blacker thought mere fantasized sex "racially suicidal"—because it did not lead to reproduction—and found in the relentless drive for actual satisfaction the "*biological* trustee of our racial future.") By profession and experience, he was too knowledgeable to accept the simplistic prejudices of mainline eugenics, or its often wrongheaded genetics, some of which he likened to taking the family pedigree of mining-accident victims—all of them, of course, male—to support a hypothesis that "the tendency to have mining accidents [is] the product of a sex-linked gene."[24]

In the nineteen-thirties, Blacker and Osborn were elected to key offices in their national eugenic societies. Both men, although genuflecting at times to the hard-core conservatives in their constituencies, steadily moved their organizations a sanitizing distance away from the right—especially the pro-Nazi right. (Blacker took special pains to prevent British eugenics from being tarred with the Nazi brush, not only because he thought the pro-Nordic, anti-Semitic policies of the Nazis "ridiculous" but because, unlike American eugenics, the British variety had attracted a number of Jews.)[25] Blacker and Osborn both sought to construct a eugenics in keeping with the known facts of heredity. To that end, both turned their societies from propaganda promising universal social redemption to sober educational efforts concerning heredity and health. The British Eugenics Society had been made well off in 1930 by a bequest valued at sixty thousand pounds from Henry Twitchin, an Australian sheep farmer, who explained to the Society that he had been "born of unsound parents and inherited their weaknesses," had himself declined to marry, and wanted to assist in discouraging the propagation of the unfit. The American Eugenics Society, impoverished by comparison, had to content itself with the sponsorship of various conferences in New York City, which, though few in number, did offer such attractions as lectures by Will Durant on eugenics and civilization; Rabbi Sidney Goldstein on eugenics and birth control; and Arthur Morgan, the head of the Tennessee Valley Authority, on the socioeconomic obstacles that eugenics faced.[26]

Blacker channeled some of his society's income into research. Osborn paid for a small research staff out of his own pocket. Both edited and wrote a variety of eugenically related books. Both painstakingly reshaped their respective societies as older members retired (or resigned in distress at the direction things were taking), in order to reduce the influence of lay eugenicists and strengthen the hold of professionals in eugenically relevant fields. Blacker proffered the hand of friendship directly to the left, inviting the participation in his society's affairs of Haldane, whom he disliked; Hogben, whom he scorned (according to Blacker, Hogben delivered his inaugural lecture as professor of social biology at the London School of Economics while wearing "a pink tie and [with] his hair arranged in such a way that three curls dangled down over his forehead, rather like what you see behind the counters in Selfridges—we were not entirely pleased"); and even Penrose, an outspoken opponent of eugenics, who accepted a lecture invitation with the warning: "You know the risks you are taking."[27] To Blacker, such risks were worth taking, since, along with Osborn, he was eager for respectable academic support. By the forties, both had rebuilt their cadres of officers and members to include a number of distinguished geneticists, physicians, psychologists, and demographers.

Many of the new visionaries of the thirties and forties happily called themselves eugenicists but stayed out of the eugenic societies. Haldane, who held Blacker in contempt, generally refused to have much to do with the British Eugenics Society; so did Hogben. However, Julian Huxley was a mainstay of the group, and like-minded biologists befriended its American counterpart. Whether in or out of organized eugenics, the Blackers and the Haldanes, the Osborns and the Mullers formed a loose coalition of what one might call reform eugenicists, who rejected in varying degrees the social biases of their mainline predecessors yet remained convinced that human improvement would better proceed with—for some, would likely not proceed without—the deployment of genetic knowledge.

What differentiated reform eugenicists from the standard reformers of the day was their conviction that biology counted—that not only did nurture figure in the shaping of man but so, significantly, did nature. Reform eugenicists tended to insist that the science of biology revealed, in Julian Huxley's phrase, "the inherent diversity and inequality of man." To eugenicists, biologically based inequality in mental capacity seemed manifest in the fact that people in the same socioeconomic class scored across a broad range in I.Q. tests. If test performance varied under presumably constant environmental conditions, this could be accounted for, or so it seemed to many eugenicists, only by variation in native ability.[28] Huxley predicted that even if environmental disparities were eliminated the genetically flawed core of what the British Joint Committee on Mental Deficiency had called "the social problem group" would remain, and that the professional classes would be revealed as "a reservoir of superior germ plasm, of high average level notably in regard to intelligence." Reform eugenicists were inclined to believe that, as Herbert Jennings put it, "on the average, a greater proportion of poor genes will be found in the delinquent group, a greater proportion of better genes in the self-controlled or self-supporting group."[29]

The reformers recognized, however, that hardly anything was known about precisely what role heredity played in the achievement, or lack of it, of the bulk of the population. Inadequate housing, medical care, education, and opportunity could just as easily as heredity account for the dissolution and physical or mental disease among lower-income groups. Until basic environmental conditions were equalized among all socioeconomic strata, reform eugenicists held, no one had any right to say that one stratum differed from another solely by the force of heredity.[30]

Quite the contrary, given the growing body of scientific evidence. Frederick Osborn stressed that I.Q. tests revealed no marked hierarchy of occupational groups. To be sure, on the average, tradesmen, clerks, and professional workers did better on such tests than skilled and semi-skilled workers, who in turn outscored unskilled, irregular workers. But the range among individuals within a given occupational group was large enough to make for a good deal of overlap between the groups; a sizable number of people in the lowest group were at least as intelligent as a sizable number in the highest. Reform eugenicists felt compelled by such evidence to break away from the identification of innate ability with race or class—from what one of them characterized as "the idea of encouraging or discouraging either Park Avenue or Hester Street"—and to concern themselves instead with the biological qualities of individuals.[31]

They also argued the importance, to both eugenics and the social welfare, of adequate diet, health care, housing, and education. They called for the abolition of slums; the creation of decent housing and of recreational and day-care centers; the right to a job and a fair wage. Yet for reform eugenicists of the left, measures of social melioration were by no means sufficient. ("Don't let's go on pretending it's all the dear old Edwardian Age!" Huxley told C. P. Blacker.) In a celebrated 1936 lecture to the British Eugenics Society, Huxley said flatly that a system based on private capitalism and public nationalism was ipso facto dysgenic: it failed to utilize existing reservoirs of valuable genes and it led to the ultimate dysgenics—war. The left mixed its eugenics with the socialist reconstruction of society. "We can't do much practical eugenics," Huxley declared, "until we have more or less equalized the environmental opportunities of all classes and types—and this must be by levelling *up*."[32]

The reformist outlook defused the long-standing eugenic concern with the differential birthrate—the tendency of lower-income groups to outreproduce the middle and upper classes. Raymond Pearl suggested in his authoritative *Natural History of Population* that the higher birthrate among lower-income groups was not the result of sexually wanton inattentiveness to the consequences of copulation but partly of a genuine desire for children, partly of contraceptive ignorance. Blacker and Osborn, both staunch advocates of making birth control available to lower-income families, could agree with Haldane that if everyone were provided with "the same economic incentives to family limitation as exist among the rich," as well as with the same access to contraception, the differential birthrate might well take care of itself.[33]

That the differential birthrate did not stand high on the reformers' scale of social anxieties bespoke their recognition that social quality was to be found in all groups, including those near the bottom of the socioeconomic scale. In part, too, the relative lack of concern derived from the conclusion of demographers during the Depression that the birthrate in the United States and England had fallen below the rate necessary to replace the existing population, and was declining even among lower-income groups. Enid Charles summarized the situation in *The Twilight of Parenthood*, published in 1934. Differential class fertility, she wrote, had been "a temporary and exceptional phenomenon" of a peculiar time in history—a time that, according to all indicators, was gone. The reproduction rate was likely "to continue to fall steeply," and within a hundred years the population of England and Wales combined could shrink to less than that of contemporary Greater London.[34]

In the reform-eugenic view, society needed the reproductive contribution of all competent people. Mainline concern with "the race" was beginning to be replaced by attentiveness to "the population." The new language was more than just a change in terminology; it reflected the reform eugenicists' belief that valuable characteristics were to be found in most social groups, and that the best in human variation was to be encouraged. Frederick Osborn, who was as eager as C. P. Blacker to save eugenics from the discredit brought upon it by the Nazis, put the the most attractive face possible on the reformist version in his *Preface to Eugenics,* of 1940:

> We cannot tell the heights to which any man may rise, until he meets the particular opportunity appropriate to his unique possibilities. . . . When personal freedom is denied, and the attempt is made, by enforcing a rigid environment, to form men in a common mold, individual variations are repressed and men lose their power of choice, tending to be pawns in the hands of circumstance. Eugenics, in asserting the uniqueness of the individual, supplements the American ideal of respect for the individual. Eugenics in a democracy seeks not to breed men to a single type, but to raise the average level of human variations, reducing variations tending toward poor health, low intelligence, and anti-social character, and increasing variations at the highest levels of activity.[35]

Chapter XII

BRAVE NEW BIOLOGY

REFORM EUGENICS WAS IN PART self-deluding; notions like "anti-social character" and "highest levels of activity" were freighted with class-dependent biases. But for the most part it was free of its predecessor's patent social prejudice, and in the thirties and forties it provided an umbrella broad enough to shelter eugenic impulses ranging from the meliorative to the utopian. Lancelot Hogben, idiosyncratic radical though he was, found common cause with Lord Horder, physician to the royal family and loyal member of the British Eugenics Society, in issues centered on health. For Frederick Osborn and C. P. Blacker, the focus of reform eugenics was the social and biological quality of the population. For Hermann Muller, J. B. S. Haldane, and Julian Huxley, reform eugenics pointed, as the original version had for Francis Galton, to a more distant goal—in Muller's words, "the conscious social direction of human biological evolution."[1]

The medically-minded reformers found allies outside the eugenic movement, among genetically oriented physicians like Lionel Penrose. Penrose, who ranked the welfare of "the race" far below that of individual patients and their families, thought that genetics might be advantageously deployed in preventive or therapeutic medicine. At the time, only a handful of medical schools taught any genetics at all, let alone the human variety. Haldane sniped that "a medical student who has attended three lectures on the entire subject of genetics is unusually well informed." Penrose remembered the "rather mysterious way" in which his Cambridge University medical class in the late nineteen-twenties had been introduced to the familial nervous diseases such as amaurotic idiocy, the contemporary term for Tay-Sachs disease. "We were told that these conditions occurred in several members of the same sibship but were given no clue to the mechanism of causation." Many doctors seemed to scorn the genetics of disease. Dr. Madge Thurlow Macklin, of the University of Western Ontario, one of the pioneers of medical genetics, reported in a 1931 issue of *Science* that during a discussion of the inheritance of clubfoot a physician "indignantly demanded to be shown 'a clubfooted ovum.' "[2] Clinical physicians perceived no value in genetic knowledge for the treatment of disease; if a malady was hereditary, the prevailing medical attitude had it, it must be neither treatable nor preventable.

For reform eugenicists, however, there was a good deal of value at least in attempting to prevent its transmission to succeeding generations. In both the United States and Britain, some of them began to offer lectures at their universities on genetics in relation to medicine. (Among them was F. A. E. Crew, a trained physician and the professor of animal genetics at the University of Edinburgh, who had told Blacker, "I hold the view that it is infinitely better to present the eugenic argument to a class of senior medical students than to spend one's time rushing round the country and talking to mothers' meetings.") Several of them turned out books designed to inform general practitioners about—to use the title of Blacker's effort—*The Chances of Morbid Inheritance.*[3] So informed, family physicians would be able to provide patients with eugenic prognoses of intended offspring, and the patients, so instructed, could then make appropriate decisions. In the case of genetically dominant diseases, such as Huntington's chorea, reform eugenicists held that both society and the families in question would be better off if the victims could somehow be persuaded to refrain from procreation. Because recessive disorders tended to occur disproportionately in consanguineous marriages, Haldane and even Penrose suggested discouraging the marriage of first cousins. Haldane estimated that stopping such marriages would reduce the incidence of amaurotic idiocy by about fifteen percent, of congenital deaf-mutism by some twenty-five percent, and of xeroderma pigmentosum, a fatal skin disease, by nearly fifty percent. Frederick Osborn went so far as to predict that as science grew better able to identify the carriers of recessive defects, restrictions on their marriages might well become an accepted public-health measure.[4]

But Penrose likened the main usefulness of human genetics to that of giving people spectacles.[5] His views were strongly shaped by his work with phenylketonuria, the recessive metabolic disorder that occurs in the liver. There, the lack of an enzyme prevents the normal metabolic processing of phenylalanine—a common constituent of ordinary foods—into tyrosine. Some of the phenylalanine is turned into phenylpyruvic acid, which is excreted and is what reveals itself under chemical test by coloring the victim's urine green. But most of the phenylalanine remains unmetabolized in the body, and, for reasons not yet understood, its abnormally high presence retards brain development beginning in the first days of life. In the mid-thirties, Penrose experimented with therapeutic diets that were nearly free of phenylalanine. He recalled the result of having administered such a diet to one of his patients at the Royal Eastern Counties' Institution, in Colchester:

> At first, phenylpyruvic acid disappeared completely from his urine and I fancied that his mental condition improved slightly. Trouble started after about two weeks, when the patient began to lose weight and, in consequence of his partial starvation [causing the body to begin metabolizing some of its own protein, which contained phenylalanine], excretion of phenylpyruvic acid began again. I consulted Sir Frederick Gowland Hopkins [the Nobel-laureate biochemist] at Cambridge. He expressed great interest in the problem and estimated that, for about £1000, it might be possible to produce enough synthetic diet, free from phenylalanine, to feed one patient adequately for a week. So the matter rested. The experiment had to be discontinued.[6]

Still, in Penrose's opinion recognition of the genetic origins of disease could permit early and accurate diagnosis—he thought the characteristics of rare recessive diseases were "just as much clinical signs as . . . the sounds heard through a stethoscope"—and thus more efficient treatment. He shared his ideas on genetic diagnosis at a 1938 meeting of the British Association for the Advancement of Science: "This aspect of prevention is less frequently emphasized, but it may become very important in the future. In some illnesses, the contribution of heredity is to increase the susceptibility or sensitivity of the individual. In such a case, the onset of illness might be prevented by warning the susceptible person to avoid types of environment . . . dangerous to him." (Penrose, who died in 1972, lived to see his therapeutic ideas made practical. By the early nineteen-sixties, a half-dozen commercially prepared low-phenylalanine diets were available which would prevent PKU retardation if administered throughout childhood to those identified at birth as victims of the disorder.)[7]

IN THE HEYDAY OF THE mainline movement, so-called positive eugenics—the encouragement of the breeding of the "better stocks"—had inspired little more than secular sermons against the use of birth control in the upper classes and trumpet calls to reproductive duty. But by the thirties, with their authoritative predictions of dire population decline, positive eugenics commanded significant attention in and out of reform circles on both sides of the Atlantic.[8] Some reform eugenicists attributed the decline in the upper-class birthrate to the reluctance of intelligent people to bring children into the world who might become fodder in future wars. Lancelot Hogben suspected that it might have resulted from changes in upper-class sex habits, including the prevalence of separate sleeping arrangements. "The use of electric light by the bedside, the possibility of having a hot bath at any hour of the day and night, scrupulous washing of the genitalia, enjoined by so many medical men, the bodily fastidiousness which asserts the demand for single beds—all these factors taken together may . . . affect the probability of conception materially," he wrote.[9] Perhaps a more pertinent factor was the concern that responsible parents were said to feel regarding the economic instability, social insecurity, and uncertain educational prospects that would confront their children. Haldane allowed that "the average doctor would probably beget at least one more child if he could be sure that his children would be satisfactorily educated at State or State-aided schools."[10]

That the concerns of women figured in the issue of upper-class fertility did not escape Hermann J. Muller, whose former wife, a mathematician, had been fired from her university job upon the birth of their son, with the admonition that motherhood and career would not mix. Muller assaulted male eugenicists for supposing that most intelligent women loved to be pregnant—that they loved "the frightful ordeal of childbirth," the demands of child care, and abstention from the stimulating life of the world outside the home. He argued that for the majority of women, especially the "more idealistic and capable . . . struggling to emerge from their slave psychology of yesterday," bearing and rearing an old-fashioned-size family was "a form of martyrdom too protracted and repeated to be endured: one quick burning at the stake would be much easier." He concluded, "On the part of a host of intelligent women, therefore, there is a growing mass strike against child-bearing."[11] Muller's analysis earned special notice on the eugenic left, but what captured attention across the spectrum of reform eugenics was the interpretation that Ronald A. Fisher advanced in his 1930 classic, *The Genetical Theory of Natural Selection.*

Fisher's own family had its share of ability (and other qualities: as a cousin once commented, "Some Fishers were brilliant, some were dull, some very sane and responsible, some were brilliant but went off the rails, some just went off the rails"). Fisher's father built a fortune as an art dealer —social London was said to rank the firm of Fisher and Robinson with Sotheby's or Christie's—but lost it all not long after the turn of the century. While an undergraduate at Cambridge University, where he went on a scholarship, Fisher had begun to concern himself with the eugenic goal of multiplying the socially strong. At a meeting of the Cambridge Eugenics Society in 1911, he marveled that "the Englishmen from Shakespeare to Darwin . . . have occurred within ten generations," and added, "The thought of a race of men combining the illustrious qualities of these giants, and breeding true to them, is almost too overwhelming, but such a race will inevitably arise in whatever country first sees the inheritance of mental characters elucidated." Young Fisher aimed to extract from research in heredity, particularly of a mathematical type, the knowledge to effect "a slow but sure improvement in the mental and physical status of the population" and to "ensure a constant supply to meet the growing demand for men of high ability."[12]

Comfortable with male intellectuals, Fisher was ill at ease with women, tongue-tied among strangers, and at times rude and irascible. In 1913, after a year of postgraduate work, he left Cambridge, a misfit without means but with a self-estimable credo. "Like all healthy philosophies, eugenics urges us to simplify our lives, and to simplify our needs. . . . We must be ready to sacrifice social success, at the call of nobler instincts." Unable to obtain suitable academic employment, Fisher worked on statistical problems for the Mercantile and General Investment Company in the City of London, then after the company director had instructed him that baggy trousers and a dirty sports coat would not do, moved into schoolteaching. When the war broke out, he volunteered for the Army, but was rejected because of poor eyesight (his thick spectacle lenses resembled beer-bottle bottoms). Avowed patriot, political conservative, and Church of Englander that he was, Fisher was deeply disheartened by the Army's refusal, and the death of his brother in France in 1915 no doubt made the pain acute.[13]

Fisher's life brightened when in 1917 he married Ruth E. Guinness, the seventeen-year-old daughter of an evangelical preacher. She had grown skeptical about the evangelical God and eagerly joined with Fisher in a faith new to her, eugenics. The Fishers quickly started to procreate and continued to do so after Fisher, in 1919, joined the staff of the Rothamstead Experiment Station. Ruth Fisher eventually bore her husband eight children. Joan Fisher Box, her father's biographer, remembers that her mother saw to the endless needs of house and garden, children, and husband on a tight budget and with little domestic help. She helped care for the groups of mice and snails that Fisher kept for genetic breeding experiments, had his boots polished, read *The Times* to him at breakfast, and in the evening discussed his diverse interests, even coming into the bathroom to listen to him while he bathed. (The marriage was fine for many years, but eventually, at least for Ruth Fisher, who came to feel used and ignored, it was fine no longer, and she divorced him.)[14]

Before the appointment to Rothamstead, Fisher had kept at his scientific pursuits as best he could, with special encouragement from Leonard Darwin, whom he had come to know through his undergraduate involvement in eugenics. Darwin appreciated Fisher not only for his brilliance and eugenic devotion but also, no doubt, for his conviction that his father's theory of evolution by natural selection, then in dispute among biologists, was correct. Darwin arranged for the society to provide Fisher one hundred pounds for 1916 to work on eugenic investigations. About that time, Fisher formally demonstrated that the biometric analysis of heredity was consistent with Mendelian genetics—that, for example, the correlations measured between the heights of parents and children could be predicted from the assumption that the trait was polygenic in origin. Although a classic—and the foundation of what became *The Genetical Theory of Natural Selection*—Fisher's paper on the subject was rejected for publication by the Royal Society of London (Fisher later attributed the rejection to its having been reviewed by "a biologist who knew no statistics and a statistician who knew no biology"). Darwin, eager that it should see print, had the Eugenics Society sponsor its appearance in the *Transactions of the Royal Society of Edinburgh,* and he urged Fisher to keep going, insisting that mathematical treatment was perhaps the only way that the difficulties in the theory of natural selection could be worked out.[15]

Rothamstead was the principal agricultural research station in England, and there in the nineteen-twenties, Fisher was able to keep going across a broad front of investigation. To help plan and evaluate plant and animal breeding experiments, he did profoundly important work in statistics, particularly in the development of methods for avoiding hidden biases in research design and for interpreting the meaning of experimental results. The breeding program, in which the analysis of the relative roles of heredity and environment figured significantly, enriched the perspective that Fisher brought to questions in the area of genetics and evolution. He steadily pursued the writing of *The Genetical Theory of Natural Selection,* sent Darwin every chapter as it was completed, and dedicated the book to him when it was finished.[16]

Fisher's treatise addressed a major conundrum in evolutionary theory that, one recalls, had vexed Charles Darwin himself as well as Francis Galton. In their day, the problem had been: Could evolution, a story of major changes in organisms, occur through the natural selection of small variations? In the period prior to the First World War, the advent of Mendelian genetics had transformed the question: Could evolution proceed from the natural selection of the minute variations that cropped up in single-gene mutations or in the genetic recombinations of sexual reproduction? That it likely could was suggested by the experimental work of Thomas Hunt Morgan with fruit flies and by the research of other geneticists in the United States and abroad. Fisher theoretically generalized such results. So, independently, did J. B. S. Haldane and the American Sewall Wright. Each of the three forged mathematical models of the evolutionary impact of various small, selective advantages in genotype upon the overall genetic makeup of a given population, and taken together, their work brilliantly demonstrated the consistency of Darwin's theory of evolution with Mendelian genetics in its rapidly developing complexity.[17]

Questions concerning human evolution had helped stimulate Fisher to consider population genetics in the first place, and he took up the issue of low upper-class fertility in the closing third of the book, having noted in the preface that "the deductions respecting Man are strictly inseparable from the more general chapters." Fisher contested the type of analysis that explained the low fertility causally in terms of possession of wealth or professional success, including excess food and leisure, the stress of brain-work, the enervating influence of comfort.[18] Fisher attributed differences in fertility to physiological factors—and also to variations in mental, behavioral, and even moral character. All added up to variations in temperament that, to Fisher's mind, figured in decisions for or against marriage and especially for or against reproduction. Some people welcomed contraception because their temperaments disposed them to favor "sexual anarchy." Others accepted it because they were temperamentally inclined to raise a few children well rather than many children poorly. Whatever their particular reasons, the behavioral outcome—a correlation of low fertility with high social position—posed what Fisher, like some eugenicists before him, considered a eugenic paradox: If success in the Darwinian sense meant high fertility rates, then in modern Western society evolutionary success went together with social failure, and social success with evolutionary defeat.[19]

Fisher resolved the paradox by invoking what amounted to a biological theory of the Protestant ethic. The theory, inchoate in his early ideas about human evolution, had first achieved explicit formulation when in 1913 he read in the *Eugenics Review* an article by J. A. Cobb that he estimated as "containing the greatest addition to our eugenic knowledge since the work of Galton." Cobb had advanced the basic point that in any society allowing greater social advantage to the members of small rather than of large families, temperamental qualities making for de facto sterility would tend to rise in the social scale. In modern industrial society, the smaller the family, the more the resources that could be accumulated, and the larger the resources, the more advantages that could be passed on to children in the form of education and capital. Fisher added to Cobb mainly the demonstration, at least to his own satisfaction, that the temperament which made for infertility was a genetic product. After all, women who came from large families tended to have large families themselves. Some analysts argued that they did so by force of tradition, but to Fisher their very receptiveness to tradition was itself a genetically determined character trait. So was the temperament —the ambition and intellect, the readiness to defer gratification and accumulate means—that in modern society made for success. Thus, the environment of modern society naturally selected for and united at the top the traits of low fertility and high ability. In Fisher's summary: "The various theories which have sought to discover in wealth a cause of infertility, have missed the point that infertility is an important cause of wealth."[20]

What most worried Fisher was the low fertility of the professional and clerical middle class. Confident that the destiny of the nation depended on the extent to which its citizens combined "enterprise with prudence, or character with intellect," he was certain that "the fate of this class is of the deepest concern for the future of our nation." But through birth control, he was reported in paraphrase to have told the Linnaean Society in London in 1932, professionals and better-paid workers were "destroying their racial stock . . . as rapidly as any communist could wish to see the intelligentsia extirpated." To Fisher, even if the birthrate of the lower classes should drop to that of the upper, Britain would remain in jeopardy: Such equalization of fertility would not take the country "a step nearer to arresting the process by which the eugenically valuable qualities of the nation are being destroyed."[21]

To reverse the trend, Fisher argued for a comprehensive scheme of state family allowances. The scheme resembled that long advanced by such social reformers as Eleanor Rathbone, but only superficially, for in Fisher's version, the allowances were to be aimed preferentially at the eugenically desirable sectors of the middle class. The government would provide an allowance for each child proportional not to the family's absolute need but to its total earned income; high-income families would receive more per child than low-income families. Regardless of the number of children, parents would thus be enabled to provide each child equally with whatever social and economic advantages befitted the family's station in life. Of course, Fisher's brand of family allowances would give more to those who already had—that is, to his own professional middle class. In Fisher's defense of the scheme, it would simply replace the principle of equal pay for equal work with that of an equal standard of living for equal work.[22]

The more C. P. Blacker thought about Fisher's genetic theory of the differential birthrate, the more did he conclude that Fisher had "practised upon the intelligentsia of this country a most interesting hoax." People had known about the dependence of fertility upon prudential considerations long before Fisher, and Blacker could not see how the situation was at all illuminated by "a lot of vague talk about hypothetical genes which are supposed to produce sterility in certain particular environmental conditions."[23] Nevertheless, Fisher's theory was persuasive to Huxley, and for a time it captured even J. B. S. Haldane (who had said of *The Genetical Theory of Natural Selection* that "no serious future discussion, either of evolution or eugenics, can possibly ignore it"). Blacker was sufficiently bothered by the declining fertility among the better sort to endorse the family allowance idea (while "eugenically valuable" people were found in all social classes, he later noted, he thought it "possible that they may be proportionately more numerous in some classes and occupations than others"). Support also came from the Eugenics Society—despite opposition from rump mainliners hostile to "state paternalism"—from other reform eugenicists, and even from Lionel Penrose. In the United States, Osborn and his allies embraced variants of Fisher's scheme, notably tax exemptions for children keyed to the actual cost of rearing them and salaries proportionate to size of family for teachers, professors, ministers, and possibly government employees. In *Tomorrow's Children*, a pamphlet published by the American Eugenics Society, Ellsworth Huntington, a demographer at Yale, summarily declared: "It is hard to see how a perfect eugenic system can prevail until every intelligent married couple is able to have as many children as it wishes without lowering its economic status."[24]

THE EAGERNESS TO FOSTER higher fertility among the eugenically valuable fortified reform eugenicists in their embrace of meliorative economic measures or socialist reconstruction. Yet to some on the eugenic left, particularly biologists, social measures would not by themselves make the eugenicists' utopian dream of man's genetic improvement a reality.

Whatever the economic system, the argument ran, if people married primarily for eugenically reproductive purposes, they might be spiritually destroyed. If they procreated solely out of love, the product of their union might not be eugenically valuable. To Hermann Muller, and to twenty-two British and American scientists who signed his "Geneticists' Manifesto," in 1939, the course was obvious: for the sake of eugenics, replace "the superstitious attitude toward sex and reproduction now prevalent" with "a scientific and social attitude." Render it "an honor and a privilege, if not a duty, for a mother, married or unmarried, or for a couple, to have the best children possible, both in respect of their upbringing and of their genetic endowment."[25] This was, of course, no new idea in the eugenics of the left. It recalled the Shavian demand that society allow able women to conceive children by able men whom they might never see again. But the early eugenic radicals had, like Shaw, advanced more of a sexual than a scientific revolution. Muller and his allies proclaimed that sexual revolution could now proceed in tandem with what was already—or was likely soon to be —known about genetics and reproduction. In their view, something akin to a utopian eugenics was, in short, becoming a scientifically practical prospect.

J. B. S. Haldane had given the utopian vision of eugenics explicit scientific statement in *Daedalus,* a slim, remarkable book he published in 1924. The technological inventor, Haldane observed, was a Prometheus whose every innovation, from fire onward, had been "hailed as an insult to some god." In contrast was the first biological inventor, Daedalus—the first genetic engineer, a later generation might say—who oversaw the procreation of the Minotaur by arranging the coupling of Pasiphaë and the Cretan bull. This "most monstrous and unnatural action in all human legend was unpunished in this world or the next," Haldane averred. But if Daedalus escaped the vengeance of the gods he suffered "the agelong reprobation of a humanity to whom biological inventions are abhorrent." Physical and chemical invention might be blasphemy; biological invention was "perversion" and to most observers appeared "indecent and unnatural," offensive not to some god but to man himself. So did Haldane expect laymen to perceive as perverse what, with evident relish, he proposed that the new Daedalus might accomplish.[26]

The proposition took the form of an essay that a Cambridge undergraduate might read to his tutor a hundred and fifty years in the future about the influence of biology on history. Reviewing the early eugenic movement, Haldane's undergraduate noted that it had provoked class hatred but had served a useful purpose in preparing the public for what was to come—the first "ectogenetic child," produced in 1951 by the fictional scientists Dupont and Schwarz. The undergraduate of the twenty-first century explained:

> Dupont and Schwarz obtained a fresh ovary from a woman killed in an aeroplane accident, and kept it living in their medium for five years. They obtained several eggs from it and fertilized them successfully, but the problem of the nutrition and support of the embryo was more difficult, and was only solved in the fourth year. Now that the technique is fully developed, we can take an ovary from a woman and keep it growing in a suitable fluid for as long as twenty years, producing a fresh ovum each month, of which 90 per cent can be fertilized, and the embryos grown successfully for nine months, and then brought out into the air. . . . As we know, ectogenesis is now universal, and in this country less than 30 per cent of children are now born of woman. The effect on human psychology and social life of the separation of sexual love and reproduction . . . is by no means wholly satisfactory. The old family life had certainly a good deal to commend it, and although nowadays we bring on lactation in women by injection of placentin as a routine, and thus conserve much of what was best in the former instinctive cycle, we must admit that

in certain respects our great-grandparents had the advantage of us. On the other hand, it is generally admitted that the effects of selection have more than counterbalanced these evils. The small proportion of men and women who are selected as ancestors for the next generation are so undoubtedly superior to the average that the advance in each generation in any single respect, from the increased output of first-class music to the decreased convictions for theft, is very startling.[27]

Haldane predicted that if reproduction were completely separated from sexual love mankind would be "free in an altogether new sense." No matter that the ultimate result would involve taking the wombs from women, mechanically fostering conception, engineering fetal development and parturition, then chemically making breast feeding possible: if biological innovation began with a perversion, it usually ended as "a ritual supported by unquestioned beliefs and prejudices." Was there not something slightly disgusting about milking cows with machines or drinking beer out of teacups? Man had grown accustomed to these innovations. Why not to innovations concerning the sexual act? He had nothing to fear from the gods, only from himself. Haldane celebrated the scientist of the future as the lonely figure of Daedalus, garbed in black robes, proud of his "ghastly mission" and singing a "song of deicides."[28]

Within a year of publication, *Daedalus* sold some fifteen thousand copies—a substantial number—and it provoked a good deal of attention among literary and leftist intellectuals who concerned themselves with science in relation to society. Ectogenesis earned none too flattering treatment in the principal book it inspired, Aldous Huxley's *Brave New World*. (Nor did Haldane come off very well in *Antic Hay*, in which he is the prototype of Shearwater, the biologist too absorbed in experiments to notice his friends bedding his wife.) Yet to geneticists on the left, Haldane's utopian speculation exemplified possibilities increasingly "less fanciful," as Enid Charles remarked, than at the time of the book's publication.[29] In fact, in the following decade Hermann Muller and the British eugenicist Herbert Brewer independently came to insist that modest first steps toward Haldane's goal might be feasible.

HERMANN MULLER WAS THE product of an intellectual family, originally refugees from the German upheavals of 1848 and, during his childhood, proprietors of an artistic-metalware business in New York City. Muller did his doctoral work at Columbia University, in the laboratory of Thomas Hunt Morgan, joining Morgan's group in the splendid research on the genetics of *Drosophila*. Muller's contributions were brilliant, and he was one of the four co-authors of *The Mechanism of Mendelian Heredity*, which the group published in 1915. In the mid-twenties, at the University of Texas, Muller inaugurated a research program that succeeded in demonstrating that genetic mutations could be induced in fruit flies by X rays—the achievement that brought him, in 1946, his Nobel Prize in physiology or medicine. When he first fully reported his results, at the Fifth International Congress of Genetics in Berlin, in 1927, his listeners realized, in the recollection of one of them, that they had been "privileged to be present at the moment of a decisive advance in man's probing of nature—the first time that he had willfully changed the hereditary material."[30]

Muller, who seems to have craved recognition, soon came to believe that the Morgan group had stolen his ideas, had failed to give him proper credit for his contributions to the early *Drosophila* work, and had blocked his professional advancement. He felt isolated at Texas. In 1932, when he had recently been passed over for election to the National Academy of Sciences, and his first marriage was on the rocks, he walked into the woods and swallowed a roll of sleeping pills. Searchers found him the next day sitting dazed under a tree. He had a suicide note in his pocket—addressed to Edgar Altenburg, a close friend from his Columbia days—which included a bitter attack on "the predatory operations of T. H. Morgan."[31]

Some years later, Altenburg recalled that at Columbia Muller had "traded in the three R's for the three S's—science, sex, and socialism." Muller's was an armchair socialism, drawn little from reading in its doctrines, imbibed mainly from his father and his own circle of friends in New York. Nevertheless, he advanced the socialist cause with a bantam outspokenness. His radicalism got him into trouble at the University of Texas. In 1933, he went to Leningrad to work in a laboratory of the Institute of Genetics, which was part of the Soviet Academy of Sciences and was then headed by the accomplished plant geneticist Nikolay I. Vavilov.[32]

Muller never became a Communist, probably because of what he witnessed in Russia: the hegemony of the plant physiologist Trofim Lysenko, and the persecution of Lysenko's scientific opponents, the advocates of the genetics of Gregor Mendel and Thomas Hunt Morgan (among them Vavilov, who in 1941 was sent to prison, where he died two years later). At a meeting of the Lenin All-Union Academy of Agricultural Sciences in 1936, Muller courageously declared that to be forced to choose between what the Soviets called Mendelism-Morganism and the Lamarckian doctrines of Lysenko was to be "confronted with a choice quite analogous to that between medicine and shamanism, between astronomy and astrology, between chemistry and alchemy."[33] Despite his growing disillusionment with the U.S.S.R., however, Muller maintained that "only the eugenics of the new society, freed of the traditions of caste, of slavery, and of colonialism, can be a thoroughgoing and a true eugenics." He was sure that eugenics as practiced by the American capitalist order would lead to a population composed of "a maximum number of Billy Sundays, Valentinos, Jack Dempseys, Babe Ruths, even Al Capones."[34]

Herbert Brewer was a socialist, too, but more in the practical, incremental vein of the white-collar clerk eager to welcome whatever socioeconomic improvements might occur. He was a postal clerk in Maldon, England—"eugenicist by profession and a post-office clerk by accident," he said in a letter to C. P. Blacker—and one of those brilliantly inventive autodidacts whose exceptional talents the British class system so often wastes. In 1911, at the age of fourteen, he had been forced by his family's desperate poverty to leave school; he seems to have discovered eugenics through voracious reading, particularly in Wells and Shaw. He developed his own ideas on the subject in the thirties while working—sometimes eleven hours a day, seven days a week—in the Maldon post office. The strain told on him, and so did the cruel lack of scientific opportunity; he suffered a nervous breakdown and bouts of severe depression.[35] Although, like Muller, Brewer believed that eugenics stood the best chance in a classless society, he judged that in the meantime everyone ought to enjoy a better genetic endowment. If the salvation of the human species required socialism "to make a better world to live in," it also required eugenics "to make better men to live in the world."[36] Early in their eugenic careers, both Muller and Brewer had flirted with negative eugenics—the mainline idea of ridding the world of the biologically unfit. In the thirties, they came to focus on positive eugenics, which to them meant the biological fostering of aptitudes and faculties that might aid in the creation of the socialist order, and forms of talent and intelligence essential to literary, artistic, and scientific achievement.

For Brewer, the strategy to be followed consisted in first "raising up the great mass of mediocrity and inferiority to the level of the best existing"; second, in "advancing from the present best to the superman." Creating a superman would be a long and difficult task, Brewer told the British Eugenics Society in 1935, but bringing the mediocre up to the level of the best might require just "a few generations," through the process of "eutelegenesis."[37] Brewer coined the word to refer to the eugenic breeding of human beings via pregnancies produced "from afar"—that is, by artificial insemination.

First successfully achieved with animals at the end of the eighteenth century, artificial insemination was by the thirties an object of practical interest to stockbreeders. Since the mid-nineteenth century, it had also been carried out, sporadically, on women whose husbands were sterile and who wanted children. According to a report in the March 1934 issue of *Scientific American*, requests for sperm donors in the United States were currently estimated to come from between one thousand and three thousand women a year. The report noted that the women usually wanted assistance from the biologically best donors, and that artificial insemination for eugenic purposes made "possible to humans a privilege, in posterity, heretofore enjoyed only by thoroughbred plants and animals," and it continued, "Some 10,000 to 20,000 babies [could] be born every year from selected sources, while less than 500 babies per year are now being born to the men of real talent in our country. What will be the eugenic effect on the race, if this same tendency grows?"[38] In fact, artificial insemination was not yet a very reliable technique. In one scientist's estimate, only about a third of the attempts in human beings resulted in pregnancies. The reasons for the low success rate lay in areas of human physiology about which little was known—among them the vitality and longevity of spermatozoa kept outside the male body, and the hormonal and chemical requirements necessary for conception in the female's. But to Brewer, who scoured the scientific literature on the subject, the success with animals implied that artificial insemination could be made to work reliably among human beings.[39]

Brewer and Muller began a correspondence in 1935, while Muller was still in Russia. They were both tantalized by recent research in the physiology of reproduction. In 1934, at Harvard University, the endocrinologist Gregory Pincus (who later became one of the principals in the development of the contraceptive pill) had managed to wash eggs out of monkey ovaries and fertilize them in vitro. In due course, he did the same with rabbits, injecting the fertilized eggs back into the female, who proceeded to bring her artificially engineered pregnancy successfully to term.[40] Both Muller and Brewer perceived eugenic implications in the work: the fertilization of a genetically "superior" human egg by a similar sperm in a test tube, and the implantation of the zygote in a third-party female—no doubt genetically inferior, but an able nurturer of the fetus. Obtaining the eggs would require certain advances in technique. Muller was encouraged by a report in 1935 from the Rockefeller Institute, in New York, that the Nobel-laureate surgeon Alexis Carrel—with the aid of Charles Lindbergh (who had designed an advanced perfusion pump in Carrel's laboratory)— had succeeded in keeping mammalian ovaries alive and growing outside the body. Brewer pondered less radical methods—including some similar in principle to the procedure developed in the nineteen-seventies by Patrick C. Steptoe and Robert G. Edwards which resulted in the birth of the famed Louise Brown. Brewer dubbed test-tube fertilization "penectogenesis," because he considered it a major step toward Haldane's ectogenesis.[41]

Yet it was artificial insemination that commanded the attention of

Brewer and—especially—Muller. To Muller, artificial insemination meant that the eugenic future need not await the distant advent of penectogenesis, let alone the still more fantastic ectogenesis. Muller had begun developing his own ideas about eugenics via artificial insemination before he went to the Soviet Union. He advanced them in *Out of the Night*, a book published in 1935 in the United States, where it sold only about a thousand copies. Muller sent a copy of *Out of the Night* to Stalin, thinking that the leader of the Soviet Union might recognize in its ideas some possible ways of accelerating Soviet socioeconomic advance. Perhaps because of Stalin's puritanism—not to mention his affinity with Lysenko—Muller succeeded only in making himself persona non grata in the Kremlin. Whatever thought Muller may have had about how Stalin might carry out a program of Soviet eugenics, he and Brewer both stressed the point that eutelegenesis would entail no compulsion. "Eugenic advance must be the voluntary adventure of free men and women, or nothing," Brewer noted in the *Eugenics Review*.[42]

Brewer publicly evaluated artificial insemination as "a simple manipulation, less painful than drawing a tooth, and no more unchaste than an ante-natal examination." Privately, he thought that problems could arise in finding an unobjectionable method of obtaining sperm. However innocuous the techniques, he expected that eutelegenesis would be "stigmatized as immoral and not respectable." It was necessary to remember, he remarked at a meeting of the Eugenics Society, echoing J. B. S. Haldane and resolutely facing the future, that often "the immorality of yesterday is the social duty of tomorrow." In *Eugenics and Politics,* a 1937 pamphlet published by the Society, he observed that the aims of eutelegenesis were not merely compatible with socialism: "They *are* socialism, biological socialism. . . . They involve nothing less than a socialization of the germ plasm, the establishment of the right of every individual that is born to the inheritance of the finest hereditary endowment that anywhere exists."[43]

Still, Brewer and Muller were resolved to see eutelegenesis proceed slowly. Eugenicists might study the results of its deployment for infertility, and then it might be tried by a few pioneers deliberately for eugenic ends. Eventually, such trials might inspire respectful acceptance of it. Muller declared in *Out of the Night* that "in the course of a paltry century or two . . . it would be possible for the majority of the population to become of the innate quality of such men as Lenin, Newton, Leonardo, Pasteur, Beethoven, Omar Khayyám, Pushkin, Sun Yat-sen (I purposely mention men of different fields and races), or even to possess their varied faculties combined."[44]

Women were noticeably absent from Muller's pantheon of talent; the role of women in eutelegenesis amounted to little more than that of conceptual vessels for the sperm of admirable men. This sexual asymmetry was dictated by physiology. Between the ages of twenty-five and fifty-five, the normal man was estimated to produce about three hundred and forty billion sperm. By comparison, women produced only a minuscule number of ova. If only one out of a thousand of the male sperm was utilized, Brewer noted enthusiastically, one man in a year could fertilize five million women. Eutelegenesis thus "immensely magnified" the reproductive power of "a few superior males." Muller and Brewer, for all their socialist principles, seemed ready to welcome what physiology dictated.[45] Brewer mused about the possibility of hiring women for the use of their bodies in an experimental eutelegenetic program. The resultant children would be adopted by worthy couples; the women themselves would get five hundred pounds and a bit of scientific glory. Muller confided to Brewer that there had been talk in the Soviet Union of crossbreeding human beings and apes, and stories of Russian women prepared to volunteer for artificial insemination with ape sperm. Brewer found the idea ghastly, but he thought the attitude of the women significant.[46] He contended at a meeting of the Eugenics Society that "the whole nature of woman is dominated by her reproductive function," and by her sense of "altruism in relation to the child." In *Out of the Night,* Muller revealed a similar cast of mind: "How many women, in an enlightened community devoid of superstitious taboos and of sex slavery, would be eager and proud to bear and rear a child of Lenin or of Darwin! Is it not obvious that restraint, rather than compulsion, would be called for?"[47]

Out of the Night appeared in England in 1936—Brewer arranged for its publication—to a glowing reception and with a sale, in connection with the Left Wing Book Club, of thirteen thousand copies. British reviewers from the left to the far left—from C. P. Snow in *The Spectator* to *The Daily Worker*—were full of admiration for the eutelegenetic idea, calling it not only socially desirable but scientifically sound. Haldane, responding to eutelegenesis partly out of the deep distress he felt over the childlessness of his marriage, told Brewer that he was prepared to supply his name, money, and gametes to the cause and predicted that the results of eutelegenesis "will be as important as those of the industrial revolution."[48] Julian Huxley doubtless spoke for much of the eugenic left when he celebrated eutelegenesis for rendering it "open to man and woman to consummate the sexual function with those they loved, but to fulfill the reproductive function with those whom on perhaps quite other grounds they admired." George Bernard Shaw urged Brewer on: "When I, who have no children, and couldn't have been bothered with them, think of all the ova I might have inseminated!!! And of all the women who could not have tolerated me in the house for a day, but would have liked some of my qualities for their children!!!" The encomium came with a check for a hundred pounds, and the Shavian signature with a phallic flourish at the tail.[49]

Reform eugenicists generally believed with Huxley that "the whole progress and stability of the collective human enterprise" depended upon the gifted capable minority who might prevail against the socially heavy "dead-weight of the dull, silly, underdeveloped, weak and aimless." Eutelegenesis, the brave new biology of the left, promised, at least in evolutionary time, to make the multiplication of the gifted minority imminent. "Not only is our genetic improvement patently possible," Muller declared, "but it is far surer and more feasible than any ultimate conquest of the atom, of interplanetary space, or of external nature in general."[50]

Chapter XIII

THE ESTABLISHMENT OF HUMAN GENETICS

M OST REFORM EUGENICISTS were aware that man as yet knew too little about human heredity to enact sweeping eugenic changes, let alone usher in a eutelegenetic utopia. They stressed that the task of eugenics had to be further research, particularly in the field of human genetics—a science that in the thirties had few practitioners and was intellectually, as Lionel Penrose remarked, "still in its early infancy."[1]

Man, Penrose noted drily, was not a laboratory animal and did not live in the conveniently circumscribed environment of a test tube. Unlike plants or other animals, human beings could not be subjected to controlled experimental breeding. Human geneticists had to obtain their data from direct clinical experience, reports in medical journals, and the records of hospitals and mental institutions, or, more generally, from surveys of patients, schoolchildren, or some other selected population. Special categories of people were held to be particularly useful, notably identical twins raised separately or, better yet, apart, since they provided an exact case of genetic comparison; "racially" mixed populations, since their variability of traits could substitute for the results of controlled hybridization; and consanguineous parents, since they were more likely than mates at random to join recessive genes in their offspring.[2]

In the thirties, students of human heredity insisted that human genetic investigations had to be emancipated from the biases that had colored mainline-eugenic research—notably the attentiveness to vague and often prejudiced behavioral categories and the assessment of traits in deceased family members on the basis of hearsay or gossip. But human geneticists of the day realized that sanitizing the data-gathering process would by no means solve all of the major methodological problems of the subject. Even with data that were unimpeachable, it was no trivial task to determine whether a trait was hereditary, and if so, in what way. "Genetic analysis of human data," Hogben noted, "is a much more subtle task than the interpretation of experimental results in animal or plant breeding, and presupposes some knowledge of the theory of probability."[3]

In sexual reproduction, the laws of probability can predict the frequency with which possible genetic combinations will occur in offspring. Recall that, for example, if two organisms mate, each containing genes for the dominant trait A and the recessive a, on average one-quarter of the offspring will be homozygous for A, another quarter the same for a, and one half will be heterozygous as Aa. Since A is dominant, it will be expressed in three-quarters of the offspring. Tests of genetic hypotheses hinged on measuring the frequency with which particular traits appeared in successive generations, but the soundness of the tests depended upon the production of a sufficient number of offspring in each generation to express all the possible genetic outcomes. Thus, plant and animal geneticists preferred to experiment with organisms that reproduced prolifically (and, preferably, rapidly, so as to have a long series of generations for analysis). Thomas Hunt Morgan's choice of *Drosophila* had endowed his genetic research program with a decisive advantage because fruit flies amply satisfied both criteria. Geneticists disliked man as a subject because he satisfied neither. He bred slowly, his families were small, and his life cycle, as Penrose remarked, was "much more lengthy than that of, say, a rabbit."[4]

In the typical human family, the offspring express only a sample—rather than the complete inventory—of the genetic combinations the parents can provide. In some families carrying recessive genes, the trait may not be expressed in any of the offspring; in others, it may be expressed with misleadingly high intensity. If both parents are heterozygous for a recessive genetic disease, the probability that any child of theirs will be homozygous for it is one in four. But if the parents were to bear, for example, two children, one of whom suffered from the disease, the observed familial incidence would be one in two; if the diseased infant were the family's sole child, it would be one in one. Thus, the smaller the family, the more the human geneticist who observed it for a given trait risked what came to be called an "ascertainment bias"—a tendency, given the genetic makeup of the parents, to find a higher frequency for the trait in families showing it than its true probability of occurrence.[5]

By the early nineteen-thirties, correction for ascertainment bias could be accomplished with mathematical procedures based on the theory of probability. The procedures took into account the fact that in genetic surveys those families in which the trait was expressed would be counted, while those containing the gene in unexpressed—and hence undetectable —combinations would not. On this basis, the frequency to be observed for the trait could be calculated and then compared with the surveyed incidence to indicate whether the trait was—or was not—truly a Mendelian dominant or recessive. Methods of this type led to a convincing demonstration that juvenile amaurotic idiocy was caused by a recessive gene and to the strong suggestion that schizophrenia was not. (An early study of schizophrenia had measured an incidence of the disorder among siblings of about the twenty-five percent characteristic of a Mendelian recessive, but correction for ascertainment bias revealed the true frequency as slightly under five percent.)[6]

A corpus of formal mathematical genetics was also available to analyze the genetic dynamics of human populations. Early in the century, the British mathematician G. H. Hardy and the German physician Wilhelm Weinberg had independently arrived at a mathematical formulation—eventually called the Hardy-Weinberg law—for the frequency with which different genotypes occurred in populations breeding at random. Fisher, Haldane, and Wright enriched the mathematical arsenal with their work on population genetics, providing methods for assessing the effects on evolutionary development of such events as dominant and recessive mutations and changes in genetic fitness. The tools of population genetics could be used with different types of human genetic data, but particularly with surveys of trait incidence in populations—as distinct from families—ranging in size from the local to, in principle, the global. Moreover, a comparison of observed trait frequencies with what theory predicted provided a way of testing hypotheses concerning the genetic basis of the traits themselves.[7]

Such an analysis had been carried out in the twenties by the German mathematician Felix Bernstein, who used the Hardy-Weinberg law to work out the genetics of the then-known human blood groups. Human blood groups were first identified by Karl Landsteiner, a Viennese physician and later Nobel laureate in physiology or medicine, when early in the century he noticed that the blood of patients contained three different isoagglutinins—substances that would react to particular antibodies—which he categorized as "A," "B," and "O." In 1911, it was demonstrated that the blood groups resulting from them—A, B, AB, and O—were inherited, seemingly in a Mendelian fashion with A and B each a dominant member of two separate pairs of genetic factors. Research on the genetics of blood groups lagged until the twenties, when, following the recognition that they varied with "race," it picked up, notably in Germany. Bernstein, taking advantage of the growing quantity of blood-group data, showed, in 1924, that the genetic factors for A, B, and O were not two separate pairs but three forms of the same gene, with A and B coequal to each other—thus the AB group—and both dominant to O. In 1927, Landsteiner, who had moved to the Rockefeller Institute for Medical Research in New York, reported, with his colleague Philip Levine, the existence of two more isoagglutinins, "M" and "N," which made for three more blood groups— M, N, and MN—that were rapidly shown also to conform, albeit in slightly different fashion, to the rules of Mendelian genetics.[8]

The progress in blood groups was rich with implications for human genetics. The problem of sorting out the relative contributions of nature and nurture to human traits was particularly acute with characters like intelligence that varied in continuous grades—that is, did not neatly segregate from one another among offspring as did, for example, blue and brown eyes. Even in cases of ungraded qualities, insufficient attention to environmental forces could lead to mistaken analysis. Dietary deficiency rather than genes was responsible for rickets, for example, yet more than one treatise on human heredity reproduced family pedigrees to demonstrate that a tendency to rickets was genetic.[9] Students of human heredity treasured well-defined, sharply segregating traits as immune as possible both to uncertainty in identification and to environmental influence. Thus the attention given in human genetics to the normalities of eye or hair color, to unmistakable physical deformities, and, at least for Penrose, to the biochemically specific phenylketonuria. And thus the considerable interest stimulated in the early thirties by the discovery that human beings possessed a heritable sensitivity to the taste of the compound phenylthiocarbamide, or PTC.

The sensitivity itself was accidentally detected in 1930 by a scientist at the Du Pont Laboratories in Ohio who was working with a compound that contained PTC. Dust from the compound tasted unpleasantly bitter to his assistant, while the dust had no taste at all to him. Laurence Snyder, a geneticist at Ohio State University, demonstrated that the lack of PTC sensitivity depended upon a single recessive gene. PTC tastes bitter to the majority of people, either heterozygous or lacking in the gene for the trait. It is tasteless to the smaller, homozygous fraction of the human population. (In the thirties, at the Edinburgh Zoo, Fisher and colleagues were delighted to find evidence that the same genetic trait occurred in man's cousin the chimpanzee. Eight of the zoo's chimps were given a series of sugar solutions progressively stronger in PTC to drink. When the concentration reached fifty parts per million, six of the eight displayed reactions ranging from apparent hurt to hostility—from retreating, with back turned, to the rear of the cage to spitting out the potion in seeming angry disgust at the onlooking scientists. The other two chimps appeared to be homozygous non-tasters for PTC.)[10]

Like PTC or PKU, the blood groups provided precisely the kind of unambiguous trait that human geneticists liked to find. They were, so far as anyone knew, wholly unaffected by environment, dietary or otherwise. Then, too, for human genetics, blood groups were superior to rare recessive traits, even if biochemical, because, unlike PKU or PTC, they were universally expressed; everyone fell into one or another blood-group category. As such, they were thought in the thirties to open the door to the genetic mapping of the human chromosomes, which aimed to determine whether the genes for two traits were linked by residence on the same chromosome and their relative distance apart. The blood groups provided a specific set of chromosomal markers—"a locus of reference," Hogben put it—to which the genes for other traits could be linked. If linkages could be found between the gene for, say, PKU and the gene for blood group ABO, then one would know that the PKU gene lay on the same chromosome as the ABO gene, and one would know the relative distance between the two genetic factors.[11]

The determination of linkage depended at bottom upon the fact that in the sexual division of the cells a segment of one chromosome could be exchanged with the similar segment of its counterpart. The farther away from each other the genes for two traits, the more likely that one would end up on an opposite chromosome as a result of segmental crossover. Once the sexual division of the cell occurred, the two chromosomes, and hence the genes for the two traits, would be separated from each other into different gametes. Thus, the closer together on the chromosome the genes for two traits, the more likely that both would be transmitted to a given offspring, while the farther apart, the less likely. Observationally, the study of linkage in human populations required measuring the frequency with which different traits did—or did not—occur together, and the transformation of the frequency measurements into proofs of linkage demanded mathematical treatment similar to, though rather more complicated than, the procedure Bernstein had used to demonstrate the inheritance of the A, B, and O blood groups. In 1931 Bernstein himself supplied a concrete algebraic method for the purpose. Hogben, who recognized the significance of Bernstein's work, called attention to it in his influential *Genetic Principles in Medicine and Social Science,* published in 1931, adding that if unambiguous markers like the blood groups or the PTC trait could be found for every chromosome, then there would be a set of socially unbiased benchmarks in connection with which the human genome could be catalogued.[12]

To Hogben, Fisher, and others, linkage studies were also promising for eugenic prognosis. Identifying the carriers of dominant traits had never been much of a problem to eugenicists; sooner or later, the traits expressed themselves. But they had long been stymied by the problem of tagging the heterozygous carriers of genes for recessive traits, which were not expressed until—too late from a eugenic point of view—they joined homozygously in offspring. (For this reason Herbert Spencer Jennings declared, in 1930, that to recognize the carriers of recessive genes for defects would be "one of the greatest biological discoveries that could be made; one of the most fruitful in immediate practical application.") Linkage studies might reveal that a deleterious recessive gene occurred on the same chromosome as did one of the blood groups; anyone who came from a family known to have the gene and who was also found to have that blood group would be spotlighted as a probable carrier of the recessive. Similarly, if the gene was a dominant, the identification of an infant's blood group would enable one to predict the probability—it would depend upon the degree of linkage— that the disease resulting from the dominant would be expressed in the child. Appropriate steps might then be taken to prevent the expression— or at least to mitigate the effects—of the disease itself. If the disease came on late in the childbearing years, people fated to contract it could be advised before they had children of the chance of transmitting it to their offspring and they might then refrain from reproduction.[13]

Whether the aim was eugenic improvement or basic understanding, Haldane, in his *New Paths in Genetics*, a benchmark book of 1941, rightly pointed out that "in the study of human genetics, statistical methods replace the various technical devices, such as milk bottles and etherizers, which are familiar to the *Drosophila* worker...," adding, "They are essential adjuncts to any study of human genetics which goes beyond the mere accumulation of pedigrees."[14]

IN 1931, HOGBEN, TAKING NOTE OF the new mathematical methods, declared the prospects of advancing the field "as an exact science" much brighter than they had been since the first flush of enthusiasm that followed the rediscovery of Mendel's papers. Still, the new methodological engine required the fuel of data and the hands of analytical operators. Hogben called for the establishment of what amounted to a multi-part human genetic research program: twin studies to sort out the relative roles of heredity and environment; measurements of variability within hybrid populations to test for "race"-specific characters; pedigree investigations, especially from medical records, for determining the genetic basis of disease; and surveys of consanguinity, to decide whether certain diseases or physical traits might be the product of homozygous recessives. With the aim of creating a human genetics devoid of social prejudice, the data were to be gathered with scrupulous care, and as much of the information as possible was to concern unambiguous traits, particularly, of course, blood groups.[15]

Both the collection and the analysis of the data promised to require the collaboration of trained geneticists, clinicians, and ethnologists. Bernstein's establishment of the genetic basis of the ABO blood groups had involved the scrutiny of information from thousands of families by numerous investigators. Hogben, preparing for an investigation of twins, had found himself filing the addresses of four to five thousand prospective subject families. Given the magnitude of the overall task, it was patently evident to Hogben that "the advancement of human genetics is extremely costly, that important contributions will not any longer be made by isolated individuals, and that the study of human inheritance imperatively demands organised team work on a very large scale."[16]

In the United States, the principal candidate to conduct large-scale inquiries in human heredity was the Eugenics Record Office at Cold Spring Harbor. By 1926, as a result of its various surveys and studies, the Office had accumulated about 65,000 sheets of manuscript field reports, 30,000 sheets of special traits records, 8,500 family trait schedules, 1,900 printed genealogies, town histories, and biographies. To gain control of the material, it had developed something akin to a Dewey decimal system for trait classification. The Office, its records spilling through the rooms of its small, crowded quarters, was by far the chief center of its kind in the United States, and a constant stream of investigators from North America and Western Europe came to Cold Spring Harbor to examine its records, techniques of data gathering, and modes of analysis.[17]

Since the late twenties, Charles Davenport had been suggesting that the Office concentrate on human genetics as such, to help raise the field to the same high level of quality as genetics proper. But the research of the Office had of course been carried out as a branch of mainline eugenics, committing all the methodological sins and biases that reform eugenicists like Hogben thought it imperative to eliminate from human genetics. In 1935, the administrators of the Carnegie Institution of Washington, long suspicious of—and at times embarrassed by—its eugenic activities, appointed a blue-ribbon committee of scientists to assess its work. The committee concluded that the thousands of records, along with the elaborate indexing system, concerning family heredity were "unsatisfactory for the study of human genetics." Among the reasons: traits such as personality, character, sense of humor, self-respect, loyalty, holding a grudge, and the like could seldom be measured, or honestly recorded if they were. The committee added that in light of events in Germany human genetics research ought not to be carried out under a eugenics rubric. In 1939, Vannevar Bush, the new president of the Carnegie Institution, persuaded the head of the Eugenics Record Office, Harry Laughlin, who was suffering from severe epileptic attacks, to follow Davenport into retirement, and in 1940 the Office was shut down entirely.[18]

In England in the early thirties, the resources for the research program that Hogben envisioned were not readily to hand. A leading feature of the work at the Galton Laboratory had remained the collection of family pedigree material and its publication in *The Treasury of Human Inheritance*, the first volume of which had in successive parts continued to record in rich detail the familial occurrence of various afflictions, including harelip and cleft palate, deaf-mutism, hemophilia, and disorders of bone development. A second volume had been inaugurated in 1922, when Julia Bell, a longtime stalwart of the Galton Laboratory, began publishing pedigree data on anomalies and diseases of the eye. But though Bell had been aided by a grant from the Medical Research Council, through the postwar decade the Galton Laboratory had insufficient funds to pay for research workers to analyze even its own valuable material, let alone to pursue the study of human genetics at a significant level. Hogben was similarly strapped at the London School of Economics, where he continued to hold the professorship of social biology that William Beveridge, the head of the school, had created with a grant from the Rockefeller Foundation for the purpose of closing the gap between the social and the life sciences by fostering studies in the biological, including the genetic, basis of human behavior. Elsewhere in Britain, especially at Cambridge and Edinburgh, reform eugenicists chafed to take advantage of the promising new methods available for human genetics yet lacked the means to do so. In July 1931, a large group of them, including Haldane, Huxley, and Hogben, along with a diverse contingent of equally stellar figures in British genetics, medicine, and psychology, convened in London and urged that something be done to eliminate the obstacles holding up progress in the field.[19]

A few months earlier, Hogben, taking matters into his own hands, had turned to the head of the Medical Research Council, Sir Walter Morley Fletcher. Fletcher, who the year before had committed the council to help sponsor Penrose's work at Colchester on the role of heredity in mental deficiency, declared at the London meeting what he evidently had already told Hogben—that the council would take an interest in whatever aspects of human genetics might be of medical importance. Early in 1932, Fletcher established a council Committee on Human Genetics under Haldane's chairmanship that included Fisher, Hogben, Penrose, and Julia Bell, and was charged with providing the council with expert advice upon the directions in which it could profitably extend work in the field.[20]

The directions were provided, the advice followed, and in the thirties the Medical Research Council came to foster much of the research program in human genetics that seemed so opportune to Hogben, Fisher, Haldane, and their reform-eugenic colleagues. The most major undertaking was a survey for a period of years of hospital patients—the total was expected to reach 500,000—to determine the incidence of various diseases in the British population and, by investigating the patients' family histories, whether the diseases were hereditary in origin. A special purpose of the survey was the identification of diseases that might be the products of homozygous recessive genes, and special pains were thus taken to measure the incidence of consanguineous parentage. The Medical Research Council Committee on Human Genetics also lent its name and advice to the Bureau of Human Heredity, established in London in 1936, to the applause of both *The Times* and *Nature,* for the purpose of providing a clearinghouse of hereditary information, a repository of pedigree data, and a center for inquiries. (Its director soon announced from the bureau's small office in Bloomsbury that inquiries were coming in from all over Britain. "Men and women are at last taking the idea of child-rearing seriously.")[21]

Analysis of the Medical Research Council's hospital survey data was assigned to Julia Bell, who was also given the council's support for a new volume—it was to center on nervous diseases and muscular dystrophies— in *The Treasury of Human Inheritance.* The council supplied a subvention to Hogben's group at the London School of Economics for analyzing the survey data particularly with regard to the genetic basis of amaurotic family idiocy and Friedrich's ataxia and to whether either was genetically linked to any of the blood groups or to PTC sensitivity.[22] But the principal recipient of largesse for blood-group work was Ronald A. Fisher, after he succeeded Karl Pearson in 1933 as the Galton Professor of Eugenics.

In 1919, Fisher had refused an offer from Pearson to join the staff of the Galton Laboratory under terms that would have permitted him to teach and publish only what Pearson approved. During the twenties, Fisher had envied Pearson the facilities of the Galton Laboratory for work in human heredity, while regretting that, despite the fine statistical methods employed there, the laboratory failed to pursue any Mendelian genetics. After his appointment to the Galton chair, Fisher got rid of Pearson's paleontological collections, artifacts, and casts of Paleolithic man. "His chief aim," Pearson privately stormed, "seems to be to cast scorn on his predecessor and all who use any of his methods." Actually, Fisher merely proposed to go beyond Pearson in ways that the study of human heredity required, to create a laboratory of mathematical genetics, with attention given to both words in the phrase. While continuing to advance mathematical statistics, Fisher made Mendelian genetics an intrinsic feature of work at the Galton. Deeming it essential to test hypotheses of human heredity via controlled breeding in man's mammalian counterparts, he brought with him to the Galton seventy cages of mice, the well-bred progeny of his years at the Rothamstead Experimental Station.[23]

In 1930, Fisher had been excited to learn that Charles Todd, a serologist with the Medical Research Council, had developed analytical methods that could detect fine differences in certain biochemical factors in the blood of fowl—fine enough to distinguish between even closely related animals. Fisher suspected that the factors were genetic in origin and that Todd's methods, if applied to the blood of human beings, might lend themselves to linkage analysis for the identification of carriers of recessive genes, including genes for mental deficiency. In 1934, he broached this idea to a representative of the Rockefeller Foundation in Europe. The Foundation promptly awarded the Medical Research Council $35,000 for the Galton Laboratory from 1935 to 1940, enough to increase its annual budget by about forty percent. The Rockefeller report for 1935 explained: "Mental defects are variable and elusive in their manifestations; those factors in the blood which are probably closely linked to mental defects promise a more direct genetic interpretation than is possible from symptoms and many types of measurements."[24]

With the Rockefeller money, Fisher established a serological research unit at the Galton and hired a small corps of able researchers, notably two serologists, the physicians George L. Taylor and Robert R. Race, as well as two women to assist them, Aileen M. Prior and Elizabeth W. Ikin. The linkage work of the staff concentrated on disorders known to be genetic, including the dominant Huntington's chorea and the recessive PKU. Race and his colleagues identified the victims of disease at various hospitals in London, while the PKU work was carried out in cooperation with Penrose at Colchester. They then traced the family members to take their blood groups, test them for PTC sensitivity, and record whatever other genetic characteristics—eye color, for example—they might express. In due course, a series of papers started coming from the Galton Laboratory under the names of Ikin, Prior, Race, and Taylor that took their place as standards in the field of blood-group genetics.[25]

Haldane pursued his own small research program in linkage with the assistance of Julia Bell at University College London, where he was appointed to the professorship of genetics in 1933, then to the new Weldon Professorship of Biometry in 1936. He focused on male sex-linked characters, since they were manifestly carried on the same chromosome. ("I am a fundamentally lazy man, and like to see definite results when I do make an effort," Haldane later said of this work.) According to a report from Bell in June 1936, her effort produced what Haldane described as "a 'sensational' pedigree showing linkage of Haemophilia and color-blindness." Bell added, "It is really very exciting." Haldane soon demonstrated that the likelihood was remote that Bell's observations had arisen by chance, thus confirming that she had achieved the first certain pedigree demonstration of linkage in human beings.[26]

Fisher's group failed to find any linkages between, on the one side,

Todd's serological factors, the blood groups, PTC sensitivity, or any other universal character and, on the other, any type of genetic disease or disorder. Hogben's team was similarly unsuccessful, and so was every laboratory in the United States and Britain that in the thirties and forties attempted the task. Almost as disappointing was the search for consanguinity, that indicator of recessive disorders, among the parents of British hospital patients. The Medical Research Council reported at the end of the thirties that the incidence of first-cousin consanguinity for some 100,000 cases seemed to be low—no more than about six tenths of a percent.[27] Nevertheless, the reform-eugenic research program of the thirties and forties yielded a great deal of fundamental importance for the science of human genetics. It produced powerful mathematical methods of linkage analysis and, for the most part, a large amount of reliable data (an exception was the material accumulated by the Bureau of Human Heredity in London, which Penrose eventually declared "quite inadequate for any scientific purpose"). It was useful to human genetic analysis to have what the Medical Research Council survey provided—measures of the national incidence of consanguinity as well as of the incidence of numerous rare diseases. It was also advantageous, Haldane noted, to know that single recessive genes did not seem to account for goiter, pyloric stenosis, harelip, or spina bifida, among other maladies.[28]

Perhaps most significant were the data gathered on blood groups, accumulated from samples of thousands of people in the United States and Britain during the thirties, then enlarged enormously during the Second World War to meet the demands of the armed services. Early in the war, Fisher's serological unit was itself moved from the Galton Laboratory to the relative safety of Cambridge University; there it was made part of the British blood transfusion service, which depended upon blood-group analysis to match the blood of donor and recipient. Blood-group research also led to the medically important identification in the United States between 1939 and 1941—by Karl Landsteiner and Alexander S. Wiener on the one hand, and by Philip Levine with collaborators on the other—of the Rh factor and its role in the hemolytic disease of newborn infants, and subsequently to the untangling by Fisher and others of Rh-factor genetics.[29]

Knowledge of blood-group genetics found increasing legal use in establishing the possibility—or impossibility—of paternity. A group M parent could not produce a group N child, for example, nor could a group M child be born of a group N parent. Laurence Snyder argued that, in the class of MN blood groups, at least half the men accused of paternal responsibility by the mother were very likely not guilty. Blood tests could also help decide the rare case of disputed maternity, as when a woman sued her husband for support of a child who she claimed was theirs but who, in fact, was not even hers (she had obtained the infant from an orphanage). Blood tests were first used in the British courts in 1932, but, contrary to the practice in many Northern European countries, American courts were at first reluctant to admit them in cases of disputed paternity; the law tended to give great weight to the word of the mother. But in the thirties a few states enacted statutes governing the use of the tests in paternity suits, and in 1940 the admissibility of the data was upheld by the U.S. Court of Appeals for the District of Columbia. By the postwar period they were commonplace items of evidence in the courtrooms of Britain and the United States (even though in 1946 the Supreme Court of California, in a widely followed case, held the actor Charlie Chaplin to be the father of a child despite the decree of blood-group genetics that he could not have been).[30]

A different use of blood-group data had been suggested during the First World War, when the Polish serologist Ludwig Hirszfeld and his wife, who were medically assisting the Serbian Army on the Macedonian front, had sampled sixteen different peoples in that polyglot area and demonstrated that the distribution of the four blood groups then known varied from one ethnic population to another. By the forties, increasing refinement in the identification of blood groups—group A, for example, was discovered to segregate into two slightly different forms—permitted drawing detailed serological profiles of distinct populations and determining their degree of intermixture. Blood-group genetics thus joined ethnography, anthropology, and demography as a valuable tool in the study of human history, particularly migrations and mixings, and ultimately of human evolution.[31]

Having reliable data on the incidence of heritable human traits assisted explorations in an area of both evolutionary and medical importance—mutation rates in man. Naturally occurring mutations were familiar to plant and animal geneticists, and after Hermann J. Muller's work in the twenties, so were the artificially induced variety. Haldane drew upon theoretical population genetics, which provided mathematical tools for dealing with mutation rates, to account for the persistence of hemophilia in human populations. The persistence was puzzling because, though hemophiliacs tended to die before fathering children—their marriage rate was only about a quarter of that of the general population—the incidence of hemophilia did not seem to diminish with time but appeared to remain relatively constant. The explanation that contemporary hemophiliacs were the residue of a historically much larger number that had steadily diminished in time led to the absurd conclusion that one thousand years earlier the entire population of Britain must have been hemophiliac. Haldane proposed, instead, that hemophiliacs who died without reproducing were constantly replaced by people who were made fresh carriers of the disorder by mutation, and in

1930 he calculated that, indeed, mutation accounted for as much as one-third of the hemophiliacs in each generation.[32]

Lionel Penrose followed a similar line of reasoning with epiloia, a disease from which some of his Colchester patients suffered. Its symptoms were idiocy, epilepsy, and tumor formation on various organs, and its victims did not ordinarily reproduce. Since the disease did not thus seem likely to arise principally from hereditary transmission, Penrose suspected that the genes for it must come from mutation, and he estimated the mutation rate as comparable to that for hemophilia. Haldane's was the first estimate of a sex-linked mutation in man; Penrose's, the first for the autosomal—that is, non-sex-linked—type.[33]

IN ITS EARLY YEARS, from 1930 to 1945, human genetics was, as new scientific disciplines tend to be, populated by a small band of pioneering enthusiasts, entrepreneurs, and evangels. Many were trained in genetics proper and then studied human biology, some even taking medical degrees, while others were physicians who picked up genetics one way or another. Medical practitioners on the whole remained skeptical of or indifferent to human genetics, but physicians formed perhaps a third of the leadership in what by the late forties was an emerging Anglo-American community of human geneticists. The community was small in size—fewer than two hundred people published any research in the field at all, while fewer than fifty published more than once—yet it had obtained footholds in a number of institutions of learning.[34]

In most scientific fields, a comparatively small fraction of people account for a disproportionately large fraction of progress, and so it was in human genetics. From 1930 to the end of the Second World War, about a quarter of the human geneticists in the United States and Britain produced more than sixty percent of the published papers, and about a tenth—a cadre of leaders totaling fewer than twenty people—were responsible for some forty percent, including the corpus of fundamental work that established the methodological foundations of the field. The large majority of this most productive tenth were British. Frederick Osborn lamented in 1940 that "the United States, which leads all other countries in most of the sciences, has lagged far behind in the study of human heredity," and for at least a decade after the Second World War the center of gravity of the nascent discipline rested on the eastern side of the Atlantic. J. B. S. Haldane was wont to say, with pardonable inaccuracy, that only about a half-dozen people in the world knew anything about human genetics, and, with one exception—Gunnar Dahlberg, a Swede—all were English.[35]

Relatively few American geneticists turned to work in the field probably because the techniques and skills of plant and animal genetics, in which most were trained, did not readily transfer to human genetics, with its reliance upon medical knowledge and clinical surveys and with its special mathematical methods. Indeed, in the United States, plant and animal geneticists tended to discourage prospective colleagues from having anything to do with human genetics, reminding them that it was associated with the racism, sterilizations, and scientific poppycock of mainline eugenics. Arthur Steinberg, who defied the obstacles and ultimately became one of the leading human geneticists in the United States, recalled having been warned that it was just too difficult to get the necessary reliable information on human heredity. "The records are poor; classification is poor. . . . Let's work with experimental organisms. The only thing you can do with human genetics is develop prejudices. And anyone who went into human genetics was immediately classified as a person of prejudice."[36]

To contest the prejudice scientifically in the United States, with its legacy of a racist eugenics, was to take up research in intelligence or fertility, subjects which fell primarily to psychologists and demographers rather than to geneticists. An exception was the genetics of I.Q., and one of the few Americans among the human-genetics leadership was the reform eugenicist Horatio Hackett Newman, the University of Chicago biologist whose work dealt with how twins raised in different environments performed on intelligence tests.[37] Cyril Burt to the contrary notwithstanding, there were hardly enough such twins—Newman had after all managed to find only nineteen pairs—to go around in either the United States or Britain. Of course, many people in both countries suffered from various types of mental disabilities. But while in Britain the 1929 report of the Joint Committee had invested the issue of mental deficiency with a sense of urgency, in the United States the "menace of the feebleminded" had been dissolved as a public issue. There was no felt need in the United States for any major program of research in the area, particularly not for one that paid special attention to the relative roles of genes and environment in the production of mental deficiency or disease.

While American authorities continued to recognize heredity as a causal factor in some forms of mental deficiency, discourse in the field seems to have swung strongly in an environmentalist direction. No doubt the trend reflected a reaction against Henry H. Goddard's extravagant claims of the dominance of nature over nurture, yet it probably expressed a good deal more—something particular to American scientific and medical culture of the era—and to account for it would require a book on why, among other things, Freudian psychiatry, with its pronounced emphasis on psychic nurture, took strong hold in the United States, while British psychiatry came to rest more on such considerations of nature as neurophysiology. It is perhaps significant that two important exceptions to the trend were both European imports. One was George A. Jervis, an M.D. and Ph.D. in psychology from the University of Milan who became director, in 1933, of the New York State Department of Mental Hygiene and who through a series of deft biochemical experiments helped to show that PKU resulted from the body's inability to metabolize phenylalanine. The other was the psychiatrist Franz J. Kallmann, who came from Germany in 1936 to the staff of the New York State Psychiatric Institute of Columbia University and brought with him the conviction that severe forms of mental disease revealed a constitutional predisposition to the disorder that could be hereditary. He held that a dominant gene predisposed people to manic depression, a recessive one in its homozygous state to schizophrenia. Kallmann detected much more of a hereditary pattern in mental disease than his American contemporaries were willing either to accept—and rightly so, since Penrose at the time and others later judged Kallmann's work unconvincing—or to pursue.[38]

In Britain, not least because of the concern stimulated by the Joint Committee report in 1929, studies in the genetics of mental deficiency flourished. Physicians who dealt with the heredity of mental disorder were compelled to turn to genetics; geneticists who confronted the problem, to medicine. Either way, work in the field naturally opened out to human genetics in general. Lionel Penrose's career in human genetics owed its origins, of course, in no small part to his research in mental deficiency, and so did the career of John Fraser Roberts, ultimately another British leader in human genetics.

The product of a prosperous North Wales farming family who started his scientific life as a sheep geneticist, Roberts was drawn into human genetics first during his postgraduate studies at the University of Edinburgh by F. A. E. Crew, then by Ronald A. Fisher, who generously supplied repeated help with statistics. Like Fisher, whose disciple he became, Roberts was a political and religious conservative—raised a Methodist, he developed strong Roman Catholic inclinations and compromised on the Church of England—and he inclined to a reform eugenics compatible with his scientific knowledge and social temperament. He embarked on work in the genetics of mental deficiency when in 1933 he was appointed principal investigator at the Stoke Park Colony for the Mentally Defective in Bristol. The funds for his post came from the Burden Mental Research Trust, a philanthropy recently endowed with ten thousand pounds for the investigation of mental diseases and disorders. A strong hand in the decision as to how the money should be spent had been taken by E. O. Lewis, a member of the advisory committee to the Trust, who argued successfully that it should be used for research complementary to that underway by

Penrose at Colchester. Roberts thus began to analyze the diverse factors that shaped the mental qualities of an apparently normal school population—he studied 3,400 children in Bath selected to represent a cross-section of I.Q. scores from the highest to the lowest—and of the institutionalized population at Stoke Park. The more Roberts got into the work, the more interested he became in human genetics as such. In the mid-thirties, he took a medical degree, not with the aim of practicing medicine but to deepen his knowledge of the human organism. "I was purely interested in human genetics," he said many years later, "so I did the absolute minimum and crawled through. I was very lucky in my final medical oral, and after that day I put my stethoscope away." While in medical school, Roberts discovered how ignorant his fellow students were of the laws of heredity, so he soon wrote *An Introduction to Medical Genetics*, an influential text when it was first published in 1940 and today, in its seventh edition, still a standard.[39]

Adding to the British edge was the support given human genetics in general from the early thirties by the Medical Research Council. In the United States by contrast, the federal government had not yet begun the munificent funding of basic science that would characterize research activities after the Second World War. The National Institutes of Health, just getting started in the thirties, did not award grants for research to universities, and it did not sponsor research anywhere on human genetics. Americans eager to pursue such research were thrown back upon the resources of state universities, which at the time were willing to invest little in the field, or upon what between the wars was the principal patron of the science in the United States, the Rockefeller Foundation, which was something of a private precursor to the National Science Foundation and National Institutes of Health combined.

The Rockefeller philanthropic interest in eugenics, dating back to before the First World War, had continued, albeit sporadically, into the late twenties, when the Foundation began to support the research of Professor C. R. Stockard, of Cornell Medical College, in "eugenics and heredity." In the early thirties, doing its part to deal with a world seemingly going out of control, the Rockefeller officers ventured a programmatic departure: to sponsor scientific research—medical, biochemical, biophysical, and psychological—in the analysis of human behavior. In the medical section of the Foundation, the mandate was interpreted to allow for the funding of investigations in the heredity of mental disease. Rockefeller monies went to efforts in human genetics at research facilities in Europe, including, of course, Penrose's at Colchester and Fisher's at the Galton Laboratory, but not at any in the United States, no doubt reflecting the fact that in America the genetics of mental deficiency now commanded little interest. The interpretation given the program in the natural sciences section of the Foundation was expressed in the annual report for 1935: "It is clear . . . that the human race needs, and needs desperately, a fuller and more useful knowledge of human genetics, and yet it is equally obvious that genetics, at least for many years to come, must base its progress upon experimentation with lower forms of life."[40]

In the United States before the Second World War, Laurence Snyder had to do his human genetics on a shoestring. His interest in human genetics had originated during two undergraduate summers that one of his professors at Rutgers University had arranged for him to spend during the First World War at Cold Spring Harbor. Snyder imbibed enough of Davenport's ideas to confess many years later that he did "grow up in the eugenics shadow, so to speak, and had to find my way out of it." Still, Snyder was never in such thrall to Davenport's eugenics as to ignore the key advances in the methods and knowledge of human heredity. At Harvard in the mid-twenties, he devoted his doctoral research to blood-group genetics, mastering Felix Bernstein's mathematical genetics, especially his use of the Hardy-Weinberg law of gene frequencies. Continuing the work in the late twenties while at North Carolina State College, he roamed the region taking blood samples on Cherokee Indian reservations and at the large family reunions common in the mountains. Years later Snyder laughingly explained that, unlike experimental animals, people "take themselves home at night, put themselves to bed, and you can conduct intelligible conversations with them," adding that his human genetics did not require "a lot of upkeep and money." With this low-budget research, Snyder added to the confirmation of Felix Bernstein's theory of the genetics of the ABO blood groups. He also used the blood-group data from Cherokees, whites, and the progeny of their intermarriages to construct a quantitative index of the degree of biological intermixture that had occurred between the two communities.[41]

In 1930, Snyder was appointed to the faculty of Ohio State University to build the genetics program, and in 1932 he was made professor of medical genetics—it was probably the first such designation in the United States—in the medical school. Snyder recruited some of his doctoral students into human genetics and onto the faculty, notably David C. Rife, a specialist in twin studies, and Charles W. Cotterman, a highly original—albeit highly eccentric—student of mathematical genetics who did brilliant work yet declined to publish most of it. Still, Cotterman kept Snyder on his methodological toes and helped him work out the mathematics for demonstrating the recessive nature of the PTC-tasting trait. The Ohio State enterprise commanded most of the subjects—blood groups, family surveys, mathematical methods, twin studies—central at the time to pioneering human genetic research. Both Snyder and Rife produced enough to earn places in the discipline's Anglo-American leadership, and Ohio State glowed with singular prominence on the American horizon of the field.[42] Yet Snyder found his efforts to build human genetics into an activity of permanent distinction repeatedly stymied by all the forces that adversely confronted the discipline in the United States.

A son of medical missionaries, Snyder was at pains to proselytize for human genetics in the medical school, but apart from a few of the physicians, the faculty there, Snyder remembered, treated him with ridicule. "I was asked publicly to explain the gene for a stomach, and to give an opinion on whether the gene for the heart was dominant or recessive." Some of the doctors insisted that the growing number of diseases that could be successfully treated must not have a genetic component. Snyder remembered, too, that difficulties arose from the identification of human genetics with eugenics, especially the Nazi variety. At Ohio State, he was unable to obtain the financial support necessary to enlarge his research group, or even enough to keep it together. He appealed for aid to the Carnegie and Rockefeller philanthropies and was turned down. In 1934 he was appointed chairman of a committee of the National Academy of Sciences–National Research Council to foster human genetics in America, but the committee was no more successful at the task than was Snyder by himself. In 1947, discouraged, Snyder left Ohio State and research in human genetics for a deanship at the University of Oklahoma.[43]

Good as it was, the Ohio State group at its best never matched the scientific power of the British school, particularly its masterful forging of the mathematical methods that were essential to the development of human genetics during this period. In the thirties, almost three-quarters of the British leadership in the discipline worked at or were affiliated with the Galton Laboratory, which meant with Fisher and Haldane. The two men disliked each other's polar-opposite politics; according to some, they also disliked each other personally. But Haldane not only had supported Fisher for appointment to the Galton Professorship but had told the selection committee that Fisher was the only possible candidate for the post. Both found common ground in reform eugenics and greatly respected each other's considerable scientific talents. At the Galton, they were energetically interactive—Haldane often joined Fisher's afternoon staff tea—in developing the special mathematical methods that human genetics required and in pooling Fisher's strength in mathematical rigor with Haldane's vast biological and physiological knowledge.[44]

From 1930 to 1945, Fisher and Haldane were the most productive pair in human genetics on either side of the Atlantic. Much of their work appeared in the *Annals of Eugenics,* a quarterly journal started by Karl

Pearson in 1926, control of which Fisher acquired when he became Galton Professor. Fisher changed the subtitle of the *Annals*—under Pearson it had been a journal "for the scientific study of racial problems"—to a journal "devoted to the genetic study of human populations," and, assisted by a subvention from the Eugenics Society, he published a wide range of articles dealing with various aspects of the subject in a predominantly mathematical fashion. Between 1930 and 1945, the largest cluster of human genetic analysis —indeed, some forty percent of the work published in Britain and the United States combined—saw the printed light of day in the *Annals*.[45] Fisher and Haldane provided intellectual guidance to the Galton staff as well as to visitors who came to work at the laboratory and, more important, through the journal they set a standard of first-class research in human genetics for scientists elsewhere to emulate.

Chapter XIV

APOGEE OF THE ENGLISH SCHOOL

B Y THE MID-FORTIES, human genetics increasingly depended upon a variety of disciplines, not only mathematical statistics and genetics proper but psychology, demography, physiology, biochemistry, and medicine. No one person in either the United States or Britain commanded such a range of specialties, not even polymaths like Haldane. But in Britain, experts in one scientific area could with relative ease obtain help from those in another. Almost all of the leading practitioners were located in the environs of Greater London or less than an hour or two away by train, and the concentration had long made for advantageous cross-disciplinary reinforcement among British human geneticists, especially among Haldane, Fisher, and Hogben. In the United States, by contrast, the work of human geneticists had suffered in the early years of the discipline from the vastness of the country, from the absence of a dominant scientific center. The pioneering Americans in the field had been located at different institutions, each of them as geographically distant from each other as Ohio State was from, say, Chicago, where Horatio Newman did his twin studies, or the New York area, where Landsteiner and Levine pursued their blood-group work.[1]

After the war, British human genetics remained advantageously centered on Greater London. Robert Race returned from Cambridge to head a new Medical Research Council unit on blood-group genetics in the old Lister Institute, a grotty building by the Chelsea Bridge. He was joined there in the late forties by Ruth Sanger, eventually his wife, who, having got interested in Rh-factor phenomena during wartime duty in blood transfusion, had come from Australia to work with him. In due course, with the aid of only a few technical staff, they forged the scientific collaboration that in the postwar era made them preeminent authorities

in blood-group genetics and their *Blood Groups in Man* in its successive editions the standard reference on the subject. Race maintained his
close relationship with Fisher, who even though he had left the Galton to
take up the professorship of genetics at Cambridge University, came to
visit them often in London, staying at their home, discoursing about
blood groups in the kitchen while Sanger, trying to prepare dinner, worried that Fisher, his vision as impaired as ever, would knock his pipe ashes
into the butter.[2]

The striking wartime progress in the understanding of Rh-factor
disease established blood-group genetics for a time as a glamorous field.
(Sylvia Lawler, who as a young physician went to work with Race after
the war, recalled that you had to have a Ph.D. even to handle the precious anti-Rhesus serum.) Numerous visitors made their way to the Lister,
and Race and Sanger were in touch with physicians and geneticists throughout Great Britain, but Sanger remembered that their principal locus of interaction was the community of human geneticists in the London region.
Members of the community kept in touch frequently via telephone, pub
chats, visits to each other's laboratories and homes. In central London itself, the postwar community now included John Fraser Roberts, who had
moved to the London School of Hygiene and was devoting some of his
effort to blood-group work; also J. B. S. Haldane, still at University College London, who applied some of his theoretical power to the puzzle of
why Rhesus hemolytic disease should have had a selective survival value
in human evolution.[3] And it was particularly enriched by Lionel Penrose
after 1945, when he returned from his wartime stay in Canada to succeed
Fisher as Galton Professor of Eugenics at University College.

HALDANE HAD ARRANGED THE matter. ("I think that you and I are the British
people under 60 who have contributed most to human genetics, and therefore one of us should have the chair. As you have specialized on man and
I have not, your claim is somewhat greater.")[4] While Haldane was a brilliant theorist, Penrose, by now a world authority in the genetics of mental
deficiency, was also a clinician, not only medically qualified but well versed
in psychology as well as psychiatry, a scientist who thrived on direct
contact with his human subjects. While at Colchester in the thirties, where
he felt somewhat isolated, Penrose had drawn considerably upon the cluster
of expertise centered on London, especially the biochemical knowledge of
Haldane and the statistical of Fisher and Hogben. Penrose was acutely
sensitive to the importance of avoiding the epistemological pitfalls that had
so distorted earlier work in human heredity. Neither a master biochemist
nor a statistician, he was nevertheless clever, clever enough to invent his

own ingenious methods of overcoming ascertainment bias and for performing biochemical assays. The more Penrose branched out into human genetics, the more he came to personify a richly multidisciplinary orientation—statistical, biochemical, medical, and genetic—to the study of human heredity.

Although unconcerned with the development of mathematical statistics for its own sake, Penrose early appointed to the Galton staff Cedric A. B. Smith, an able statistician from Cambridge University (who piqued Penrose's interest during his job interview with the revelation that he was a Quaker convert and had spent the war on hospital duty). But while maintaining the Galton's biometric tradition, Penrose shifted the emphasis of the laboratory in a medical and biological direction, establishing ties with hospitals, medical schools—especially the University College Hospital complex just across Gower Street—and mental institutions, which supplied data on the diverse physiological characteristics and afflictions found among their patients.[5] He also reached out to the overall University College London Department of Biometry, Genetics, and Eugenics, of which the Galton was a part and which was headed by Haldane, who continued to hold the Weldon Professorship of Biometry.[6]

Penrose's wife, Margaret, had known Haldane since her girlhood, when their fathers were both fellows of New College, Oxford. Haldane was one of the few scientists in the world who enjoyed Penrose's unreserved admiration, and Haldane repaid the compliment; the two were warm friends. At the Galton, even more than in Fisher's day, Haldane played the role of theoretical gadfly, goad, and collaborator to the laboratory staff. He suffered neither fools nor shoddy work. He was mercurial to the point of explosiveness, and sometimes brutally tactless, once telling a staff member who had just completed the manuscript of a textbook on human genetics that the publication of the book would be "harmful to yourself, to the science of genetics, and to the department of which you are a member." C. A. B. Smith, who liked and respected Haldane, came to consider it a blessing that Haldane's office was at the south end of University College while his was at the north, because Haldane's temper would tend to abate while he stormed across the distance between the two.[7]

Yet Haldane was on the whole generous with accolades, even though the Galton staff used to say that you could get more praise from him if you were his enemy than if you were his friend. University College people would gather in what is now called the Haldane room—it was then termed the "mixed" common room, because both men and women were permitted entry—to listen to Haldane, sprawled in an easy chair, discourse on science, politics, or anything else people wanted to argue about. He lit up the Galton with the force of his awesome intelligence and the surprise of his irrepressible wit. Penrose would tell people that his own lectures should be billed: "Text by Lionel Penrose, jokes by J. B. S. Haldane." Some of Haldane's genetic theorizing proved to be wrong because he often relied on other people's data. Although Haldane fell away from the Communist Party, he insisted upon maintaining an open mind about whether some of T. D. Lysenko's ideas might be correct and suggested that support for the possibility might be found in aspects of recent research in biochemical genetics. Still, he declared himself unconvinced by Lysenko's sweeping contention —that environmental modifications of organisms were genetically transmissible. Haldane supplied Penrose's people with a brilliant command of genetic theory, especially in its mathematical formulation, with illuminating hypotheses as to how widely disparate phenomena might fit together —and with the force of his long-standing belief, dating at least from the discovery of phenylketonuria, in the essential importance to human genetics of biochemistry.[8]

HALDANE HAD RECOGNIZED phenylketonuria as another in the class of biochemical abnormalities to which Archibald Garrod had drawn attention early in the century. Garrod, an eminent British physician, brought to medical research a combination of skills and insight rare for his day—not only considerable clinical powers but also wide knowledge in biology and biochemistry. To him, the physician who would cure must first understand. That cast of mind informed Garrod's fundamental work on alcaptonuria, done in London at the turn of the century mainly at the Hospital for Sick Children in Great Ormond Street. Signaled by the blackening of an infant's urine soon after birth, the disease was harmless to the young but, as the years passed, produced a blackening of the cartilages along with a tendency to certain arthritic lesions. Garrod not only demonstrated that the condition was attributable to a recessive Mendelian character but joined the Mendelian hypothesis to what was known—partly as a result of his own labors in the laboratory—about the biochemistry of the condition.[9]

What blackened the urine of alcaptonurics was homogentisic acid, an intermediate product of the body's metabolism. Metabolic processes could be likened to biochemical pathways along which proteins, fats, carbohydrates, and the like were changed into successive intermediate products. At each step, an assist was given by an enzyme—an organic catalyst essential to the biochemical transformation in which it was involved. In normal protein metabolism, homogentisic acid was oxidized and the process moved to the next transformational step on the pathway. Garrod argued that in alcaptonurics the normal metabolic pathway was blocked, leaving the homogentisic acid intact to be excreted in the urine. The reason that the metabolic process halted was the lack of an enzyme necessary to catalyze the normal oxidation, and this enzyme deficiency, Garrod speculated, resulted from the homozygous expression of a recessive Mendelian character.[10]

In succeeding years, Garrod pondered the biochemical and genetic evidence of albinism, cystinuria, and other conditions associated with abnormal metabolites and concluded that all derived from blocked metabolic pathways—that is, from "inborn errors of metabolism," to use the title phrase of the classic book that he published on the subject in 1909. He summarized the general idea in 1923, in the second edition: "If any one step in the process fail the intermediate product in being at the point of arrest will escape further change, just as when the film of a biograph is brought to a standstill the moving figures are left with foot in air."[11]

Haldane appreciated Garrod's theory as one of the great speculative insights in the history of biochemical genetics, and he was prone to follow William Bateson's dictum: "Treasure your exceptions," especially such scientifically suggestive exceptions as Garrod's rare inborn errors. At the end of the twenties, Haldane was godfather to the renewal of a research program on the genetics of plant colors that William Bateson had helped foster in the first decade of the century but that had been dormant for twenty years. The revived effort involved collaboration between scientists at the John Innes Horticultural Institute, where Haldane was a consultant, and at the biochemical laboratory of Frederick Gowland Hopkins at Cambridge University, where he was then on the faculty. The work proceeded by breaking down the plant pigments into their different biochemical constituents and, through experimental breeding, locating the sources of the constituents in different genes.[12]

Outside the small band of scientists around Haldane, the genetic significance of Garrod's ideas went largely unrecognized far into the interwar period. Biochemists appreciated Garrod for his work in metabolism but had little interest in heredity. Physicians paid little attention to the medical conditions arising from Garrod's inborn errors because they were thought to be rare and, consequently, unimportant diseases. Most geneticists apparently knew nothing about Garrod, not least because, one suspects, they were disinclined to take seriously theories of heredity concerning human disorders. Besides, neither physicians nor geneticists knew much biochemistry.[13]

Garrod's work became known to the Americans George W. Beadle and Edward L. Tatum at Stanford University about the end of the thirties, shortly after they embarked on the course of research in biochemical genetics that would lead to the Nobel Prize. Beadle had come to the subject via fruit-fly investigations at the California Institute of Technology and then in Paris with the European geneticist Boris Ephrussi; their collaboration had led to the hypothesis that genes somehow shaped the biochemical pathways which produced the insect's different eye colors. At Stanford in 1940, Beadle, the geneticist, and Tatum, a young biochemist, began to pursue the hypothesis with *Neurospora*, an ordinary bread mold, which reproduced rapidly and about which a good deal was known. Garrod had taken metabolic variations found among human beings and searched for genetic differences; Beadle and Tatum triggered genetic mutations in the mold with X rays and analyzed the resultant metabolic variations. They found that, with a specific gene bred into it, the mold could metabolize a given substance, while with the gene bred out, it could not—in short, that the absence of the gene forced the mold into a metabolic error. On receipt of the Nobel Prize in 1958, Beadle would declare that he and Tatum had only "rediscovered what Garrod had seen so clearly," adding, "We were working with a more favorable organism [than man] and were able to produce, almost at will, inborn errors of metabolism."[14]

Beadle spelled out the striking import of the rediscovery when, in 1945, he reviewed the general implications of recent work in biochemical genetics: ". . . that to every gene it is possible to assign one primary action and that, conversely, every enzymatically controlled chemical transformation is under the immediate supervision of one gene, and in general only one." In 1948 that idea was distilled down to an apothegm—the "one gene–one enzyme hypothesis." A powerful guide for research, the phrase added force to Haldane's assertion in *New Paths in Genetics*—the book had called attention to the work of Beadle and Tatum as well as Garrod—that henceforth the "geneticist cannot possibly neglect biochemistry."[15]

Among the human geneticists who paid a lot of attention to Haldane was Harry Harris, who would eventually succeed Penrose in the Galton chair. Harris is endlessly amused by the vagaries of chance in life, including the chain of chances that led him to a career in biochemical genetics. He comes from a family of Eastern European Jewish immigrants who worked in the needle trades in Manchester, and he earned a medical degree at Cambridge, with vague ambitions of going into psychiatric research. During the war, a stint of house duty at Sunderland, a mental hospital on the north coast of England, convinced him that he could contribute little if anything significant to the understanding of mental illness, but while there he became fascinated with the news then appearing in medical journals about the hereditary dynamics of the Rh factor. He studied Haldane's *New Paths in Genetics*—"a beautiful book," he reflected later—fascinated by its contents, which he little understood, and drawn by its author's left-wing politics, which he knew a good deal about and which his own resembled. Soon Harris inaugurated a modest genetics research project on premature baldness, which he thought might be hereditary because it had happened to his father and to all his uncles. Harris submitted a paper on the subject to *Annals of Eugenics,* which he had discovered one day in London on a visit to the library of the British Medical Association, and, much to his delighted surprise, learned in what he remembered as a "sweet note" from Lionel Penrose that it would be published.[16]

Shortly afterward, Harris, activated as a medical officer in the Royal Air Force, was sent to Burma, where he continued his postdoctoral self-education from, among other sources, a copy of Ronald A. Fisher's *Genetical Theory of Natural Selection* and Lionel Penrose's Galton inaugural lecture: "Phenylketonuria: A Problem in Eugenics," which was published in *The Lancet.* Here was a type of disease, Penrose said, which suggested that biochemistry surely had "a great contribution to make towards the understanding of human inheritance." Harris was fired by the biochemical theme but, more important, by Penrose's sophistication in dealing with disparate human genetic issues. In 1946, back in England attending sick quarters for the Air Force at a base near London, Harris dropped in at the Galton to meet Penrose, who spent three hours talking with him and encouraged him to come to work in the laboratory.[17] Of course, there was no job available, but perhaps Harris could obtain a fellowship from somewhere. Harris managed to garner a stipend from the Royal College of Physicians to work on diabetes. Penrose thought it a useful subject to study genetically, but he also stressed to Harris that he could explore other things, too.

For Harris the autodidact, curious, imaginative, and resourceful, the Galton was an excellent place to be. Although concentrating on the diabetes work, he pursued the various approaches to human genetics practiced at the laboratory, including group surveys, individual family analyses, and statistical assessments. He absorbed Penrose's eagerness to fasten on problems that could be rendered objective—clinically, biochemically, or otherwise—and quantified. He started to search for new sharply defined characters that obeyed Mendel's laws. He collaborated with Hans Kalmus, a Viennese refugee on the Galton staff, in investigating PTC taste sensitivity, finding Mendelian patterns of responsiveness to other substances with the same chemical grouping.[18] Then Harris met Charles Dent, a physician across the street at University College Hospital who knew about the recently developed method of paper chromatography for the separation and identification of biochemical compounds.

The method started with drying the sample of compounds—a common hair dryer would do the trick—on an area near one end of a strip of filter paper. This end would be placed in a small cup filled with a solvent and set at the top of a container, while the other end of the paper would be permitted to hang down toward the bottom. Gradually, the solution from the cup would diffuse along the paper. The diverse compounds in the sample would migrate along with the solvent, but at different rates, and thus they would spread apart from each other. Once the diffusion ceased, the paper would be sprayed with a reagent, then baked dry in an oven. The dry paper would be freckled with a series of spots, each arising from the presence of the specific biochemical compound that had reached that point, and the analysis of the sample would depend upon the separation of the spots. The separation could be increased by doing paper chromatography in two dimensions—that is, by turning the paper ninety degrees and repeating the process with new diffusing chemicals. The compounds could also be qualitatively identified simply by comparing the positions of the spots with those resulting from a sample with known constituents.

Charles Dent, a chemist before turning to medicine, had recognized the value of paper chromatography and after the war had begun to use it to assay the amino acids in urine. During Dent's first efforts with the urines of seemingly normal people, he detected in some of them a spot that did not seem to be characteristic of any of the twenty amino acids that are the building blocks of proteins. (The spot happened to occur in the urine of a colleague, Robert Trotter, so for a long time Dent called it the "T-spot.") Harris, having heard about what Dent was doing, persuaded Dent to teach him paper chromatography. Harris was to search normal urines for the T-spot, then attempt through family studies to determine whether the odd amino-acid excretion signified a genetic condition. Only some progress was made along these lines—although the T-spot was biochemically identified, the reason for its appearance in certain urines remains in doubt—but in short order Dent invited Harris to pursue a similar research program with the urines of his patients who suffered from cystinuria.

Cystinuria is marked by the excretion in the urine of large quantities —much more than the forty to eighty milligrams a day that healthy people excrete—of the amino acid cystine, often in the form of stones. Garrod had suspected that the condition arose from another inborn error of metabolism, but neither the biochemical nor the genetic evidence for such a theory was as clear-cut as for alcaptonuria. Particularly confusing was the presence in the urine of a spectrum of amino acids that varied inconsistently from one cystinuric victim to another. Harris and colleagues—at the Galton and at London Hospital Medical College, whose staff he joined in 1953—cleared up a good deal of the confusion through paper chromatography and family analysis. They successfully distinguished between cystinuria and other diseases which also yielded abnormal amounts of cystine in the urine. They also showed that cystinuria occurred in two main forms. One was accompanied by excessive excretion of cystine and the amino acids lysine, ornithine, and arginine; the other by excessive excretion of cystine and lysine.

Harris concluded that the first form was the product of a homozygous condition—that is, it was caused by the presence of two recessive genes, one from each parent. The second form arose from what Harris termed an "incompletely recessive" gene, in its single—or heterozygous—state.[19]

Harris and Dent's finding of two types of cystinuria where only one had been believed to exist stimulated Haldane to recall the state of botany and zoology before Linnaeus' eighteenth-century classification of all living nature into genus and species and to grumble that, except for the blood groups, human genetics was in "a pre-Linnaean stage." But while his work did suggest that there was a good deal more to be learned about human traits, Harris had also established the subject of the aminoacidurias as an important new branch of human biochemical genetics.[20]

IN THE SPRING OF 1945, when Penrose was trying to find a way back to England from Canada, Haldane had noted "how hard it must be to get a passage across unless you are a politician (of one of the acceptable brands), a financier, or a physicist." After the Second World War, human geneticists possessed neither the glamour nor the power of physicists, those emperor scientists who had forged radar and the atomic bomb and won the war. Geneticists nevertheless benefited from the general upsurge in the funding of scientific research, especially by governments. Throughout Penrose's tenure, the Galton was well supported by the Rockefeller Foundation and modestly assisted by the Medical Research Council. The permanent staff, including affiliates like Haldane, was comprised of perhaps eight to ten people (in the Boston physician Park Gerald's recollection of his visitor's impression of the mid-fifties, there seemed to be more geneticists at the Galton than in all of New England). Still, by the standards of post-1945 science, the Galton was neither munificently funded nor heavily staffed. Sylvia Lawler, who moved from the Lister to the Galton, remembered her experimental equipage: a few deep freezes, some pipettes, and a "sort of old microscope that Pasteur would have thrown out." For the most part, people sat at tables and desks working with numbers and papers.[21]

Penrose stretched the available resources to the limit. Positions were funded, usually temporarily, on a catch-as-catch-can basis, with a fellowship here or an assistantship there. A number of the make-do posts were held by women, who of course had been employed in abundance at the Galton since Karl Pearson's day and constituted a relatively cheap supply of trained —often highly trained—scientific labor. Some of the women at the Galton felt themselves unfairly relegated to positions inferior to those held by men, and a few became lastingly bitter about it. But at the time only a small number of permanent career opportunities in the laboratory—or in human genetics, for that matter—were available for anybody, male or female. Around 1950, Harry Harris asked Penrose what his future might be. Penrose replied that he didn't know, that employment was a problem in this business of human genetics. In fact, people at the Galton were waiting to see what happened to Harris.[22] The Galton was a work-hard place, but it was also lively, congenial, and stimulating. Sylvia Lawler recalled that there were considerable compensations for enduring a woman's position at the laboratory, not least the sheer excitement of being there.[23]

Penrose remained much as Ruth Darwin had described him in 1930— not much of an administrator, but a first-class thinker. He brought to the Galton that cast of mind which made no distinction between serious science and scientific play. Unlike Pearson and Fisher, Penrose was a decidedly laissez-faire director. He did not run the laboratory so much as preside over it. "Anyone who managed to get a Ph.D. there had to have a streak of originality," Sylvia Lawler later noted. "There was no spoon feeding. Penrose would take people in, shut them in a room, and let them get on with it." Unlike Haldane, Penrose was not ordinarily generous with praise. At times some unfathomable insecurity led him to disparage or ignore the qualities of colleagues, especially those outside the Galton, and he was no more capable of extending direct human encouragement to the Galton staff than he was to his own children. Still, he usually found time for people with results or problems that interested him.[24] Rarely saying much, he tended to respond to queries with an intuitive judgment of what was likely to be scientifically right or wrong, and when pressed, he could be perplexingly elliptical. However, since Penrose did not explain the probable flaw in a piece of work, people had to figure it out for themselves. In Sylvia Lawler's judgment, the staff were also made to use their heads because the technological opportunities were limited by the lack of sophisticated equipment. When Park Gerald arrived from Boston and discovered he would be unable to pursue laboratory work extensively, he went, he recalled, "into a panic for a few months and then finally managed to settle down," adding, "And because I couldn't do anything else, I started to think. And I had the best thought that I ever had—actually conceived the relationship between the various hemoglobin genes."[25]

If Penrose inculcated anything explicitly, it was the essential importance of quantification. He found in measurement, whether of biochemical excretions or of developed physical characteristics, the best possibility of enlarging the scope of certainty in human genetics. No pure Cartesian rationalism for him. He used to snipe at French scientists: "The reason they get it wrong is that they're so logical." Declining to take anything on pure trust, he always wanted to do his own calculations, in his own way. Still, by example Penrose taught that measurement and mathematics had to be tempered by scientific experience and judgment. Alexander Bearn, an American physician who spent time at the Galton, recalled that he would sometimes show Penrose a set of data to which he had applied some statistical calculations. "He would do funny little scribbles on backs of envelopes and say, 'That's about right,' and as an afterthought, 'If you want to check it, you can always give it to C. A. B. Smith—he is very good at these things and always gets the decimal point right.' "[26]

Smith in fact provided essential statistical aid to the entire staff, performing complicated calculations concerning pedigree analysis that might take weeks. He was indispensable in helping Sylvia Lawler establish, by the use of blood-group markers, two of the first three autosomal linkages found by 1960. In many other subjects, including biochemical, statistical, and clinical genetics, the Galton was a groundbreaker. In the postwar years, enlarging upon his long-standing interest in Down's syndrome, Penrose himself devoted a major part of his own effort to the investigation of fetal malformations, both congenital and hereditary, and in 1949 he published *The Biology of Mental Defect*, a classic work, widely hailed for giving scientific rigor and credibility to the subject, and unrivaled in its successive editions on either side of the English-speaking Atlantic.[27]

When the American human geneticist James V. Neel first visited the Galton in the mid-fifties, he was struck by the fact that the famous laboratory had few experimental facilities and basically consisted of three offices —one of them Penrose's, ten feet square and lined with books. Neel was reminded of a proverb his professor liked to quote: "It's not the size of the cage that determines how sweetly the canary sings." The Galton sang the songs of human genetics with exquisite sweetness and power. Its preeminence rested on neither size nor money; it hinged, rather, on the high quality of its diverse staff, above all Penrose and Haldane, and on what both fostered, particularly an offbeat, skeptical esprit and an incisive style of thought that attracted original men and women and permitted them to thrive. Between 1945 and 1965, when Penrose left the directorship, the Galton was a mecca for aspiring human geneticists from England, the Empire, the United States, and the Continent, and a list of the postgraduate visitors to Gower Street reads like a later Who's Who in the field.[28]

Chapter XV

BLOOD, BIG SCIENCE, AND BIOCHEMISTRY

In 1950, in the United States, a corps of enthusiasts formed the American Society of Human Genetics, and in 1954 they established the *American Journal of Human Genetics*. The meetings of the society were tiny, and it was difficult to get enough good articles to fill an entire issue of the journal —though the editors could usually rely on James Neel, whose work was held in high regard even at the Galton.[1]

Neel first learned about genetics in the early thirties at the College of Wooster, in Ohio, in the last chapter of his first-year biology textbook. "It was . . . not quite a religious conversion," he remembered, "but that was just the most fascinating thing I'd ever read." In 1935 he embarked on work for a Ph.D. at the University of Rochester, concentrating on *Drosophila* genetics under Curt Stern, a recent German refugee and leading fruit-fly geneticist. Increasingly interested in human heredity, Neel sat in on most of the courses a first-year medical student would take. He acquainted himself with the statistical methods necessary for the study of heredity in man, using such writings as Hogben's and Fisher's, and in his last year of graduate work he took a new seminar in human genetics that Stern, at his suggestion, agreed to offer. At the time, Neel recalled, going into human genetics seemed "a pretty lonely gamble." Nevertheless, after three years of temporary positions and further research with fruit flies, he returned to Rochester in 1942 as a second-year medical student, receiving his M.D. in 1944 and remaining there to do his internship and residency.[2]

While pursuing his medical studies, Neel kept in touch with Curt Stern, who was helping to carry out at Rochester part of the research for the Manhattan Project on the biological and health effects of radiation. Most of what was reliably known about the subject derived from work with fruit flies; there was little data concerning mutation rates, either spontaneous or induced, in mammals, especially man. Neel was interested in the subject, and after the end of the war, he argued to local military officials that studies of mutation ought to be carried out in Hiroshima and Nagasaki. In the fall of 1946, now fully qualified as a physician, Neel went to Japan as a member of an official American scientific and medical survey team; he remained in the country to oversee the establishment of the Atomic Bomb Casualty Commission, and, in 1947, he inaugurated a study of the genetic impact of the atomic blasts on the populations of Hiroshima and Nagasaki.[3]

Neel set up headquarters in Hiroshima in a large building near the bay called the *Gai-Sen-Kan*—the "House of Triumph"—where the Japanese Second Army had given combat-bound troops send-off parties. Medical field surveys soon revealed that almost half of the births in Hiroshima and Nagasaki were to parents who either had not been in the city at the time of the bombing or else had been so far from the hypocenter that they had suffered no substantial radiation exposure. Recognition of that fact eased the task of establishing control groups for the two cities (to assess the impact on people exposed to radiation, it was essential to know birth patterns among people unexposed to it). Neel's investigators expected any radiation effects to manifest themselves in congenital defects, stillbirths, abnormal birth weights, sex ratios, and survival rates of live-born infants. All were impure indicators of mutation, since they could result from socioeconomic conditions, but they were the best his team could hope to get under the circumstances. The group gathered extensive background information on the parents to prevent social bias from creeping into their analyses.

Amid the shortages of postwar Japan, women who registered their pregnancies in the fifth lunar month received a food card—an incentive that greatly facilitated Neel's task, since it brought virtually all pregnant women in the two cities within his team's investigative reach. When they registered, they were asked to fill out a duplicate form, one to keep for themselves and complete at the end of the pregnancy—which usually occurred at home with the assistance of a midwife. Neel later wrote: "In Japan the social stigma attached to the birth of a malformed child is rather considerable. Every effort had to be made to develop a program which would not antagonize the mothers of malformed children by exposing them to what they considered undue publicity."[4] The midwives were essential in overcoming this obstacle. Not only did a very high percentage of pregnant women register, but the project received a comparably high percentage of returns, including notice if anything unusual happened during the pregnancy. Entirely a field operation in the beginning, the genetic program soon acquired a permanent clinic to which a sample of nine-month-old children were brought for careful examination, and Neel eventually added a laboratory to back up the clinic and an autopsy program for children who were stillborn or who died postnatally.

In 1948, with the program well underway, Neel shifted to a consultative involvement in it and returned to the United States to resume a joint faculty appointment—which he had accepted and briefly occupied in 1946 —in the Medical School and the Laboratory of Vertebrate Biology of the University of Michigan. Initiative for the creation of the post seems to have come from Lee R. Dice, an ecologist and head of the laboratory who was eugenically inclined and who had persuaded the university to establish a small outpatient heredity clinic to help people learn whether they might have "bad" genes. Neel, who was responsible for the clinic, had begun in 1946 to explore how the carriers of genetic disorders might be detected.[5] "When I came into human genetics," he recalled, "I had one, I guess absolute, guiding principle: Try to be as rigorous as I would have been had I remained with *Drosophila.* That meant picking problems carefully, problems where we could get solid scientific evidence about inheritance in man." Neel's search for solid scientific evidence—and for indicators of deleterious genetic carriers—had focused his attention, like that of others before him, on human blood. "You spread it out, you look at it, you treat it objectively," he remarked. Blood was known to consist partly of red cells containing hemoglobin, partly of white cells, and partly of serum, which was largely water but was believed to include at least one protein, albumin. What caught Neel's scientific eye was not the blood groups but blood disorders, particularly two related to the red-cell hemoglobins—thalassemia and sickle-cell anemia.

Thalassemia—anemia of the sea—was so called because it was most commonly found among people of the Mediterranean region. In 1940, at the Johns Hopkins Hospital, the hematologist Max Wintrobe had shown that the disease was the same as that known in the United States as Cooley's anemia, after the Detroit physician Thomas Cooley, who in 1925 had clinically differentiated it from various other childhood blood disorders. Cooley had thought the anemia, which seemed to occur in both borderline and fatally gross forms, congenital rather than hereditary. Wintrobe suspected that it might well be genetic, because the parents of children with gross thalassemia were often themselves borderline cases.[6] Neel, while completing his graduate studies in medicine at Rochester University in the early forties, took time to probe the genetics of the disorder among the numerous people in the Rochester area who were of southern Italian and Greek extraction. He recalled that he and a colleague "pretty well nailed down that there were two kinds of thalassemia, the very severe and the very mild" —thalassemia major and minor, he dubbed them. Thalassemia major resulted from the homozygous, and thalassemia minor from the heterozygous, expression of a recessive gene. "At that time," Neel went on, "it didn't take too much imagination to think that there might be other such blood diseases."

Sickle-cell anemia was a prime candidate. Ordinarily lethal, and disproportionately common among blacks, Greeks, and inhabitants of the Indian subcontinent, the disease takes its name from the shape that the victim's red blood cells periodically assume. A normal red blood cell resembles a disc that is concave on both sides; a diseased cell tends to curl up to look something like a sickle. Sickling of the red blood cells impedes the flow of the blood and can also lead to their destruction. It had long been known that the red blood cells from certain people could be made to sickle in the laboratory under reduced oxygen pressure, but that not everyone with such cells suffered from the anemia. In the prevailing medical understanding, all susceptibility to sickling was transmitted by a dominant gene—one that expressed itself differently in different individuals, causing a condition that varied in intensity from the harsh to the benign. Neel, however, suspected that, like thalassemia, sickle-cell anemia might be a recessive disorder, and after his arrival at the University of Michigan he determined to settle the matter. He found twenty-nine sickle-cell anemics among the black population in the Detroit area and examined forty-two of their parents. He calculated that if the dominant-gene hypothesis was correct, the red blood cells should be susceptible to sickling in only about three-quarters of the parents. In 1949, he reported that the cells could be made to sickle in every parent tested—a highly improbable outcome in terms of the dominant-gene hypothesis, but one that fitted the recessive hypothesis nicely. A single recessive gene for the disorder made people carriers of an apparently harmless sickle-cell trait, while a homozygous dose of the gene made them victims of the sickle-cell disease.[7]

That same year, Linus Pauling and several of his postdoctoral research fellows at the California Institute of Technology completed an independent inquiry into the physical properties of sickle-cell hemoglobin. The Pauling group employed the technique of electrophoresis, which had been pioneered early in the century and brought to a high degree of effectiveness in the thirties by the Swedish physical chemist Arne Tiselius. Electrophoresis relied on the fact that substances of different molecular makeup, if dissolved in a liquid and then subjected to the force of an electric voltage, would migrate through the liquid at different speeds. Tiselius's apparatus permitted the measurement of these variant speeds—the observation, in a sense, of the substance's molecular signature. By electrophoresis, Tiselius had been able to determine that blood serum contained, in addition to albumin, at least three additional, hitherto unrecognized proteins, which he designated the alpha, beta, and gamma globulins. (This achievement, together with the development of the electrophoresis apparatus, earned him the 1948 Nobel Prize in chemistry.)[8]

Using electrophoresis, Pauling's group discovered that the hemoglobin molecule in sickling cells differed physically from that in the normal type. More striking, in people with sickle-cell trait, about forty percent of the hemoglobin displayed the abnormal molecular properties, whereas in people with sickle-cell anemia all of it showed the abnormality. The Pauling group, reinforcing Neel's conclusion, interpreted their results to mean that the trait and the disease derived from a particular recessive gene involved in the synthesis of the hemoglobin molecule.[9] Neel recalled, excitement filling his voice, that Pauling's people "had no genetics in their paper," and continued, "They had the biochemistry. I had no biochemistry. I had the genetics." The genetic and biochemical results matched convincingly. "Our two papers just fitted together."

NEEL CONTINUED TO WORK with the genetics program in Hiroshima and Nagasaki, spending several weeks each year in Japan as the investigation proceeded. By February 1954, the combined Allied and Japanese staff had surveyed 76,626 pregnancies and examined almost 20,000 nine-month-old babies. About that time it was decided to bring the original project to an end, since eighty percent of the offspring likely to come from parents who had been heavily exposed to the radiation of the bombs had already been born and the rate at which additional birth data could be obtained was thus rapidly diminishing. That year Neel and William J. Schull, a colleague at the University of Michigan, co-authored *The Effect of Exposure to the Atomic Bombs on Pregnancy Termination in Hiroshima and Nagasaki*, a report whose results were as prosaic as the title. With regard to stillbirths, neonatal deaths, birth weight, or any other indicative category, the survey found no statistically significant genetic damage. Neel and Schull hastened to add that the study could "in no way be interpreted to mean that there were no mutations induced in the survivors of the atomic blasts." Everything known about radiation genetics argued that mutations must have occurred, but the frequency with which they had was no doubt too small to be detected with the techniques of the survey. To Neel the outcome was no surprise. He recalled that, on the basis of what was known of radiation genetics, "none of us who were professionals in the field had expected major findings out of Hiroshima and Nagasaki. We anticipated that they would be quite borderline."[10]

Nevertheless, Neel and his colleagues had thought the project would prove important for what it would show about the biological impact of radiation in general and about special aspects of the genetics of large human populations. He later pointed out that in Hiroshima "there were wards where cousin-marriage rates were very high, and these happened to be either close to the hypocenter or far away. . . . So early on it was clear that in the course of doing the radiation study we could set the stage for . . . the definitive study of consanguinity effects." A consanguinity study would be especially warranted, Neel thought, because Japanese vital statistics since the Meiji restoration had been organized around the family. Using these records, Neel said, you could start with an individual and go back as much as six generations.[11]

Neel had long been interested in why deleterious genes persisted in human populations, and particularly in whether they could be accounted for by natural mutation or not. After 1954 the genetic follow-up studies in Japan were extended into new subject areas. Neel, who remained involved with them, brought Japanese scientists to Ann Arbor for training in human genetics, and he drew upon the Japanese survey for material bearing upon questions concerning mutations. He also took a strong interest in the paper concerning sickle-cell genes in the malarial regions of Africa that was published in a 1954 issue of the *British Medical Journal* by Anthony C. Allison, a medical biologist at Oxford University.

Allison had himself contracted malaria while a child in Kenya, where he had been raised until his departure for boarding school in England. At Oxford University he worked in population genetics, took a doctorate in biochemistry, and then, in 1949, began medical training. During the summer of that year he went to Africa with an Oxford exploring club as their medical anthropologist, to survey blood-group variations and genetic markers in local populations. He noticed that sickle-cell trait occurred more frequently in low, wet regions, where the incidence of malaria was high, than in elevated, dry ones, where it was not. Evidence cropped up from other scientists as well that the frequency of sickle-cell trait was relatively more intense in malarial regions elsewhere, and on a return visit to Africa, Allison himself observed that it reached as high as forty percent in some tribes. In a restricted population, so high a frequency would ordinarily lead to many cases of sickle-cell anemia, diminished reproduction, and hence steady elimination of the sickle-cell gene. Allison judged that the high frequencies could not be maintained by mutational replenishment of the gene in each generation, since the necessary mutation rate would have to be three thousand times greater than that generally believed to occur in human beings. He came to suspect, therefore, that the trait persisted with such force because it conferred, upon those who possessed it, a resistance to strains of the malarial parasite—and thus a reproductive advantage.

In 1953, Allison tested this idea by examining two hundred and ninety very young African children: two hundred and forty-seven lacking sickle-cell trait, forty-three possessing it. The children were from rural areas around Kampala, Uganda, and ranged in age from five months to five years —an age group especially vulnerable to malaria because its members are old enough to have lost their neonatal immunity to the disease but too young to have begun to develop acquired immunity. He found infections with malarial parasites in about forty-six percent of the children without the trait but in only twenty-eight percent of those with it. In his 1954 report of these results, Allison concluded: "In areas where malaria is hyperendemic children having the trait will tend to survive, while some children without the trait are eliminated before they acquire a solid immunity to malarial infection. The protection against malaria might also increase the fertility of possessors of the trait." Similar reasons, he suggested, might account for the relatively high incidence of such disorders as thalassemia.[12]

In the fifties, James Neel embarked on surveys of the geographical distribution in Africa of certain abnormal hemoglobins, and he stepped up what he had early begun in Michigan: research in human population genetics. He mounted extensive genetic field studies throughout the state to determine the frequency of specific medical syndromes, estimate mutation rates, and assess the rapidity with which deleterious genes might be accumulating in the population. He also explored the genetic outcome of consanguinity in Japan. All the field studies were backed up in his laboratory, particularly through electrophoretic studies of hemoglobin variants. Like Penrose, Neel surrounded the human genetics work with research in genetics proper, including mouse and fly genetics.[13]

All the while, Neel rose rapidly up the Michigan academic ladder, expanding his department with no less managerial skill and entrepreneurial energy than he had brought to the postwar task in Japan. Located originally in a small white house where Dice had established the heredity clinic, the department steadily acquired unused space in several other older laboratories, then, in the mid-fifties, moved into a large new building of its own. The annual department budget, about $30,000 when Neel first came to Ann Arbor, climbed in tandem with the physical expansion. By the late fifties a growing number of doctoral and postdoctoral students were coming to Ann Arbor both from the United States and abroad.[14]

State and local philanthropies paid much of the bill in the early days, but less so as the department obtained a growing quantity of the funds that the federal government was now providing the nation's colleges and universities for research in a wide range of scientific fields. Human genetics qualified for the federal largesse that came to the life sciences as such, and it also enjoyed a degree of support for its connection, via the genetic effects of atomic radiation, to national security in the nuclear age. Attention to the matter escalated considerably once the issue of atmospheric nuclear testing erupted in the mid-fifties, though as early as 1949 the Atomic Energy Commission was devoting, in the words of its report for that year, "a major part of its biological research to the effects of radiation on heredity." The direct genetic effects of radiation were studied experimentally with lower organisms, notably mice; no laboratory could deliberately irradiate human beings. As radiation research subjects, human beings were to be found among those who had been exposed to radiation outside the laboratory— not only the people who had been atomic-bombed at Hiroshima and Nagasaki but also, for example, women who had undergone pelvic X rays during pregnancy or shoe buyers who had been fitted with the aid of the fluoroscopes common in the stores of the era. Yet it was a point of science policy in the United States—and in Great Britain—that reliable understanding of the impact of radiation on the human genetic complement— the "genome"—required supporting the advancement of knowledge in human genetics as such.[15]

Penrose privately reflected that Neel's group could "get as much money as they like from the government for human genetics because of their direct connections with the Atomic Bomb Casualty Commission in Japan." True enough, Neel obtained funds from the Atomic Energy Commission for his investigations of human mutation rates, the genetic outcome of consanguinity in Japan, and the rate at which deleterious genes might be accumulating in the general population. Nevertheless, the Public Health Service supported the sickle-cell work; the Rockefeller Foundation supplied some funds to help train new human geneticists and, beginning in the late fifties at a munificent level, so did the National Institutes of Health. "We used the grant system," Neel said, smiling. "Believe me."[16]

The amount of money, and the way Neel used it, made for sharp differences in style, scope, and size between the energetically expansive Department of Human Genetics at Michigan and the Galton, with its three rooms. In 1954 one of the young Galton staff members, then spending some time with Neel, remarked upon how at the Michigan laboratory genetic data gathered from patients was centrally and systematically organized, a sharp contrast to the Galton, which had no system. In 1958 Penrose himself visited Ann Arbor and confided the experience to his private notes: "Immediately I am swept off to the great Institute of Human Genetics and shown superb maps of Michigan with dots and flags for various kinds of cases and the perfect filing system with cross references of diseases and relatives. . . . There is no lack of intellect in this Ann Arbor department. . . . In spite of all their excellent work I have a feeling that we could do much more with the same opportunities or rather, I should say, more interesting things."[17]

Perhaps, perhaps not, but Neel's research program was unquestionably imaginative and adventurous. He and Penrose were both honored in 1960 by selection for one of the Joint Awards given by the American Public Health Association and the Albert and Mary Lasker Foundation—one of the most prestigious American prizes in medical research.[18]

IN THE FIFTIES, human genetics in the United States attracted a number of new recruits, both Ph.D.s and, increasingly, M.D.s, aided and abetted by the opportunities for study and research available because of the government's interest. Neither group's professional training prepared them to deal with the special requirements—particularly the mathematical and biochemical demands—of the subject. Like Neel before them, the new entrants introduced themselves to the field autodidactically, using the works of Fisher, Hogben, Haldane, Crew, Penrose, Roberts, and Race and Sanger. By 1954 they also had James V. Neel and William J. Schull's *Human Heredity*, whose authors had been at pains to introduce their readers to the mathematical methods of human genetics. And everyone seemed to read Curt Stern's *Human Genetics*, first published in 1949, though it was not as much to the taste of the physicians as Fraser Roberts's. ("A lovely book," one of them said, adding, however, that Stern was not a medical man and his text "wasn't bedside genetics.")[19]

About twenty of the neophytes took part of their education as visitors to the Galton Laboratory. They remember that Penrose's people tended to cast regular animadversions against many practitioners of human genetics in the United States, partly because they thought their work shoddy and overlaid with eugenics. (During his 1958 visit to North America, Penrose noted his opinion of two postprandial lectures by an officer each from the American eugenics and human genetics societies: "When not offensive they showed gross ignorance and stupidity.") The Galton staff, tilted so much to the political left, also disliked U.S. cold war policies. Barton Childs, who was at the Galton during the Korean War and became one of the pioneers of human genetics at the Johns Hopkins Medical School, remembered that two of the staff members would get together at tea "and shred another American reputation each day." Nevertheless, the Americans at the Galton generally thought the staff from Penrose on down hospitable enough, and most had ample opportunity to absorb—in the osmotic way one did at the Galton—the Penrose-Haldane way of approaching human genetics.[20]

Geneticists with Ph.D.s were drawn to human subjects via work on the national security issue of radiation effects or because, willy-nilly, they found themselves affiliated with medical laboratories. A number took it up if only because, like James Neel, they wanted to help capture the science of human heredity from the oppressive hand of mainline eugenicists. Arthur Steinberg had set out to be a fruit-fly geneticist. In the thirties, during his postgraduate days at McGill University, the faculty included a plant geneticist, C. Leonard Huskins, who had an interest in human genetics and introduced a great deal of eugenics into his courses, one of which Steinberg helped teach. Steinberg remembered that one day when he was not there, Huskins told a class of about a hundred and fifty undergraduates, in so many words: "Because Dr. Steinberg is a Jew, he believes that genetics has relatively little to do with intelligence and character. . . . Because I'm an Englishman, I believe that heredity has much to do with it." Steinberg repeatedly argued with Huskins and others about human heredity, growing ever more interested in—and critical of—prevailing beliefs on the subject. After the war, he decided that maybe he should do something about the state of the field, "and that's when I changed to human genetics."[21]

Among physicians, recruitment to human genetics tended to originate with their noticing familial patterns in areas of clinical research, often pediatrics. Barton Childs recounted the beginnings of his interest: "I was in charge of pediatric outpatients here at Hopkins and was aware of the number of infants that turned up with congenital malformations. No one knew much about the causes of those things. There seemed to be two ways to study them. One was teratology"—the study of major deformities—"which consisted in taking something out of every bottle on the shelf and giving it to some poor pregnant rat and then observing what happened to her fetuses. That seemed to me about as gross as hitting somebody over the head with a sledgehammer and devoid altogether of scientific elegance. The other tack"—the one Childs chose—"was to look at family aggregations of cases and see whether one could learn something about genes and what they might be doing in these disorders."[22]

Victor McKusick, one of Childs's colleagues on the Johns Hopkins medical faculty, came to human genetics through his research on disorders of connective tissue, notably Marfan's syndrome, which includes long spindly legs—it has been speculated that Abraham Lincoln suffered from the disease—and among whose victims McKusick noticed familial patterns of occurrence. McKusick had learned biostatistics while in medical school at Hopkins from Raymond Pearl, Karl Pearson's early American acolyte, and he had followed the subject further under the epidemiologist Abraham M. Lilienfeld. In the early fifties, he helped form the Galton-Garrod Society at Hopkins, a small club devoted to human genetic studies that included Barton Childs, Lilienfeld, and the geneticist Bentley Glass, whose interest in human heredity derived in part from his concern with racial equality and with the nuclear arms race.[23]

Tantalized by what he learned, McKusick increasingly specialized in human genetics of a clinical type. It was exciting to him because, he remarked, just as heritable disorders of connective tissue led to the eye, heart, nervous system, and bones, clinical genetics allowed one "to swashbuckle through different fields." Hopkins was a good place to do clinical genetics because, with its various specialty hospitals, it covered the entire medical waterfront. In 1957 McKusick was appointed director of the new Moore Clinic for Chronic Diseases and head of a brand-new Division of Medical Genetics at the Medical School. His department, originally specializing in heritable disorders of connective tissue and in cardiovascular disease, steadily branched out into other areas, including linkage studies, and through some of its first staff established ties with Neel's laboratory in Ann Arbor and Penrose's in London. It was the first—and the leading—program of clinical genetics in the United States.[24]

By 1959, the landscape of human genetics in America was a good deal more populated than it had been in 1945, with membership in the American Society of Human Genetics having reached almost five hundred men and women. In the prewar era, the absence of a scientific center may have diminished the vitality of the discipline in the United States, but the expansionist postwar circumstances of American science turned the institutional pluralism to advantage, producing several centers, each of sufficient size to include the multidisciplinary expertise so important to research in human heredity. Neel's and McKusick's laboratories loomed particularly large on the landscape, but peaks of quality could be seen in most regions of the country. At the end of the fifties, Americans accounted for about half of the Anglo-American leadership that had developed since 1945.[25]

A GROWING FRACTION OF that leadership was drawn to biochemical subjects under the stimulus of the complementary advances in such disorders as the aminoacidurias and the blood anemias. "When the biochemical wave began to gather momentum," Neel later remarked, "there were not very many reputations being made at the bedside."[26]

The reputations being made in Britain tended to come from work in the aminoacidurias, no doubt reflecting the influence of Harry Harris. The overall British attitude toward research in abnormal hemoglobins was perhaps summarized by Anthony Allison when, in 1955, he complained to Penrose: "Most of the Oxford medical people think that I have been wasting my time working on sickle cell anaemia—a rare disease in a far off country of which they know little!" Research in abnormal hemoglobins tended to concentrate in the United States, whose population, much more ethnically and racially diverse than Britain's, was drawn from different regions of the world and included relatively high incidences of sickle-cell trait, thalassemia, and other possible blood disorders. Besides, as Neel once remarked, no doubt with the hemoglobins in mind, "because of the favorable funding situation . . . American investigators have been especially prominent in undertakings that required large laboratories or extensive field surveys."[27]

Yet the two national strands of human biochemical genetics gradually overlapped as practitioners in both countries drew upon the results of work then underway in the biochemical branch of plant, animal, and, increasingly, bacterial genetics and upon the rapid growth of knowledge concerning the biochemistry of the human body. The merger was also fostered by the spread of such new technical methods as paper chromatography in the analysis of hemoglobins, amino acids, and other biochemical compounds.

Paper chromatography had a distinct advantage over the Tiselius "moving boundary" electrophoresis that Pauling had used to differentiate sickle-cell from normal hemoglobin. In the Tiselius apparatus, the liquid with the substance to be analyzed was admitted to a tube which already contained a similar liquid free of the substance. In the region of contact between the two liquids, a boundary layer would form, and the measurement to be taken after the application of the electrical voltage was of the speed with which this layer moved. The trouble was that the detection of the boundary required an elaborate optical system that occupied a lot of laboratory space and cost a good deal of money. Paper chromatography demanded, besides the filter paper, only a tall, tabletop-sized container, some chemicals and water, and the organic sample with its diverse compounds to be analyzed. Compared to moving-boundary electrophoresis, which was, to be sure, good for substances of high molecular weight like proteins, it was quick, cheap, convenient, and also effective for low-weight substances like amino acids.[28]

In the early fifties, paper chromatography was joined by the similarly low-cost and efficient paper electrophoresis and, in 1955, by Oliver Smithies's invention of starch-gel electrophoresis. Smithies, an Oxford-trained biochemist then at the Connaught Medical Research Laboratories in Toronto, was looking for a way to separate insulin from proteins related to it. Paper electrophoresis would not do the job because the insulin kept sticking to the filter paper. On a visit to another laboratory in Toronto, Smithies happened to see a type of electrophoresis that successfully separated proteins using a slurry—a watery mixture—of starch grains. The proteins did not stick to the grains but migrated around them at rates that depended on their different molecular compositions. However, the detection of the proteins required cutting the slurry into thick slices and chemically analyzing each one—a time-consuming process that Smithies, who had no laboratory assistance, could not afford. Smithies hoped to identify the proteins by staining them, but realized that he could not use stain with a watery slurry. Then he remembered from his childhood days of helping his mother with the laundry that starch could be cooked into a thick liquid that would set into a gel upon cooling. He quickly made a starch gel and, as he recalled, found that proteins, including insulin, "migrated through it as beautiful sharp bands which could be stained."[29]

The new chromatographic and electrophoretic methods made it possible for many laboratories, unable to support the costly and complicated Tiselius moving-boundary method, to get into the business of searching for biochemical variants, not only among diseased people who showed up in clinics but among the much larger normal population. Testing his starch-gel method, Smithies promptly discovered that the proteins in human blood sera from different people, all previously thought to be the same, were not. Starch-gel electrophoresis also helped reveal that blood sera contained more than twice as many proteins—at least twenty—than had previously been known. Both chromatography and electrophoresis were indispensable in Frederick Sanger's disentanglement, done at Cambridge University and completed in 1955, of the amino-acid sequence that composed bovine insulin —a feat in sharp confirmation of the theory that proteins consisted of chains of amino acids. And both were crucial in the research that Vernon Ingram began in 1956 to see whether there might be a specific chemical difference between normal hemoglobin and the fateful sickle-cell variant.[30]

Ingram was a protein chemist working in the Cambridge University laboratory of Max Perutz, which was devoted to figuring out the structure of the hemoglobins and was one of the places in England with an interest in the abnormal varieties. Anthony Allison had recently visited the laboratory and left behind some sickle hemoglobin, which Perutz suggested that Ingram might want to analyze. Perutz's interest piqued Ingram's. So did the likely utility of the project for a line of inquiry of concern to Francis Crick, who with James D. Watson in 1953 had published the double-helical structure of the genetic material—deoxyribonucleic acid, or DNA. Since then, Watson, Crick, and other scientists had been forging ideas about how the information contained in DNA was translated into the development of organisms. Crick had gotten Ingram interested in trying to test experimentally a key implication of these ideas—namely, that a protein produced by a mutant gene must differ in its amino-acid sequence from one produced by a normal gene. Ingram had already looked, unsuccessfully, for such differences in a few proteins. The protein of sickle-cell hemoglobin, which was known to differ from the normal version because of a change in a single gene, provided a neat opportunity for looking again. And Frederick Sanger's work with insulin in the nearby biochemistry department—which Ingram knew about—suggested how to look effectively.[31]

To analyze the sickle hemoglobin, Ingram first broke up the chain of amino acids of which it was composed into about thirty short pieces, each containing about ten amino acids. Such chains are also called "peptides" or, if the chain is relatively long, "polypeptides," since one amino acid is connected to another by a so-called peptide bond. Ingram then subjected the soup of peptides to paper electrophoresis and chromatography. Neither technique used by itself yielded anything of interest, but then Ingram deployed them together, in sequence, to force a larger separation between the peptides. Now the telltale spots on the filter paper—the "fingerprint" of the sample, Ingram called them in the paper he published on the work in October 1956—revealed that sickle hemoglobin differed in only one peptide spot (peptide number 4) from the pattern that occurred with normal hemoglobin.[32]

After more months of laborious work, Ingram managed to identify chemically the sequence of amino acids in each peptide number 4—that is, in the one from the normal hemoglobin and in the one from its sickle-cell counterpart. In 1957, he reported that the sickle-cell variant differed chemically in only one regard from three hundred amino acids that were estimated to compose the normal half-hemoglobin molecule: at the point where the normal chain contained a glutamic-acid link, the sickle chain contained a link of valine. "It is remarkable," Harry Harris remarked with understatement a few years later, "that such a subtle difference in molecular structure should have such profound pathological consequences."[33]

Ingram's work, Harris added, had "opened up an entirely new chapter in human genetics." By the late fifties, a large number of clear-cut biochemical variations were known, including more than a dozen inborn errors of metabolism arising from probable enzyme deficiencies and numerous polymorphisms—that is, traits that occurred in a population in different forms, each with a frequency of at least a few percent—among the hemoglobin and blood-serum protein variants, knowledge of which was accumulating from research around the world. Not all these variations seemed likely to have originated genetically in the same way. For example, a mutant gene could result in a failure of protein synthesis, as with certain red-blood-cell con-·stituents whose absence brought on particular anemias; or it could produce abnormal proteins like the sickle-cell hemoglobins and possibly even abnormal enzymes. Harris was tempted to speculate that variations in the fine structure of human enzymes might yield drastic changes in their activity —an effect that had earlier been demonstrated in *Neurospora* and *E. coli* bacteria—and that these might well lie at the base of many inborn metabolic errors.[34]

Whatever the case, when in Naples in 1959 Penrose opened a conference on human genetics, he was rightly moved to declare: "At the present time the application of mathematical methods is no longer a dominating factor. Biochemical methods are now in the ascendant." But Penrose also wanted to say at length—only the biochemical focus of the conference prevented him from doing so—that equally in the ascendant were methods concerning human chromosomes.[35]

Chapter XVI

CHROMOSOMES—
THE BINDER'S MISTAKES

I N AUGUST 1955, Joe-Hin Tjio, a young Indonesian who was then working in Zaragoza, Spain, came to Lund, Sweden, for one of his periodic collaborations with Albert Levan. Both were primarily plant cytologists, but now their attention was turned to the chromosomes in the human cell. The nucleus of the normal human cell contains two sex-determining chromosomes—XX for females and XY for males—plus twenty-two pairs of autosomes—that is, chromosomes unrelated to sex. The total comes to forty-six. That fundamental number of human cytogenetics was established by Tjio and Levan during Tjio's visit in 1955—long after cytologists had started counting the chromosomes of man in the eighteen-nineties.[1]

The very early counts had yielded numbers that varied around twenty-four, which was consistent with those obtained for other mammals. The trouble then was that cytologists made their counts with tissue taken from corpses, often those of executed criminals; upon the death of mammalian cells, the chromosomes tend to clump together rapidly, thus deceiving even the microscope-aided eye into falsely low counts. Recognizing the problem, the Belgian cytologist Hans von Winiwarter used fresh tissue obtained during surgery and immediately fixed with a chemical preparation. In 1912, he reported the human chromosome number to be forty-seven for males and forty-eight for females. Von Winiwarter explained the sexual difference by arguing that while the human female had two sex chromosomes —a double X—the human male must have only one, a single X.[2]

Von Winiwarter's result, neither confirmed nor rejected, was evidently regarded as an anomaly by most cytologists, but at the beginning of the nineteen-twenties his use of fresh tissue caught the attention of Theophilus S. Painter, a cytogeneticist at the University of Texas. One of Painter's former students happened to be practicing medicine at the state

mental institution in Austin. Painter obtained the testes from three patients
—one white, two black—all of them castrated, Painter reported, because of
"excessive self-abuse coupled with certain phases of insanity." Within a few
minutes of their removal from the blood supply, the specimens were slit
into multiple sections and dropped into a fixing solution. In mid-1921,
Painter reported to a colleague that "my best counts now give me 48
chromosomes for both the Negro and white man . . . and [I] feel confident
that this is correct."[3] Perhaps his confidence derived from the fact that the
figure squared with von Winiwarter's for females. More important, as in
other mammals, the total included the male sex-chromosome combination,
X and *Y*. It was also consistent with his counts in spermatocytes, which,
as the products of sexual division, should have contained half the number
in non-sex cells, and, so far as Painter saw, did have twenty-four. After
Painter published a full report of his work in 1923, other cytogeneticists
confirmed his count. For the next thirty years, just about everyone believed
the human chromosome number to be forty-eight, for both sexes.[4]

In retrospect, the reasons for the persistent miscounting are clear
enough. Normally, the chromosomes lie in a region of the cell nucleus that
takes on a deep color upon staining. In the quiescent cell, the individual
chromosomes cannot be visually differentiated from the region. They can
only be seen—and counted—in the process of cell division, when they
emerge as separate, colored—hence the name—rodlike entities. To obtain
a chromosome count, human cells had to be captured and fixed at the
moment of division. The more cells in a state of division, the better the
prospect for chromosomal observations. Particularly suitable were tissues
with rapidly proliferating cells, notably embryos or testes, which are sites
of constant cellular division.[5]

Such material, obtained fresh from living bodies, was, to say the least,
difficult to come by. Many more human chromosome counts seem to have
been done with testes than with ovaries for the simple reason that the taking
of ovarian tissue required a major surgical procedure. The human
cytogeneticist often had to wait, ready to fix his specimens, outside operating rooms or, in the case of a team that confirmed Painter's count, literally
at the foot of the gallows. Once obtained and fixed, the specimens were
sliced into thin sections with a fine blade—the blade cutting through the
nucleus of a given cell as a knife might cut through an egg in the middle
of a meat loaf. Just as successive sections of meat loaf would contain successive slices of egg, successive sections of cell—perhaps two or three—would
include serial slices of the complete nucleus. Since the chromosomes were
spread through the nucleus, some would wind up in one section, some in
the next. The cytologist added the number found in each section to reach
the total in the cell. But because of imprecision in where the blade happened

to cut, fragments of a chromosome located—and already counted—in one section might turn up as candidates for counting in the next. Then, too, compared to fruit flies, which have four pairs of chromosomes, the human cell nucleus is small and the number of chromosomes large. Even when separated and fixed during cell division, human chromosomes are crowded together. They appeared to cytologists of Painter's era as something like the noodles suspended in a soup—some lying beneath others and difficult to count accurately. It was not easy to decide whether the noodle that resembled an "L" under the microscope was a single bent chromosome or two straight ones.[6]

The cytologist Tao-Chiuh Hsu, who once saw a slide of one of the human testicular sections that Painter had prepared, later wrote: "I failed to make any sense of the twisted, crowded, stacked chromosomes. It's amazing that [Painter] even came close!" Every enumeration of human chromosomes required judgment, and judgment left room for conformation to orthodoxy. Human chromosomal counts sometimes suggested a figure different from forty-eight, but most cytologists, expecting to detect Painter's number, virtually always did so.[7] Indeed, the preconception in favor of forty-eight was so powerful that it operated on Hsu himself when, in 1952, he set off the train of experimental work that led to the revision down to forty-six.

Hsu had come from Chekiang University in China in 1948 to take a Ph.D at the University of Texas; now a postdoctoral fellow in human cytology at the medical branch of the university in Galveston, he was looking at cell nuclei in preparations of fetal spleen tissue. It was with distinct incredulity, Hsu recalled, that he saw in one of the preparations "some beautifully scattered chromosomes." Similar pretty pictures appeared in other slides, but when he examined additional preparations, the chromosomes "resumed their normal miserable appearance." Hsu guessed that something about the original preparations must have been special. For some months, he sought assiduously to find out what. There was no need for him to hover outside some operating-room door to obtain fresh spleen cells. Plenty were available because the original sample had been subjected to tissue culture—the technique by which cells are kept alive and multiplying in vitro with suitable nutrients. Tissue culture had come into use in cytology laboratories after the Second World War, and provided a continuous supply of dividing cells. Hsu systematically altered the preparation procedure of one sample after another of the abundant embryonic spleen cells. Nothing worked until April 1952, when he added distilled water to the balanced salt solution commonly used to rinse the tissue specimens before fixation.[8]

This so-called hypotonic solution liberated the chromosomes from the cell spindle—a warp of fibers that form during cell division to guide them on their journey—and it also swelled the cell volume, which allowed the chromosomes more room to separate. Hsu guessed that the preparations in which he had seen the chromosomes so clearly must have been accidentally washed in hypotonic solution before being fixed. Turning accident to advantage, he proceeded to look closely at the human chromosomes—not to check the number but to examine their structure. In many cells, he recalled with some irony, "I had difficulty in getting the count to equal forty-eight." Nevertheless, his vision filtered through the prevailing preconception. Hsu managed to count to Painter's figure. He later confessed to feeling like a football player who returns an interception forty yards only to find himself "fumbling the ball at the three-yard line."[9]

Hsu's metaphor did him a disservice; at the time, he did not know that he was in a contest with nature for the correct human chromosomal count. Neither, three years later, did Tjio and Levan when they found the right number: their aim had been to explore in detail the morphology of human chromosomes in lung tissue taken from legally aborted embryos. The difference between their work and that of all previous analysts of human chromosomes was its reliance not only on tissue culture and hypotonic treatment but on two other techniques newly deployed in human cytology. One was the pre-treatment of the cells with colchicine, an alkaloid extracted from the seeds of a crocuslike herb. Colchicine arrests cell division midway through its course, thus providing many more cells to be observed in the process of splitting. It does so in a way that further frees the chromosomes to disperse throughout the cellular volume. And it tends to contract chromosomal size, thus diminishing the likelihood of confusing overlaps. The other was the "squash technique," so named because, instead of being sectioned, the cells to be examined were literally squashed with the thumb under a thin glass plate. With the cell thus flattened into something resembling a pancake, the chromosomes are spread onto a single plane of optical focus. Once Tjio and Levan applied all four techniques in combination to their embryonic lung cells, they immediately saw an unambiguous forty-six human chromosomes. Further experiments in the fall and winter of 1955 yielded the same count with high consistency, and in 1956 they published their results, though not without residual anxiety about challenging Painter's much-confirmed number.[10]

WITHIN DAYS OF ITS publication, Tjio and Levan's article was read in England by Charles E. Ford, a cytogeneticist in a radiobiological research unit of the Medical Research Council located at the British Atomic Energy Research Establishment at Harwell, near Oxford. In connection with stud-

ies in leukemia, Ford had worked with mouse and, recently, human cytogenetics. Already adept at the essential techniques of the field, he had in fact
helped alert Tjio and Levan to the value of treating specimens with colchicine and hypotonic solution. An Oxford University surgeon, impressed
with the clarity of Ford's cytological preparations, had offered to send
human testicular material for chromosomal analysis. Ford had passed up the
opportunity and, as he read Tjio and Levan, wished he hadn't. Now Ford
and John Hamerton, a colleague at Harwell, swiftly confirmed the count
of forty-six, using fresh human tissue supplied by the Oxford surgeon.[11]
The work brought Ford to the attention of the human geneticists in London, where interest in human cytogenetics was rising rapidly.

Among those concerned with the subject was Paul E. Polani, a physician at Guy's Hospital on the south side of the Thames, on a sight line from
St. Paul's Cathedral. Polani had started in genetics during his undergraduate days in Italy just before the Second World War, and from 1948 to 1950,
while on a fellowship, he had spent part of his time at the Galton Laboratory with Penrose. In 1954, in the course of his research on the causes of
congenital heart disease, Polani came across three women who suffered
from an aortal defect usually found among males but who also had Turner's
syndrome, a condition found almost exclusively among females. Given the
characteristics of Turner's syndrome—a thick, webbed neck, shortness of
stature, and, especially, rudimentary ovarian and mammary development—
Polani wondered whether the Turner's patients might genetically resemble
males. At this time, indications of human genetic sex were beginning to be
obtained by using the 1949 discovery of Murray L. Barr, a cytologist at the
University of Western Ontario: routine staining revealed a small satellite
(eventually called a "Barr body") near the nucleolus in the cells of females
but not usually of males. Females were thus classified as "chromatin positive," males as "chromatin negative." Polani tested his Turner's females and
found that all three were chromatin negative.[12]

This outcome stimulated Polani to further research into human "intersexes"—people of one sex who displayed some characteristics of the other
—and he gathered information on twenty-five more women, about half
with Turner's syndrome and the rest with simply no ovarian development.
He found twenty of the twenty-five to be chromatin negative. There was,
however, scientific doubt that chromatin negativity could be taken as a
definite sign of genetic maleness, particularly among abnormal human
beings. Pondering how alternatively to determine the genetic sex of the
women, Polani hit upon the ingenious idea of surveying them for a sex-
linked trait and, following a discussion of the matter with Penrose, he
resolved to test them for the predominantly male trait of red-green color
blindness. He observed this trait in four out of the twenty-five women—

a frequency significantly higher than expectation in such a group of genetic females, but one consistent with expectation in a comparably sized sample of genetic males. In his report of these results in *The Lancet,* in July 1956, Polani suggested that the Turner's women might be chromosomally *XO*— that is, might have only one *X* chromosome, instead of the normal female's two.[13]

Polani enlarged his work on color blindness in the human intersexes to include males with Klinefelter's syndrome—a condition with the symptoms of tallness, minor mammary development, and, often, testicular atrophy and mild mental deficiency. Barr and a colleague had just found that Klinefelter's males were chromatin positive—that is, they displayed the nuclear staining feature characteristic of normal females. In October 1958, Polani reported that color blindness occurred among such Klinefelter's with a frequency characteristically observed among females, and he suggested that, like females, Klinefelter's males must have two *X* chromosomes. The question was whether they had a *Y* chromosome, too. There was no way to determine the answer without looking directly at the karyotypes—the word comes from *karyon,* the Greek for "kernel," and signifies the display of chromosomes in the cell nucleus.[14]

In 1955, Polani had tried to determine the genetic sex of a few of his Turner's patients by looking at their karyotypes with the aid of Gordon Thomas, an anatomist at Guy's Hospital who knew how to do tissue-cultures. Inexperienced at working with human chromosomes, they obtained—from three Turner's women and seven normal people used as controls—only a handful of complete cell samples, and none of sufficient quality to assess what sex chromosomes the cells contained. (They did manage to count forty-five chromosomes in one of the karyotypes but mistrusted the result, partly because the number did not square with the prevailing belief in a normal total of forty-eight chromosomes, even if the cell was one *X* chromosome short.) In February 1956, Polani attempted to persuade a practiced cytogeneticist to help him; the man declined because he was unconvinced by Polani's arguments that the Turner's women might be *XO.* But in the fall of 1958, now eager to examine the karyotypes of Klinefelter's males, Polani turned with success to Charles Ford, whom he had met the year before at a conference on sex and the cell nucleus at King's College Hospital in London.[15]

Ford had recently perfected a method for treating bone marrow— another source of rapidly proliferating cells—in a way that yielded a large number of cells in a state of mitosis within a matter of hours. The method reduced to virtually nil a then-presumed risk of long-term tissue culture: that it could result in chromosomal changes of a misleading kind because they occurred not in the body but in the process of cell division in the culture itself. Early in 1958, Ford had used the bone-marrow technique to scrutinize a Klinefelter's karyotype in collaboration with Lazlo G. Lajtha, a hematologist at the Churchill Hospital, Oxford, and Patricia A. Jacobs, a young cytogeneticist from Edinburgh who had come to Harwell for a few months to learn the techniques of bone-marrow preparation. They had counted forty-six chromosomes, including two X's, which was consistent with the chromatin-positive reading characteristic of females. They had not found a Y chromosome. Even though the Klinefelter's was an apparent male, this was no surprise at the time. Fruit flies with an XO complement of sex chromosomes were males, while those with an XXY complement were females. The prevailing extrapolation from these data had it that the Y sex-chromosome played no role in the determination of maleness, even in human beings. Still, the examination of one Klinefelter's karyotype hardly settled the matter, and late in 1958 Polani sent a sample of Klinefelter's bone marrow for analysis to Ford at Harwell.[16]

Unknown to Ford, the chromosomes of a Klinefelter's male had been under scrutiny in Edinburgh since the early summer by Patricia Jacobs and John A. Strong, a local physician. Jacobs had returned to her Medical Research Council Unit, which specialized in radiation genetics and where she had been examining the karyotypes of human beings with radiation-induced leukemias. Unable to find more than a few such people, Jacobs had decided to apply her newly mastered bone-marrow techniques in a resumption of the Klinefelter's work she had begun with Ford. Though she did not at first believe what the Klinefelter's karyotype revealed, Jacobs was compelled to the identical conclusion that Ford at Harwell, still ignorant of her investigations, reached when he scrutinized the sample from Polani: The Klinefelter's male karyotype contained not two but three sex chromosomes—two X's plus the Y of the normal male. Jacobs and Strong published their results in January 1959. At the time, as Lionel Penrose later wrote to Haldane, who had moved to India, the discovery of the extra Klinefelter's chromosome "astonished everyone." Not the least astonishing feature of the new knowledge was that human beings differed from fruit flies in the role played by their sex chromosomes: In *Homo sapiens*, the Y determined maleness, even if in *Drosophila* it did not.[17]

The Klinefelter's results set Penrose to thinking. Early in the thirties, the Dutch physician P. J. Waardenburg and the St. Louis pediatrician Adrien Bleyer had independently suggested that Down's syndrome might be the product of a chromosomal anomaly, and by the end of the decade Penrose had come to embrace the suspicion. In 1952, at his urging, Ursula Mittwoch, a member of the Galton staff, scrutinized the sex-cell karyotype of a Down's male. Though inexperienced at cytology, she managed to count twenty-four chromosomes, half of the forty-eight that one would then expect to find in a normal cell after meiotic division—which implied that Down's syndrome was not the result of a chromosomal disorder. For Penrose, the Klinefelter's results reopened the question. Penrose knew of a Klinefelter's Down's at the Harperbury Hospital, identified in a search he had initiated there in the fall of 1958 for chromatin-positive males and chromatin-negative females. In his letter to Haldane a few months later, Penrose recounted, "Naturally, I wanted at once to try our luck with the Klinefelter mongol."[18]

Charles Ford was ready and eager to do the karyotype analysis, but it took time to get the relatives' consent for the removal under anesthetic of the bone-marrow cells. Then, for three weeks or so from late February 1959, a virulent Asian flu epidemic completely tied up the hospital facilities. In the meantime, reports filtered into England that Jérôme Lejeune, a young French human geneticist, had learned something of consequence about Down's syndrome karyotypes.[19]

LEJEUNE'S CAREER IN genetics started in 1952, when, as a recent graduate in medicine, he returned from military service to work with Raymond Turpin at the Hospital Saint-Louis, in Paris. Turpin, a professor of pediatrics at the University of Paris, was one of the very few people in France at the time interested in human genetics. His hospital practice included a group of Down's syndrome patients, and he turned over responsibility for them to Lejeune.[20] Neither Turpin nor Lejeune believed John Langdon Down's original hypothesis that victims of the condition were throwbacks to some atavistic Mongolian "race." In his clinical work, Lejeune saw a Down's child from Indochina whose appearance differed sharply from that of normal children of the region; the syndrome stood out among Orientals as well as among Caucasians. Lejeune suspected that Down's syndrome had something to do with hereditary mechanisms. Like a number of physicians elsewhere confronted with such inklings, he embarked on a postmedical course of study toward a doctorate in science with emphasis on biochemistry and genetics. Postwar French austerity made the task of research less straightforward: Lejeune had no laboratory, no microscope, only a single room without running water. Pondering what experimental research he might pursue under those conditions, he decided to concentrate on the palm prints of Down's victims.[21]

In 1953, Lejeune scrutinized the configurations of lines on the palms of ninety-three Down's patients, two hundred and forty-six members of their families, and two large control groups drawn at random—except that one group was evenly divided for sex—from the Parisian population. Lejeune assessed the configurations quantitatively and arrived at a numerical index of the degree to which, on a given palm, they occurred in association with each other. He found that the Down's patients had a strikingly higher associative frequency of abnormal palm lines than did the people in either of the control groups. To Lejeune, this signified that Down's syndrome must involve some deep genetic change from the normal. One of the palm lines found in the syndrome was the so-called simian crease. Lejeune knew very little about primatology, but it occurred to him that a clue to the deep change might be found in the palm configurations of apes and monkeys—especially the lower-order monkeys from which the simian crease took its name.[22]

At the Natural History Museum in Paris, he measured the configuration of palm lines on the skins of the apes and monkeys preserved there. The palm lines of normal human beings showed no resemblance to those of either the lower-order monkeys or the anthropoid apes—orangutans, gorillas, and chimpanzees. But there were extraordinary similarities between the Down's palms and those of the inferior monkeys—for example, mangabeys and macaques.[23] Lejeune supposed that the distinction between the palm lines of anthropoid apes and those of the lower-order monkeys must have resulted from the accumulation of numerous single-gene changes over evolutionary time. He speculated that the Down's palm lines, too, must arise from a polygenic difference between the Down's victims and normal human beings—occurring, obviously, not over evolutionary time but in one generation, from parent to child. Lejeune reasoned that the necessary change had to involve the only genetic material then known to be large enough to carry a polygenic message—a chromosome.[24]

At this point, Lejeune's mind turned to the haplo-four fruit fly. (Cytogeneticists designate as "haploid" those cells—for example, mammalian gametes—that contain only half the normal number of chromosomes. The haplo-four takes its name from the fact that it possesses only one member of the fourth chromosomal pair found in normal *Drosophila*.) The haplo-four fruit fly has various abnormal characteristics, including thinner bristles, a shortened body, and a prolonged larval stage. No one of these characteristics announces the haplo-four; they declare themselves as an ensemble—a syndrome. Lejeune thought of the haplo-four as a kind of "mongol fly." Just as the "mongol fly" was missing a chromosome, Lejeune came to think, in 1954, that the victims of Down's syndrome must lack a chromosome, too.[25]

Lejeune had by this time moved with Turpin's group to the Hospital Trousseau. He wanted to look at the chromosomes of his Down's patients, but he was not familiar with human cytogenetic techniques and was unable to find anyone in Paris who was. Besides, there was not much money for research and only limited laboratory facilities at the hospital. He therefore turned to various other subjects—mainly radiation genetics, for which Turpin, like many biologists, was able to raise funds in the mid-fifties. All the while, however, he had his chromosomal hypothesis in mind and kept hoping to test it, especially after the work of Tjio and Levan was published.

The opportunity arose in 1957, with the arrival in Turpin's clinic of Marthe Gauthier, a cardiologist who had recently learned the technique of tissue culture; Turpin authorized her to use it in collaboration with Lejeune.[26] Sometime about the spring of 1958, Gauthier cultured tissue taken from the fascia lata—the smooth connective tissue that covers muscle—of three Down's patients at the Hospital Trousseau. Lejeune, using the newly developed cytogenetic techniques, prepared karyotypes and examined them through a microscope discarded by the hospital's bacteriology laboratory; it was so worn that he had to stabilize its adjustment gears by inserting between them a piece of tinfoil from a candy wrapper. He photographed the karyotypes with equipment borrowed from the pathology department, expecting them to show, like those of the "mongol fly," the absence of a chromosome. Instead, they showed that the Down's patients had forty-seven chromosomes rather than forty-six.[27]

Lejeune wondered whether the extra chromosome was typical of the Down's patients or an artifact of the tissue culturing. Aging cultures were known to produce chromosomal anomalies. But the cultures had been no more than a month old before he obtained the karyotypes—too short a time, Lejeune thought, for the aging phenomenon to occur. More troubling to him was a recent paper by Masuo Kodani, an American cytogeneticist then working with the Atomic Bomb Casualty Commission in Japan, claiming that in some normal human beings the chromosome number might be forty-seven. If Kodani was correct, then the "extra" chromosome Lejeune had detected in his patients might not be extra at all and might have nothing to do with Down's syndrome. In a lecture at McGill University in September 1958, just after the Tenth International Congress of Genetics, in Montreal, Lejeune swallowed his doubts enough to show the photographs of the three Down's karyotypes and advance his belief that the cause of the syndrome was an extra chromosome. His audience seemed for the most part unconvinced.[28]

After he returned to Paris, Lejeune prepared karyotypes of cells from eight non-Down's patients at the Hospital Trousseau. Each of the karyotypes showed forty-six chromosomes. Though still somewhat anxious about putting his Down's results into print, he finally published the work in the *Comptes Rendus* of the French Academy of Sciences in January 1959. In the same journal, in mid-March, he reported the results of an examination of nine Down's karyotypes and argued with greater confidence that the extra chromosome was the cause of the syndrome.[29]

IN ENGLAND BY NOW, the crowding of Harperbury Hospital had eased enough to take the bone-marrow sample from the Klinefelter's Down's (Orlando J. Miller, a young American physician then on a Population

Council fellowship at the Galton Laboratory, dates the event between March 19 and March 23, 1959). Half the sample was sent to Ford at Harwell, who recalls finding the extra Down's chromosome (plus, of course, the extra X for the Klinefelter's character) just two days after hearing about Lejeune's results. At the Galton, Miller and Ursula Mittwoch detected the identical chromosomal anomaly in their half of the bone-marrow sample. Additional confirmation came from Edinburgh, where Jacobs and her co-workers, also without knowing about Lejeune, had begun to look at the chromosomes of Down's victims because they tended to suffer from a high incidence of leukemia.[30] News of the Down's results moved the provost at University College London in May to send Penrose a note: "It must be one of the most important things that has happened in genetical studies for a long time." And it was. Penrose remarked some months later that the events of the past year amounted to "a major breakthrough in the science of human genetics," adding that he found "the photograph of the cell from the man with two extra chromosomes from which the intelligence level, the behavior and sexual characters can be confidently predicted, just about as astonishing as a photograph of the back of the moon."[31]

However, there was still doubt about the nature of the extra Down's chromosome. Penrose thought that it was a member of a trisomy—that is, the occurrence of one of the twenty-two autosomal chromosomes as a triplet rather than as a pair. Lejeune had not been certain—and neither had the other investigators—whether it was that or a supernumerary chromosomal piece of unknown origin. But within a year the abnormality was demonstrated to be indeed a trisomy—of the chromosome designated No. 21 by agreement at a genetics conference in Denver, Colorado, in April 1960. (The agreement assigned numbers to the chromosomes in order of descending size.)[32]

Also in 1960, investigators in Sweden, in addition to Polani and Ford, and Penrose and others in England, concluded that a particular form of this trisomy accounted for the small number of cases of familial occurrence of Down's syndrome. It arose from the presence in some people of what is called a translocation—in this case, the attachment of one of the 21-chromosomes to the 14-chromosome. If a gamete containing the 14-21 combination plus the other 21-chromosome was passed on to a fetus, the offspring would possess two regular 21-chromosomes plus the 21 on the No. 14. If a gamete transmitted the 21- and 14-chromosomes only in their hybrid form, the child would be normal. But because these chromosomes were attached to each other the child would be a carrier, and his or her children would be at risk for trisomy-21.[33] The detection of the cause of "mongolism" in such cellular accidents finished off—or should have—its vestigial association with some kind of atavism. Lejeune, Penrose, and others publicly urged that the racially tinged nomenclature of the condition be abandoned in favor of different terms, including "Down's syndrome" or "trisomy-21."[34]

The sharp turn of events in human cytogenetics originated in different approaches—particularly in the Cartesian rationalism of Lejeune on the one side of the Channel and British step-by-step empiricism on the other, but they joined incandescently to light up a vast unexplored region on the human cytogenetic map. Charles Ford had analyzed a Turner's bone-marrow sample sent him by Polani and had reported in 1959 that, as Polani suspected, Turner's females were missing a second sex-chromosome. In 1960 other birth defects were shown to result from chromosomal anomalies, and it was demonstrated that lymphocytes in the blood could be cultured for karyotype analysis—a technical advance that put human chromosomal studies within reach of any scientist or physician who wanted to undertake them. Penrose later remarked of the hereditary mechanism that "the instructional errors, when single genes are involved, are too small to be seen. They are like mistakes made by an imaginary printer whereas chromosome aberrations are like the mistakes of a binder."[35] By the early sixties, human geneticists were equipped with the cytogenetic techniques essential to seeing the binder's mistakes.

THE EXPLORATION OF THE new regions—not only human cytogenetics but human biochemical genetics—surged ahead with remarkable force, drawing people in steadily increasing numbers, enlarging what was by now a flourishing international community in the discipline—the First International Congress of Human Genetics had been held in 1956—that included scientists from most of the nations of Western Europe as well as from Japan and Latin America. In the United States and Britain, and no doubt elsewhere, a significant fraction of the new practitioners were physicians. Victor McKusick wryly observed that cardiologists had long had the heart, neurologists the nervous system, and nephrologists the kidney. The discovery of trisomy-21 gave medical geneticists the chromosome—"our organ."[36] Yet the enterprise of human genetics was also populated by an army of specialists, scientists in the variety of disciplines upon which research in the subject had come to depend.

J. B. S. Haldane was elected to be president of the Third International Congress of Human Genetics scheduled for 1966 in Chicago, but by the time of the meeting he was dead of cancer. The office devolved upon Penrose, who took the occasion to deliver a tribute to his old friend and to mark the change in the field they had so long cultivated together: "Before I worked at University College, I imagined that a laboratory for studying human genetics would have to contain experts in anthropometrics, statistics, clinical pathology, cytology, biochemistry, and serology. Now an ideal center would contain teams of people from all these disciplines and also include biophysicists, enzymologists, embryologists, and electron microscopists. Human genetics has grown from being a quiet hobby, involving merely the collection of pedigrees of rare diseases and deformities, to one of the most complicated and demanding disciplines in the whole of science. When I was asked by prospective research workers, thirty years ago, whether it was worthwhile studying the subject at all, I used to reply that though at the time things seemed to be developing very slowly, there would soon be an explosion. The explosion has now taken place."[37]

Chapter XVII

A NEW EUGENICS

HUMAN GENETIC RESEARCH may have been spurred in part by reform-eugenic goals, but the more that was revealed about the complexity of heredity in human beings, the less did eugenics—even much of the reform variety—appear defensible in principle, or even scientifically within reach. The dozens of variations discovered in hemoglobins, metabolic processes, and, in the sixties, enzymes made it evident that human beings were infinitely differentiable in their biochemistry. No stigma could be attached to the impersonal substitution of a single amino acid that produced sickle-cell anemia. In 1966, Lionel Penrose observed, "The social and biological values of hereditary differences are continually altering as the environment changes. . . . At the moment . . . our knowledge of human genes and their action is still so slight that it is presumptuous and foolish to lay down positive principles for human breeding. Rather, each person can marvel at the prodigious diversity of the hereditary characters in man and respect those who differ from him genetically. We all take part in the same gigantic experiment in natural selection."[1]

Moreover, the revelations of the Holocaust had all but buried the eugenic ideal. After the Second World War, "eugenics" became a word to be hedged with caveats in Britain and virtually a dirty word in the United States, where it had long been identified with racism. In their 1954 textbook, *Human Heredity*, James Neel and William J. Schull censured the eugenics of the past, warned against the extremes to which its biases could lead, and, while endorsing reform eugenics, did so in a gingerly fashion and with an insistence that the first order of business was to continue advancing the science of human genetics.[2]

Penrose proclaimed in his inaugural lecture as Galton professor that the only "racial" issues with which human genetics ought to be concerned were those relating to the human race as a whole. The staff of the Galton Laboratory bristled with contempt for the country's remaining eugenic activists, lumping reformists like C. P. Blacker with earlier mainliners like Leonard Darwin. The Galton's institutional identification with eugenics made Penrose cringe. He told the University College provost in 1961 that since the war the work of the Galton had been seriously handicapped by "the stigma of eugenics," and that he found it a "continual embarrassment" to have to explain that both his laboratory and the professorial chair were "wrongly named." In 1954, Penrose had changed the name of the laboratory's principal publication, the *Annals of Eugenics,* to the *Annals of Human Genetics,* and now he succeeded in persuading the authorities of University College to rename his chair the Galton Professorship of Human Genetics.[3]

In the offices of the Eugenics Society, a few blocks from Victoria Station on Eccleston Square, there was no affection for Penrose on the part of either Blacker, who regarded Penrose's occupancy of the Galton Chair as an offense to its intent, or the human geneticists who continued to associate with the Society. R. A. Fisher, long estranged from Penrose, privately remonstrated to Blacker, in 1951, on his successor's attitudes: "The coincidence that opponents of Eugenics in this century have been almost always Communists, or fellow travellers, cannot . . . be overlooked." The British and American eugenic societies, recognizing that the opposition extended far more widely than Fisher's simplistic characterization of it, had to concede, in the phrases of the 1947 minutes of the American group, that "the time was not right for aggressive eugenic propaganda."[4]

Both societies continued, discreetly, to follow the course they had begun in the thirties, attracting to membership or involvement in their scientific activities various distinguished geneticists from across the political spectrum. Frederick Osborn noted of one of his conferences, esoteric with computerized models of human evolution, "This is a far cry from the propagandist eugenics of Madison Grant and my dear uncle." Their scientific efforts, however, were steadily overwhelmed by the vast outpouring of work that developed in human genetics, demography, the field of human reproduction, and the like. (The British society, fueled by its modest endowment, survives to this day as a minor learned society, in musty offices on Eccleston Square. In 1972, its American counterpart became the Society for the Study of Social Biology, a vestige of the original organization. The year before, Osborn had ruefully anticipated the transformation in a brief, unpublished history, in which he lamented that "the American public . . . does not care to envisage the possibility that individuals are born with different genetic potentials, with different possibilities for defect, for happiness, or for service to the community.")[5] Nevertheless, even though the eugenic ideal had gone out of fashion, a variety of scientists pursued one element or another of the reform-eugenic program.

. . .

IN ITS EFFORTS TO encourage the use of genetics for medical purposes and to improve the biological quality of human populations, reform eugenics had helped lead to the opening of facilities devoted explicitly to genetic advisory services. In the United States, perhaps the first was the Heredity Clinic at the University of Michigan, which opened its doors in 1940, and which James Neel headed from 1946 to 1981. The second was probably the Dight Institute, which was established in 1941 at the University of Minnesota. (Charles F. Dight, an eccentric insurance company medical examiner who lived in a tree house, left a sizable legacy—accumulated through a combination of shrewd investments, acute frugality, and failure to file income-tax returns—to the university for the establishment of a clinic "to promote biological race betterment—betterment in human brain structure and mental endowment and therefore in behavior.") The first genetic advisory clinic in Britain was established in 1946 by John Fraser Roberts at the Hospital for Sick Children, on Great Ormond Street, in London. The clinical offerings went by different names, including "genetic hygiene"— a term that Sheldon Reed, the director of the Dight Institute from 1947 to 1977, objected to because it connoted toothpastes, deodorants, and the like. It was Reed who invented the term that eventually prevailed—"genetic counseling."[6]

People tended to seek genetic counseling either because they wondered about a seemingly hereditary pattern of disease or deficiency in their families or because a child already born to them was afflicted with what they or their physician suspected was a genetic disorder. By the fifties, for a tiny number of disorders, genetic counselors could tell from biochemical tests whether either potential parent carried the deleterious recessive gene. For the most part, they could also provide informed estimates of the risk that another child might be born with the same disorder as its sibling. If, for example, the parents had previously conceived a child with Tay-Sachs disease, the fatal disorder of the nervous system which is caused by the homozygous occurrence of a recessive gene, the laws of genetics dictated that the odds of their conceiving another were one in four. If they had previously conceived a child with a disease that was not conclusively identified as genetic—as was the case with a wide variety of disorders, ranging from anencephaly to schizophrenia—the calculation of risk was based upon statistical summaries of known family data. According to the empirically determined risk, if a woman had already borne a child with Down's syndrome the chances were four in a hundred that a succeeding child would be similarly afflicted. Genetic counselors of these early years had to agree with William Schull, who emphasized at a 1958 conference on the subject that genetic counseling was "certainly a very imperfect art even in the hands of the very best of us."[7]

In the forties and fifties, given the fact that counseling could provide prospective parents with little more than odds, there was no great demand for it. In 1951, there were ten genetic-counseling clinics in the United States, and perhaps three or four in Britain, counting the consultation that Lionel Penrose was willing to give at the Galton. By the end of the decade, there were perhaps thirty in the two countries combined, representing a steady but by no means striking increase. In Britain, such clinics were staffed by physicians. But, on the whole, physicians seemed to be indifferent to the kind of advisory services that genetic clinics could offer—particularly in the United States, where a large fraction of the counselors were Ph.D.s. Although a little over half the medical schools in this country and Canada in 1953 offered some instruction in genetics, full courses in medical genetics were offered only by seven of them—less than a tenth of the total. The others offered a few hours in the subject in such standard courses as anatomy.[8] But attitudes in the Anglo-American medical community toward the utility of human genetics were changing. The shift resulted in part because of the proselytizing activities of the two eugenic societies, still more because some exposure to genetics was included in an increasing number of medical school curricula. Specialists in blood diseases were beginning to recognize that diagnostic assistance could be provided by human geneticists working with hemoglobin disorders. Particularly important in dramatizing the medical value of human genetics was the discovery that Down's syndrome was the result of a chromosomal anomaly. (Victor McKusick recalled that after the announcement of trisomy-21 "doctors would notice that disorders ran in families, so they would send the patients over to have us look at their chromosomes.")[9]

In 1960, at McKusick's instigation and with the financial support of the National Foundation–March of Dimes, a summer course in human genetics aimed mainly at medical school faculty was established at the Jackson Laboratory in Bar Harbor, Maine. (A success from the outset, the program continues to thrive, teaching mouse genetics and human genetics to about one hundred people each year.) By 1972, courses in genetics were required in half of all American medical schools and offered in more than three-quarters, and there were at least five chaired departments of human genetics in their British counterparts. Pediatricians, in particular, had come to understand that human genetics was an important tool of postnatal diagnosis, which would permit the early commencement of proper care in cases of disease. While genetic disorders seemed to occur in fewer than one out of fifty births, they accounted for one out of eight infant deaths.[10]

The paradigm therapeutic case was, of course, PKU. Contrary to

Lionel Penrose's original belief, the detection of phenylpyruvic acid in the urine did not definitely indicate the presence of the disease. But in the early sixties, Robert Guthrie, a physician at the School of Medicine of the State University of New York at Buffalo, perfected a far more reliable test, using a strain of *Bacillus subtilis* which would grow in a cultured sample of blood only if the sample was abnormally rich in phenylalanine. Even before the new test was devised, Britain had established a program to screen newborn children for PKU. After the advent of the Guthrie test, a number of American states and municipalities set up such programs. Between 1966 and 1974, a screening program in New York City identified fifty-one PKU infants. The screening cost came to less than a dollar per child, and over the eight years the total spent on the city's program came to no more than a million dollars, as compared to the cost—then estimated to be some thirteen million dollars—that would have been required to keep these children in institutions.[11]

The evident cost-effectiveness of the PKU programs strongly suggested widening the scope of postnatal screening. By 1971, almost nine hundred maladies had been definitely identified as single-gene disorders—the phrase denotes a disorder caused either by a dominant gene, a sex-linked gene, or two recessive ones—and a thousand more were suspected of having single-gene origins. At least a hundred of the former could be treated. Many single-gene disorders did not lend themselves to mass testing, but with no great additional effort or cost the existing PKU programs could be modified to test for some fifteen additional inborn errors of metabolism —galactosemia, for example, which arises from the inability of the body to process galactose, a substance derived from milk. The disorder's consequences—liver enlargement, cataracts, mental retardation, and, not uncommonly, infant death—could be avoided with early detection and the administration of a galactose-free diet.[12]

The detection in the newborn of single-gene disorders was increasingly complemented by the ability to recognize the recessive potential for them in prospective parents. By the early seventies, carriers of at least fifty genetic disorders could be identified.[13] No one argued seriously for the screening of every possible parent, but some did urge the screening of people from groups at comparatively high risk for particular genetic diseases, notably blacks—in the United States, one in twelve has a recessive sickle-cell gene. Demand for sickle-cell screening arose around 1970, partly from within the black community. The *Journal of the American Medical Association* endorsed the view of Dr. Robert Scott, a hematologist at the Medical College of the Virginia Commonwealth University, who called for the establishment of a screening program for blacks of marriageable age. If both members of a couple were discovered to carry the sickle-cell trait, they could shape their reproductive plans in the awareness of the one-in-four chance that their baby would suffer from sickle-cell anemia. The screening idea, supported in both lay and medical quarters, caught on rapidly. Beginning in 1971, sickle-cell-screening laws were enacted in seventeen states, often under the sponsorship of black legislators. In 1972, with the blessing of the Nixon administration—the President had proposed reversing the nation's "sad and shameful" neglect of the disease—Congress passed the National Sickle Cell Anemia Control Act, which provided for research, screening, counseling, and education.[14]

Blacks, of course, were not the only ethnic group susceptible to a genetic disease. Americans of Mediterranean extraction were at high risk for thalassemia (Cooley's anemia), and Ashkenazic Jews for Tay-Sachs disease. In 1969, Tay-Sachs was shown to result from yet another inborn metabolic error, the absence of the enzyme hexosaminidase A, and in short order a simple test—an examination of blood serum to determine whether the Hex A activity was lower than normal—was developed to identify heterozygotic carriers of the gene. In 1971, Dr. Michael Kaback, of The Johns Hopkins University School of Medicine, and Dr. Robert Zeiger, of the National Cancer Institute, mounted a pilot program of voluntary screening for Tay-Sachs carriers among a total of eighty thousand Ashkenazic Jews of childbearing age in the Greater Baltimore area. In two years, ten thousand people volunteered to be tested, and the Tay-Sachs gene was found in about one out of thirty of them.[15]

In 1972, the National Cooley's Anemia Control Act, which contained provisions similar to those in its sickle-cell predecessor, was signed into law. In the wake of the success of the Tay-Sachs program in Baltimore, similar screening efforts began in many other American cities, and Senator Jacob Javits, of New York, introduced a bill for a National Tay-Sachs Control Act. Medical, lay, and political commentators promptly called a halt: the Congress seemed headed for regular action on the ethnic genetic disease of the month. Javits agreed that a more general approach was preferable, and so did the spokesmen for other genetic disease interest groups. In the spring of 1976, Congress passed the National Genetic Diseases Act, which absorbed its two predecessors and provided for research, screening, counseling, and education in Tay-Sachs and various other disorders, including cystic fibrosis, Huntington's disease, and muscular dystrophy. Assisted by federal dollars, a number of states enlarged their postnatal screening programs to encompass tests for such additional inborn metabolic errors as galactosemia. By 1975 almost half a million people had been screened for sickle-cell trait and tens of thousands more had been tested for Tay-Sachs or thalassemia.[16]

By this time, genetic prognosis had gone beyond counseling prospective parents on the odds of conceiving a genetically diseased child. The procedure known as amniocentesis had come into widespread use by the late sixties. In amniocentesis, a long needle is inserted in the uterus, and fluid containing fetal cells is withdrawn from the amniotic sac. This procedure had first been used to assess whether an infant would suffer from Rh-factor disease and thus require transfusion immediately after birth; it was later used to identify fetal sex—primarily to determine whether the fetus might suffer from a sex-linked disease. In the late sixties, fetal cells were cultured to diagnose the rapidly growing list of chromosomal disorders, and of genetic disorders detectable by biochemical means. By the mid-seventies, virtually all of the hundred or so known chromosomal disorders could be detected in utero, and so could twenty-three inborn errors of metabolism, including the error that produced Tay-Sachs disease; almost forty more seemed potentially detectable.[17] If the fetus was found to suffer from a disorder, the prospective parents could elect abortion. The right of abortion had been secured in Britain in 1967, by an act of Parliament that permitted the termination of pregnancy for various reasons, including— according to what has sometimes been called "the eugenic clause"—substantial risk that the child would be seriously handicapped as a result of physical or mental abnormalities.[18] In the United States, the right of abortion was rendered constitutional by the Supreme Court's 1973 ruling in the case of *Roe* v. *Wade.*

Amniocentesis and legalized abortion together stimulated a major boom in prenatal genetic diagnosis. Prior to 1976 only some five thousand prenatal diagnoses of genetic disorders seem to have been carried out in the United States, and about seventy-five hundred were conducted in Great Britain. After that date, the number rose rapidly in both countries, reaching at least twenty thousand annually in the former and seven thousand in the latter. In 1960, there had been between thirty and forty clinics and counseling centers in the United States. By 1974, the number had jumped to about four hundred. Almost a quarter of these were established and maintained with assistance from the National Foundation–March of Dimes, which, its war against polio having been won, had turned in the sixties to the subject of congenital muscular-skeletal disorders, and in 1970 had decided to mount a major effort in the general area of birth defects.[19] While in the United States genetic counseling was unregulated—a number of the "centers" were small units manned by Ph.D.s in genetics—in Britain it was, for the most part, practiced by qualified physicians and was governed as a branch of medicine. In the mid-seventies, British physicians began to integrate their genetic counseling units, including prenatal diagnostic services and laboratories, into the National Health Service, planning to provide, ultimately, at least two clinical geneticists in each of the country's health service regions of three to five million people.[20]

In the early years of genetic counseling, some geneticists had sought to turn the practice to eugenic advantage—to reduce the incidence of genetic disease in the population, and by extension to reduce the frequency of deleterious genes in what population geneticists were coming to call the human gene pool. To that end, some claimed that it was the counselor's duty not simply to inform couples about the possible genetic outcome of their union but also to instruct them whether or not to bear children at all. Through the fifties, however, the standards of genetic counseling had turned strongly against eugenically oriented advice—that is, advice aimed at the welfare of the gene pool rather than that of the family. The standards also had it that no counselor had the right to tell a couple not to have a child, even for the sake of the couple's own welfare.[21] Whatever the standards, Lionel Penrose noted in 1969 that a large fraction of the patients who sought genetic advice acted in a way that "would be considered generally to be reasonable"—that is, "they avoid risks which are serious and accept those which are only moderate"—and he predicted that "the result of skillful counselling, over a long period of years, will undoubtedly be to diminish, very slightly but progressively, the amount of severe hereditary diseases in the population." Perhaps so, but James Neel observed, in a paper for a 1971 symposium on ethical issues in human genetics, "Any population policy—or for that matter, no population policy—may have implications more far-reaching for the gene pool than all the genetic counseling of the next hundred years."[22]

THE POSTWAR POPULATION explosion had mocked prewar demographic predictions. "Thirty years ago," J. B. S. Haldane ruefully remarked in 1963, "statisticians were writing about 'the twilight of parenthood,' 'les berceaux vides'. . . and I was fool enough to believe them." Frederick Osborn found no cause for anxiety in the American statistics. They revealed that the middle and upper middle classes were contributing mightily to the baby boom, and that educated groups appeared to be reproducing at a rate sufficient to replace themselves. In fact, according to an influential series of studies published in the mid-sixties by the young population geneticist Carl Jay Bajema, the net reproduction rate in the United States tended to be higher among people with above-average I.Q.s than among those below the average. In Britain, the 1951 census hinted that the educated classes, like their American counterparts, were coming closer to the mark of their reproductive duty.[23]

In the postwar era, Anglo-American eugenic attention extended to the global population explosion—particularly in what was then called the underdeveloped world. No doubt for some reform eugenicists the rapidly multiplying populations of Asia, the Middle East, Africa, and Latin America represented some sort of immense "social problem group." Yet it required no race prejudice to find a good deal that was dysgenic in the proliferation of people in environments that offered inadequate food, housing, education, and medical care.[24] After the war, eugenicists renewed their advocacy of contraception, finding in it the principal instrument for dealing with the population "crisis," including eugenically adverse differential birthrates at home (which Huxley and Osborn both perceived in the comparatively high fertility of groups at the social bottom, particularly the lowest-income blacks in the United States). A number of them—a leading example was Osborn, who organized the Population Council in 1952—became prominent population-control activists. But although a diversion for eugenic energies, population control, focused as it was on the issue of quantity, addressed only grossly the eugenic interest in improving the human gene pool. That interest remained vital to the reformist outlook.[25] The matter of the quality of the gene pool was explicitly dealt with by Hermann Muller in his 1949 presidential address to the American Society of Human Genetics: "Our Load of Mutations."

Muller's analysis proceeded from the fact that the individual human genome was, like those of all other species, constantly subject to change by mutation. The mutation could be spontaneous or, as Muller had demonstrated in his own work, could be induced by radiation. Some mutations might lead to genetic improvement; most, it was believed, were deleterious, even deadly. In the main, mutations were recessive. But Muller drew upon recent research with fruit flies and that of Neel, among others, with human beings to point out that many mutant genes, although recessive, behaved as partially dominant genes which—as, for example, those for thalassemia minor—were partly expressed in the organism. In human beings, these mutant genes might make for greater susceptibility to cancer, diabetes, hypertension, or any number of infectious or mental disorders. Though a given single mutation might not be lethal, others might eventually crop up in the same individual genome. The gradual accumulation of these mutations, which would be spread through breeding, constituted the "genetic load" of the human race—the total number of potentially lethal genes in the human gene pool.[26]

According to Muller, genetic load reduced evolutionary fitness. An individual's load would be eliminated from the gene pool by his or her death before reproduction. But while the genetic load might diminish with pre-reproductive death, the loss was constantly offset by fresh mutations. In a stable species like man, the degree of load was assumed to be the amount that the species could tolerate at equilibrium—that is, the point where the rate at which disadvantageous mutations were created equaled the rate at which they were eliminated. Muller put the accumulated load at an average of eight genes per person, out of the tens of thousands each individual was estimated to carry. He further speculated that in primitive man this genetic load was sufficient to bring about the death from genetic causes of twenty percent of the human race in each generation.[27]

In this regard, modern man was no different from his prehistoric forebears. "Most of us have a nearly twenty-per-cent chance of death or of reproductive inefficacy from genetic causes," Muller declared, but he pointed out that mankind had recently ceased to live "under those comparatively primitive conditions . . . to which a rough genetic equilibrium must have become established."[28] Modern man, of course, benefited from improved sanitation, nutrition, housing, and medical care; and post-Hiroshima man might well find himself living in a higher radiation environment. Thus, deleterious genes were no longer being eliminated at the prehistoric rate, and an increase in radiation would speed up the rate at which mutations were induced. The human genetic load was getting bigger and would continue to grow. The greater the effectiveness of medicine, the greater the load that might be tolerated.

Muller estimated the degree of load that might be reached in eight generations (about two hundred and forty years), assuming a continued advance in medical technology: it would be the same as that expected from the absorption by all the parents in one generation of two hundred roentgens of gamma radiation—a dose comparable to the average at the surface within two kilometers of Ground Zero at Hiroshima. The greater the genetic load, Muller warned, the more pitiful and less recognizable as human would our descendants be. Instead of struggling with "external enemies of a primitive kind such as famine, climatic difficulties, and wild beasts," the human beings of the future "would be devoted chiefly to the effort to live carefully, to spare and to prop up their own feeblenesses, to soothe their inner disharmonies, and, in general, to doctor themselves as effectively as possible." He concluded that "everyone would be an invalid, with his own special familial twists."[29]

Muller's belief that the therapeutic powers of modern civilization were working dysgenic effects echoed early-twentieth-century eugenics. Time and the cold war had tempered his socialism. Still, his theory differed from the mainline creed in that it did not identify dysgenic trends with race or class—mutations occurred in all sectors of society—and was couched in socially antiseptic, genetic language.[30] Advanced with a Nobelist's authority, the specter of genetic load pervaded the debates over the genetic effects of atomic radiation. It also formed a central tenet in the reform-eugenic response to the population explosion. Julian Huxley fretfully declared in the course of a 1963 London symposium on the future of the human species: "The population explosion is making us ask . . . What are people for?

Whatever the answer . . . it is clear that the general quality of the world's population is not very high, is beginning to deteriorate, and should and could be improved. It is deteriorating, thanks to genetic defectives who would otherwise have died being kept alive, and thanks to the crop of new mutations due to fallout. In modern man, the direction of genetic evolution has started to change its sign from positive to negative, from advance to retreat: we must manage to put it back on its age-old course of positive improvement."[31]

To Muller, meeting the mutation problem required the exercise of eugenic reproductive control. He explained to a physician in 1954 that "the fact that the so-called eugenics of the past was so mistaken . . . is no more argument against eugenics as a general proposition than, say, the failure of democracy in ancient Greece is a valid argument against democracy in general." In the standard eugenic vein, Muller argued for both diminishing reproduction among high-load people—presumably identifiable by their genetic diseases or disorders—and increasing it among those blessed with especially valuable genes. He recognized that in the wake of the Nazis people would not tolerate compulsory interference with human reproduction. "I think much of 'negative eugenics,' such as compulsory sterilization of alcoholics or criminals, is definitely out," he wrote to a correspondent in California. Muller expected people with high genetic loads to refrain voluntarily from procreation out of a sense of social duty. Similarly, it would be considered "a social service for those more fortunately endowed to reproduce to more than the average extent."[32]

At a 1959 conference at the University of Chicago celebrating the centenary of Darwin's *Origin of Species*, Muller presented a paper reviving the old idea of eutelegenesis as his positive-eugenic method for offsetting the effects of increased genetic load. (He had no ally in Herbert Brewer this time. After the war, Brewer had abandoned eugenics in revulsion at what the Nazis had made of it.) He soon came to call the latter-day version of the plan "germinal choice." In Muller's view, recent developments in the field of artificial insemination—particularly the demonstrated success of freezing and storing sperm, then thawing them for vaginal injection—enhanced the plan's prospects for success. The preservation techniques—dry ice had been used first, then the much colder liquid nitrogen—allowed for the accumulation of sperm from a given donor, and as the number of sperm increased so did the chance of producing a pregnancy. More important to Muller's positive-eugenic purpose, the frozen sperm might well be stored until, say, twenty years after the death of the donor. By that time, it could be better judged whether the donor, outstanding in life, seemed truly outstanding in calm retrospection. Thus the effort to guide man's evolution could be kept to the highest standard.[33]

Muller recognized that germinal choice, neat as it seemed, raised a number of vexing difficulties. People might well confuse choice with coercion. While the divorce of sex from procreation had taken strong social hold, the emancipation of procreation from sex was as yet hardly consonant with prevailing values. Artificial insemination was practiced mainly to overcome marital infertility. The donor's identity was normally kept secret from the prospective parents, and this worked squarely against the idea of knowingly choosing a superior father. Besides, exemplary donors might well carry their share of genetic load. And since the genetic basis for superior traits was hardly known there was no way to predict the outcome of any particular conception. Germinal choice would merely weight the results in favor of the preferred procreative consequence, not guarantee it.[34] Yet Muller felt that high-minded couples would be willing to forgo the guarantee to themselves as parents for the sake of what the collective process would yield—an increase in mankind's genetic quality. Eventually, the wise use of selection could breed out the load of disadvantageous genes from the limited fund of advantageous ones. Muller recognized that fears of coercion might arise, but he insisted that germinal choice would be strictly voluntary. And surely in the beginning a few couples would be willing to break with social convention and pioneer the procreational revolution.[35]

In the early sixties, with the aim of getting that revolution started at least on a modest scale, Muller looked into the establishment of a Foundation for Germinal Choice. Some of his old allies responded with advice or encouragement, among them C. P. Blacker, Frederick Osborn, and J. B. S. Haldane. So did some new ones, including a claque of a different cut. One of these was Robert K. Graham, a millionaire who had pioneered the development of shatterproof plastic eye-glasses and was the president and chairman of the board of the Armorlite Lens Company, in Pasadena, California. At a meeting with Muller in June 1963, Graham agreed to provide a thousand dollars to establish and about three hundred dollars a year to maintain a liquid-nitrogen repository for the sperm of outstanding men. High intelligence and altruism were to be among the primary criteria for donors. Muller thought that Julian Huxley would be an ideal donor. Graham suggested Muller himself, who, however, stipulated that his sperm not be used until twenty-five years after his death.[36]

But Graham, a political conservative, put too much emphasis for Muller's old-socialist taste on the genetic increase of intelligence and too little on the genetic increase of altruism. Graham's views, Muller thought, smacked enough of the old eugenics to jeopardize the germinal-choice project. Muller dissociated himself from Graham and abandoned plans for the foundation. Nevertheless, in 1971, four years after Muller's death and despite objections from his widow, Graham created the Hermann J. Muller Repository for Germinal Choice. A few years later, he began to collect donations of sperm, exclusively from Nobel laureates—the physicist William Shockley was a donor (the only one to reveal his name)—and to look for healthy and intelligent female recipients. Now housed in an office building in Escondido, California, and formally titled the Repository for Germinal Choice—but with Muller's name listed in its brochure as a co-founder—the Sperm Bank, as it is commonly known, has relaxed the Nobel requirement for donors. Its frozen deposits include, however, only the gametes of scientists. The Repository claims that fifteen offspring now owe their paternity to it.[37]

Germinal choice stimulated a good deal of ridicule at an anticipatory distance. Members of the Anglo-American genetics community tended to judge it either socially impractical or scientifically unworkable, or both. (There was word of a telling endorsement for germinal choice from a special, non-scientific quarter. Aldous Huxley was all for the scheme, according to the report of one of Muller's acolytes, who had chatted with the author of *Brave New World.* Huxley considered it far superior to the approach of the activists in the early eugenic societies who had wanted to sterilize their genetic inferiors.) Yet well into the sixties Muller and his ideas occupied center stage at scientific symposia, and he saw several versions of his 1959 paper into learned print. Julian Huxley invoked Darwin's success at forging a theory of evolution despite his ignorance of genetics to scoff at the claim that one had to know more about human heredity before one could think of human biological improvement.[38]

Support of the principle of germinal choice came, in varying degrees, from points across the spectrum of evolutionists—from the Harvard University systematist Ernst Mayr to the University of Wisconsin population geneticist James F. Crow. Natural selection, Crow observed, was "cruel, blundering, inefficient," while deliberate human selection could be based on criteria of "health, intelligence, or happiness." Francis Crick pronounced himself in agreement with "practically everything" that Muller had to say, and went on to wonder "why people should have the right to have children." (Perhaps, Crick mused, one might have a "licensing scheme," so that "if the parents were genetically unfavorable, they might be allowed to have only one child, or possibly two under certain special circumstances.")[39] In 1969, in a *Life* feature entitled "The Second Genesis," the respected science reporter Albert Rosenfeld declared, "We are now entering an era when, as a result of new scientific discoveries, some mind-boggling things are likely to happen. Children may routinely be born of geographically separated or even long dead parents, virgin births may become relatively common, women may give birth to other women's children, romance and genetics may finally be separated, and a few favored men may be called upon to father thousands of babies."[40]

No DOUBT THE ATTENTION given germinal choice reflected the concern within the biological community and its thoughtful public over the impact of the population explosion. No doubt for a time it reflected the apprehension over the genetic effects of increased radioactivity. Yet germinal choice drew more widespread and serious consideration than had the eutelegenesis of the thirties probably because Muller's advocacy of it coincided with the arrival—or, at least, proclamations of the arrival—of the new biological revolution, genetic engineering.

The term "genetic engineering" was coined in 1965 and rapidly came to denote a cluster of micro-manipulations of the reproductive or hereditary process, some of which, like cloning, had little to do with genetics. Cloning was originally a botanical technique of asexual reproduction by means of a "cutting."[41] In the genetic-engineering sense, the idea was to select an individual with desirable traits, remove from one of his or her cells the nucleus with its full set of chromosomes, and substitute it for the nucleus in a recently fertilized egg. The remodeled egg, implanted in a uterus, would then develop into a fully formed fetus with precisely the same genetic complement as the original individual. To some, cloning seemed eugenically preferable to germinal choice. Germinal choice, joining in the traditional fashion the paternal seed with the maternal egg, allowed for the reproduction in the offspring of only half the genes of the treasured donor. Cloning produced the donor's exact genetic duplicate.[42]

By the end of the sixties, only an amphibian—a frog—had ever been successfully cloned. The cloning of embryonic mammals appeared in principle within reach, but the cloning of, say, a grown human being, whose genome would have lost its developmental potency, seemed to lie somewhere in the science-fiction future.[43] Not so the test-tube fertilization of human embryos. In 1970, Robert G. Edwards, a physiologist at Cambridge University, reported in London to a symposium on the social impact of modern biology, "We can now recover eggs from women some two or three hours before ovulation, fertilize them in culture using ejaculated spermatozoa from the husband, and grow some of them into blastocysts" —that is, the embryonic bundle of cells ready to implant itself on the uterine wall. Edwards and his collaborator, the gynecologist Patrick Steptoe, had so far failed to accomplish implantation and thus create a pregnancy; they intended to keep trying.[44]

Steptoe and Edwards's aim was to enable women who were infertile by reason of blocked oviducts to have natural children, but there was speculation that the technique could be extended to genetic engineering. Echoes of ectogenesis were heard in predictions that science might soon learn how to manage gestation from conception to term outside the wombs of women, in the laboratory. Edwards himself suggested that embryos could be typed, and that those found with the genes or chromosomes for disease could be rejected prior to implantation—a procedure that most couples would surely consider preferable to abortion. Or, he added, embryos might even be modified, to rid them of defect.[45]

The prospect of genetic modification stemmed mainly from the triumphs of molecular biology—especially the working out of the genetic code. By the mid-sixties, a satisfactory theory had been formulated of how the genetic information contained in DNA governed the making of proteins in organisms. The theory started with the structure of DNA, with its double helical strands joined at regular intervals across the distance between them by one of either two nucleotide base pairs—adenine with thymine, or cytosine with guanine. The genetic information resided in the sequence of nucleotides along the helix, with particular sequences of three nucleotides representing particular amino acids. Through a complicated biochemical mechanism, a series of such triplets was translated at a cellular site into a chain of amino acids, which enfolded themselves into a specific protein—for example, a constituent of the eye—involved in the organism's structure or, as in the case of an enzyme, figuring in one of its processes, like metabolism. There were triumphs in the laboratory, too—notably, in 1967, the duplication of DNA in a test tube by Arthur Kornberg and collaborators at Stanford University; in 1969, the isolation of a gene from a natural cell by Jonathan Beckwith with colleagues at the Harvard Medical School; and, in 1970, the first synthesis of a gene by H. Gobind Khorana, who led a group at the University of Wisconsin in constructing a specific strand of DNA by linking one nucleotide base-pair to another.[46]

The rapid forward march of genetics opened stunning new vistas of human biological engineering. Genes could in principle be repaired simply by modifying a few nucleotides along the DNA molecule or fashioned *de novo* by stringing together the right nucleotide sequence. Genetic manipulation, Albert Rosenfeld told readers of his 1969 article, might well "bring into being new species of creatures never before seen or imagined in the universe."[47]

Of course, virtually all molecular genetic triumphs in the laboratory were brought about with microorganisms, particularly *E. coli*, the workhorse bacterium of the field; what might be accomplished in principle with human beings was at the opening of the seventies distant from practical achievement. It might be known that the human genetic code was carried in the coils of DNA compacted in the chromosomes of the cell nucleus; it was unknown where each gene was situated in the tangle, and what function each fulfilled. Human geneticists, and even most molecular biologists, recognized these difficulties, but in the opinion of the latter group especially they were difficulties that molecular genetics would eventually overcome.

Assessments of relative size differentiated one chromosome from another only in a rough fashion and permitted the definite determination of only five of the twenty-two human autosomes (that is, the chromosomes independent of sex). But in 1968 Torbjörn Caspersson, of the Karolinska Institute in Stockholm, advanced a new, more precise method of identifying chromosomes—by banding. The method involved treating a chromosome with a fluorescent substance, usually quinacrine mustard, which marked it with a particular pattern of fluorescent bands. In the eventual explanation of why the process worked, the quinacrine molecules slipped themselves between the nucleotide pairs along the axis of the DNA helix, and their fluorescence was quenched when they were adjacent to the pair containing guanine while it was unaffected when they were next to that containing adenine. Since the distribution of guanine along the helix varied from one chromosome to another, the distribution of fluorescence—the result of the position, width, and brightness in each of the bands—did, too, thus providing each chromosome with an unambiguous label.[48]

In the early sixties, various biologists had begun to develop the techniques of "somatic-cell genetics"—the genetics to be learned from the culturing of ordinary body cells. The techniques included the culturing of hybridized cells made by fusing the cells of two closely related species. The mouse had long been used as a reliable biological proxy for human beings. Mice were genetically close enough to man—they have forty chromosomes, compared with man's forty-six—to make the hybrid viable. In 1967, at New York University, Howard Green and Mary Weiss managed to culture mouse/human hybrid cells through a number of generations. They found that as the cell generations accumulated, the number of human chromosomes gradually diminished—a fact with major implications for human genetic research.[49]

Consider, for example, that the production of enzymes is controlled by genes. Researchers would monitor the enzymatic activity in each generation of hybridized cells. The activity coming from genes on the mouse chromosomes could be differentiated from that coming from genes on the human ones. If the activity governed by one of the human enzymes disappeared in the same generation as one of the human chromosomes, then researchers would know that the vanished chromosome had contained the gene for that enzyme. If the activity for two always disappeared together, then they would know that the genes for the two were linked. In 1968, Green and Weiss, among others, demonstrated that the gene for the enzyme thymidine kinase was situated on the 17-chromosome. This was the first assignment of a single gene to one of the autosomal chromosomes. Other assignments followed rapidly, and in 1974 James Neel noted that there were reports of "a new linkage every month." Identifying the chromosomal seat of a specific human gene did not reveal exactly where on the chromosome that gene—one of a thousand or more—was situated; nevertheless, the cartography of the human genome had begun.[50]

The methods of recombinant DNA, more far-reaching in their implications, developed from the deployment in the early seventies of "restriction enzymes." These remarkable proteins will cut a given strand of DNA at particular points in its sequence of nucleotides. By appropriate manipulation of the enzymes, a fragment can be snipped from the DNA of one organism—say, a human being—and spliced into that of another, like a bacterium. Inserted in *E. coli,* the recombined DNA will reproduce as rapidly as its bacterial host. The process can thus provide numerous copies of the original fragment—and of whatever biochemical products the code on the fragment would produce.

Recombinant DNA offered startling new means of biochemical synthesis and the design of new microorganisms for the performance of specific biochemical tasks. These promises figured in the so-called biotech boom. Others added fresh vigor to the speculations concerning the possibility of manipulating the human genome, for recombinant DNA techniques promised a quantum leap over one of the major obstacles in the way of human genetic engineering—the difficulty of isolating particular human genes. Once isolated, it was said, a given gene or set of genes could be identified, produced in quantity, and, in principle, inserted in the malfunctioning cells of a person suffering from a genetic disease. An adult diabetic, for example, could then obtain his own insulin internally rather than from daily injections. Genetic surgery could be performed on infants, or even on blastocysts. Indeed, an entire human genome could perhaps someday be tailored to whatever specifications might be desired.[51]

Even before the advent of recombinant DNA, Robert L. Sinsheimer, the distinguished molecular biologist at the California Institute of Technology, reached back to Francis Galton to wrest a deeper social meaning from the trends in molecular biology. In 1969, he declared, "A new eugenics has arisen, based upon the dramatic increase in our understanding of the biochemistry of heredity and our comprehension of the craft and means of evolution." This new eugenics was not to be confused with the old version of Galton's, Sinsheimer continued. It would require no large-scale social program over many generations, and no pervasive program of social control. The new eugenics could be accomplished on an individual basis: "The old eugenics would have required a continual selection for breeding of the fit, and a culling of the unfit. The new eugenics would permit in principle the conversion of all of the unfit to the highest genetic level. The old eugenics was limited to a numerical enhancement of the best of our existing gene pool. The horizons of the new eugenics are in principle boundless—for we should have the potential to create new genes and new qualities yet undreamed."[52]

Sinsheimer thought the possibility "potentially one of the most important concepts to arise in the history of mankind," and he concluded, "Indeed, this concept marks a turning point in the whole evolution of life. For the first time in all time, a living creature understands its origin and can undertake to design its future. Even in the ancient myths man was constrained by his essence. He could not rise above his nature to chart his destiny. Today we can envision that chance—and its dark companion of awesome choice and responsibility."[53]

Chapter XVIII

VARIETIES
OF PRESUMPTUOUSNESS

THE ARRIVAL OF THE new eugenics coincided with a sea change in the Anglo-American sociopolitical environment. What had long been assumed—namely, that the principal cause of social pathology was nurture rather than nature—was once again under challenge in the United States and Britain. The challenge was implicit in the so-called white backlash in America; in the declaration of Sir Keith Joseph, who eventually became Margaret Thatcher's Minister for Education and Science, that the poor, proliferating to excess, were leading Britain to "degeneration"; in the budgetary neglect of education in both countries; in the political trend that eventually put Ronald Reagan and Thatcher in office.[1] In the academic world, the challenge was made explicit by the revival of attention given to the issue of race and intelligence.

No single publication did more to precipitate the revival than Arthur R. Jensen's 1969 article in the *Harvard Educational Review*, "How Much Can We Boost IQ and Scholastic Achievement?" Jensen, a professor of education and psychology at the University of California at Berkeley, later insisted that attention to his title question had been "all but completely stifled" by the "zeitgeist of environmentalist egalitarianism." Jensen declared himself an egalitarian of a certain type: he stood staunchly for complete equality of social, economic, and educational opportunity for individuals. His quarrel with the zeitgeist was that it stifled attention to his title question and that it inclined analysts and policymakers to treat people not as individuals but as groups. Currently, for example, the zeitgeist was encouraging a nurture-oriented response—compensatory education and the like—to the problem that blacks as a group did not do as well in standard academic competition as whites as a group. This response presumed that in the sort of abstract abilities which figured in academic work blacks as a

group were by nature as capable as whites. Yet the data of black performance on I.Q. tests were well known: on the average, blacks scored about
fifteen points lower than whites. Environmental deprivation was the prevailing explanation of the difference, but the prevailing explanation, Jensen
declared, could be wrong.[2]

"The possible importance of genetic factors in racial and behavioral
differences has been greatly ignored," Jensen asserted, "almost to the point
of being a tabooed subject, just as were the topics of venereal disease and
birth control a generation or so ago." They ought not to be ignored, he
went on. He noted that races differed physiologically, anatomically, and
biochemically, and declared, "There is no reason to suppose that the brain
should be exempt from this generalization." It was reasonable "to hypothe-
size that genetic factors may play a part" in racial differences in I.Q.-test
performance.[3]

Jensen had steeped himself in the principles, methods, and results of
human population genetics. His article, a hundred and twenty-three
pages long, reviewed the extensive literature on the subject which explored the issue of heredity and intelligence—principally twin and foster-
child studies extending back forty years or more. Jensen advanced his
assessments with the standard hedges ("there might be") and double
negatives ("it seems not unreasonable") of scholarly discourse. Yet Jensen
was an intellectual disciple of the British hereditarian psychologist Cyril
Burt—it was through Burt's works that he had first approached the issue
of heredity and intelligence. For all the hedges, there was no mistaking
his belief: the average difference in I.Q. scores between blacks and whites
indicated a highly probable average difference in native scholastic intelligence.[4]

On both sides of the Atlantic, Jensen's writings invigorated the
hereditarian school of thought on intelligence, including the wing that was
little if at all concerned with race. Richard Herrnstein, professor of psychology at Harvard, called attention in *The Atlantic* to the considerable data
suggesting that the occupational hierarchy in American society was
strongly correlated with grades of intelligence, and he went on to reason
that, assuming intelligence to be strongly heritable, the United States might
be turning increasingly into a "hereditary meritocracy." Though he found
the prospect troubling, the syllogisms of social and genetic science compelled him to observe that in the future, for example, "the tendency to be
unemployed may run in the genes of a family about as certainly as bad teeth
do now." The most prominent champion of hereditarianism in England
was Hans J. Eysenck, who had been a protégé of Burt's as an undergraduate
and was now a professor of psychology in the University of London
Institute of Psychiatry. As a result of immigration from regions of the

former empire, Britain was beginning to experience the kind of racial strains that had long afflicted the United States. While, however, Eysenck noted that research was "revealing considerable scholastic backwardness and low IQ scores among colored children," what preoccupied him was not race but, to use the title of a book he published in 1973, "the inequality of man."[5]

Eysenck faulted Herrnstein for giving insufficient weight in his model of a hereditary meritocracy to the phenomenon of regression. Like Herbert Spencer Jennings a half-century earlier, he argued that genetic recombination tended to give children at the lower end of society more, and at the upper end fewer, capabilities than their parents, and that such reassortment could make for mobility up and down the social ladder. Nevertheless, Eysenck insisted that low social class was no handicap to performance on I.Q. tests, that I.Q. in part determined position on the social ladder, and that any child's I.Q. was "largely inherited." Such logic led Eysenck to the same educational implications that Jensen had drawn: school policy ought to aim at a diversified curriculum, which meant a scholastic course for those children—even if, in the main, socially advantaged—who could benefit from it, and something different for those children—even if, for the most part, socially deprived—who by virtue of their genes could not. Like Jennings and Herrnstein, Eysenck counted himself a social liberal, yet in consonance with them he predicted that social reform was doomed to fail unless it "takes into account limitations set by inexorable biological"—meaning, genetic—"facts."[6]

Jensen coupled black-white genetic differences with the high inner-city birthrate to raise the hoariest of mainline-eugenic issues—the possibility of dysgenic trends in urban slums. For several years, William Shockley had been arguing that the failure to explore fully the subject of race and intelligence was keeping society ignorant of the knowledge to combat such trends. In October 1969, apparently reenergized by Jensen, Shockley urged the National Academy of Sciences to encourage research into the possibility that the quality of the United States population was deteriorating genetically—a proposal that the Academy considered but that it eventually went no further toward endorsing than to recommend greater interdisciplinary cooperation between behavioral genetics and such fields as psychology and education. In the spring of 1971, during the Academy's annual meeting, Shockley explained to reporters: "Diagnosis will, I believe, confirm that our nobly intended welfare programs are promoting dysgenics—retrogressive evolution through the disproportionate reproduction of the genetically disadvantaged." Shockley thought that such reproduction was "so much more severe" among blacks than among whites that the disparity threatened the next generation of blacks with "genetic enslavement"—a predestined subordination akin to that of one of the lower-ranking breeds in *Brave New World.*[7]

Edward O. Wilson went far beyond genetic theories of intelligence in the sweeping hereditarianism with which he interpreted human social behavior. Wilson, a professor of zoology at Harvard University and one of the world's leading authorities on insect societies, was a principal figure in the newly emerging discipline of sociobiology. Sociobiology, as he put it, was concerned with "the study of the biological basis of social behavior in every kind of organism, including man." Its practitioners took an interdisciplinary approach to their subject, and—what made it new, according to Wilson—sought to forge the multidisciplinary insights they attained into a coherent structure that was consistent with the principles of ecology and genetics.[8] Wilson brilliantly summarized a great quantity of research that had been done in the field—virtually all of it on non-human species—in a 1975 treatise, *Sociobiology: The New Synthesis.* Despite its seven hundred pages and half-million words, the book provoked considerable attention in lay as well as professional journals, and in 1977 the subject of sociobiology reached the cover of *Time* magazine. What stimulated most of the popular interest was the last of Wilson's twenty-seven chapters, which was devoted to a speculative analysis of human social behavior. He subsequently expanded upon the human aspects of the subject in various articles, including one in *The New York Times Magazine,* and then in his 1978 book *On Human Nature.*[9]

In these writings, Wilson explained how sociobiologists approached the problem of identifying behavioral traits in which genes played a role. Each living form could be viewed as "an evolutionary experiment, a product of millions of years of interaction between genes and environment." Through the close comparison and contrasting of different "experiments," it was possible to construct the principles of a genetics of behavioral evolution. The extension of the analysis to man was admittedly difficult and tricky, but it could be accomplished by observing species closely related to human beings, and taking into account what was known about human hunter-gatherer societies. Traits that were consistent throughout the order Primates, Wilson wrote, were "likely to have persisted in unaltered form into the evolution of *Homo.*" (In contrast, traits that were not thus consistent across the order could not safely be extrapolated.) Chimpanzees were, in fact, "close enough to ourselves in the details of their social life and mental properties to rank as human in certain domains," and such findings added weight to "the hypothesis that human social behavior rests on a genetic foundation—that human behavior is, to be more precise, organized by some genes that are shared with closely related species and others that are unique to the human species."[10]

Centrally important to sociobiology was a behavioral trait that evolutionary biologists had long noticed, and had come to call "altruism." Self-sacrificial acts seemed to be commonplace among many animals—for example, honeybee workers, which would sting at mortal cost to themselves for the sake of the hive, or certain small birds which would whistle upon the approach of a hawk, placing themselves in jeopardy to warn the rest of the flock. Yet the prevalence of altruism—and of the presumed genes for such behavior—posed a paradox for evolutionary theory: on the one hand, natural selection was supposed to favor traits that assisted individuals to survive and reproduce, thus spreading their genes into the next generation; on the other, there seemed to be a selective advantage in a trait of self-sacrifice that reduced the individual's chances of survival and reproduction. To evolutionists, the key to the paradox possibly lay in the fact that every individual shared some fraction of its genes with its relatives, and the closer the relative, the greater the fraction, on the average, of genes that were shared. The individual honeybee worker might sacrifice its own genes in the defense of the hive, but its act would help to perpetuate the gene pool of its kin group, of which some of its own were a part. Many evolutionary biologists thus resolved the paradox with the theory of "kin selection"—the idea that natural selection favored kinship groups comprised of at least some individuals with a behavioral tendency to surrender themselves to the Darwinian good of the whole.[11]

The hypothesis of kin selection figured significantly in shaping Wilson's speculative extrapolations to man of the genetic traits that sociobiologists had found in animals. Salient among the traits were aggressiveness (protection of the group) and territoriality (safeguarding its ecological niche). There was also male dominance over females—to the end of maximizing the proliferation of favorable genes, since a single dominant male could repeatedly impregnate many different females, while a single female could herself be impregnated only periodically. (Polygyny, Wilson noted, was permitted in about three-fourths of all human societies; the taking of multiple husbands, in only about one in a hundred.) And there was the familiar altruism, which Wilson utilized to suggest, among other things, that homosexuality might have a genetic basis. (Homosexuality was strongly expressed in the most intelligent primates; and in primitive human societies, Wilson reasoned, homosexuals could have helped other members of the same sex in their tasks, perhaps benefiting their survival and reproduction rates, with the effect of increasing "the genes these individuals shared with the homosexual specialists.") Genetic drives, Wilson argued, might even lie at the emotional source of certain ethical propositions that mankind regarded highly—the "oughts" that parents should sacrifice for children, or citizens for the nation.[12]

Wilson considered himself politically and socially a liberal. From his genetic hypothesis of homosexuality he inferred that homosexuals ought not to be discriminated against "on the basis of a religious dogma supported by the unlikely assumption that they are biologically unnatural." He insisted that no genetic heritage of male dominance could justifiably be used "to argue for anything less than sex-blind admission [policies in colleges and professional schools] and free personal choice." He warned that while mankind might be genetically programmed to warlike behavior and to the maximizing of reproductivity, acting out the program could lead to global disaster. He declared that no genetic "is" should be confused with any moral "ought."

Nonetheless, Wilson wondered about man's ability to control himself for the greater moral good. Like Francis Galton, he ruefully perceived modern humanity as suffering from deep inner contradictions—eager to build a better world, yet bedeviled by the gene-based behavioral impulses of its prehistoric forebears. "To chart our destiny," he declared in a Galtonian vein, "means that we must shift from automatic control based on our biological properties to precise steering based on biological knowledge." The time had come for "ethics to be removed temporarily from the hands of philosophers and biologized." With greater self-knowledge, man could "hope to decide more judiciously which of the elements of human nature to cultivate and which to subvert, which to take open pleasure with and which to handle with care." Of course, knowledge by itself would not alter the fundamental constraints on human behavior, but human genetics would progress, knowledge about human behavior would accumulate, and genetic engineering might ultimately render the human genome alterable. "At the very least, slow evolutionary change will be feasible through conventional eugenics," Wilson predicted. In some distant future, the human species could "change its own nature."[13]

Wilson's sociobiological writings on man contained no Galton-like celebrations of the biological merits of any given social class, nor did they ratify the claims of Jensen or Shockley. Indeed, Wilson considered his work in no way an endorsement of the long history of outrages committed in the name of eugenics, and he dissociated himself entirely from genetic theories of racial differences in intelligence.[14] And yet, his books and essays were in fact salted with statements that could be—and in practice were—taken in a sense contrary to his good intentions. For example, even as he opposed discrimination against women, he declared in virtually the same breath that in hunter-gatherer societies a genetic bias had led men to hunt and women to stay at home; that this bias might still be "intense enough to cause a substantial division of labor even in the most free and most egalitarian of future societies"; and that "even with identical education and equal access to all professions, men are likely to continue to play a disproportionate role in political life, business and science."[15]

Wilson, moreover, seemed to contradict himself, at the very least implicitly, on a crucial issue: whether genes determined particular behaviors or simply made a range of behaviors possible. The lack of clarity on this point clouded his argument that greater genetic self-knowledge was a prerequisite for social self-control, for if behavior was genetically determined in some close sense, then it could hardly be subject to the governance of reason. Thus, short of changing the human genome, the broad spectrum of social behaviors that Wilson regarded as possibly genetic in origin—not only aggressiveness, xenophobia, and sexism but conformity, spite, and genocide—might well be fixed and unalterable, and so might the social practices and arrangements to which they gave rise. To many analysts, Wilson's sociobiology appeared to be a revival of the social Darwinism of the late nineteenth century, a biological sanctification of the social status quo—in all, as *Time* summed up the opposition's case, a "reactionary political doctrine disguised as science."[16]

MEANWHILE, PEOPLE WHO HAD no interest in the gene pool, in the abstractions of sociobiology, or in alleged black-white genetic differences complained bitterly about rising welfare costs, juvenile delinquency, and the high birthrate among lower-income groups. Whatever their concern, many were rediscovering a simple countermeasure in sterilization. In 1971, Shockley suggested to the annual convention of the American Psychological Association that the sterilization of persons of low intelligence might be encouraged through a system of financial incentives, the amount of payment to be proportional to the number of points below 100 that the candidate for sterilization scored on an I.Q. test. Mental-health administrators could deploy more persuasive methods of encouragement. In the sixties, state legislatures had begun repealing compulsory sterilization laws in favor of statutes that authorized sterilization on a voluntary basis—in some cases for classic eugenic purposes, in many others for the benefit of mentally retarded patients. But in a number of state mental institutions, inmates, especially women, were not released unless they first submitted to "voluntary" sterilization. Administrative practice de facto frequently required the same of mentally deficient people, institutionalized or not, who wanted to marry.[17]

If federal funds were to be used for sterilizations in health and welfare programs, consent was required, but in 1976 the requirement was reported to have been utterly ignored in the sterilization of thousands of women by the Indian Health Service. And the requirement had been interpreted with deplorable latitude in dealing with other minority-group women. In 1973, national attention was given to the news that two southern black sisters, aged twelve and fourteen, had been sterilized under the auspices of the

Office of Economic Opportunity. "Consent" had been obtained from the mother, in the form of her mark "X," but she later said that she thought her daughters were simply going to be given some shots. In a 1975 study prepared for the President's Committee on Mental Retardation, Monroe Price and Robert Burt, professors at the U.C.L.A. and University of Michigan law schools, respectively, declared, "We are not too far removed, in time or in ideology, from Justice Holmes and *Buck* v. *Bell.*" People were "too sophisticated to talk eugenics, at least out loud," the professors noted, and they continued, "The language of 'fiscal responsibility' and 'parenting environment' [makes] a more appealing case than the rhetoric of 'wards of the state' and 'menace to society.'"[18]

Indeed, in the renewed claims of a dependence of intelligence upon race or class, the popularity of human sociobiology, the revived apprehensions about a differential birthrate, the fresh resort to sterilizations, there was a good deal that would have warmed the heart of an early-twentieth-century mainline eugenicist. There was nothing in any of it that a Robert Sinsheimer could approve; his "new eugenics" was an extension of the reformist program that had colored the social connotation of human genetics since the nineteen-thirties. All the same, to a number of observers even the eugenics of chromosomes and enzymes, of medical applications and utopian visions, threatened in practice to shade over into some of the old mainline sins.

In the early sixties, British geneticists had noticed that male inmates of prisons or mental institutions displayed a higher incidence of a particular sex-chromosome anomaly—the so-called XYY anomaly—than did males in the general population. A research team in Edinburgh headed by Patricia Jacobs reported in *Nature* in 1965 that it had found males with an extra Y chromosome in disproportionately high numbers among inmates of a Scottish hospital for the treatment of patients with "dangerous, violent, or criminal propensities." Jacobs raised the speculation that males with an extra Y were disposed to unusually aggressive behavior. (In fact, in a study of some forty-one hundred men in Denmark which was published in 1976, a team of Danish and American researchers found that XYY males were more than twice as likely as XY males from the same socioeconomic group to be convicted of crimes. However, the evidence suggested that the higher conviction rate among the XYY males was the result of lower-than-average intelligence, not of an unusual propensity to aggressiveness. They had none, the team concluded, since crimes of violence were no more frequent among the XYY males than among the XY ones.) Jacobs' speculation was advanced with strong caveats about the actual role, if any, of the extra Y in aggressive behavior, but all caveats were lost in the explosion of publicity that followed. Various scientists, legal scholars, and public officials argued that *XYY* males were almost certainly prone to criminal violence, and suggested that they be identified through screening programs and kept under scrutiny.[19]

Echoing Francis Galton, some biologists supposed that a new morality might soon arise which would encourage a couple to forswear the birth of children with genetic defects. To help the new morality along, British and American geneticists suggested the establishment of computerized genetic information banks so that people could be informed if they were at high risk for transmitting a serious hereditary disorder to offspring. Orlando J. Miller, now a professor of human genetics at Columbia University, proposed in the mid-seventies that attempts even to legislate eugenic programs might not lie too far in the future, noting, "Individuals in a society which is willing to allow even normal fetuses to be aborted simply at the request of the parents are not likely to be very tolerant of a known abnormal fetus."[20]

In the United States, a former health-systems analyst in the office of the Surgeon General was quoted in a 1974 *Fortune* article as saying that some five billion dollars could be profitably spent over twenty years to reduce the incidence of Down's syndrome by a program of voluntary diagnosis and abortion; if the reduction amounted to fifty percent, society would save some eighteen billion dollars. He also estimated that similar programs aimed at other genetic diseases might bring the saving to between seventy-five and a hundred billion dollars. "If we allow our genetic problems to get out of hand," the analyst declared, ". . . we as a society run the risk of overcommitting ourselves to the care of and maintenance of a large population of mentally deficient patients at the expense of other urgent social problems."[21]

Paul Ramsey, a professor of religion at Princeton University, devoted considerable thought to the ethics of human genetic control. He saw no difference in principle between requiring premarital blood tests and premarital genetic tests, and in 1970, in his book *Fabricated Man*, he proposed that the state might use its marriage-licensing power to prevent the transmission of "grave dominantly inherited diseases," explaining, "The freedom of parenthood is a freedom to good parentage, not a license to produce seriously defective individuals to bear their own burdens." As though taking a leaf from Ramsey's book, the Chicago Bar Association urged that the Illinois marriage laws require premarital tests for "diseases or abnormalities causing birth defects." The Bar Association lawyers also suggested that Illinois might, if and when the feat became technically possible, require from applicants the correction of the genes for certain race-specific maladies—for example, Tay-Sachs or sickle-cell anemia—before it issued a marriage license.[22]

In a number of states, genetic screening programs became compulsory. They raised constitutional questions about the right of privacy, and those for sickle cell were said to strike at the right of equal protection of the law because the principal group at risk for the disease was black. In the sixties, major American corporations had begun to develop screening programs for prospective employees, and these programs, underway by the early 1970s, threatened to restrict employment opportunities for people alleged to be genetically susceptible to hazards in the workplace.[23] As it happened, the programs were often marred by technical confusion or ignorance. The preamble of the National Sickle Cell Anemia Control Act, for instance, opened with the blatantly erroneous statement that two million Americans suffered from sickle-cell "disease." The fact was that two million carried the harmless sickle-cell trait; fewer than a hundred thousand had the disease. In practice, the sickle-cell programs, many of them short on follow-up counseling, often left people detected as carriers unnecessarily anxious about their procreational futures.[24]

Public attention was given to the deaths, in 1968 and 1969, of four seemingly healthy black Army recruits during basic training at Fort Bliss, in El Paso, Texas. Autopsies showed that all four had severe sickling of the red blood cells. The sickling could have been a consequence of death, but it was judged to have been the cause: it was thought to have occurred because of an oxygen deficit brought about by physical stress at the camp's thirty-seven-hundred-foot altitude. Subsequently, people with sickle-cell trait—because of the fear that their red blood cells might sickle at high altitudes—were prohibited from entering the Air Force Academy, restricted to ground jobs by various major commercial air carriers, and often charged higher premiums by insurance companies. By late 1972, not long after the sickle-cell act was passed, some spokesmen of the black community in the United States were indicting sickle-cell screening programs as racially discriminatory, a form of anti-black eugenics, and even a step toward genocide.[25]

Postnatal screening programs for PKU might not be racially charged, but they appeared to rest on undue technical confidence. Critical assessments revealed that the Guthrie test for PKU produced some positive results that were false, and that if it was administered too soon after birth it failed to identify some infants who in fact had the disorder. The consequences of false diagnosis could be dire: a phenylalanine-deficient diet given to non-PKU children could result in disability, and PKU infants fed on a normal diet would likely develop mental retardation. Marc Lappé—a biologist in the California State Department of Health Services who became a sharp and persistent critic of genetic screening—pronounced the PKU picture decidedly "blemished," not least because of the "overly sanguine and simplistic views of the first screeners." Genetic-screening laws struck Jonathan Beckwith, a professor of microbiology and molecular genetics at Harvard Medical School, as potentially "the opening wedge for a eugenics program." In 1974, in an article in *Psychology Today,* Beckwith and two younger co-authors warned that "in the age of the technological fix, this country is heading for genetic and behavioral control of society." They continued, "Who will exercise the control? Who will make the decisions about which genes are defective, and which behavior abnormal? Who will make the decisions about the genetic worth of prospective human beings?"[26]

Such concerns, arising during the Vietnam War—which helped spark a general revolt against expert authority that rendered most declarations of scientific imperatives widely suspect—stimulated the formation of a diverse coalition of dissenters from the new eugenics. As in the interwar years, the coalition included civil libertarians, members of the political left (the so-called New Left of the day), and minority-group leaders, especially prominent Catholics. It also included a new identifiable group—the professional ethicists, some of them secular scholars and intellectuals, others clerics of various faiths. Particularly influential was the small corps gathered by Daniel Callahan, an intellectually adventurous Catholic layman who heads the Institute for Society, Ethics, and the Life Sciences, also known as the Hastings Center, in Hastings-on-Hudson, New York, where in the early seventies affiliates like Marc Lappé began to explore the social and moral conundrums arising from the uses of human genetics. The demand for social responsibility on the part of science was particularly strong among younger members of the medico-scientific community who had been radicalized or moved to dissent by the Vietnam War, but there also appeared a socially critical—and self-critical—trend among many of their older colleagues. In the late sixties and early seventies, numerous conferences were held on ethics and the new genetics, virtually all of them including a complement of concerned biomedical scientists.[27]

The conferences were more numerous in the United States than in Britain, where—apart from the linkage of I.Q. and educational policy with race—the new eugenics, like the old, seemed to stimulate less controversy. To be sure, the advent of the new immigrants had increased the frequency in the country of such genetic diseases as thalassemia, but while screening of newborns for PKU was mandatory, there were no such programs that disproportionately affected any racial or ethnic minorities. Then, too, in Britain most matters that concerned the application of genetics to human beings were commonly understood to fall, along with genetic counseling, under the authority of the medical profession, and the debates about human genetic engineering tended to be confined to committees of the National Health Service or of professional societies rather than fought out in the general press.[28] Yet the gravity of the issues moved some of the British dissidents to join with their American counterparts in mounting a powerful critique of the new eugenics, including a number of its moral, social, and technical premises.

Many of the dissident scientists aimed withering fire at the renewed hereditarian claims for I.Q. and then at those of human sociobiology. The publication of Jensen's 1969 article stimulated an astonishing outpouring of critical debate and writings. (Richard Lewontin, the Harvard University population geneticist, remarked in 1970 that "Professor Jensen has surely become the most discussed and least read essayist since Karl Marx.") Some of the fire aimed at Jensen resembled storm-trooper harassment, including the disruption of his lectures and classes. The *Harvard Educational Review* responded to the tumult by stopping sale of the issue in which Jensen's article appeared and at one point refusing to send him the reprints of the piece usually made available to a scholarly author. The editors, under fire from parts of the academic community, eventually halted the suppression; they had defended it at the time on the ground that the article could not properly be circulated without accompanying rebuttals.[29]

Edward Wilson complained of intimidation at Harvard and of personal abuse for his allegedly conservative politics. At a daylong panel discussion of sociobiology, at the February 1978 meeting of the American Association for the Advancement of Science in Washington, D.C., Wilson rose to respond after a series of speakers had criticized his work. Before he could begin, he was confronted by about a dozen young men and women loudly demanding that he be denied the podium and shouting, "Fascist!" "Nazi!" "Racist!" "Sexist!" One of them poured a bucket of water on him, declaring, "You're all wet, Wilson!"[30]

Such responses both to Wilson and to Jensen made for one of the more deplorable episodes in the annals of academic freedom—one of the more counterproductive, too, since for a time it clouded the exposure of the many serious intellectual flaws that numerous critics found in the theories of both men.

In his 1969 article, Jensen had contended that intelligence tests, though they did not assess all mental ability, did measure a type of intelligence that was as objectively real as atoms or genes. From his review of the literature on heredity and intelligence he extracted a fundamental claim, which he based on the technical concept of "heritability." In its technical sense, "heritability" is defined by geneticists as that fraction of the variance in a characteristic—say, I.Q.—within a given group which is accounted for by genetic differences. The studies that Jensen had reviewed concluded that heritability in I.Q. scores was high. This conclusion led Jensen to the most critical step in his argument: blacks and whites differed on the average in the measure of an entity—intelligence—that was largely "heritable" within each of the two groups; it was therefore highly probable that the difference between the two groups was based in part on a hereditary biological difference—a difference in genetic makeup.[31]

Jensen's assertion of the high heritability of intelligence rested in part on the studies of twins conducted by Cyril Burt, which were revealed as fraudulent in the early seventies by the Princeton psychologist Leon Kamin. Intellectually, the debunking of Burt undermined but did not devastate Jensen's claims.[32] Far more telling was the scrutiny given his arguments by critics who indicted him for various methodological and epistemological sins. Most had been committed by his forebears in the annals of genetics and intelligence, a number of whose works he relied on, and the charges against Jensen strongly echoed those brought by prior critics—notably Lancelot Hogben—against the hereditarian school.[33] The most penetrating arguments—their force was recognized even by some leaders of the hereditarian school—were marshaled in articles by Richard Lewontin, himself something of a latter-day Hogben—an outspoken and idiosyncratic Marxist, a polymath, and an eminent population geneticist, who mixed brilliance with a remarkable mastery of biological facts and statistical inference.

To Lewontin, the evidence that Jensen offered in support of his case was "irrelevant." Part of the variation in I.Q. scores within a group actually tested might be attributable to heritability; this did not mean, however, that the variation in every similar group arose to the same degree from a hereditary factor. Heritability estimates were specific to the particular group surveyed, and to the particular environment in which that group was found. Hence, they were not measures of universal cause but indicators of local environmental and genetic circumstance. Lewontin pointed out the "fundamental error" of Jensen's argument: heritability estimates applied only to the analysis of differences *within* groups; Jensen had erred in using heritability to help account for the difference in scores *between* two groups—in this case, whites and blacks—a use of the concept which, Lewontin said, was technically "meaningless."[34] Moreover, like Hogben, Lewontin stressed the pitfalls involved in comparing any two groups: environmental differences could make genetically identical organisms seem genetically unequal, or genetically disparate organisms seem genetically similar. It was true that Jensen, recognizing the importance of adjusting for environmental disparities, had made a point of observing that blacks scored lower than whites of the same socioeconomic status, but in this regard Lewontin thought him "biologically naive." How did Jensen know the major environmental sources of difference in I.Q.-test performance?[35]

Luigi Luca Cavalli-Sforza and Walter Bodmer, in an influential article in the October 1970 issue of *Scientific American,* pointed to the same flaw. Bodmer, a professor of genetics at Oxford, and Cavalli-Sforza, a population geneticist at the University of Pavia, had recently completed their joint classic treatise *The Genetics of Human Populations.* Both had, at different times, been students of Ronald Fisher's; they met in the late nineteen-fifties in Fisher's laboratory at Cambridge University, and they became friends and collaborators in 1962, when they taught a joint course at Stanford University. Both brought to the issue of race and intelligence not only enormous expertise but also the sharp perspective of foreigners acquainted with American culture. They added their own echoes of Hogben to the debate by emphasizing a number of elements that had a possible role in the production of apparent intelligence differences, including variations in the uterine environment; protein-deficient diets; and cultural legacy.[36] They also queried Jensen's assumption that for blacks and whites similar socioeconomic status meant similar environments:

> Black schools are well known to be generally less adequate than white schools, so that equal numbers of years of schooling certainly do not mean equal educational attainment. Wide variation in the level of occupation must exist within each occupational class. Thus one would certainly expect, even for equivalent occupational classes, that the black level is on the average lower than the white. No amount of money can buy a black person's way into a privileged upper-class white community, or buy off more than 200 years of accumulated racial prejudices on the part of the whites, or reconstitute the disrupted black family, in part culturally inherited from the days of slavery. It is impossible to accept the idea that matching for status provides an adequate, or even a substantial, control over the most important environmental differences between blacks and whites.

The difficulty of establishing such methodological control by any other means was immense—indeed, very likely insurmountable. Bodmer and Cavalli-Sforza concluded, in an apparent rebuttal of William Shockley's repeated calls for research into racial differences in intelligence, "For the present at least, no good case can be made for such studies on either scientific or practical grounds."[37]

In 1973 the Genetics Society of America resolved to issue a statement on genetics, race, and intelligence. A draft statement, drawn up by a small committee and ready by January 1975, elicited overwhelming support from the membership, but also stimulated more than eighty letters of critical comment. An important objection held that no scientific society should take an official position on an issue like the role of genetics in intelligence, not least because scientific disputes could not wisely be decided by majority vote. The most frequent criticism was aimed at one particular sentence in the draft statement: "In our view, there is NO CONVINCING EVIDENCE OF GENETIC DIFFERENCE IN INTELLIGENCE BETWEEN RACES." To some of the dissenters, the statement came too close to implying that there were definitely no such differences. Among them was Norman H. Horowitz, a prominent geneticist and professor of biology at the California Institute of Technology, who held that human populations that had evolved in geographically isolated regions doubtless differed in the gene frequencies for a variety of traits; the Society's statement, he argued, "should explicitly recognize the possibility—many geneticists would say the probability— that there are racial differences in the gene frequencies for these [mental] abilities. . . ." Horowitz added that the statement should also "take the position that such considerations have nothing to do with the value of an individual human being, because they are only statistical considerations; and, in any case, the value of a person rests on much more than his IQ alone." He concluded: "In short, it should distinguish clearly between scientific questions and moral ones."[38]

Few of the critics, however, went so far as to declare it probable, as distinct from merely possible, that races differed genetically in intelligence. The expressed objections to the disputed sentence pivoted mainly on the point—a corollary to the proposition of Bodmer and Cavalli-Sforza—that just as the presence of such a difference could not be demonstrated, neither could its absence. Prompted by the letters, the drafting committee drew up a revised statement that won the endorsement of many of the dissenters. It was issued in 1976, not in the name of the Society but in the names of the 1,390 people—over half the membership—who endorsed it. Reflecting the moral issue that Horowitz had spotlighted, it declared: "We deplore racism and discrimination . . . because they are contrary to our respect for each human individual. Whether or not there are significant genetic inequalities in no way alters our ideal of political equality, nor justifies racism or discrimination in any form." Reflecting the principal scientific dissent, it replaced the controversial upper-case sentence with the judgment: "In our views, there is no convincing evidence as to whether there is or is not an appreciable genetic difference in intelligence between races."[39]

. . .

RICHARD LEWONTIN was no friend of human sociobiology and neither was one of his professorial colleagues at Harvard, the paleontologist and evolutionary biologist Stephen Jay Gould. Both belonged to the Sociobiology Study Group, which comprised some thirty-five scientists and students—Jonathan Beckwith was also a member—in the Boston area, and was affiliated with the radical-left-oriented organization Science for the People. In the *New York Review of Books*, in November 1975, and then elsewhere, Lewontin, Gould, and other members of the Group scathed Wilson and his claims with a barrage of political, scientific, and *ad hominem* arguments.[40] "Our rhetoric was at fault," Gould later said, but he insisted that the opposition to Wilsonian sociobiology was not at its foundation political. Gould awarded "high praise" to most of Wilson's *Sociobiology* for its "lucid discussion of evolutionary principles" and its "indefatigably thorough discussion of social behavior among all groups of animals." His opposition rested on an issue of scientific methodology that permeated Wilson's Chapter 27, dealing with human social behavior. Wilson himself later recalled that he had been entirely unaware that Chapter 27 might be taken as a political statement—and a conservative one, at that—and in *On Human Nature* he cautiously tried to defuse the political explosiveness of his speculations. Gould commended the attempt, but saw the book's hedging as exposing the methodological flaws of human sociobiology all the more baldly. "We may have been more sensitive to the flaws because we disliked the implications; but we didn't make them up," he wrote.[41]

Gould himself spotlighted the flaws, notably in two essays for general audiences published in 1976 and 1978. As an evolutionary biologist, he readily conceded that Darwinian processes could "work on behavior as well as form"; that biology could "abet our Socratic search to know ourselves"; and that "genes have something to do with human uniqueness." But here the methodological issue intruded: How could one determine just what role genes played in human behavior? Wilson, precluded from performing breeding experiments with human beings, had been compelled to resort to a set of indirect analytic strategies. Gould found all of them dubious. He disputed Wilson's claim that certain behavioral traits—for example, susceptibility to indoctrination—were universal in man. He went on to strike at two arguments that were central to Wilson's case. One held that if altruistic acts in animals expressed the natural selection of genes for the trait, then by some principle of continuity altruism in human beings must also be genetically grounded; the other insisted that if a given form of social behavior was adaptive in an evolutionary sense, then its origins were genetic. Both claims, Gould stressed, foundered in principle on a simple proposition: Similar behaviors in man and primates could proceed from dissimilar causes—in primates from genes but in man, whom natural selection had equipped with the potential for a vast range of behavioral patterns, from culture.[42]

Although Wilson acknowledged that human behavior was shaped more—perhaps a great deal more—by culture than by genes, he was for the most part unclear about where the role of genes ended and that of culture began. Instead of coming to grips with the issue, he resorted mainly to a number of imaginative tales—"just-so stories," Gould called them—of how behavioral patterns of interest could be accounted for by the natural selection of genes. Virtually all of these tales, Gould noted, could be replaced by equally plausible stories that hinged on cultural explanations. In most types of human social behavior, there was simply no decisive method for choosing between a genetic or a cultural story. "What is the direct evidence for genetic control of specific human social behavior?" Gould asked. "At the moment, the answer is none whatever."[43]

Now, AS IN THE interwar period, there was no evidence that the higher birthrate of blacks or other lower-income groups was polluting the human gene pool. Many of the new dissidents argued that it was arbitrary, too, to talk about "genetic load" without, as Lionel Penrose had once noted, having some notion of what constituted genetic "fitness." (Theodosius Dobzhansky, of Rockefeller University, a renowned evolutionist and population geneticist, had reminded biologists, at a 1961 conference that took up H. J. Muller's ideas of genetic load and germinal choice, that "usefulness and harmfulness are not the intrinsic properties of a variant gene; genes are useful, neutral, or harmful only in a certain environment," and he had continued, "What is good in the Arctic is not necessarily good on the equator; what was good in man in the ice age is not necessarily good now; what is good in a democracy is not necessarily good under a dictatorship.")[44] So the public biologists of the interwar years had asserted, but now the point could be substantiated by, for example, the resistance to malaria conferred by the sickle-cell gene in its heterozygous state. The human race, it was increasingly clear, was genetically polymorphic to a dizzying degree. Eliminate that immense variation, and the race would be "genetically frozen," as Lionel Penrose put it. Enlarge it—as seemed to be happening in the modern age, with the mixture and interbreeding of previously separated human groups—and, in Penrose's words, there would be an increase in "the number of man's possible inborn reactions, whether physical or psychological, to his rapidly changing civilized environment."[45]

Indeed, the use of genetic screening and selective abortion could well bring about dysgenic results. In the absence of amniocentesis, a couple who had, say, a Tay-Sachs child might choose to have no more children. They would thus not transmit their recessive Tay-Sachs genes to the next genera-tion. With amniocentesis, the couple could successively abort Tay-Sachs fetuses until they bore as many normal children as they desired, practicing what human geneticists call "reproductive compensation." But there would be a two-in-three chance that the recessive Tay-Sachs gene would be trans-mitted, via each normal offspring, to the succeeding generation. Instead of decreasing the incidence of deleterious genes in the population, amni-ocentesis and abortion would over time likely increase it. In fifty genera-tions, it was calculated, the frequency of carriers of the recessive cystic-fibrosis gene would rise from five in a hundred to seven and a half in a hundred—an increase of fifty percent.[46]

For some, that might be reason enough to oppose abortion, but more powerful arguments against the taking of prenatal life came from a special wing of anti-eugenicists—the right-to-life movement. The wing was made up largely of Catholics and fundamentalist Protestants, but it also had allies among secular ethicists and others unable to reconcile abortion with the sanctity of human life. Paul Ramsey, generally a dissident despite his views on the eugenic regulation of parenthood, indicted the language used with abortion—such words as "therapy" and "treatment"—as "a logical and moral contradiction," a kind of medical doublespeak. But critics like Ram-sey differed in significant ways from the right-to-life movement, not least in a lack of doctrinaire zealousness.[47]

A number of anti-abortionists insisted with scientific certitude that abortion is murder because human life begins at conception. However, along with many other scientists and physicians, James Neel has held that scientific knowledge does not authorize anyone to "pontificate on the pre-cise moment at which the developing embryo becomes human," adding, "I cannot resist pointing out that quite clearly the early embryo has gill slits, and has an appendage labeled a tail in every textbook on embryology. Is it human at the time it's exhibiting gill slits and a tail? The necessary definition is philosophical or legal, not scientific." Harry Harris, since 1976 a professor of human genetics at the University of Pennsylvania, has de-plored as "self-righteous" attacks against the right to abort even severely disordered fetuses and likened the attitudes that energize such attacks to those of the eugenicists who "two generations ago might have said, Why don't we just slaughter all these . . . parasites on society."[48] Mainline eugenicists, obsessed with the procreational practices of others, claimed that certain people, because of alleged genetic inadequacies, must not procreate. Right-to-life advocates demanded that children once conceived must be born, no matter what pain their genetic disendowments will force them to bear and no matter what the emotional or financial cost to their families.

Prenatal diagnosis has actually fostered births. Likely birth defects have been detected in only a few percent of the fetuses carried by women who undergo amniocentesis—which is to say that amniocentesis has provided the vast majority of couples compelled to use it with the knowledge that their fetus is normal and with the reassurance to bring it to term. The number of abortions carried out as a consequence of adverse prenatal diagnosis has constituted a minuscule fraction of all legal abortions each year in the United States and Britain. Nevertheless, a number of right-to-life advocates have equated genetic counseling with abortion and have sought to intimidate private organizations that support such counseling.[49]

The National Foundation–March of Dimes, while denying that the severe fire from the movement influenced its policies, disclosed in 1978 that it intended to reduce its considerable support of genetic-services programs. In 1976, a major appropriation bill was amended to limit the use of Medicaid funds for abortions to the termination of pregnancies that endangered the mother's life or resulted from promptly reported rape or incest. In 1980, the Supreme Court upheld the amendment by a narrow majority, despite arguments that it violated the right of equal protection of the laws in that it singled out abortion as a medical service to be denied to lower-income women. The omnibus National Genetic Diseases Act of 1976 languished unfunded for two years, perhaps in part because of right-to-life pressure. Although it enjoyed limited funding thereafter, in 1981 the Reagan administration virtually abolished the federal role in genetic programs by burying the money for them in an omnibus block grant to the states for maternal and child health. There, in competition with numerous other programs for the reduced public-health dollar, genetic services have fared poorly.[50]

Ironically, the principal figure in the discovery of trisomy-21—the leading cause of post-amniocentesis abortion—emerged in the seventies as a prominent spokesman of anti-abortion groups on both sides of the Atlantic. From the rostrum of the American Society of Human Genetics, Jérôme Lejeune made his position on abortion clear: if the American biomedical community truly sought to decide which embryos were not worthy of eventual birth, it should establish a new research entity, "the National Institute of Death." Lejeune, a Catholic, stressed in a recent conversation that his uncompromising objection to abortion proceeds as much from medical as from religious principle. "Amniocentesis and abortion injure the practice of medical science," he argues. "They have transformed the traditional goal of medicine from a cure to an attack on the patient. Young medical-genetics students ask me these days why I continue to work on trisomy-21—after all, Down's fetuses can be discarded. I think of trisomy-21 as a symptom of disease. The students think of it as a symptom of death."[51]

Such attitudes, Lejeune claims, diminish biomedical research on Down's syndrome and on other forms of congenital mental deficiency, and that is medically dangerous. "Only a very small number of Down's syndrome children are detected by amniocentesis," he explains. "Most physicians do not recommend the procedure for women under thirty-five. The probability of giving birth to a Down's child is lower for a younger woman, but, since they bear so many more children than those over thirty-five, they produce a majority of all Down's syndrome offspring." In the seventies, Lejeune began to explore the biochemical nature of Down's syndrome in what most geneticists regard as a quixotic attempt to understand why a third 21-chromosome yields such debilitating results. But he looks forward, he says, "to the day when a mongolian idiot, treated biochemically, becomes a successful geneticist."[52]

Divided though they were on the abortion issue, the dissidents of the seventies tended to agree with Daniel Callahan's judgment that "we will indeed have descended into the pit if we make genetic perfection a condition for the right to exist." They denied that any increase in genetic load would unduly burden the health-care system. The calculations concerning load were far from precise, and the rate of medical progress was far outpacing its seeming rate of increase. Calculations of what might be required to reduce it significantly led to drastic conclusions—as Penrose had once estimated, the sterilization of one percent of the British population to rid that gene pool just of the recessive for PKU. Kurt Hirschhorn, a medical geneticist at Mount Sinai Medical School in New York City, pressed the issue to a *reductio ad absurdum*: the interference with everyone's reproduction to halt the propagation of the load that on the average everyone was said to carry.[53]

It was hardly sensible to base reproductive decisions, let alone public policy, in the present on uncertain predictions of consequences a millennium in the future. Moreover, some genetic disorders could already be dealt with therapeutically, through diet, vitamins, surgery, enzyme induction, or drugs. Still more might be prevented by cleansing the physical environment of pollutants. (Joshua Lederberg, who had studied medicine as well as earned a Nobel Prize for his work in bacterial genetics, estimated that environmental factors—drugs, food additives, unclean air, and the like—might well account for eighty percent of the prevailing human mutation rate.) There was no sentence of perpetual doom in every current gene-environment mismatch. Though man might fail, he might also succeed in fashioning an environment salutary to his genetic endowment.[54]

He seemed more likely to be stymied by the reverse—the positive-eugenic goal of engineering an improved or superior genetic endowment. Arno G. Motulsky, professor of medicine and head of the Division of Medical Genetics at the University of Washington Medical School, summarily declared: "The possibility of *safely* placing the right gene into its right place within a human chromosome (particularly in the germ tissues) in the foreseeable future seems far-fetched. . . . We should not hold up false hopes to our patients." Even with all that had been learned about human heredity, positive eugenics still appeared thoroughly problematic. The arguments for maintaining human genetic variation worked as powerfully against positive-eugenic schemes as against negative-eugenic ones. To design an angel it remained necessary to know the specifications of heaven. Critics of Muller's germinal-choice scheme liked to point out that, as his social ideas had changed, his nominees for the reproductive pantheon had changed, too. No one knew what combinations of genes produced genius. Perhaps valuable traits were reinforced by unattractive or debilitating counterparts. Might not Dostoevsky's genius, Theodosius Dobzhansky asked, have been in some way conditioned by his suffering from epilepsy?[55]

Moreover, artificial insemination raised difficult legal and ethical issues, ranging from legitimacy and child support to liability for birth defects. So did in-vitro fertilization. If the embryo was faulty, should it be discarded? Leon R. Kass, actively concerned with the social issues raised by the life sciences, inquired, "Who decides the grounds for discard? What if there is another recipient available who wishes to have the otherwise unwanted embryo? Whose embryos are they? The woman's? The couple's? The geneticist's? The obstetrician's? The Ford Foundation's? . . . Shall we say that discarding laboratory-grown embryos is a matter solely between a doctor and his plumber?" Kass, extending his animadversions to cloning, doubted whether Mozart, Newton, or Einstein would have consented to be genetically duplicated. "Indeed, should we not assert as a principle that any so-called great man who *did* consent to be cloned should on that basis be disqualified, as possessing too high an opinion of himself and of his genes? Can we stand an increase in arrogance?"[56]

James Neel, though welcoming recombinant DNA as a tool of basic human genetics, attacked as presumptuous the talk of using the method to improve upon what so little was yet known about: "the single most precious possession man has—the double-stranded helix which, against all odds, makes us human." Lay critics warned against subverting the "mystery" of human existence. A 1969 *Life* magazine poll by Louis Harris revealed that only one out of three people in the United States approved of artificial insemination to assist childless couples to become parents; a majority objected to any sort of genetic methods aimed at producing superior human beings.[57] Thoughtful people on both sides of the Atlantic agreed with Lionel Penrose, who declared a few years before his death that he "would rather live in a genetically imperfect society which preserves human standards of life than in one in which technological standards were paramount and heredity perfect."[58]

The President's Commission for the Study of Ethical Problems in Medicine and Biomedical and Behavioral Research, set up in 1980, reported two years later that public concern expressed about the recombinant splicing of human genes seemed "to reflect a deeper anxiety that work in this field might remake human beings, like Dr. Frankenstein's monster." The commission thought such concerns exaggerated. Nevertheless, it observed that "genetic-engineering techniques are not only a powerful new tool for manipulating nature—including means of curing human illness—but also a challenge to some deeply held feelings about the meaning of being human and of family lineage."[59]

Chapter XIX

SONGS OF DEICIDE

I N 1983, AT A CONFERENCE on gene therapy, Ola Huntley, the mother of three sickle-cell anemic children and a counselor of sickle-cell patients, declared, "I am angry that anyone presumes to deny my children the essential genetic treatment of a genetic disease. I see such persons as simplistic moralists." For all the vociferous objections to the new eugenics, most authorities on health, disease, and reproduction have tended to side with Huntley and to believe that though there may be problems with genetic screening, counseling, and therapy the lesson to be drawn was surely not to proscribe them. People need and deserve to have whatever information may be available concerning genetic risks, genetic disorders, and modes of treatment.[1]

In the United States and Britain, genetic disorders are now known to occur in between three and five percent of all live births, and chromosomal disorders—for example, Down's syndrome—in at least a half-percent. The percentages may be small, but the absolute annual numbers suggest a wrenching magnitude of individual afflictions—in the United States, up to one hundred and sixty-five thousand abnormal infants, including from six to eight thousand with neural-tube defects like spina bifida, five thousand cases of Down's syndrome, fifteen hundred of cystic fibrosis, at least a thousand of sickle-cell anemia. Genetic and chromosomal illnesses or malformations are reported to account for between twenty and thirty percent of all pediatric hospital admissions. Twelve percent of all adult hospital admissions are said to involve illnesses with a significant genetic component. At least fifteen percent of all diagnoses for mental retardation report it as unambiguously hereditary.[2]

Despite the increasing recognition of genetic risk, it is estimated that in the United States in 1979 and 1980 only half the pregnant women who were deemed medically appropriate for amniocentesis underwent the procedure, and it was being performed on only ten percent of the comparable group in Britain. Practicing physicians were said to be failing to refer patients for prenatal diagnosis, partly because many of them still lacked adequate genetic knowledge. Then, too, in Britain the plan to staff each National Health Service region with two genetic consultants was, in 1983, far short of realization for want of funds and of an adequate number of genetically trained physicians; in a number of regions genetic counseling was being provided on an ad hoc basis. Nevertheless, physicians in both countries were increasingly sensitive to the value of genetic knowledge in general practice; courses in various aspects of human genetics were offered in the vast majority of medical schools; and a growing number of people were entering training programs in medical genetics. In 1980, an American Medical Board—the mechanism for certifying practitioners in different specialties—was established for medical genetics and promptly qualified by examinations about five hundred fifty people in various subspecialties of the field. About five hundred genetic-counseling facilities were operating in the United States, perhaps one hundred fifty of them connected with major teaching and research hospitals and providing full diagnostic services. In Britain, the Royal College of Physicians accredited medical geneticists. By 1984, some type of genetic counseling was available in every National Health Service region, and a system of regional and national laboratories provided analyses requiring chromosomal, biochemical, and recombinant DNA techniques.[3]

In the United States, legal obligation entered the practice of medical genetics when, in the mid-1970s, the question began to be raised in the lower courts of a number of states whether damages could be sought from physicians who failed to provide their patients with appropriate genetic counseling. In 1978, the New York State Court of Appeals—becoming the first higher state court in the nation to deal with the issue—ruled on two companion suits that had been brought independently by Dolores Becker and Hetty Park and their husbands, all of Long Island, New York, against their respective obstetricians. Becker had become pregnant at the age of thirty-seven and given birth to a Down's syndrome daughter. According to the Beckers' complaint, their physician had not informed them of the sharply higher frequency of such births among women over thirty-five, nor had he offered Dolores Becker amniocentesis. Park had borne a child with polycystic kidney disease who had died five hours after birth, then produced a second with the same genetic disorder who had died at the age of two and a half. According to the Parks' suit, Hetty Park had consulted her doctor before conceiving the second child and he had advised that the hazard of such a repeat outcome was virtually nil. In the judgment of the Appeals Court, the parents had a right to sue the physicians for not having warned them of the risks in their pregnancies and that the obstetricians, if found at fault, could be held financially responsible for the extraordinary costs of caring for an offspring with a genetically predictable disease or disability.[4]

The Beckers had also wanted to call their physician to account on behalf of their Down's syndrome daughter for "wrongful causation of life," on the grounds that she had been denied the "fundamental right of a child to be born as a whole, functional human being." The Court of Appeals, following the prevailing legal standard, disallowed that claim as a basis of suit, the majority holding that "whether it is better to have been born at all than to have been born with even gross deficiencies is a mystery more properly to be left to the philosophers and the theologians." However, new ground was broken in this area too when, in 1980, the California State Court of Appeals ruled on a "wrongful life" action brought by Temar Curlender on behalf of his daughter Shauna against Bio-Science Laboratories and Automated Laboratory Sciences. Curlender and his wife, Phyllis, had consulted the laboratories in 1977 to determine whether either of them carried the recessive gene for Tay-Sachs disease. Believing from the report of the laboratory tests that they had nothing to fear, they had conceived Shauna, who was diagnosed in 1978 as a Tay-Sachs baby. Their suit sought punitive damages, compensation for the pain and suffering to be endured during Shauna's expected four-year lifespan, and additional compensation for her having been deprived of 72.6 years of normal life. Although the court denied standing to the claim concerning deprivation of normal life expectancy, it did hold that Shauna had "the right . . . to recover damages for the pain and suffering to be endured during the limited lifespan available . . . and any special pecuniary loss resulting from the impaired condition." The court explained its ruling: "The reality of the 'wrongful life' concept is that such a plaintiff both exists and suffers, due to the negligence of others. It is neither necessary nor just to retreat into meditation on the mysteries of life. . . . The certainty of genetic impairment is no longer a mystery."[5]

One of the most powerful sources of pressure for further research and treatment in medical genetics has come from the victims of genetic diseases and their families. Many are organized in private foundations, including the National Genetics Foundation, the Hereditary Disease Foundation, the National Hemophilia Foundation, the Cooley's Anemia Foundation, the Cystic Fibrosis Foundation, and the Huntington's Disease Foundation of America. They not only support research but also lobby for their constituencies. Not surprisingly, they tend to take a skeptical view of the distress voiced in recent years over interference with the human genome, and they welcome the powerful new tools for prenatal diagnosis emerging from the accelerating advance of biomedical knowledge and techniques, especially the methods of recombinant DNA.[6]

By the early eighties, amniocentesis could detect likely, though not conclusive, signs of at least two polygenic disorders—the common neural-tube defects anencephaly and spina bifida. Both were signaled by high concentrations of alpha fetoprotein, a substance that was presumed to leak from the fetus because of the failure of the spine to close and that appeared not only in the amniotic fluid but also in the mother's blood, which made finding first indications of its presence relatively simple. Amniocentesis could also reveal about one hundred chromosomal anomalies and about as many genetic disorders of a molecular type.[7]

Prenatal biochemical tests depend upon detection of the protein associated with the defective gene. The protein, however, cannot be easily or safely detected in numerous cases—notably sickle-cell anemia. (The telltale hemoglobin can be obtained only by direct extraction of fetal blood—a procedure extremely hazardous to the fetus.) With recombinant DNA technology, the examination of the protein can be sidestepped and the relevant DNA itself analyzed directly. The trick relies on choosing a restriction enzyme that will cut from the DNA chain a strand containing or adjoining the gene of interest. The points cut by the restriction enzyme are at a known distance from each other on the normal chain; any abnormality in their neighborhood will cause these points to shift, and thus a strand containing or adjacent to an abnormal gene will not be the same length as a strand taken from a normal chain. A comparison of the fragment taken from cells in the amniotic fluid with a normal strand indicates the presence or absence of the trait. At the end of the seventies, several investigators reported successful exploitation of this technique for the prenatal detection of sickle-cell anemia, and the comparison of DNA fragments is now said to be one of the most promising methods of genetic diagnosis.[8]

Another striking diagnostic tool has come from the employment of cell-hybridization techniques to identify the chromosome containing a particular gene, then the use of recombinant methods to reveal where on the chromosome it is located. By the early eighties, such mapping of the human chromosomes was well begun, yielding for each a rapidly expanding library of gene locations. Many of the genes do not cause disease but are responsible for one or another of a growing list of biochemical polymorphisms—enzymes, blood products, antigens, and the like. Widely present in the human population, these genes can be used as universal landmarks on each chromosome and, as such, indicators of the presence of whatever genes may be regularly found nearby them on their strand of DNA. A high frequency of association between the occurrence of such a landmark and that of a genetically based disease would imply—just as in classical linkage analysis —that the genes for the landmark and the disease were close to each other on the strand. Thus, by detecting the landmark, one could know to a certain probability that the gene for the disease was present, even without knowing the nature of the gene or its biochemical product.[9]

At the end of 1983, a team of American scientists completed an extensive study of more than five hundred Venezuelan families of victims of Huntington's disease (formerly called Huntington's chorea) and reported the detection of an exceedingly close linkage between a restriction enzyme marker and the gene for the disease. Nancy Wexler, head of the Hereditary Disease Foundation and a collaborator in the work, declared, "This has *radically* changed the face of Huntington's disease research." The dominant Huntington's gene remains unknown, but the identification of the marker—it was found, together with the gene, on the 4-chromosome—has signaled the neighborhood in which the gene resides. Eventually, the gene itself will likely be pinpointed and reveal the biochemical origins of the fatal disease. In the meantime, pedigree studies can identify similar markers specific to families at risk for Huntington's disease and thus detect whether individual members carry the gene long before they reach thirty-five to forty-five, the average age of onset. A person found to have it could refrain from procreation or, after conceiving a child, have the fetus examined for the marker via amniocentesis, with the option of abortion if it was detected.[10]

Recombinant DNA techniques are thus adding dramatically to the rapidly growing arsenal applicable not only to prenatal analysis but also to genetic screening and to postnatal diagnosis and therapy. The more that is learned about every individual's genetic makeup, the more it will be possible to determine what constitutes, for him or her, a salutary environment. And as the genetic profile becomes more specific, the knowledge about susceptibilities increases. Harry Harris likens the medical prospects to custom tailoring, explaining, "If you go [to a doctor] with some illness, you can get the standard treatment, but if you go to a custom doctor, by knowing your genetic constitution he can treat you according to your real needs. . . . In the long term, the most important thing about human genetics research is that it may enable us to tailor-make every individual's environment."[11]

Other prognostications have tended toward the medically roseate and on to the utopian: sufficient use of amniocentesis and abortion may eventually prevent the expression in each generation of single-gene disorders; genetic research may lead to the control of widespread polygenic disorders, such as heart disease, lung cancer, and atherosclerosis; if gene therapy can be practiced successfully on the ill perhaps it can enhance the lives of the healthy, by enriching their intelligence or physical strength; if somatic cells can be genetically manipulated perhaps reproductive cells can be made to pass selected enhancements on to offspring.[12]

Most such predictions have aroused expectations—and fears—that, given the current state of genetic knowledge, are exaggerated. The predictions usually originate among advocates of genetics within the biomedical community and are diffused by an often uncritical press. They stem from some mixture of genuine scientific vision with professional self-interest and an eagerness to justify the enormous contemporary investment in genetic research. Similar processes were at work in the heyday of mainline eugenics. It is fortunate that biology is far from the public weakling that it felt itself to be in the early twentieth century, when some of its practitioners, hoping for professional power, argued for a reign of eugenic expertise, and others, fearing that they might lose what power they had, were unwilling to decry publicly what they knew to be technically false.

A new generation of public biologists has emerged in the United States and Britain, far more numerous than the generation that earlier attacked mainline eugenics and, as a result of the recombinant DNA revolution, far more pluralist. A divergence of professional—and financial—self-interest has scattered the scientific dissenters of the early to mid-seventies to different points on the social responsibility spectrum. But while today's public biologists display little correlation between their general political inclinations and their views toward the new eugenics, they help provide experts aplenty ready to take up the cudgels against abuses proposed, fostered, or imagined in the name of genetic imperatives. During the last decade, the counterattack of technically knowledgeable dissidents has in fact somewhat subdued the new eugenics. In response to the outcry against the early practices characteristic of the sickle-cell-screening programs, reforms were incorporated into the National Genetic Diseases Act: at the state and local levels, screening and counseling programs were made voluntary rather than mandatory; eligibility for other federal services was not to hinge on participation; and the results were to be kept confidential. In 1978, the outrage over the use of federal money to sterilize lower-income and minority women stimulated the Department of Health, Education, and Welfare to include strict safeguards in its sterilization guidelines.[13]

To date, the most powerful restraint on the revival of eugenics has been nature itself. Single genes account for only a small fraction of human traits, disorders, and diseases. Like intelligence, most human characters are polygenic, and therefore are not even genetically understood, let alone subject to manipulation. There is widespread agreement among geneticists that, with a few exceptions, gene therapy is distant for single-gene disorders and beyond sight for the polygenic variety. The President's commission on ethics in biomedical research summarized the outlook for gene therapy: "The technology . . . involves four steps: cloning the normal gene, introducing the cloned genes in a stable fashion into appropriate target cells

. . . regulating the production of the gene product, and ensuring that no harm occurs to the host cells in the patient. Only the first step—cloning a normal counterpart of a defective gene—is a straightforward matter with current knowledge and technology."[14] Louise Brown, born fully formed and healthy in England in 1978 after her test-tube conception had been engineered by Steptoe and Edwards, was a Brave New World baby only as far as her conception was concerned; she was implanted in her mother's uterus shortly thereafter and carried naturally to term. Considerable scientific doubt remains that a developed human being can ever be cloned. Even with the powerful methods of recombinant DNA, the genetic engineering of new men or women at the zygote stage looms at this point as a science-fiction speculation—tantalizing, as always, but a speculation nonetheless.[15]

Yet, as the President's commission noted of the pace of advance in human molecular biology, "time and again in the past ten years, the speed with which events have unfolded has taken well-informed observers by surprise." The commission was itself surprised by how much closer human genetic therapy seemed when its own study was completed than it had seemed when the commission was set up. By the early eighties, at least a hundred and fifty restriction enzymes were known, and some eight hundred human genes—about one percent of the estimated human genetic complement—had been identified and mapped on their chromosomal sites. Francis Ruddle, a professor of biology and human genetics at Yale University, predicted that the major outline of the human gene map would be known by the year 2000.[16]

Recently, a recombinant gene was made to function in a multicellular animal, and a genetic defect was repaired in a fruit fly. Some predict gene therapy for man by the end of the century. The genetic design of man? Perhaps in a hundred years, Robert Sinsheimer suggests, in defiance of the numerous biologists and medical scientists who stress the complexity of the human organism. In 1983, he told an audience at the Massachusetts Institute of Technology that the human race has just entered what its descendants a century hence will regard as "the Stone Age of genetic engineering." Of course man is complex, but he is also, "not surprisingly, very divisible," Sinsheimer elaborated in a recent conversation. "Man is a product of evolution. He was built up one mutation or genetic recombination at a time. It stands to reason that given enough time we can analyze him right down to his last genetic and biochemical brick."[17]

THE ADVANCE OF GENETIC knowledge has already increased the range of medical and procreative opportunities, and the choices raised by their advent can be discomfiting. Genetic screeners worry that the publicity given screening programs may cause needless apprehension among people whom the roll of the genetic dice has favored, and that the genetic information obtained may lead to unrelievable anxiety among those whom it has not. Many more genetic diseases can now be identified than can be cured or even treated. Someone with the gene for Huntington's disease might well prefer not to know it, since the knowledge that he or she will fall victim to it would mean having to live under a sentence of certain debilitation and doom. However, the Boston physician and medical geneticist Park Gerald recalled having to tell a woman that her husband, the father of her three children, was himself the son of a man with Huntington's disease. "The husband at age thirty-five still was at risk. The woman was raging: Why didn't somebody tell me this?" Even after she came to terms with her husband's jeopardy, Gerald's patient declined to inform her three teenage children of the risk they faced. The revelation of genetic hazard has been observed to result not only in repression but in anxiety, depression, and a sense of stigmatization.[18]

Some genetic counselors report that their patients show no difficulty in comprehending the information they are given, but various studies by psychologists and psychiatrists have concluded that a large fraction of counselees are likely not to understand, assimilate, or remember analyses relevant to their own genetic constitutions. Jack Singer, a physician at the Genetics Screening Unit at Guy's Hospital in London, has declared, "The issues involved are just too emotionally charged for the parents to take anything like an objective attitude." Only a small percentage of women who seek genetic counseling are advised to undergo amniocentesis. The procedure is carried out usually when the expectant mother is known to be at risk for bearing a child with a genetic disorder, and one of the major signals of risk is that the mother has already given birth to such an offspring. Just considering the abortion of a similarly afflicted second child may well affect her attitude toward the first, Park Gerald notes, adding that she inevitably wonders how she could tell her living child with spina bifida that she chose to kill its unborn sibling. The child might ask, "Why don't you kill me, Mommy?" When couples abort a fetus after discovering that it has a birth defect, they have often suffered severe guilt reactions, termination of sexual relations, and even divorce.[19]

One is led to recall Haldane's observation in *Daedalus* that to humanity biological innovation is initially abhorrent, a perversion, an offense not to some god but to man himself. History suggested to Haldane that, led by the scientist with his songs of deicide, man might slay his inner demons, come to terms with the seeming perversions, and transform unnatural innovations into natural, humanly advantageous customs. Twentieth-century history has certainly suggested that among significant fractions of the population sharp change is possible in standards of sexual behavior and reproductive practice. C. P. Blacker told Hermann Muller in 1961 that opposition to germinal choice might well go the same way as opposition to birth control. "It is surprising," Blacker added, in what turned out to be something of an understatement, "how quickly new ideas can percolate nowadays." Even mainliners had recoiled at using abortion as a tool of eugenics, but very few couples seem to reject it after amniocentesis reveals a defective fetus, and in the United States in 1982 there were (mainly for non-medical reasons) more than one and a half million legal terminations of pregnancy—almost one for every two live births.[20]

Dr. Wayne Decker, of the Fertility Research Foundation of New York, remarked to a reporter for *The New York Times Magazine* in 1974, "A lot of things we wouldn't do a few years ago, we no longer think twice about. For instance, I do forty or fifty artificial inseminations a week, whereas a few years ago we would do ten or twelve a year. The repellent connotations of artificial insemination are almost nonexistent now. Couples not only accept it but seem often to regard it as more natural than adoption." Amid the gathering force of the women's movement, it seemed to some people a natural way to become a single parent. Among them was Afton Blake, a Los Angeles psychologist and the second woman to have a baby with the assistance of Robert Graham's Sperm Bank. Blake wanted, she explained, to raise a child "without conflict from a spouse," adding, "An unborn child should be guaranteed the best genetic material." By 1984, babies conceived by artificial insemination were being born to surrogate mothers, the test-tube fertilization of ova was becoming a clinical commonplace, and an embryo produced by artificial insemination in the womb of one woman had been successfully implanted and carried to term in the womb of another.[21]

Given that changes in individual attitudes inevitably affect the scope of institutional action, both public and private, history surely teaches that serious attention is owed the warnings, however shrill they may sometimes be, of the dissenters from the eugenic revival. Early in this century, nascent genetic theory was invoked to bear a weighty load of human social claims. Biology still knows little about the role of genetics in behavior, but it might someday learn—or claim to have learned—more. In that event, the definition of "defect" might become once again a hereditarian cloak for social prejudice.[22] One can hardly be confident that principles of political and social equality will, as a practical matter, remain unscathed by scientific contentions of racial differences in such traits as intelligence. The ancient impulses setting group against group survive in the views of a William Shockley and, if not in the intent of an Arthur Jensen or Hans Eysenck, certainly in the way what they have said may be used. (Daniel Patrick

Moynihan remarked in 1969, when he was a White House adviser, that "the winds of Jensen" were storming through the Capitol.) In 1981, the Air Force Academy ended its ban on cadets with sickle-cell trait, but at least six major American corporations genetically screen employees for sensitivity to toxic substances, and almost five dozen other firms, most of them in the Fortune 500, reported in 1982 that within five years they expected to put a similar policy into practice.[23] Hemophiliacs may not have an unequivocal right to employment as butchers; still, in some untold fraction of cases the burden of workplace safety could well come to fall less on the company than on the employees—a circumstance that would particularly affect ethnic or racial groups among whom the incidence of, say, sickle-cell trait or thalassemia is disproportionately high.

The willingness of individuals to use rapidly developing genetic and reproductive knowledge may have more subtle effects. Genetic screening and counseling, amniocentesis and abortion, and attempts at genetic therapy will probably long remain matters of private, voluntary choice, to be arrived at by consultation between individual families and their physicians. But the more that medical science can assist people with genetic disease to survive, the greater the cost that will be placed upon the socio-medical system. The more that people with heritable disorders can and do practice "reproductive compensation," the higher will rise the frequency of the genes for such disorders in the human gene pool. Private decision-making in the realm of genetic disorder and disease may ultimately lead to public consequences, and thus to demands for public regulation of reproductive behavior. A sizable number of people may argue that the right to have genetically diseased children, or even to transmit deleterious genes to future generations, must be limited or denied. Dissenters such as Daniel Callahan maintain that the resolution of such public problems must turn on "the willingness of society to bear the social costs of individual freedom."[24] Yet that willingness has varied enormously with history. How the public, or politically powerful public coalitions, will respond to the steady pressure of problems raised by the advance of genetics depends upon what reconciliation society chooses to make between the ancient antinomies—social obligations as against individual rights, and reproductive freedom and privacy as against the requirements of public health and welfare.

The criteria of choice are currently clouded, and they are not likely to be compellingly delineated by assertions of righteous certitudes on the one hand or invocations of genetic imperatives on the other. People may perhaps be tempted to seek rules of decision in some renewed version of Francis Galton's secular faith, and urge courses of action in the name of eugenics. It bears remembering that eugenics has proved itself historically to have been often a cruel and always a problematic faith, not least because it has elevated abstractions—the "race," the "population," and more recently the "gene pool"—above the rights and needs of individuals and their families.[25] Galton, obsessed with original sin, had expected that the ability to manipulate human heredity would ultimately emancipate human beings from their atavistic inclinations and permit their behavior to conform to their standards of moral conduct. But in fact, the more masterful the genetic sciences have become, the more they have corroded the authority of moral custom in medical and reproductive behavior. The melodies of deicide have not enabled contemporary men and women to remake their imperfect selves. Rather, they have piped them to a more difficult task: that of establishing an ethics of use for their swiftly accumulating genetic knowledge and biotechnical power.

Notes

1. Francis Galton, *Inquiries into the Human Faculty* (Macmillan, 1883), pp. 24–25.
2. J. B. S. Haldane, "The Implications of Genetics for Human Society," in S. J.

Geerts, ed., *Genetics Today: Proceedings of the XIth International Congress of Genetics . . . , 1963* (Pergamon Press, 1964), pp. xcviii–xcix.

Chapter I:
FRANCIS GALTON, FOUNDER OF THE FAITH

1. Karl Pearson, *The Life, Letters, and Labours of Francis Galton* (3 vols. in 4; Cambridge University Press, 1914–30), IIIA, 348.
2. Francis Galton, "Hereditary Talent and Character," *Macmillan's Magazine*, 12 (1865), 157–66, 318–27; Francis Galton, *Hereditary Genius: An Inquiry into Its Laws and Consequences* (Macmillan, 1869). Unless otherwise noted, references in *Hereditary Genius* are to the 1869 edition.
3. Galton, *Hereditary Genius*, pp. 37–38.
4. *Ibid.*, pp. 1–2 and *passim*.
5. *Ibid.*, p. 1.
6. Pearson, *Galton*, IIIA, 91; Galton, "Hereditary Talent and Character," p. 165; *Hereditary Genius*, pp. 345, 362; Francis Galton, "Hereditary Improvement," *Fraser's Magazine*, 87 (1873), 125–28.
7. Galton, *Hereditary Genius*, pp. 40–43.
8. Derek W. Forrest, *Francis Galton: The Life and Work of a Victorian Genius* (Taplinger, 1974), pp. 2–4, 7; Pearson, *Galton*, I, 66; II, 255; IIIB, 446; Francis Galton, *Memories of My Life* (Methuen, 1908), p. 13. Galton noted, "It . . . appears to be very important to success in science, that a man should have an able mother. . . . A child so circumstanced has the good fortune to be delivered from the ordinary narrowing partisan influences of home education." Galton, *Hereditary Genius*, pp. 195–97.

9. Raymond E. Fancher, "Biographical Sources of Francis Galton's Psychology" (unpublished manuscript), pp. 4–6; Theodore M. Porter, "Galton and Correlation: A Census of Gemmules" (unpublished manuscript), pp. 5–6.
10. Ruth Schwartz Cowan, "Nature and Nurture: The Interplay of Biology and Politics in the Work of Francis Galton," *Studies in the History of Biology*, 1 (1977), 148. Galton complains of the headaches in letters to his father, Dec. 5, 1838, and Nov. 10, 1838, Francis Galton Papers, University College London, file 108/c. Galton assured his parents that the digression from medical preparation would only be temporary, adding that cousin Charles Darwin urged the move on grounds that "the faculty of observation rather than that of abstract reasoning tends to constitute a good Physician," and that the higher parts of mathematics were entwined with chemical and medical phenomena, which were of necessity empirical. Pearson, *Galton*, I, 110.
11. Galton, *Memories*, pp. 78–79; Pearson, *Galton*, I, 173; Ruth Schwartz Cowan, *Sir Francis Galton and the Study of Heredity in the Nineteenth Century* (University Microfilms, 1969), pp. vii–viii.
12. Quoted in Pearson, *Galton*, I, 203.

13. Forrest, *Galton*, pp. 30–32, 37; Pearson, *Galton*, I, 203.

14. Report of the phrenologist, Donovan, 1849, Francis Galton Papers, University College London, file 81.

15. Galton, *Memories*, pp. 150–51; Forrest, *Galton*, p. 56.

16. Quoted in Pearson, *Galton*, I, 200.

17. Forrest, *Galton*, pp. 55–58, 60; Pearson, *Galton*, II, 11–12; Galton, *Memories*, p. 159.

18. Galton to Edgar Schuster, Feb. 3, 1906, Francis Galton Papers, file 245/21.

19. Quoted in Pearson, *Galton*, IIIA, 124; I, 231–32.

20. Quoted in Pearson, *Galton*, II, 340; I, 157, note 1; IIIA, 248; II, 277, note 1.

21. Forrest, *Galton*, pp. 69, 116–17.

22. While plotting wind and pressure charts, Galton noticed that centers of high pressure were associated with clockwise directions of the wind around a center of calms, a system that he named an "anticyclone" (in distinction from "cyclones," the name given to centers of low pressure associated with counterclockwise directions of the wind around a center of calm). Pearson, *Galton*, II, 39, 42; Forrest, *Galton*, pp. 213–21.

23. Galton, *Memories*, pp. 288–89; Galton, *Hereditary Genius*, p. v.

24. Raymond E. Fancher, "Francis Galton's African Ethnography and Its Role in the Development of His Psychology," *British Journal for the History of Science*, 16 (March 1983), 79.

25. Forrest, *Galton*, pp. 95–96; Galton, *Hereditary Genius*, p. 14; Pearson, *Galton*, II, 74, 96; IIIA, 90; Porter, "Galton and Correlation," p. 13.

26. Galton, *Hereditary Genius*, pp. 360–61; Galton, "Hereditary Talent and Character," p. 325.

27. Galton, "Hereditary Talent and Character," p. 164; Galton, "Studies in National Eugenics," in Francis Galton, *Essays in Eugenics* (Eugenics Education Society, 1909), p. 66; Francis Galton, *Inquiries into the Human Faculty* (Macmillan, 1883), p. 207; Galton, *Hereditary Genius*, pp. 356–57; Forrest, *Galton*, p. 85; Pearson, *Galton*, IIIB, 441.

28. Pearson, *Galton*, I, 173, 116–17; Forrest, *Galton*, p. 46; Galton recalled of his breakdowns, "A mill seemed to be working

inside my head; I could not banish obsessing ideas." Galton, *Memories*, pp. 78–79.

29. Fancher, "Biographical Sources of Francis Galton's Psychology," p. 7. Pearson, *Galton*, I, 171; II, 396.

30. Pearson, *Galton*, II, 11–12; Cowan, *Sir Francis Galton*, pp. 88–89.

31. Pearson, *Galton*, IIIA, 32.

32. John Stuart Mill, *Autobiography* (Henry Holt, 1874), pp. 137, 143; Galton, *Memories*, pp. 154–56; Pearson, *Galton*, II, 282; Forrest, *Galton*, pp. 147–48. Galton supposed that "delight in self-dissection must be a strong ingredient in the pleasure that many are said to take in confessing themselves to priests." Galton, *Inquiries into the Human Faculty*, p. 60.

33. Galton, *Memories*, pp. 154–55; Forrest, *Galton*, pp. 84–85, 242–43; Francis Galton, *English Men of Science, Their Nature and Nurture*, Ruth Schwartz Cowan, ed. (Frank Cass, 1970), pp. 258–59.

34. Pearson, *Galton*, IIIB, 448; Galton, *Hereditary Genius*, pp. 358–59.

35. Galton, *English Men of Science*, pp. 259–60; Pearson, *Galton*, I, 207.

36. Pearson, *Galton*, I, 203. Many years later, Galton's housekeeper and companion told Karl Pearson that she and one of Galton's surviving sisters "think it best not to take any notice of those blank years," adding, "I expect his mother tore up any letters of that date." L. E. Biggs to Professor Pearson, Sept. 15, 1912, Karl Pearson Papers, University College London, list 598.

37. Forrest, *Galton*, pp. 33–34; report of the phrenologist, Donovan, 1849, Francis Galton Papers, file 81.

38. Forrest, *Galton*, pp. 33–34.

39. Forrest, *Galton*, p. 183; Hesketh Pearson, *Modern Men and Mummers* (George Allen and Unwin, 1921), p. 71.

40. Pearson, *Galton*, II, 131; IIIA, 67, 422–23; Forrest, *Galton*, p. 227.

41. Pearson, *Galton*, I, 207; Galton, "Hereditary Talent and Character," p. 327; Galton, *Hereditary Genius*, pp. 349–50.

42. Galton, *Essays in Eugenics*, pp. 42, 24–25.

43. Galton, *Memories*, pp. 310–11; Galton, "Restrictions in Marriage," Francis Galton Papers, file 138/9; Mark Haller, *Eugenics: Hereditarian Attitudes in American Thought* (Rutgers University Press, 1963), p. 18.

44. Pearson, *Galton*, II, 120–21, 364.

45. Victor L. Hilts, *Statist and Statistician* (Arno Press, 1981), pp. 15, 265–71; Victor L. Hilts, "Statistics and Social Science," in R. N. Giere and R. S. Westfall, eds., *Foundations of Scientific Method: The Nineteenth Century* (Indiana University Press, 1973), p. 223.

46. Important among the authorities whom Galton consulted was the Belgian astronomer Adolphe Quetelet, who, finding that the distribution of such human quantities as height conformed to the law of error, hoped to use the law to construct a sort of social physics centered on the "average man." Hilts, "Statistics and Social Science," pp. 38–49, 207–8, 219.

47. Francis Galton, *Natural Inheritance* (Macmillan, 1889), p. 66.

48. Galton, *Hereditary Genius*, p. vi and *passim*.

49. Forrest, *Galton*, pp. 107–8; Porter, "Galton and Correlation," pp. 44, 49–50; Galton to E. H. J. Schuster, April 22, 1905, Francis Galton Papers, file 245/21.

50. Cowan, *Sir Francis Galton*, pp. 171, 179–81.

51. Galton, "Typical Laws of Heredity," *Proceedings of the Royal Institution*, 8 (Feb. 9, 1877), 9.

52. Galton, "Typical Laws of Heredity," p. 10; Ruth Schwartz Cowan, "Francis Galton's Statistical Ideas: The Influence of Eugenics," *Isis*, 63 (1972), 516–18; Hilts, "Statistics and Social Science," p. 225. At the time, Galton reported that the offspring derived from seven sets of the seven groups of seeds—i.e., from 490 parental seeds. Almost ten years later, he provided additional data on the filial generation in "Regression Towards Mediocrity in Hereditary Stature," *Journal of the Royal Anthropological Institute of Great Britain and Ireland*, 15 (1886), 258–61.

53. Galton, "Typical Laws of Heredity," p. 7; Cowan, "Galton's Statistical Ideas," p. 520.

54. Cowan, "Galton's Statistical Ideas," pp. 522–24; Porter, "Galton and Correlation," pp. 28–29. Galton used the heights of ninety-three children and of their two hundred and five parents.

55. Francis Galton, "Regression Towards Mediocrity in Hereditary Stature," pp. 246–63; Francis Galton, "Family Likeness in Stature," *Proceedings of the Royal Society of London*, 40 (1886), 42–73; Cowan, "Galton's Statistical Ideas," pp. 522–24.

56. Francis Galton, "Co-relations and Their Measurements, Chiefly from Anthropometric Data," *Proceedings of the Royal Society of London*, 45 (1888), 135–45; Pearson, *Galton*, II, 384; IIIA, 5; Cowan, "Galton's Statistical Ideas," p. 526.

57. Galton, *Natural Inheritance*, pp. 62–63.

58. Porter, "Galton and Correlation," pp. 16–17; Pearson, *Galton*, IIIA, 56–57.

59. Galton, *Natural Inheritance*, pp. 12, 14–15, 155, 157–58, 162, 175–76, 180–81.

60. Cowan, *Sir Francis Galton*, pp. 249–50.

61. Porter, "Galton and Correlation," pp. 46–47, 52. Part of Galton's contribution to the statistics of heredity stemmed in fact from his insightful explanation of why, despite familial regression to the mean, the distribution for a character was the same in the generations of parents and offspring. Consider, as he did, the heights of all the children in a co-family—i.e., all the offspring produced by all parents in the population of a given stature: while the average height of the offspring would regress toward the mean of the population, the heights of the individual children would be dispersed around the co-family's average. The dispersions of all the co-families would together offset the regression enough to maintain the same distribution for height over the population in the children's as in the parents' generation. Galton, "Regression Towards Mediocrity in Hereditary Stature," pp. 255–58.

62. Francis Galton, Preface to the 2nd edition of *Hereditary Genius* (London: Macmillan, 1892; reprinted, 1925), p. xviii.

63. Cowan, "Nature and Nurture . . . in the Work of Francis Galton," pp. 167–73.

64. Galton, *Natural Inheritance*, p. 28; Galton, Preface to the 1892 edition, *Hereditary Genius*, pp. xiv–xix, and the last chapter in the first edition.

65. Galton, Preface to the 1892 edition, *Hereditary Genius*, pp. xx, xxvii.

Chapter II:
KARL PEARSON FOR SAINT BIOMETRIKA

1. Karl Pearson, *The Life, Letters, and Labours of Francis Galton* (3 vols. in 4; Cambridge University Press, 1914–30), I, 6, 7.

2. Ruth Schwartz Cowan, *Sir Francis Galton and the Study of Heredity in the Nineteenth Century* (University Microfilms, 1969), p. 14; Bernard Norton, "Karl Pearson and the Galtonian Tradition: Studies in the Rise of Quantitative Social Biology" (unpublished doctoral dissertation, History of Science, University College London, 1978), p. 29; Pearson, *Galton*, I, 210.

3. Mark Haller, *Eugenics: Hereditarian Attitudes in American Thought* (Rutgers University Press, 1963), p. 24. Jane Hume Clapperton, the social-health advocate, declared that the true motive for marriage must lie in love, not eugenics, adding that the unfit might be persuaded to limit their procreation voluntarily. Jane Hume Clapperton, *Scientific Meliorism and the Evolution of Happiness* (Kegan Paul, Trench, 1885), pp. 333–35.

4. Reviews of Galton's *Inquiries into the Human Faculty*, in *The Academy*, July 14, 1883, and Manchester *Guardian*, Aug. 6, 1883, copies in Francis Galton Papers, list 152/9c; Donald Pickens, *Eugenics and the Progressives* (Vanderbilt University Press, 1968), pp. 35–36.

5. Hal D. Sears, *The Sex Radicals: Free Love in High Victorian America* (The Regents Press of Kansas, 1977), pp. 120–21, 243–44; Victoria Woodhull, *The Scientific Propagation of the Human Race* (pamphlet, 1893), pp. 5, 13, 19, 35, 39.

6. Raymond Lee Muncy, *Sex and Marriage in Utopian Communities* (Penguin Books, 1974), pp. 186–87; *The Circular*, March 27, 1865; Maren Lockwood Carden, *Oneida: Utopian Community to Modern Corporation* (Harper Torchbooks, 1977), p. 61. Noyes referred to his program of better breeding through marriage selection as "stirpiculture," a word that he seems to have coined independently of Galton in 1865. See "Stirpiculture," *The Circular*, April 3, 1865.

7. Maurice Valency, *The Cart and the Trumpet: The Plays of George Bernard Shaw* (Oxford University Press, 1973), p. 219; Galton to William Bateson, June 12, 1904, Francis Galton Papers, 245/3.

8. Donald MacKenzie, "The Development of Statistical Theory in Britain, 1865–1925: A Historical and Sociological Perspective" (doctoral thesis, University of Edinburgh, 1977), pp. 126–27; Pearson, *Galton*, IIIA, 327–28; Norton, "The Galtonian Tradition," pp. 75–76.

9. Arthur's life is limned in his letters to his mother, in the Arthur Beilby Pearson-Gee file, Karl Pearson Papers, University College London.

10. Pearson to mother, Oct. 7, 1873, Karl Pearson Papers, Cabinet II, D1/e.

11. MacKenzie, "Development of Statistical Theory," pp. 141–42; G. Udny Yule, "Karl Pearson, 1857–1936," *Obituary Notices of the Royal Society*, 2 (1936), 73–74.

12. Pearson to mother, May 1 or 7, 1880; Feb. 23, 1880, Karl Pearson Papers, Cabinet II, D1; Pearson's literary favorites are evident in his Commonplace Book, October 1877, Karl Pearson Papers, Cabinet II, D1. Pearson also told his mother that the Germans were deplorably enamored of the Americans, who could "almost beat them in expectorating" and who impertinently quoted "the English poets through their noses."

13. Pearson to Robert Parker, Jan 12, 1880; Sept. 19, 1884, Karl Pearson Papers, Cabinet II, D1 and D2.

14. Norton, "The Galtonian Tradition," pp. 83–84; Yule, "Karl Pearson, 1857–1936," p. 103; Karl Pearson, *The Chances of Death and Other Studies in Evolution* (Edward Arnold, 1897), I, 194–95.

15. Donald MacKenzie, *Statistics in Britain, 1865–1900: The Social Construction of Scientific Knowledge* (Edinburgh University Press, 1981), pp. 81–84; Pearson to Robert Parker, Jan. 12, 1880, Karl Pearson Papers, Cabinet II, D1.

16. Pearson went on to say of the students, "Poor fellows, they go and listen attentively to the possibility of producing a permanent race of mules, as if that could be any cure for tea at six shillings a pound and no marmalade at one and four a pot." Pearson to mother, Nov. 29, 1879, Karl Pearson Papers, Cabinet II, D1.

17. MacKenzie, *Statistics in Britain*, pp.

81–84; Pearson asserted: "You cannot get a strong and effective nation if many of its stomachs are half fed and many of its brains untrained. . . . The true statesman has to limit the internal struggle of the community in order to make it stronger for the external struggle." Karl Pearson, *National Life from the Standpoint of Science* (Black, 1901), pp. 51–52.

18. MacKenzie, "Development of Statistical Theory in Britain," pp. 161, 168–69, 146–48, 150.

19. Shaw to Pearson, March 24, 1890, Karl Pearson Papers, file 627; MacKenzie, *Statistics in Britain*, pp. 77–78.

20. When the senior Pearson suggested that the law was a most interesting science, Pearson could only cry, "God!" and wonder how anyone could think that an interesting science might be made of dealing with rape, murder, and petty theft. Pearson to Robert Parker, Oct. 24, 1881; Pearson to G. U. Yule, April 30, 1898, Karl Pearson Papers, Cabinet II, D2, and Cabinet I, D6.

21. Yvonne Karp, *Eleanor Marx* (2 vols.; Pantheon Books, 1976), II, 82–83; Pearson to Maria Sharpe, no date [1887] and Dec. 14, 1886, Karl Pearson Papers, Cabinet II, D3; MacKenzie, "Development of Statistical Theory," pp. 166–67; Bernard Norton, "Karl Pearson and Statistics: The Social Origins of Scientific Innovation," *Social Studies of Science*, 8 (Feb. 1978), pp. 27–28; Maria Sharpe, "Autobiographical History of the Men and Women's Club," p. 1, Karl Pearson Papers, Cabinet II, D3/k. Detailed treatments of the Club, its attitudes, and its members and guests, on which part of the following is based, are in Ruth First and Ann Scott, *Olive Schreiner* (Schocken Books, 1980), pp. 144–72; and in Phyllis Grosskurth, *Havelock Ellis: A Biography* (Alfred A. Knopf, 1980), pp. 93–106.

22. Sharpe, "Autobiographical History," pp. 15–17, 42; Pearson to Sharpe, Jan. 6, 1888, Karl Pearson Papers, Cabinet II, D3.

23. Schreiner to Pearson, Jan. 30, 1887; Oct. 12, 1886, Karl Pearson Papers, Cabinet I, D2.

24. Pearson to Maria Sharpe, Sept. 8, 1889, Karl Pearson Papers, Cabinet II, D4/B.

25. Schreiner to Pearson [Dec. 14, 1886]; Schreiner to Havelock Ellis, Feb. 1, 1887; Mrs. Cobb to Pearson, Dec. 14, 1886; Donkin to

Pearson, Dec. 1886, Karl Pearson Papers, Cabinet I, D2.

26. Maria Sharpe to Papa, March 15, 1868, Sharpe Family Papers, University College London, file 102; Sharpe, "Autobiographical History," pp. 2–3, 5–6.

27. Sharpe, "Autobiographical History," pp. 43–44, 51–54, 64–65; Sharpe to Pearson, Nov. 15, 1885; March 31, 1887, Karl Pearson Papers, Cabinet II, D3.

28. Pearson to Sharpe, Oct. 22, 1888; June 14, 1887; Oct. 11, 1887; August 14, 1889, Karl Pearson Papers, Cabinet II, D3; Karl Pearson, *The Chances of Death and Other Studies in Evolution*, I, 253–54, 240; Karl Pearson, *The Ethic of Freethought* (Unwin, 1888), pp. 376, 384–87, 392, 432–33, 440–45.

29. In the spring of 1889, Sharpe, who had previously studied the works of Ibsen with discomfort, recalled that she could then bear to have the playwright's portrait in her bedroom, "even love to have it there, and listen to him with confidence when in mental distress." Sharpe, "Autobiographical History," pp. 90–91; Pearson to Maria Sharpe, Aug. 14, 1889; Pearson to mother, Aug. 18, 1889; Pearson to R. J. Parker, Oct. 14, 1889; Sharpe to Pearson, March 26, 1890, Karl Pearson Papers, Cabinet II, D4.

30. Schreiner to Havelock Ellis, Jan. 27, 1888, in *Letters of Olive Schreiner, 1876–1920,* S. C. Cronwright-Schreiner, ed. (Little, Brown, 1924), pp. 129–30; Shaw to Pearson, March 24, 1890, Karl Pearson Papers, file 627.

31. Pearson to Galton, Jan. 24, 1895, Francis Galton Papers, file 293A; Pearson said in 1889, "I shall never do scientific work of the order that leaves a name on the book of revelations." Pearson to Maria Sharpe, Sept. 21, 1889, Karl Pearson Papers, Cabinet II, D4; Karl Pearson, *The Grammar of Science* (Walter Scott, 1892).

32. Lyndsay Farrall, *The Origins and Growth of the English Eugenics Movement* (University Microfilms, 1970), pp. 60–61.

33. Garland Allen, *Life Science in the Twentieth Century* (John Wiley, 1975), pp. 1–8; Lyndsay Farrall, "W. F. R. Weldon, Biometry, and Population Biology" (unpublished paper, 1975), pp. 5–6, 8.

34. William Provine, *The Origins of Theoretical Population Genetics* (University of Chicago Press, 1971), p. 30; Farrall, "Weldon," pp. 8–10.

35. Farrall, "Weldon," pp. 10–14. Norton, "The Galtonian Tradition," p. 70.

36. Bernard Norton, "The Biometric Defense of Darwinism," *Journal of the History of Biology*, 6 (Fall 1973), 293; Farrall, *English Eugenics Movement*, p. 66.

37. Karl Pearson, "On the Laws of Inheritance According to Galton," March 11, 1889; Pearson to Maria Sharpe, Feb. 24, 1889, Karl Pearson Papers, Cabinet V, D6, and Cabinet II, D3; Bernard Norton, "Karl Pearson and Statistics," *Social Studies of Science*, 8 (Feb. 1978), 16–17.

38. Henry Adams, *The Education of Henry Adams* (Houghton Mifflin, Sentry Edition, 1961), p. 450; Norton, "The Galtonian Tradition," pp. 94, 127; Karl Pearson, *The Grammar of Science* (2nd ed.; Black, 1900), pp. 6, 36, 332. Succeeding references in this chapter to *The Grammar* are to the 2nd edition.

39. Karl Pearson, *The Grammar of Science*, pp. 373–74; Norton, "The Galtonian Tradition," p. 126; Pearson said that if something came of the collaboration with Weldon, it would be "the first step to making biology a mathematical science." Pearson to mother, Feb. 25, 1894, Karl Pearson Papers, Cabinet VI, D1; Norton, "Karl Pearson and Statistics," pp. 3, 5–6, 20–21.

40. Robert Parker to Mrs. Cobb, Oct. 26, 1889; Schreiner to Pearson, Nov. 11, 1890, Karl Pearson Papers, Cabinet II, D4; Cabinet I, D2; Pearson, *The Grammar of Science*, pp. 34–35; Shaw to Pearson, Nov. 29, 1892, Karl Pearson Papers, Shaw file.

41. Weldon to Mama, May 28, 1879; to W. A. Dante Weldon, March 9, 1879, and Nov. 23, 1879, Karl Pearson Papers, list 633/3.

42. Pearson to Mrs. W. F. R. Weldon, April 12, 1907, Karl Pearson Papers, Cabinet I, D6. The description given here of the relationship between Pearson and Weldon is based on the extensive correspondence between Weldon and Pearson in the Pearson Papers.

43. Karl Pearson, "Walter Frank Raphael Weldon," *Biometrika*, 5 (1906), 18–20; Weldon to Pearson, Dec. 3, 1892, Karl Pearson Papers, file 624; E. S. Pearson, "Some Incidents in the Early History of Biometry and Statistics, 1890–94," in E. S. Pearson and M. G. Kendall, eds., *Studies in the History of Statistics and Probability* (Griffin, 1970), p. 326.

44. Weldon to Galton, Feb. 11, 1895; March 6, 1895, Francis Galton Papers, file 340D; Weldon to Pearson, Dec. 21, 1896, Karl Pearson Papers, file 624.

45. Pearson, "Weldon," p. 12; Churchill Eisenhart, "Karl Pearson," *Dictionary of Scientific Biography* (16 vols.; Scribner's, 1970–80), X, 450, 465.

46. Pearson, "On the Laws of Inheritance According to Galton," March 11, 1889, Karl Pearson Papers, Cabinet V, D6; Pearson, *Galton*, IIIA, 48–49, 60; Karl Pearson, "Mathematical Contributions to the Theory of Evolution—III: Regression, Heredity, and Panmixia," *Philosophical Transactions of the Royal Society*, A. 187 (1896), 307; Norton, "Biometric Defense of Darwinism," p. 293; Eisenhart, "Pearson," pp. 461–62.

47. Eisenhart, "Pearson," pp. 461–62; Pearson, "Mathematical Contributions to the Theory of Evolution: On the Law of Ancestral Heredity," *Proceedings of the Royal Society*, 62 (1898), 412.

48. Galton to Pearson, Jan. 4, 1898, Francis Galton Papers, file 245/8c.

49. Pearson, "On the Inheritance of the Mental and Moral Characters in Man, and Its Comparison with the Inheritance of the Physical Characters," *Journal of the Royal Anthropological Institute of Great Britain and Ireland*, 33 (1903), 182.

50. Pearson called the measure of the relationship between the nominal variables the "tetrachoric" correlation coefficient, because the data was arrayed in a fourfold table. Pearson, "Inheritance of the Mental and Moral Characters in Man," p. 193; MacKenzie, *Statistics in Britain*, pp. 171–73.

51. Pearson, "Inheritance of the Mental and Moral Characters in Man," p. 204.

52. Galton to Pearson, Nov. 17, 1901, Francis Galton Papers, file 245/18e.

53. Pearson, *Galton*, I, 1. The description of the evolution of the Galton-Pearson relationship is based on the correspondence between the two men in the Karl Pearson Papers.

54. As he began to think about marriage, Pearson came to believe that women's rights had to be delimited by their reproductive responsibilities to society. If Maria wondered about the troublesome implications of that view for the emancipation of her sex, Pearson told her before the marriage: "We

want the best and stablest state, not the strongest individuals. . . . There are times when Spartan discipline will be of more value than Athenian genius." After the marriage, Pearson wrote of a developing "sexual instinct" among middle-class women and declared the reconciliation of maternity with career to be *"par excellence* the women's problem of the future." Pearson to Sharpe, March 19, 1888, Karl Pearson Papers, Cabinet II, D3; Pearson, *The Chances of Death,* I, 238–39, 251.

55. Karl Pearson, *National Life from the Standpoint of Science,* pp. 43–44; Karl Pearson, *The Scope and Importance to the State of the Science of National Eugenics* (Dulau, 1909), p. 41; Pearson, "On the Inheritance of Mental and Moral Characters in Man," pp. 206–7.

56. Karl Pearson, *Nature and Nurture: The Problem of the Future* (Dulau, 1910), pp. 14–15; Karl Pearson, *The Grammar of Science,* pp. 466–67; Pearson, "On the Inheritance of Mental and Moral Characters," pp. 206–7; Pearson, *The Scope and Importance to the State of the Science of National Eugenics,* p. 40; Pearson, *National Life from the Standpoint of Science,* pp. 26–27.

57. Pearson, *The Problem of Practical Eugenics* (Dulau, 1912), pp. 22–23.

58. Pearson, *Galton,* I, 60–61; IIIA, 368; Pearson, *The Problem of Practical Eugenics,* pp. 21–22, 24–25.

59. Since capitalism permitted the working classes few opportunities, Pearson remarked, commenting on the high working-class birthrate, one could not really blame them for indulging in one of the few pleasures they could afford. Pearson, *Ethic of Freethought,* pp. 336–40.

60. Pearson, *The Problem of Practical Eugenics,* pp. 36–37; Pearson, "On the Inheritance of Mental and Moral Characters," p. 207; Pearson to Galton, Jan. 10, 1901, and Jan. 2, 1896, Francis Galton Papers, file 293B; Pearson, *The Groundwork of Eugenics* (Dulau, 1909), pp. 20–21; Pearson to Yule, Jan. 25, 1895 [?], Karl Pearson Papers, Cabinet I, D6.

61. Pearson, *Ethic of Freethought,* p. 433.

62. Pearson, *The Problem of Practical Eugenics,* pp. 21–22, 24–25; Pearson, *Eugenics and Public Health* (Cambridge University Press, 1912), pp. 32–34.

63. Pearson to Galton, Oct. 14, 1908, Karl Pearson Papers, Cabinet VI, D6; Pearson, *The Scope and Importance to the State of the Science of National Eugenics,* pp. 14–15.

64. Pearson to wife, April 19, 1903; Pearson to W. P. Elderton, March 28, 1911; Pearson to Galton, Oct. 27, 1907, Pearson Papers, Cabinet II, D7; Cabinet VI, D7 and D6.

65. The attitude toward science at Oxford was generally unsympathetic. Weldon complained to Pearson that if a boy was found wanting in Greek, he was turned next to mathematics, and then, after learning "the anatomy of the frog, and a shoddy hypothesis about the pedigree of animals," was given a science scholarship as a last resort. Weldon had failed to get "one man to care for anything I say outside a textbook! Their tutors all tell them one is an amiable crank." Weldon to Pearson, May 19, 1899; July 11, 1900, Karl Pearson Papers, Weldon file.

66. Pearson to his mother, Nov. 5, 1893; Pearson to Yule, Dec. 8, 1900; Pearson, "History of the Biometric and Galton Laboratories," c. 1920; "Report by Professor Karl Pearson . . . Worshipful Company of Drapers . . . (1903–1909)," p. 2, Karl Pearson Papers, Cabinet VI, D1; Cabinet I, D6; list 247; list 233.

67. Pearson to W. F. R. Weldon, July 13, 1900; Pearson to Galton, April 8, 1907, Karl Pearson Papers, Cabinet VI, D2 and D6. The American biologist Raymond Pearl, while spending 1905 at Pearson's laboratory, wrote: "The great biometric laboratory of University College is all comprised in one room with two windows [and] with six or seven other people, one of whom is Dr. Alice Lee, whose most settled conviction is that the proper temperature of a room is *not over 58 degrees."* Raymond Pearl to Herbert S. Jennings, Oct. 8, 1905, Raymond Pearl Papers, American Philosophical Society Library, Philadelphia, Jennings file; Pearson to Galton, Jan. 11, 1898, in Bernard Norton, "Karl Pearson and the New Discipline of Statistics" (unpublished manuscript), note 96.

68. Pearson to Major Greenwood, Sept. 22, 1904, Karl Pearson Papers, Cabinet VI, D5; Pearson to Galton, Dec. 13, 1900 [Jan. 3, 1901]; Feb. 1901; March 14, 1902; March 5, 1903, Francis Galton Papers, files 293/d, e, and f.

69. Pearson to Galton, Jan. 28, 1902, Francis Galton Papers, file 293/e; Eisenhart, "Karl Pearson," p. 454; F. Yates, "George Udny Yule," *Obituary Notices of the Royal Society,* 8 (1952–53), 315–16.

70. Yule, "Karl Pearson, 1857–1936," p. 102.

71. Pearl to Jennings, Oct. 8, 1905; Jennings to Pearl, March 5, 1910, Pearl Papers, Jennings file; Daniel J. Kevles, "Genetics in the United States and Britain, 1890–1930: A Review with Speculations," *Isis*, 71 (1980), 448–49.

72. Pearson to Mrs. Weldon, Oct. 19, 1906, Karl Pearson Papers, Cabinet I, D6; G. Udny Yule to Greenwood, May 18, 1936, G. Udny Yule Papers, Royal Statistical Society, London, Box 2, Pearson file. One bruised ex-collaborator of Pearson's later remarked to another, "As you say, K.P. is morally a first-class shit." Major Greenwood to Raymond Pearl, July 24, 1926, Pearl Papers, Greenwood file.

73. Yule, "Karl Pearson, 1857–1936," p. 103; Yule to Miss Elderton, May 24, 1936, G. Udny Yule Papers, Box 2, Pearson file.

74. J. B. S. Haldane, "Karl Pearson," in Pearson and Kendall, eds., *Studies in the History of Probability and Statistics*, p. 427.

75. Pearson, *Galton*, IIIA, 312. Pearson lamented to Weldon, "I wish we had both been born Germans; we should have established a new 'discipline' by now and have had a healthy supply of workers." Pearson to Weldon, Oct. 9, 1900, Karl Pearson Papers, file 263.

76. Galton's will is in the Francis Galton Papers, file 131/1; Farrall, *English Eugenics Movement*, pp. 130–32.

77. Galton to Pearson, Nov. 15, 1906, Karl Pearson Papers, Cabinet VI, D6; Pearson, *Galton*, IIIA, 222–23, 258; Farrall, *English Eugenics Movement*, pp. 106, 108, 110–12; Pearson to Galton, Dec. 20, 1904; June 23, 1905, Francis Galton Papers, file 293F; W. D. M. Paton and C. G. Phillips, "E. H. J. Schuster, 1897–1969," *Royal Society of London Notes and Records*, 28–29 (1973–75), 111.

78. Pearson to Greenwood, June 24, 1911, Karl Pearson Papers, Cabinet VI, D7; Norton, "The Galtonian Tradition," p. 105; University College London Applied Statistics file, Estate Committee Resolution, No. 31, Dec. 1904, Drapers Company Records, The Worshipful Company of Drapers, London.

79. Documents in re anonymous gift, 1912; subscription list, in Karl Pearson Papers, Cabinet IV, D7; Cabinet I, D7/2; list 239; E.

S. Pearson, *Karl Pearson: An Appreciation of Some Aspects of His Life and Work* (Cambridge University Press, 1938), p. 77.

80. E. S. Pearson, *Karl Pearson*, p. 71; notes by Ethel Elderton, attached to Elderton to Yule, May 22, 1936, G. Udny Yule Papers, Box 2, Pearson file; Pearson to Galton, Nov. 5, 1908; Pearson to Egon Pearson, May 19, 1931, Karl Pearson Papers, Cabinet VI, D6 and D8.

81. A list of students and research fellows is in "Report to the Court of the Worshipful Company of Drapers . . . ," 1918, Karl Pearson Papers, list 233.

82. "Report of the Francis Galton Laboratory Committee for Presentation to the Royal Commission on University Education in London (Jan. 1911)"; J. B. S. Haldane to E. S. Pearson, May 8, 1936; "Report by Professor Karl Pearson . . . Worshipful Company of Drapers . . . (1903–1909)," p. 2; Karl Pearson Papers, list 237; list 233; Cabinet I, D9.

83. Pearson, *Galton*, IIIA, 358–59; Pearson to Julia Bell, Oct. 21, 1910, and Sept. 4, 1912; Pearson to Elderton [Aug. 1913]; Pearson to David Heron, July 10, 1915; Pearson, "Report on the Galton . . . especially with regard to Their Income and Expenditures," [1926 or 1927]; Farewell Presentation to Ethel M. Elderton . . . ," 1935, Karl Pearson Papers, Cabinet I, D7/1; Cabinet VI, D7; list 233; Cabinet VI, D8; Galton to Pearson, Jan. 4, 1898, Francis Galton Papers, file 245/18c; Ethel M. Elderton, *The Relative Strength of Nurture and Nature* (Dulau, 1909), pp. 6–7; Rosealeen Love, " 'Alice in Eugenics-Land': Feminism and Eugenics in the Scientific Careers of Alice Lee and Ethel Elderton," *Annals of Science*, 36 (1979), 145–58.

84. "Report to the Court of the Worshipful Company of Drapers . . . ," 1918, Karl Pearson Papers, list 233. This discussion is generally based on the successive annual reports of the Galton and Biometric laboratories in the Pearson papers.

85. Karl Pearson, *Tuberculosis, Heredity, and Environment* (Dulau, 1912), pp. 42–43.

86. The statistician G. Udny Yule, one of the first people to work with Pearson, called *Biometrika* "the most personally edited journal that was ever published." Yule, "Karl Pearson, 1857–1936," p. 100.

87. Elderton, *The Relative Strength of Nurture and Nature*, p. 33.

Chapter III:
CHARLES DAVENPORT AND THE WORSHIP OF GREAT CONCEPTS

1. V. Kruta and V. Orel, "Johann Gregor Mendel," *Dictionary of Scientific Biography* (16 vols.; Scribner's, 1970–80), IX, 277–83.
2. *Ibid.*, p. 281.
3. Alfred R. Wallace to Galton, Feb. 3, 1891; Feb. 7, 1891; Galton to Members of the Committee of the Royal Society for the Measurement of Plants and Animals [1897]; W. Thiselton-Dyer to Galton, Dec. 9, 1896; Francis Galton Papers, University College London, files 142/28; 142/1; 142/2B; Charles B. Davenport, "A Summary of Progress in Experimental Evolution," p. 5, Charles B. Davenport Papers, American Philosophical Society Library, Philadelphia; Henry F. Osborn, E. Wilson, and W. K. Brooks, "Report of the Advisory Committee on Zoology," Oct. 15, 1902, Records of the Carnegie Institution of Washington, Carnegie Institution of Washington, Washington, D.C., Advisory Committee on Zoology file.
4. Bateson to Galton, Aug. 9, 1900, Galton Papers, file 198; Daniel J. Kevles, "Genetics in the United States and Britain, 1890–1930: A Review with Speculations," *Isis*, 71 (1980), 452–53.
5. Weldon to Pearson, Nov. 1901, Karl Pearson Papers, University College London, Kevles, "Genetics in the United States and Britain," pp. 444–45; W. F. R. Weldon, "Mendel's Laws of Alternative Inheritance in Peas," *Biometrika*, I (1901–2), 236–38.
6. Garland E. Allen, "Edmund Beecher Wilson," *Dictionary of Scientific Biography*, XIV, 431.
7. Garland E. Allen, "Thomas Hunt Morgan," *Dictionary of Scientific Biography*, IX, 520–22.
8. Karl Pearson, *The Life, Letters, and Labours of Francis Galton* (3 vols. in 4; Cambridge University Press, 1914–30), IIIA, 324–25; Alexander G. Bearn, "Inborn Errors of Metabolism, I: Archibald Garrod and the Birth of an Idea," *Transactions of the Medical Society of London*, 93 (1976), 47–53; Charles E. Rosenberg, "Charles Davenport and the Beginning of Human Genetics," *Bulletin of the History of Medicine*, 35 (1961), 270.
9. Davenport assured his sister that he was "not fired" from Harvard. Davenport to Frances G. Davenport, Nov. 19, 1899, Charles B. Davenport Papers, Frances G. Davenport file; E. Carleton McDowell, "Charles Benedict Davenport," *Bios*, 17 (1946), 12–14; Charles E. Rosenberg, "Charles Benedict Davenport and the Irony of American Eugenics," in Charles Rosenberg, *No Other Gods: On Science and American Social Thought* (Johns Hopkins University Press, 1976), p. 90; Davenport to Galton, Oct. 11, 1902, Francis Galton Papers, file 235.
10. Daniel J. Kevles, *The Physicists: The History of a Scientific Community in Modern America* (Alfred A. Knopf, 1978), p. 69; Davenport to Professor Franklin Hooper, April 21, 1902, Charles B. Davenport Papers, Cold Spring Harbor, Beginnings file.
11. Franklin N. Hooper to Charles D. Walcott, July 24, 1902; Davenport to John C. Merriam, Feb. 26, 1929, Charles B. Davenport Papers, Cold Spring Harbor, Beginnings file; Cold Spring Harbor Series #1; the budgets are taken from the annual *Yearbooks* of the Carnegie Institution of Washington; Bateson to Davenport, April 25, 1906, Charles B. Davenport Papers, Bateson file.
12. Charles B. Davenport, "Station for Experimental Evolution" [1919]; "Memorandum for President Woodward" [1919], Charles B. Davenport Papers, Cold Spring Harbor Series #1; Mark Haller, *Eugenics: Hereditarian Attitudes in American Thought* (Rutgers University Press, 1963), p. 64; McDowell, "Davenport," p. 26.
13. "Report of Howard J. Banker" [192?], Charles B. Davenport Papers, Cold Spring Harbor Series #2.
14. Davenport to John C. Merriam, April 17, 1931; Davenport to Alexander Graham Bell, April 19, 1909; "American Breeders Association, Section on Eugenics"; Davenport to the Editor of *Science*, Oct. 12, 1909, Charles B. Davenport Papers, Cold Spring Harbor Series #1; Bell file; Breeders Association file; James McKeen Cattell file.
15. Charles B. Davenport, *Heredity in Relation to Eugenics* (Henry Holt, 1911), pp. 66–67, 72–74, 79–80, 102, 157, 126; Charles B.

Davenport, *Eugenics—The Science of Human Improvement by Better Breeding* (Henry Holt, 1910), pp. 11–12; Davenport, "Euthenics and Eugenics," Charles B. Davenport Papers, Breeders Association file.

16. Davenport, *Heredity in Relation to Eugenics*, pp. 77, 79–80, 93.

17. Davenport, *Heredity in Relation to Eugenics*, p. 80; Davenport, "Euthenics and Eugenics," Charles B. Davenport Papers, Breeders Association file; Haller, *Eugenics*, p. 68. Holding that "blood" prevailed against even salubrious climate, Davenport declared, "The rising generation in California is characterized by disease of the mucous membranes, because a generation ago much weak protoplasm was attracted to this State as a Sanatorium." Davenport to Joseph F. Gould, Oct. 17, 1913, Charles B. Davenport Papers, Gould file.

18. Davenport, *Heredity in Relation to Eugenics*, pp. 216, 218–19, 221–22.

19. *Ibid.*, pp. 248–49. After the First World War, Davenport worried about Britain's losing the "reproductive capacity" of her wounded soldiers, since "the next generation of Englishmen largely lies in the loins of these excellent young men." Davenport to R. S. Woodward, Feb. 11, 1919, Charles B. Davenport Papers, Cold Spring Harbor Series #1.

20. Davenport, *Heredity in Relation to Eugenics*, pp. 221–22, 224; Davenport to Joseph F. Gould, Feb. 17, 1914, Charles B. Davenport Papers, Gould file.

21. *Heredity in Relation to Eugenics*, pp. 256–59; Davenport, *Eugenics*, p. 16.

22. Davenport to W. P. Draper, March 23, 1923, Charles B. Davenport Papers, Draper file; Davenport to John C. Merriam, April 4, 1930, Charles B. Davenport Papers, Cold Spring Harbor Series #1; Davenport, *Heredity in Relation to Eugenics*, pp. 92, 102, 104, 157.

23. Pearson to Galton, July 14, 1906, Karl Pearson Papers, Cabinet VI, D6.

24. Weldon to Pearson, Jan. 9, 1904, Karl Pearson Papers; Kenneth Ludmerer, *Genetics and American Society* (John Hopkins University Press, 1972), p. 60; Rosenberg, "Davenport and the Beginning of Human Genetics," p. 271.

25. The young American geneticist Raymond Pearl, a eugenicist, too, pronounced Davenport's research "very careless and slipshod." Pearl added, "Much of Davenport's poultry work is, in plain language, 'no good.' . . . Bateson . . . could not see how Davenport ever got any reliable results with the quality of stock he used." Raymond Pearl to Herbert S. Jennings, July 19, 1910, Raymond Pearl Papers, American Philosophical Society Library, Philadelphia, Jennings file; Smith Ely Jelliffe to Davenport, Dec. 26, 1910; Dec. 28, 1910, Charles B. Davenport Papers, Jelliffe file.

26. Davenport to Jelliffe, Dec. 27, 1910, Charles B. Davenport Papers, Jelliffe file.

27. "Record of Family Traits" for Davenport, Aug. 5, 1913, Davenport, "My Autobiography"; Davenport, Diary, 1878, 1881; Davenport, "Henry E. Davenport," Charles B. Davenport Papers, Family Records file; Davenport Autobiographical file; Diary file; MacDowell, "Davenport," pp. 4–7, 9.

28. MacDowell, "Davenport," pp. 5–6; Davenport Family Records, Dec. 15, 1914; Davenport, "Ich Dein [*sic*]", Theme 1885/86, Charles B. Davenport Papers, Family Records file.

29. Davenport to A. B. Davenport, March 18, 1885, A. B. Davenport to Davenport, May 5, 1885, "Autobiography of C. B. Davenport," Charles B. Davenport Papers, A. B. Davenport file; Autobiography file; MacDowell, "Davenport," pp. 8–12.

30. Davenport to Mrs. A. B. Davenport, June 3, 1888; July 22, 1888; Davenport, "Autobiography, 1916"; Davenport to Gertrude C. Davenport, Sept. 13, 1893, Charles B. Davenport Papers, Mrs. A. B. Davenport file; Autobiography file; Gertrude Crotty Davenport file; MacDowell, "Davenport," pp. 14–15.

31. Davenport to Gertrude Davenport, Jan. 24, 1904, Charles B. Davenport Papers, Gertrude Crotty Davenport file; MacDowell, "Davenport," p. 35.

32. MacDowell, "Davenport," p. 34; Rosenberg, "Davenport and the Beginning of Human Genetics," p. 274; Davenport to Editor, New York *Evening Post*, 1915; Davenport, "A Constructive Program for Eugenics Work in Nassau County," Oct. 1913; Davenport to Madison Grant, Feb. 10, 1917, Charles B. Davenport Papers, Cold Spring Harbor Series #2; Nassau County Association file;

33. Haller, *Eugenics*, p. 91; MacDowell, "Davenport," p. 37.

34. MacDowell, "Davenport," pp. 33–35; Davenport to Gertrude C. Davenport, Dec. 12, 1913, Charles B. Davenport Papers, Gertrude Crotty Davenport file.

35. Davenport to Gertrude Davenport, Dec. 10, 1918; Davenport to Millia Davenport, July 3, 1923; Billy [Millia] Davenport to Davenport [no date], "Tuesday night," Charles B. Davenport Papers, Gertrude Crotty Davenport file; Millia Davenport file. Billy Davenport added that she was quite as willing for a friend's father "to have been an Irish workman, who made money in real estate . . . and sent his sons to college as for him to be a small middle western farmer who sent his sons to Oberlin."

36. Anne Lutz to Davenport, April 23, 1909; Dec. 2, 1909, Charles B. Davenport Papers, Cold Spring Harbor Series #2.

37. Davenport to John J. Burke, Feb. 20, 1926; "Record of Family Traits" for Davenport, Charles B. Davenport Papers, John J. Burke file; Family Records file.

38. *Heredity in Relation to Eugenics*, p. 216; Davenport to John J. Burke, Jan. 21, 1926; to R. L. Dickinson, Aug. 13, 1923, Charles B. Davenport Papers, Burke and Dickinson files.

39. Rosenberg, "Davenport and the Irony of American Eugenics," p. 93; Rosenberg, "Davenport and the Beginning of Human Genetics," p. 272; Davenport at "Second Field Workers Conference," June 20, 21, 1913; Davenport to Starr J. Murphy, June 24, 1913, Charles B. Davenport Papers, Cold Spring Harbor Series #2, Eugenics Record Office file; Murphy-Rockefeller file; Davenport, "The Feebly Inhibited: I. Violent Temper and Its Inheritance," *Journal of Nervous and Mental Diseases*," 42 (Sept. 1915), 608.

40. Davenport, "A Constructive Program for Eugenics Work in Nassau County," Oct. 1913, Charles B. Davenport Papers, Cold Spring Harbor Series #2; *Heredity in Relation to Eugenics*, p. 258.

41. *Heredity in Relation to Eugenics*, pp. iii, 4; Davenport, "Eugenics, a Subject for Investigation . . ." [c. 1907], Charles B. Davenport Papers, Breeders Association file; Davenport, *Eugenics*, pp. 30–34.

42. Davenport, "Memorandum on History of the Eugenics Record Office," Dec. 26 [1920?], Charles B. Davenport Papers, Cold Spring Harbor Series #1.

43. H. E. Crampton to C. B. Davenport, June 3, 1905; Davenport to Leonard Darwin, April 18, 1931, Charles B. Davenport Papers, Crampton file; Darwin file; entries for Mrs. E. H. Harriman and Mary Harriman Rumsey, in *Notable American Women: A Biographical Dictionary, 1607–1950* (3 vols.; Harvard University Press, 1971); MacDowell, "Davenport," p. 29.

44. Davenport to Starr J. Murphy, Jan. 20, 1911, Charles B. Davenport Papers, Cold Spring Harbor Series #2. Surely, Davenport thought, no one could justifiably deny society the "right to question me—who is merely a sample of . . . universal germ plasm." *Heredity in Relation to Eugenics*, pp. iv–v, 269.

45. Davenport to Galton, Oct. 26, 1910, Francis Galton Papers, file 235; Davenport to Starr J. Murphy, March 22, 1911; Davenport to Mrs. Harriman, April 28, 1911; Davenport to William Henry Welch, March 1, 1915; "Notes on the History of the Eugenics Record Office," Charles B. Davenport Papers, Cold Spring Harbor Series #2; Mrs. E. H. Harriman file; Davenport Family—Notes file; Davenport, "History of Eugenics Record Office," 1929, Records of the Carnegie Institution of Washington, Harry Laughlin file.

46. Davenport, "Memorandum on History of the Eugenics Record Office," Dec. 26 [1920]; Davenport to R. S. Woodward, Oct. 18, 1919, Charles B. Davenport Papers, Cold Spring Harbor Series #1; Davenport to Starr J. Murphy, July 1910; Jan. 20, 1911; July 11, 1911; Jan. 9, 1912; July 7, 1913; Nov. 7, 1914; Feb. 11, 1915; Davenport to Board of Scientific Directors . . . , Dec. 13, 1913; "The First Quarterly Report of the Eugenics Record Office . . . ," Charles B. Davenport Papers, Cold Spring Harbor Series #2, Eugenics Record Office file; Percy Collins, "The Progress of Eugenics," *Scientific American*, 109 (Dec. 13, 1913), 459.

47. "Minutes of the Meeting of Field Officers on Human Heredity . . . , Oct. 14, 1910," Charles B. Davenport Papers, Breeders Association file; Haller, *Eugenics*, pp. 65–66, 73; MacDowell, "Davenport," p. 30; Davenport to John C. Merriam, April 17, 1931,

Charles B. Davenport Papers, Cold Spring Harbor Series #1.

48. Davenport to Alexander Graham Bell, Jan. 9, 1913; Davenport, "Eugenics Record Office" [1919]; "Memorandum on History of the Eugenics Record Office," Dec. 26 [1920?]; Davenport to Mrs. E. H. Harriman,

Jan. 5, 1913, Charles B. Davenport Papers, Bell file; Cold Spring Harbor Series #1; Mrs. E. H. Harriman file.

49. Davenport to Mrs. Harriman, Feb. 21, 1911, Charles B. Davenport Papers, Mrs. E. H. Harriman file.

Chapter IV:

THE GOSPEL BECOMES POPULAR

1. Pearson to Galton, Jan. 10, 1901, Francis Galton Papers, University College London, file 293E.

2. Geoffrey Searle, *Eugenics and Politics in Britain, 1900–1914* (Noordhoff International Publishing, 1976), p. 9; Karl Pearson, *The Life, Letters, and Labours of Francis Galton* (3 vols. in 4; Cambridge University Press, 1914–30), IIIA, 226, 235, 261; Victor Branford to Galton, June 1904, Francis Galton Papers, file 138/9.

3. Pearson to Galton, June 20, 1907; R. J. Ryle to Pearson; Oct. 15, 1913, London *Daily Sketch,* Oct. 3, 1913, Karl Pearson Papers, University College London, Cabinet VI, D6; Cabinet IV, D7; list 23B; Press Clippings Scrapbook, Papers of the Eugenics Society, Eugenics Society, London.

4. Matthew J. Bruccoli, *F. Scott Fitzgerald in His Own Time: A Miscellany* (Kent State University Press, 1971), pp. 18–19; Charles B. Davenport, "History of Eugenics Record Office," 1929, Records of the Carnegie Institution of Washington, Laughlin file; Davenport to Merriam, Nov. 7, 1932; Harry Laughlin, "Budget," 1929, Charles B. Davenport Papers, Cold Spring Harbor Series #1 and #2; Bernard Norton, "Karl Pearson and Statistics: The Social Origins of Scientific Innovation," *Social Studies of Science,* 8 (Feb. 1978), 9; Eugenics Education Society, *Second Annual Report, 1909/10,* pp. 19–24; *First Annual Report, 1908/09,* pp. 16–18; Sybil Gotto to Galton, Dec. 4, 1908; Gotto to Galton, Jan. 5, 1909; Dec. 11, 1909, Francis Galton Papers, file 240/7.

5. A. B. Wolfe, "Literature of Eugenics," *American Economic Review,* 3 (March 1913), 165; A. E. Hamilton to Davenport, March 6, 1916; Davenport to W. E. D. Stokes, April 8, 1913, Charles B. Davenport Papers,

Cold Spring Harbor Series #2. On the eve of the First World War, American magazines carried more articles on eugenics than on slums, tenements, and living standards combined. John Higham, *Strangers in the Land: Patterns of American Nativism, 1860–1925* (2nd ed.; Atheneum, 1963), pp. 150–51.

6. Havelock Ellis, "Eugenics in Relation to War," in Ellis, *The Philosophy of Conflict and Other Essays in Wartime* (2nd series; London: Constable, 1919), pp. 125–26; Vernon L. Kellogg, "The Bionomics of War," *Social Hygiene,* 1 (Dec. 1914), 52; Albert G. Keller, "Eugenics: The Science of Rearing Human Thoroughbreds," *Yale Review,* 17 (Aug. 1908), 141; Roland C. Macfie, "The Selective Effects of War," *The New Statesman,* 8 (Feb. 10, 1917), 442; Vernon L. Kellogg, "Eugenics and Militarism," *Atlantic Monthly,* 112 (July 1913), 108.

7. Albert E. Wiggam, *The New Decalogue of Science* (Bobbs-Merrill, 1923), pp. 109–11; *National Cyclopedia of American Biography,* XLIV, 542–43; Davenport to Robert S. Woodward, Dec. 21, 1920; Davenport to Leon F. Whitney, Aug. 26, 1926, Davenport to Merriam, Jan. 25, 1921; Charles B. Davenport Papers, Cold Spring Harbor Series #1; Series #2, Eugenics: Wiggam et al. file; Mark Haller, *Eugenics: Hereditarian Attitudes in American Thought* (Rutgers University Press, 1963), p. 72.

8. J. W. Slaughter to Galton, Jan. 11, 1908, Francis Galton Papers, file 318; Haller, *Eugenics,* p. 20; Lyndsay Farrall, *The Origins and Growth of the English Eugenics Movement* (University Microfilms, 1970), pp. 209–10.

9. Haller, *Eugenics,* pp. 62–63, 73–74; the composition and purpose of the Galton Society are revealed in Charles B. Davenport Papers, William K. Gregory file; Davenport to

colleague, Feb. 11, 1913, Charles B. Davenport Papers, Cold Spring Harbor Series #2, Eugenics Record Office file; President of the American Eugenics Society to James R. Fryer, June 18, 1930, American Eugenics Society Papers, American Philosophical Society Library, Philadelphia, box 12.

10. J. W. Slaughter to Galton, Jan. 11, 1908, Francis Galton Papers, file 318; Geoffrey Searle, *Eugenics and Politics in Britain*, pp. 13–14; Pauline Mazumdar, "Eugenists, Marxists and the Science of Human Genetics" (unpublished manuscript), p. 69; Richard Allen Soloway, "Feminism, Fertility, and Eugenics in Victorian and Edwardian England," in Seymour Drescher et al,. *Political Symbolism in Modern Europe* (Rutgers University Press, 1982), p. 123; Richard Allen Soloway, *Birth Control and the Population Question in England, 1877–1930* (University of North Carolina Press, 1982), pp. 122–23; Frederick Osborn, "History of the American Eugenics Society," p. 7, American Eugenics Society Papers, box 15. Both societies were comparable in size to the Fabian Society, which had some eight hundred members at the turn of the century. Farrall, *The English Eugenics Movement*, p. 212.

11. Eugenics Society budgets are taken from the annual reports; "Third Meeting of the Board of Directors of the American Eugenics Society," Jan. 3, 1927; "Ninth Meeting of the Board of Directors of the American Eugenics Society," Nov. 16, 1929, American Eugenics Society Papers, box 6; Davenport to George Eastman, Nov. 23, 1925; Leon Whitney to Davenport, Nov. 21, 1925; July 18, 1925, Charles B. Davenport Papers, Eastman file; Cold Spring Harbor Series #2, Eugenics: Wiggam et al. file.

12. C. B. Hodson to C. C. Hurst (no date), C. C. Hurst Papers, Cambridge University Library, entry 7955/16/63; Pearson, *Galton*, IIIA, 380; Eugenics Education Society, *Second Annual Report, 1909/10*, p. 1; Gotto to Galton, Dec. 11, 1909, Francis Galton Papers, file 240/7.

13. Eugenics Society, *Annual Report, 1930/31*, p. 5; *1925/26*, p. 2; *1916/17*, pp. 5–6; *1926/27*, pp. 4–6; *1927/28*, p. 6; *1928/29*, p. 4; *1932/33*, p. 4; *The American Eugenics Society* (American Eugenics Society, 1927), p. 7, American Eugenics Society Papers, box 9. Julian Huxley extolled the cause in talks on heredity broadcast over the wireless.

14. *A Eugenics Catechism* (American Eugenics Society, 1926), pp. 2–3, 10.

15. American Eugenics Society, Southern California Branch, "Minutes of Regular Meeting, April 7, 1930," American Eugenics Society Papers, box 11; "Conditions of the Awards for the Best Sermons on Eugenics," American Eugenics Society Papers, box 9; Leon F. Whitney to John Gerould, Dec. 11, 1928, American Eugenics Society Papers, box 13; "Eugenics Sermon," 1926, American Eugenics Society Papers, box 11.

16. "Eugenics Sermon," 1926; Harry H. Mayer, "Eugenics, A Sermon for Mothers Day," May 9, 1926, American Eugenics Society Papers, boxes 11, 14.

17. Kenneth C. MacArthur, "Eugenics Sermon," June 20, 1926, L. F. Whitney to MacArthur, July 20, 1927, American Eugenics Society Papers, box 14.

18. A. E. Hamilton, "Eugenics," *Pedagogical Seminary*, 21 (March 1914), 36; *The Fitter Families Eugenic Competition at Fairs and Expositions* (American Eugenics Society, 1922); Mary T. Watts to Davenport, July 28, 1923; Watts to Davenport, 192?, Charles B. Davenport Papers, Cold Spring Harbor Series #2, Eugenics: Wiggam et al. file; "Meetings of the Eugenics Committee of the U.S.A., March 28, 1925, American Eugenics Society Papers, box 4; "Fifth Meeting of the Board of Directors of the American Eugenics Society," Jan. 25, 1928; "Fifteenth Meeting of the Eugenics Committee of the United States of America," Oct. 31, 1925, American Eugenics Society Papers, box 6; Leon F. Whitney, "Memoirs," p. 195, Charles B. Davenport Papers.

19. *The Fitter Families Eugenic Competition at Fairs and Expositions;* "Department S —Eugenics," from the Premium Book of the Kansas Free Fair, Topeka, Sept. 8 to 13, 1924, Charles B. Davenport Papers, Cold Spring Harbor Series #2, Eugenics: Wiggam et al. file. The medal was inscribed with the circumferential caption: "I have a goodly heritage." See photograph of the medal in the American Eugenics Society Papers.

20. *Suggested Programs for Clubs* (Committee on Popular Education; American Eugenics Society, 192?), American Eugenics Society Papers, box 9; *The American Eugenics Society*, 1927, pp. 11–12. The descriptions of the

graphic demonstrations are taken from photographs of the exhibits, American Eugenics Society Papers. British health week exhibitions often included "heredity stalls," which showed the eugenic film, sometimes to women only. Eugenics Society, *Annual Report, 1926/27*, pp. 5–6.

21. From photographs of the exhibits, American Eugenics Society Papers.

22. Edgar Schuster, *Eugenics* (Clear-Type Press, 1912), pp. 54–55; "The First International Eugenics Congress," *Nature*, 89 (Aug. 1, 1912), 558; Raymond Pearl to Davenport, Feb. 24, 1913, Charles B. Davenport Papers, Pearl file; list of vice-presidents in announcement of the Congress, attached to Gotto to Hurst, March 27, 1912, Hurst Papers, entry 7955/12/57. A second international eugenics congress was held in 1921 in New York City at the American Museum of Natural History. Frederick Osborn, "History of the American Eugenics Society," Jan. 20, 1971, p. 3, American Eugenics Society Papers, box 15.

23. "Civilization," *Encyclopaedia Britannica* (11th edition, 1910–11), VI, 410.

24. Garland E. Allen, "Eugenics and Population Control: The Transformation of a Scientific Ideology in the Work of Raymond Pearl" (unpublished manuscript), p. 20; W. D. M. Paton and C. G. Phillips, "E. H. J. Schuster, 1897–1969," *Royal Society of London Notes and Records*, 28–29 (1973–75), 112.

25. Hesketh Pearson, *Modern Men and Mummers* (George Allen and Unwin, 1921), p. 203; Inge to Galton (no date), Francis Galton Papers, file 264.

26. Farrall, *English Eugenics Movement*, pp. 213–18, 225–27, 221–22; Donald MacKenzie, "Eugenics in Britain," *Social Studies of Science*, 6 (1976), 503–5; Haller, *Eugenics*, pp. 92–93; Kenneth Ludmerer, *Genetics and American Society* (John Hopkins University Press, 1972), p. 16.

27. James Reed, *From Private Vice to Public Virtue: The Birth Control Movement and American Society Since 1830* (Basic Books, 1978), pp. 55–56; Eugenic Education Society, *Third Annual Report, 1910/11*, p. 23; Soloway, "Feminism, Fertility, and Eugenics in Victorian and Edwardian England," p. 133; Ursula Grant Duff, "Notes on the Early Days of the Eugenics Education Society," Eugenics Society Records, Wellcome Institute for the History of Medicine, London; "Summary of

Lectures by A. E. Hamilton," 1914, Charles B. Davenport Papers, Cold Spring Harbor Series #2. A prime mover in the founding and activities of the Eugenics Education Society was Mrs. Sybil Gotto, an admiral's daughter and at the time recently widowed at the age of twenty-one.

28. Blanche Eames, *Principles of Eugenics: A Practical Treatise* (Moffat, Yard, 1914), pp. 54–56. Such physiology may have been bizarre, but reputable medical authorities of the day were also concerned with whether the chemical poisons of modern society might foster congenital disease. The British physician and eugenicist Caleb W. Saleeby reckoned that mercury, arsenic, phosphorus, and other metals might well act as "racial poisons" by passing through the placental filter, and so might nicotine, a highly diffusible alkaloid. He insisted that until the contrary was proved, "the expectant mother should not smoke." Caleb William Saleeby, *The Progress of Eugenics* (Funk & Wagnalls, 1914), p. 230.

29. Carl Degler, *At Odds: Women and the Family in America from the Revolution to the Present* (Oxford University Press, 1980), pp. 290–92, 287, 271–72; Linda Gordon, *Woman's Body, Woman's Right: A Social History of Birth Control in America* (Grossman, 1976), p. 119; Reed, *From Private Vice*, pp. 35, 42, 57; Soloway, *Birth Control and the Population Question in England*, pp. 134–35.

30. Eames, *Principles of Eugenics*, pp. 84–85; Mary Ries Melendy, *Sex-Life, Love, Marriage, Maternity* (Vansant, 1914), p. 313.

31. "Eighth Annual Meeting of the American Eugenics Society," May 12, 1933, American Eugenics Society Papers, box 6; "Minutes of Meetings, Southern California Branch, American Eugenics Society," Jan. 6, 1930, May 12, 1933, American Eugenics Society Papers, box 11.

32. Phyllis Grosskurth, *Havelock Ellis: A Biography* (Alfred A. Knopf, 1980), pp. 220, 228–31; Havelock Ellis, *The Task of Social Hygiene* (Constable, 1912), pp. 126–27; Havelock Ellis, *Studies in the Psychology of Sex* (2 vols.; 1911), II, part three: Sex in Relation to Society, pp. 583, 622–23; Gordon, *Woman's Body, Woman's Right*, pp. 121–22, 127–28; Hal D. Sears, *The Sex Radicals: Free Love in High Victorian America* (The Regents Press of Kansas, 1977), p. 121.

33. Prince A. Morrow, "Eugenics and Venereal Diseases," *Child Conference for Research and Welfare*, 1910, pp. 200–1.

34. Sears, *The Sex Radicals*, pp. 123–24; Eugenics Education Society, *Annual Report, 1911/12*, p. 24; *1913/14*, p. 7; Grosskurth, *Havelock Ellis*, p. 238; Morrow, "Eugenics and Venereal Diseases," p. 200; Eames, *Principles of Eugenics*, p. 7. It was no accident that women's clubs and salons frequently provided eugenic forums; Eugenics Education Society lecture topics included "Eugenics and Womanhood," "The New Woman and Race Progress," "The Biological Aspect of Women," "Women and Eugenics." See, for example, the Society's *Annual Report, 1910/11*, pp. 34–35; Reed, *From Private Vice*, pp. 54–55.

35. Michael F. Guyer, *Being Well-Born: An Introduction to Eugenics* (Bobbs-Merrill, 1916), pp. 159–60; Charles E. Rosenberg, "The Bitter Fruit: Heredity, Disease, and Social Thought in Nineteenth Century America," *Perspectives in American History*, VIII (1974), 213.

36. Guyer, *Being Well-Born*, p. 194.

37. *Ibid.*, pp. 167–68; Mary Dendy to Karl Pearson, Aug. 20, 1910, Karl Pearson Papers, Cabinet IV, D6.

38. In Great Britain, the various books that made knowledge of the heredity of disease available included J. Arthur Thomson's *Heredity*, which went through three editions before the First World War. Thomson was the Regius Professor of Natural History at the University of Aberdeen.

39. Percy Collins, "The Progress of Eugenics," *Scientific American*, 109 (Dec. 13, 1913), 459; Stoddard Goodhue, "Do You Choose Your Children?" *Cosmopolitan*, 55 (July 1913), 155; Lucy Moss to Pearson, May 26, 1909, Karl Pearson Papers, Cabinet IV, D6. No doubt typical of other subjects were the queries: If conception took place within seven days after menstruation, a time during which according to the 15th chapter of Leviticus a woman was "unclean," would the resulting offspring be unclean, too? Or, if the father or mother were inebriated at the time of conception, would the child be degenerate? When should conception best take place, and under what circumstances? Van E. Kasparek to W. F. Snow, Oct. 19, 1916, Charles B. Davenport Papers, Cold Spring Harbor Series #2, Eugenics file.

40. Charles A. L. Reed, *Marriage and Genetics: Laws of Human Breeding and Applied Eugenics* (The Galton Press, 1913), pp. 14–15.

41. *Eugenics at Work*, 1931, American Eugenics Society Papers, box 9; "Joint Session of the Eugenics Research Association and the American Eugenics Society," June 1, 1929, American Eugenics Society Papers, box 6; *Morning Leader*, March 8, 1910, copy in Eugenics Society Press Clippings Scrapbook.

42. Wiggam, *The New Decalogue of Science*, pp. 17–18.

43. Haller, *Eugenics*, p. 76.

44. Ludmerer, *Genetics and American Society*, pp. 34–35, 42–43, 82; W. Bateson, "Discussion of Galton's Paper, 1904," Francis Galton Papers, file 138/9.

45. Herbert S. Jennings, *The Biological Basis of Human Nature* (W. W. Norton, 1930), p. 203.

Chapter V:
DETERIORATION AND DEFICIENCY

1. Alfred Russel Wallace, "Human Selection," *Popular Science Monthly*, 38 (Nov. 1890), 93.

2. *Ibid*, p. 94; Henry Fairfield Osborn, "The Present Problem of Heredity," *Atlantic Monthly*, 67 (March 1891), 354.

3. Leonard Darwin declared, "The poorest classes contain a larger proportion of the naturally unfit than do the richer classes." Leonard Darwin, "Presidential Address," Eugenics Education Society, *Fourth Annual Report, 1911/12*, p. 7; Mark Haller, *Eugenics: Hereditarian Attitudes in American Thought* (Rutgers University Press, 1963), pp. 16–17; Charles E. Rosenberg, "The Bitter Fruit: Heredity, Disease, and Social Thought in Nineteenth Century America," *Perspectives in American History*, 8 (1974), 219–20; Arthur

E. Fink, *The Causes of Crime: Biological Theories in the United States, 1800–1915* (University of Pennsylvania Press, 1938), pp. 176, 223–35. The conflation of physical and moral traits led Galton to propose that "the worse the criminal the less sensitive he is to pain, the correlation between the bluntness of the moral feelings and those of the bodily sensations being very marked." Punishments, he concluded, should therefore be meted out not in so many lashes but in so many units of pain. Karl Pearson, *The Life, Letters, and Labours of Francis Galton* (3 vols. in 4; Cambridge University Press, 1914–30), II, 408.

4. Elof Axel Carlson, "R. L. Dugdale and the Jukes Family: A Historical Injustice Corrected," *BioScience*, 30 (Aug. 1980), 535–39; Charles B. Davenport, "Preface," in Arthur H. Estabrook, *The Jukes in 1915* (Carnegie Institution of Washington, 1916), p. iii.

5. W. C. D. Whetham and C. D. Whetham, *The Family and the Nation: A Study in Natural Inheritance and Social Responsibility* (Longmans, Green, 1909), pp. 87–88; Arnold White, *Efficiency and Empire*, G. R. Searle, ed. (The Harvester Press, 1973), pp. 119–20, 100; "Report of a Meeting at the London School of Economics," July 1931, pp. 11–12, C. C. Hurst Papers, entry 7955/18/189; Richard Allen Soloway, *Birth Control and the Population Question in England, 1877–1930* (University of North Carolina Press, 1982), pp. 17–29.

6. In the United States between 1890 and 1904, the population of institutionalized "feebleminded" more than doubled. Between 1874 and 1896, one of the leading British texts on heredity reported, the incidence of so-called defectives—including deaf and dumb, lunatics, epileptics, paralytics, crippled and deformed, debilitated and infirm—more than doubled, to almost 12 per 1,000 of population. The British eugenicist Robert R. Rentoul was appalled to find that "on one day in the United Kingdom there were 60,721 idiots, imbeciles, and feeble-minded, and of the number 18,900 were married or widowed. Here we have—under clerical blessing—a veritable manufactory for degenerates." J. Arthur Thomson, *Heredity* (2nd ed.; John Murray, 1912), pp. 528–29; Robert Rentoul, "Proposed Sterilization of Certain Mental Degenerates," *American Journal of Sociology*, 12 (Nov. 1906), 326; Haller, *Eugenics*, p. 129.

7. Haller, *Eugenics*, p. 144; Linda Gordon, *Woman's Body, Woman's Right: A Social History of Birth Control in America* (Grossman, 1976), p. 138.

8. William Henry Carmalt, review of *Heredity in Relation to Eugenics* in *Yale Review*, 2 (July 1913), 797; Whetham and Whetham, *The Family and the Nation*, p. 126; Michael F. Guyer, *Being Well-Born: An Introduction to Eugenics* (Bobbs-Merrill, 1916), p. ii; *The American Eugenics Society* (American Eugenics Society, 1927), pp. 15–16.

9. Kenneth Ludmerer, *Genetics and American Society* (Johns Hopkins University Press, 1972), pp. 32–33, 20, 25; Pearson, *Galton*, IIIA, 405.

10. Geoffrey Searle, *Eugenics and Politics in Britain, 1900–1914* (Noordhoff International Publishing, 1976), pp. 2, 32–33; Arnold White, *Efficiency and Empire*, pp. 102–3; Pearson *Galton*, IIIA, 251; Richard Allen Soloway, "Feminism, Fertility, and Eugenics in Victorian and Edwardian England," in Seymour Drescher et al., *Political Symbolism in Modern Europe* (Rutgers University Press, 1982), pp. 124–25.

11. Whetham and Whetham, *The Family and the Nation*, pp. 181–82, 148, 130–32; Soloway, *Birth Control and the Population Question*, pp. 20, 65–66, 120.

12. Davenport to W. P. Draper, April 10, 1929, Charles B. Davenport Papers, Draper file; G. Stanley Hall, "Eugenics, Its Ideals and What It Is Going to Do," *Religious Education*, 6 (June 1911), 156.

13. Lester F. Ward, "Eugenics, Euthenics and Eudemics," *American Journal of Sociology*, 181 (May 1913), 751; Guyer, *Being Well-Born*, p. 304

14. Whetham and Whetham, *The Family and the Nation*, p. 10; David Heron, *On the Relations of Fertility in Man to Social Status and on the Changes in This Relation That Have Taken Place During the Last Fifty Years* (London: Draper's Company Research Memoirs, #1, 1906). Soloway, *Birth Control and the Population Question*, pp. 30–31. Dean Inge found England "threatened with something much worse than a regression to healthy barbarism. Let anyone contrast the physique of a Zulu or an Anatolian Turk with that of our slum population, and we shall realize that we are breeding not vigorous barbarians but a new type of sub-men, abhorred by nature,

and ugly as no natural product is ugly." Inge, "Eugenics," in William Ralph Inge, *Outspoken Essays (Second Series)* (Longmans, Green, 1927), p. 266.

15. Sidney Webb, *The Decline in the Birth Rate* (Fabian Tract No. 131; London: The Fabian Society, 1907), pp. 16–17.

16. Pearson, *Galton*, II, 288; Searle, *Eugenics and Politics*, pp. 39–40.

17. Davenport to John J. Carty, Dec. 23, 1925, Davenport Papers, Carty file. The Whethams pointed with satisfaction to a report from the Poor Law Commission that illegitimate children born in areas with many Irish Roman Catholics were mostly backward and badly formed; illegitimates produced in Hampstead, Chelsea, and Kensington were "often the most refined, well-built, and promising." Whetham and Whetham, *The Family and the Nation*, pp. 136–37, 147.

18. Samuel J. Holmes, in *The Commonwealth*, 10 (Jan. 9, 1934), 2, clipping in Davenport Papers, Holmes file; Guyer, *Being Well-Born*, p. 297; Davenport to Steggerda, April 8, 1927, Davenport Papers, Series #2; Charles B. Davenport and Morris Steggerda, *Race Crossing in Jamaica* (Carnegie Institution of Washington, Publication No. 395, 1929), pp. 472–73, 468–69. For Davenport's prior views on black-white crosses, see Davenport to Madison Grant, July 1, 1914, Davenport Papers, Madison Grant file. Herbert S. Jennings argued that race crossing would produce harmonious new genetic combinations, and natural selection would eliminate whatever disharmonious combinations might occur in the process. William E. Castle of Harvard added that the "poor results" of interracial mating—disease, licentiousness, feeblemindedness—said more about the type of people, i.e., social outcasts, who tended to enter such unions than about any biological hazards inherent in the practice itself. When disease in the parents was absent, Castle noted, "racial crosses of Europeans with native people have been observed to produce offspring of complete vigor and fertility." Herbert S. Jennings, *The Biological Basis of Human Nature* (W. W. Norton, 1930), pp. 284–85; William E. Castle, *Genetics and Eugenics* (2nd ed.; Harvard University Press, 1920), pp. 265–66. See also Edwin Grant Conklin, *The Direction of Human Evolution* (Scribner's, 1922), pp. 50–52, and Jacques

Loeb, "Science and Race," *The Crisis*, 9 (1914), 92–93.

19. Ludmerer, *Genetics and American Society*, pp. 25–26; John Higham, *Strangers in the Land: Patterns of American Nativism* (2nd ed.; Atheneum, 1963), pp. 143–44, 271–72. In *The Great Gatsby*, Tom Buchanan expounded on a proxy for what was no doubt Madison Grant's book: "The idea is if we don't look out the white race will be—will be utterly submerged. It's all scientific stuff; it's been proved. . . . It's up to us, who are the dominant race, to watch out or these other races will have control of things." F. Scott Fitzgerald, *The Great Gatsby* (Scribner's, 1925, 1953), p. 13.

20. Pearson to Maria Sharpe, Nov. 4, 1887; Pearson to G. M. Morant, Dec. 27, 1924, Karl Pearson Papers, Cabinet II, D3; Cabinet VI, D8 (biometric series); Pearson, *Galton*, IIIA, 369.

21. Searle, *Eugenics and Politics in Britain*, pp. 41, 44.

22. The Whethams allowed that, though classical training was inadequate to the modern technical age, men educated scientifically were often ill equipped for important posts because "their heredity and training leave them unfit to deal with men, especially with workmen, foreigners and natives. Moreover," they added, "from the employer's point of view they often lack the guarantees of character and the intuitive sense of masterfulness that are the usual concomitants of the man of good family." Elsewhere, the Whethams declared that "scholarships have their dangers when used to raise those who win them too suddenly and completely out of their natural class," and they held that "the entrance into the upper ranks of our people of increasing numbers of persons, of various nationalities as well as our own, whose newly acquired wealth is not associated with definite territorial or local associations, has doubtless affected, and to some extent demoralized, the habits of those sections of the community who have had the misfortune to come in contact with them." Whetham and Whetham, *The Family and the Nation*, pp. 188–89, 194–96, 186.

23. Henry Fairfield Osborn, "Address of Welcome," in *Eugenics, Genetics, and the Family* (Scientific Papers of the Second International Congress of Eugenics; 2 vols.;

Williams & Wilkins, 1923), I, 2. Osborn's friend Lothrop Stoddard, the author of the pro-eugenic and widely read *Revolt Against Civilization*, classified Bolshevism, the revolt of the underman, as a denial of the natural hereditary order in man, declaring, "Against this formidable adversary stands biology, the champion of the new." Lothrop Stoddard, *The Revolt Against Civilization: The Menace of the Underman* (Scribner's, 1922), p. 238. Long after Galton's death, in the nineteen-twenties, Pearson wondered in print whether Galton had fully appreciated what follows "when, as is the usual case, a democracy *starts* with a majority of incapable citizens? . . . The incapables care nothing for the future of the race or nation." Pearson, *Galton*, IIIA, 349.

24. Thomas Pogue Weinland, *A History of the I.Q. in America, 1890–1941* (University Microfilms, 1973), pp. 66–68. The young psychologist James McKeen Cattell wrote to Galton in 1889, "We are following in America your advice and example." James McKeen Cattell to Galton, Oct. 22, 1889, Francis Galton Papers, file 220.

25. Jan Sebastik, "Alfred Binet," *Dictionary of Scientific Biography*, II, 131–32.

26. Haller, *Eugenics*, pp. 96–97; H. H. Goddard, "Four Hundred Feeble-Minded Children Classified by the Binet Method," *Journal of Psycho-Aesthenics*, 15 (Sept. and Dec. 1910), 18–19, 23–24, 26–29; Henry H. Goddard, "Two Thousand Children Tested by the Binet Measuring Test for Intelligence," *National Education Association*, 1911, pp. 870–78; Goddard to Davenport, Oct. 10, 1910, Davenport Papers, Goddard file; Stephen Jay Gould, *The Mismeasure of Man* (W. W. Norton, 1981), pp. 164–67.

27. Henry H. Goddard, *Feeble-mindedness: Its Causes and Consequences* (Macmillan, 1914), p. 4; Peter Tyor, *Segregation or Surgery: The Mentally Retarded 'in America, 1850–1920* (University Microfilms, 1972), p. 193.

28. "Minutes of the Meeting of Field Officers on Human Heredity . . . Oct. 14, 1910"; Davenport to Starr J. Murphy, Dec. 19, 1912, Davenport Papers, American Breeders Association, Eugenic Section file; Peter L. Tyor, "Henry H. Goddard: Morons, Mental Defect, and the Origins of Mental Testing" (unpublished manuscript), p. 15; Goddard to Davenport, July 25, 1912, Da-

venport Papers, Cold Spring Harbor Series #2.

29. John McPhee, *The Pine Barrens* (Ballantine Books, 1968), pp. 53–56; Tyor, *Segregation or Surgery*, pp. 206–7; Goddard remarked, "A dentist assures me that the finest set of teeth he has ever seen is in the mouth of one of our morons." Goddard, *Feeble-mindedness*, pp. 504, 508–9.

30. To Goddard, by revealing the existence of morons, the Binet-Simon tests pointed to a hitherto unrecognized group who fitted the definition for "feeblemindedness" advanced by the Royal College of Physicians in England: "One who is capable of earning his living under favorable circumstances, but is incapable from mental defect existing from birth or from an early age (a) of competing on equal terms with his normal fellows or (b) of managing himself and his affairs with ordinary prudence." Goddard, *Feeble-mindedness*, pp. 4, 7–9, 14, 17–19, 504, 514.

31. *Ibid.*, pp. 437, 547.

32. Ellis, *The Task of Social Hygiene* (Constable, 1912), pp. 32–33; Guyer, *Being Well-Born*, pp. 264–65; Fink, *The Causes of Crime*, pp. 236–37; Tyor, *Segregation or Surgery*, pp. 199–201; Haller, *Eugenics*, pp. 109–10.

33. Raymond E. Fancher, *Pioneers of Psychology* (W. W. Norton, 1979), p. 346; Terman to Robert M. Yerkes, Sept. 18, 1916, Robert M. Yerkes Papers, Yale University, box 13, folder 447; Russell Marks, *Testers, Trackers, and Trustees: The Ideology of the Intelligence Testing Movement in America, 1900–1954* (University Microfilms, 1972), pp. 23–24.

34. Weinland, *A History of the I.Q. in America*, pp. 124–25; Daniel J. Kevles, "Testing the Army's Intelligence," *Journal of American History*, 55 (Dec. 1968), 566.

35. Weinland, *A History of the I.Q. in America*, p. 107; *The New York Times*, July 19, 1916, p. 7; July 20, 1916, p. 10.

36. Yerkes to Davenport, May 27, 1941; Yerkes to Amram Scheinfeld, Dec. 19, 1950, Yerkes Papers, box 4, folder 115; box 12, file 399; Yerkes remarks at "Second Field Workers Conference," June 20, 21, 1913, Charles B. Davenport Papers, Cold Spring Harbor Series #2, Eugenics Record Office file.

37. Kevles, "Testing the Army's Intelligence," pp. 565–66; Yerkes to A. Lawrence

Lowell, Feb. 17, 1914, Yerkes Papers, box 9, folder 288; Yerkes, "Testing the Human Mind," *Atlantic Monthly*, 131 (March 1923), 367.

38. Yerkes to the General Education Board, Jan. 23, 1919; Yerkes to C. E. Seashore, Feb. 18, 1916; Yerkes to Terman, March 21, 1917; Yerkes to Raymond Dodge, Aug 4, 1921, Yerkes Papers, box 6, folder 166; folder 408; box 13, folder 447; box 4, folder 126; Kevles, "Testing the Army's Intelligence."

39. Kevles, "Testing the Army's Intelligence," pp. 571–75.

40. *Ibid.*, pp. 575–76.

41. *Ibid.*, pp. 577–78. Florence L. Goodenough, *Mental Testing: Its History, Principles, and Applications* (Staples Press, 1949), p. 67.

42. Yerkes to Abraham Flexner, Jan. 17, 1919, Yerkes Papers, box 6, folder 166; Weinland, *A History of the I.Q. in America*, pp. 248–49, 140–43; Kevles, "Testing the Army's Intelligence," pp. 578–81; Frank S. Freeman, *Mental Tests: Their History, Principles, and Applications* (Houghton Mifflin, 1939), p. 3; Yerkes, "Practical Mental Measurement: Intelligence Tests for Elementary Schools," *Scientific American Monthly*, 1 (March 1920), 270–73. According to various estimates, intelligence tests were administered annually to a few million primary and secondary students. Mental ages and I.Q.s, an observer later said, "were joyfully entered on children's permanent record cards by teachers and school principals with as much assurance as their grandfathers had placed in the skull maps drawn up by their favorite phrenologist." Goodenough, *Mental Testing*, p. 68.

43. Robert M. Yerkes, ed., *Psychological Examining in the United States Army* (National Academy of Sciences–National Research Council, 1921).

44. Haller, *Eugenics*, pp. 109–10, 123; U.S. War Department, *Annual Reports, 1919* (Government Printing Office, 1920), I, 2791; *The American Eugenics Society*, 1927, p. 16; Weinland, *A History of the I.Q. in America*, pp. 159–61.

45. Henry H. Goddard, "Mental Tests and the Immigrant," *The Journal of Delinquency*, 2 (Sept. 1917), 244, 249, 268; "Two Immigrants Out of Five Feeble-minded," *Survey*, 38 (Sept. 15, 1917), 528–29; Carl Campbell Brigham, *A Study of American Intelligence* (Princeton University Press, 1923), pp. xx, 197; Brigham to Davenport, Dec. 21, 1922, Davenport Papers, Carl Campbell Brigham file; Gould, *The Mismeasure of Man*, pp. 164–68.

46. U.S. War Department, *Annual Reports, 1919*, I, 2791; Weinland, *A History of the I.Q. in America*, pp. 179–80; Marks, *Testers, Trackers, and Trustees*, pp. 38–39; "Grouping Pupils by Intelligence," *School Review*, 28 (April 1920), 249–52.

47. John C. Almack, James Almack, "Gifted Pupils in the High School," *School and Society*, 14 (Sept. 24, 1921), 227–28; Weinland, *A History of the I.Q. in America*, pp. 172–77.

48. Leslie S. Hearnshaw, *Cyril Burt, Psychologist* (Cornell University Press, 1979), pp. 7–8, 23, 21, 30; Edgar Schuster, *Eugenics* (Clear-Type Press, 1912), pp. 153–56.

49. Cyril Burt to Yerkes, Jan. 23, 1923, Yerkes Papers, box 3, folder B; Hearnshaw, *Burt*, pp. 30, 39, 77–78, 122.

50. Henry H. Goddard, *Human Efficiency and Levels of Intelligence* (Princeton University Press, 1920), p. 1.

Chapter VI:
MEASURES OF REGENERATION

1. Victoria C. Woodhull Martin, *The Rapid Multiplication of the Unfit* (pamphlet; London, 1891), p. 38. According to C. P. Blacker, later head of the Eugenics Society in Britain, the terms "positive" and "negative" eugenics were coined by C. W. Saleeby with the approval of Francis Galton. C. P. Blacker, *Eugenics in Prospect and Retrospect* (Hamish Hamilton, 1945), p. 17.

2. *Nature*, 84 (Oct. 1910), 431; Donald K. Pickens, *Eugenics and the Progressives* (Vanderbilt University Press, 1968), p. 121. Bell thought, "In the case of men and women who are thoroughbred . . . it is obvious that

their descendants, spreading out among the population and marrying into average or inferior families, would prove prepotent over their partners in marriage in affecting the offspring, thus leading to an increase in the proportion of superior offspring produced from the average or inferior with whom they have mated." Like many eugenicists, Bell mistakenly assumed that "thoroughbred" qualities in human beings were hereditarily determined by unit characters and that these characters were dominant, to use the Mendelian equivalent of prepotent. Alexander Graham Bell, "A Few Thoughts Concerning Eugenics," *National Geographic*, 11 (Feb. 1908), 122.

3. Various press clippings, March 4, 5, and 6, 1909, Eugenics Society Press Clippings Scrapbook; Shaw to Pearson, Oct. 22, 1901, Karl Pearson Papers, Shaw file; Havelock Ellis, *The Problem of Race Regeneration* (New Tracts for the Times; Cassell, 1911), pp. 27, 34–35, 49–51, 65, 67, 70.

4. Shaw, "Preface to *Man and Superman*," in *The Complete Prefaces of Bernard Shaw* (Paul Hamlyn, 1965), p. 159; Shaw, "Discussion of Galton's Paper, 1904," Francis Galton Papers, file 138/9.

5. Kingsley Martin, *Harold Laski, 1893–1950: A Biographical Portrait* (Viking Press, 1953), pp. 8–9; Karl Pearson to Mrs. Weldon, April 10, 1913; Galton to Pearson, July 11, 1910; Aug. 4, 1910, Pearson Papers, Cabinet I, D7/1; Cabinet VI, D6; Harold Laski, "The Scope of Eugenics," *Westminster Review*, 174 (July 1910), 25–26, 31, 34. Laski ably attacked the biologist Redcliffe Salaman's attempt to demonstrate the Mendelian inheritance of a Jewish physical type, by arguing both that there was nothing physical that could be universally classified as "Jewishness" and that, in any case, any such demonstration would have to be biometric. H. J. Laski, "A Mendelian View of Racial Heredity," *Biometrika*, 8 (1911–12), 424–30.

6. Havelock Ellis, *The Task of Social Hygiene* (Constable, 1912), pp. 19–20. Shaw also noted, "Equality is essential to good breeding; and equality, as all economists know, is incompatible with property." Shaw, "Preface to *Man and Superman*," pp. 169–70. An American eugenicist remarked that barriers of class and wealth confined matrimonial choice within arbitrary hereditary limits and

kept "possible mates in widely distant spheres," adding, "Such luxuries as the parlor car, the country estate and the many-barriered ocean steamer have fixed a gulf between the millionaire and the lower middle class that is seldom traversed matrimonially except through the medium of the stage." Roswell H. Johnson, "The Evolution of Man and Its Control," *Popular Science Monthly*, 76 (Jan. 1910), 62–63.

7. *Sporting Set*, March 9 [1910]; *The Globe*, March 4, 1910, Eugenics Society Press Clippings Scrapbook. In a comment on Galton's 1904 address, Shaw called for "freedom to breed the race without being hampered by the mass of irrelevant conditions implied in the institution of marriage . . . ," adding, "What we need is freedom for people who have never seen each other before and never intend to see one another again to produce children under certain definite public conditions, without loss of honor." Shaw, "Discussion of Galton's Paper, 1904," Galton Papers, file 138/9.

8. Ellis, *The Task of Social Hygiene*, pp. 46–47. The American radical and eugenic sympathizer Scott Nearing declared that "only the woman who is a human being, with power and freedom to choose, may teach the son of a free man." Scott Nearing, *The Super Race: An American Problem* (B. W. Huebsch, 1912), p. 82.

9. Havelock Ellis, "Birth Control and Eugenics," *Eugenics Review*, 9 (April 1917), 35; Ellis, *The Task of Social Hygiene*, pp. 16–17.

10. Beatrice Webb said that she felt grateful to Shaw for *Man and Superman* because she shied as a matter of delicacy from publicly discussing the subject of human breeding. Angus McLaren, *Birth Control in Nineteenth-Century England* (Croom Helm, 1978), p. 190; Linda Gordon, *Woman's Body, Woman's Right: A Social History of Birth Control in America* (Grossman, 1976), pp. 88–89, 108–9.

11. Henry Fairfield Osborn, "Birth Selection vs. Birth Control," *Forum*, 88 (Aug. 1932), 79; Gordon, *Woman's Body, Woman's Right*, p. 306; Richard Allen Soloway, "Neo-Malthusians, Eugenists, and the Declining Birth-Rate in England, 1900–1918," *Albion*, 10 (Fall 1978), 275–76; Richard Allen Soloway, *Birth Control and the Population Question in England, 1877–1930* (University of North Car-

olina Press, 1982), pp. 145, 153; Leonard Darwin to Mrs. Hodson, June 24, 1927, Eugenics Society Records, Wellcome Institute for the History of Medicine, London, file K. 1. J. B. S. Haldane, noting that upper-class women probably suffered unduly in childbirth while lower-class ones knew little about contraception, quipped: "Rich women need more exercise, and poor women more education." J. B. S. Haldane, "Eugenics and Social Reforms," *The Nation and the Athenaeum*, 35 (May 21, 1924), 292.

12. The British physician and eugenicist Caleb W. Saleeby, sure that women's rights should be subordinated to the obligations of parenthood, inveighed against the "incomplete and aberrant women" who were "ceasing to be mammals." McLaren, *Birth Control in Nineteenth-Century England*, p. 147; *Morning Post*, March 8 [1910], Eugenics Society Press Clippings Scrapbook; W. C. D. Whetham and C. D. Whetham, *The Family and the Nation: A Study in Natural Inheritance and Social Responsibility* (Longmans, Green, 1909), pp. 198–99; Michael F. Guyer, *Being Well-Born: An Introduction to Eugenics* (Bobbs-Merrill, 1916), pp. 206–7; Edwin Grant Conklin, *Heredity and Environment in the Development of Men* (Princeton University Press, 1915), pp. 484–85; Richard Allen Soloway, "Feminism, Fertility, and Eugenics in Victorian and Edwardian England," in Seymour Drescher et al., *Political Symbolism in Modern Europe* (Rutgers University Press, 1982), p. 131. Interestingly enough, the average number of progeny among the officers of the Eugenics Education Society was 2.3, and a quarter of the group had no children at all. Soloway, *Birth Control and the Population Question in England*, p. 186. See also *ibid.*, pp. 138–39, 142.

13. Soloway, "Feminism, Fertility, and Eugenics in Victorian and Edwardian England," pp. 125, 131; Whetham and Whetham, *The Family and the Nation*, pp. 143–45; Edward L. Thorndike to Galton, Dec. 4, 1901, Galton Papers, file 327/1; A. E. Wiggam to Davenport, July 28, 1921, Charles B. Davenport Papers, Cold Spring Harbor Series #2, Eugenics: Wiggam et al. file; W. C. D. and C. D. Whetham, "Decadence and Civilisation," *The Hibbert Journal*, 10 (Oct. 1911), 193–94; Roswell H. Johnson, "The Evolution of Man and Its Control," *Popular Science*

Monthly, 76 (Jan. 1910), 68; Osborn to Carrie Chapman Catt, Dec. 28, 1917, Henry Fairfield Osborn Papers, American Museum of Natural History.

14. Hamilton Cravens, *The Triumph of Evolution: American Scientists and the Heredity-Environment Controversy, 1900–1941* (University of Pennsylvania Press, 1978), p. 53; Karl Pearson, "Eugenics, A Lecture," Dec. 7, 1923, p. 32, Pearson Papers, list f/9; Ellis, *The Task of Social Hygiene*, p. 205; Guyer, *Being Well-Born*, pp. 190–91; Charles H. Robinson, ed., *The Science of Eugenics and Sex Life, from the Notes of Walter J. Hadden* (Vansant, 1914), editor's note; Eugenics Education Society, *Annual Report, 1917/18*, p. 14. The Whethams looked forward to the day when "a comparison of scientific pedigrees" would figure significantly in preliminary marital discussions; "when birth and good breeding (in its wide sense), character and ability will be the qualities most prized in the choice of mates." Whetham and Whetham, *The Family and the Nation*, p. 223.

15. Havelock Ellis supposed that the time would come when, should an epileptic conceal his or her condition prior to matrimony, "it would generally be felt that an offence had been committed serious enough to invalidate the marriage." Ellis, *The Task of Social Hygiene*, p. 211; J. Arthur Thomson, *Heredity* (2nd ed.; John Murray, 1912), pp. 305–6.

16. *Mother Earth*, 11 (1916–17) (Greenwood, 1968), 459. Many of the women in the leadership of the Eugenics Society were suffragists and single, and no doubt constituted something of a lobby for feminist positions. Soloway, "Feminism, Fertility, and Eugenics in Victorian and Edwardian England," pp. 133–36, 126.

17. Gordon, *Woman's Body, Woman's Right*, pp. 281–82, 284; James Reed, *From Private Vice to Public Virtue* (Basic Books, 1978), p. 136; Ruth Hall, *Passionate Crusader: The Life of Marie Stopes* (Harcourt Brace Jovanovich, 1977), pp. 112–13, 175–76, 180–82, 326.

18. In the nineteen-twenties, the Council of the American Birth Control League came to include such prominent eugenicists as Lothrop Stoddard, while birth-control clinics distributed eugenically oriented advice. Similar linkages were established in England, where the Eugenics Education Soci-

ety cooperated with Marie Stopes's clinics, hosted sessions on birth control, and fostered research on contraception. Gordon, *Woman's Body, Woman's Right*, pp. 283, 286–87; Thomas Pogue Weinland, *A History of the I.Q. in America, 1890–1941* (University Microfilms, 1973), p. 232; Soloway, "Neo-Malthusians, Eugenists, and the Declining Birth-Rate in England, 1900–1918," pp. 276–77, 284–86; Soloway, *Birth Control and the Population Question in England*, pp. 201–2, 235–36; Leonard Darwin, "Presidential Address," Eugenics Education Society, *Third Annual Report, 1910/11*, p. 17; *Annual Report, 1927/28*, p. 5; Ellis, *The Task of Social Hygiene*, p. 401.

19. Ellis, *The Task of Social Hygiene*, pp. 45–46. Galton had cautioned that eugenic advocates "should move discreetly and claim no more efficacy on its behalf than the future will confirm." Otherwise, he said, "a reaction will be invited. A great deal of investigation is still needed to shew the limit of practical Eugenics." Galton, "Preface," *Essays in Eugenics* (London: Eugenics Education Society, 1909), p. ii.

20. Benjamin Kidd, "Discussion of Galton's Paper," *American Journal of Sociology*, 10 (July 1904), 13; Alfred Russel Wallace, "Human Selection," *Popular Science Monthly*, 38 (Nov. 1890), 106; "Prospectus for the Eugenics Education Society," Pearson Papers, Cabinet IV, D6; Francis Galton, *Memories of My Life* (Methuen, 1908), pp. 310–11.

21. Edgar Schuster, *Eugenics* (Clear-Type Press, 1912), pp. 77–78.

22. Perhaps, Shaw added, "we must establish a State Department of Evolution, with a seat in the Cabinet for its chief, and a revenue to defray the cost of direct State experiments, and provide inducements to private persons to achieve successful results. It may mean a private society or a chartered company for the improvement of human live stock." Shaw, "Preface to *Man and Superman*," pp. 185–86; Julian Huxley, "The Vital Importance of Eugenics," *Harper's*, 163 (Aug. 1931), 329; Donald MacKenzie, "The Development of Statistical Theory in Britain, 1865–1925: A Historical and Sociological Perspective" (doctoral thesis, University of Edinburgh, 1977), p. 101; Schuster, *Eugenics*, pp. 250–51; Leonard Darwin, "Presidential Address," Eugenics Education Society,

Third Annual Report, 1910/11, pp. 13–17; W. C. D. Whetham, *Eugenics and Unemployment* (Bowes and Bowes, 1910); G. Stanley Hall, "Eugenics: Its Ideals and What It Is Going to Do," *Religious Education*, 6 (June 1911), 156; Whetham and Whetham, *The Family and the Nation*, pp. 201–2; John F. Bobbitt, "Practical Eugenics," *Pedagogical Seminary*, 16 (Sept. 1909), 393.

23. J. Arthur Thomson, "Eugenics and War," *Popular Science Monthly*, 86 (May 1915), 424; Soloway, "Neo-Malthusians, Eugenists, and the Declining Birth-Rate in England, 1900–1918," pp. 280–82; Eugenics Education Society, *Annual Report, 1914/15*, pp. 4–5.

24. Guyer, *Being Well-Born*, p. 338. The Whethams were notably bothered by Galton's finding that heiresses tended to come from families of low fecundity, especially how "precarious must be a line of descent through heiresses," since one out of five in Galton's sample produced no male offspring. The temptation of accomplished men to marry heiresses, the Whethams insisted, had to be offset, likely by combining a hereditary title with a grant of heritable lands or a substantial sum of money. Whetham and Whetham, *The Family and the Nation*, pp. 118, 200–1.

25. Roswell H. Johnson, "The Evolution of Man and Its Control," p. 57; McLaren, *Birth Control in Nineteenth-Century England*, pp. 191–92; Ellis, *The Task of Social Hygiene*, pp. 402–3.

26. The Yale psychologist Arnold L. Gesell reckoned that society "need not wait for the perfection of the infant science of eugenics before proceeding upon a course . . . which will prevent the horrible renewal of this defective protoplasm that is contaminating the stream of village life." Arnold Gesell, "The Village of a Thousand Souls," *The American Magazine*, 76 (Oct. 1913), 15. The Whethams said bluntly that in dealing with people of "incurably vicious tendencies, we must, as with the feebleminded, organize extinction of the tribe." Whetham and Whetham, *The Family and the Nation*, p. 214.

27. Alexander Johnson, "Race Improvement by Control of Defectives," *Annals of the American Academy*, 34 (July 1909), 24–25; Henry Smith, *A Plea for the Unborn* (Watts, 1897), pp. 29–30; Simeon Strunsky, "Race-

Culture," *Atlantic Monthly*, 110 (Dec. 1912), 851; Robert C. Bannister, *Social Darwinism* (Temple University Press), p. 175; *A Eugenics Catechism* (American Eugenics Society, 1926), p. 10.

28. G. Archdall Reid, "Discussion of Galton's Paper," 1904, Galton Papers, file 138/9; William J. Hyde, "The Socialism of H. G. Wells," *Journal of the History of Ideas*, 17 (April 1956), 220–21; Johnson, "The Evolution of Man and Its Control," pp. 66–67; William J. Robinson, *Eugenics, Marriage, and Birth Control (A Practical Guide)* (The Critic and Guide, 1917), pp. 55–60. Robinson, a pro-eugenic physician and birth-control advocate, said that no test for venereal disease would be necessary for women until they became as sexually emancipated as men. At the time, the proportion of women to men afflicted with such disease, at least among the respectable classes, was one in one hundred, Robinson said, so it made no sense to him "to subject a thousand women to vaginal examinations in order to perhaps find one [ten?] who is infected." Robinson also noted that men might find ways to avoid eugenic marriage restrictions, but he thought that public opinion would demand the passage of appropriate enforcement laws in all states. After all, the better families already expected prospective bridegrooms to take out a life insurance policy—not only to protect the bride's financial future, Robinson remarked, but also so that the bride's family could ascertain the health of their daughter's betrothed. Robinson, *Eugenics, Marriage, and Birth Control*, pp. 55–60.

29. Henry H. Goddard, *Feeble-mindedness: Its Causes and Consequences* (Macmillan, 1914), pp. 565–66; Karl Pearson, *The Life, Letters, and Labours of Francis Galton* (3 vols. in 4; Cambridge University Press, 1914–30), IIIA, 373–74; Whetham and Whetham, *The Family and the Nation*, p. 212. It should be pointed out that, in the context of his day, Goddard was something of an advocate of reforming the treatment of the mentally retarded. Although morons lacked normal competitive abilities and prudence, he held, they were not hopeless but educable. With proper training, they could lead useful, productive lives. But without training that took

into account the limits of their capabilities, they could easily turn into difficult social problems. The bulk of social effort, he argued, had aimed to force delinquents, paupers, criminals, and prostitutes to behave as people believed they must be morally capable of behaving. Since by reason of their mental deficiency they were without such capacity, they ought to be dealt with not by penal coercion but with understanding care. Goddard, *Feeble-Mindedness*, pp. 4–5, 569–70, 582–83; "Report of Committee H of the Institute of Criminal Law and Criminology: Sterilization of Criminals," *Journal of Criminal Law*, 5 (Nov. 1914), 519; Henry H. Goddard, "Causes of Backwardness and Mental Deficiency in Children," *Journal of Education*, 74 (July 27, 1911), 125.

30. *A Eugenics Catechism*, pp. 8–9.

31. Arthur E. Fink, *The Causes of Crime: Biological Theories in the United States, 1800–1915* (University of Pennsylvania Press, 1938), pp. 188–89, 196; "Is Vasectomy Cruel and Unusual Punishment?" *Journal of Criminal Law, Criminology, and Police Science*, 3 (Jan. 1913), 784–85; Julius Paul, "State Eugenic Sterilization History: A Brief Overview," in Jonas Robitscher, ed., *Eugenic Sterilization* (Charles C Thomas, 1973), pp. 28–29.

32. Robinson, *Eugenics, Marriage, and Birth Control*, pp. 74–76; Pickens, *Eugenics and the Progressives*, pp. 93–94, 125; Robert R. Rentoul, "Proposed Sterilization of Certain Mental Degenerates," *American Journal of Sociology*, 12 (Nov. 1906), 327; Galton to Pearson, Jan. 6, 1907, Pearson Papers, Cabinet VI, D6; H. G. Wells, "Discussion of Galton's Paper, 1904," Galton Papers, file 138/9.

33. Guyer, *Being Well-Born*, p. 282; Yerkes, typescript introduction to Brigham's *Study of American Intelligence*, Robert M. Yerkes Papers, box 3, folder 62; Kenneth C. MacArthur, "Eugenics Sermon," June 20, 1926, American Eugenics Society Papers, box 14; Davenport to William E. Davenport, Feb. 11, 1924, Davenport Papers, W. E. Davenport file.

34. *A Eugenics Catechism*, p. 9; "Fourth Report of the Committee on Selective Immigration of the American Eugenics Society," June 30, 1928, pp. 15–16, American Eugenics Society Papers, box 9.

Chapter VII:
EUGENIC ENACTMENTS

1. Lyndsay Farrall, *The Origins and Growth of the English Eugenics Movement* (University Microfilms, 1970), pp. 242–43.

2. Kenneth Ludmerer, *Genetics and American Society* (Johns Hopkins University Press, 1972), pp. 95–96.

3. *Ibid.*, pp. 105–7; Yerkes to Paul G. Tomlinson, Oct. 31, 1922, Robert M. Yerkes Papers, box 3, folder 62; Frances J. Hassencahl, *Harry H. Laughlin, 'Expert Eugenics Agent' for the House Committee on Immigration and Naturalization, 1921 to 1931* (University Microfilms, 1971), p. 224; Franz Samelson, "World War I Intelligence Testing and the Development of Psychology," *Journal of the History of the Behavioral Sciences*, 13 (1977), 278.

4. Calvin Coolidge, "Whose Country Is This?" *Good Housekeeping*, 72 (Feb. 1921), 14; Clarence G. Campbell, president of the Eugenics Research Association, exulted: "It is a matter for congratulation that the United States now bases its immigration policy upon the eugenic quality of the racial stock." Ludmerer, *Genetics and American Society*, p. 107.

5. Sir Frank Newsam, *The Home Office* (The New Whitehall Series; George Allen and Unwin, 1954), pp. 95–97; Sir William R. Anson, *The Law and Custom of the Constitution* (4th ed., vol. II, *The Crown*, part I; by A. Berriedale Keith; Clarendon Press, 1935), pp. 322–23.

6. Pearson to Maria Sharpe, Jan. 5, 1888, Karl Pearson Papers, Cabinet II, D3; Kathleen Jones, *A History of the Mental Health Services* (Routledge and Kegan Paul, 1972), p. 187; *Nature*, 84 (Oct. 6, 1910), 431; "Dame Ellen Frances Pinsent," *Dictionary of National Biography*, volume for 1941–50, p. 673. Pinsent approached the problem of mental deficiency somewhat in the vein of Goddard. She told a church congress in 1910, "We talk of 'equality of opportunity' and forget the radical inequalities which are born with us." Mrs. Pinsent, "Social Responsibility and Heredity," Sept. 1910, copy in Pearson Papers, list 628.

7. Jones, *History of the Mental Health Services*, p. 197; Harvey G. Simmons, "Explaining Social Policy: The English Mental Deficiency Act of 1913," *Journal of Social His-*

tory, 11 (1978), 396–97; Michael Freeden, *The New Liberalism: An Ideology of Social Reform* (Clarendon Press, 1978), pp. 188–89.

8. Jones, *History of the Mental Health Services*, pp. 199–203, 208–9; Freeden, *The New Liberalism*, pp. 190–92.

9. Farrall, *English Eugenics Movement*, p. 244; Geoffrey Searle, *Eugenics and Politics in Britain, 1900–1914* (Noordhoff International Publishing, 1976), p. 111.

10. Jesse Spaulding Smith, "Marriage, Sterilization and Commitment Laws Aimed at Decreasing Mental Deficiency," *Journal of Criminal Law*, 5 (Sept. 1914), 364–66; Mark Haller, *Eugenics: Hereditarian Attitudes in American Thought* (Rutgers University Press, 1963), p. 47.

11. Smith, "Marriage, Sterilization and Commitment Laws," p. 366; Editorial, *The Nation*, 97 (Aug. 1913), 137; "Restrictions on Marriage," 192?, American Eugenics Society Papers, box 16.

12. Haller, *Eugenics*, p. 50; Adolph Meyer to Davenport, Jan. 2, 1910, Charles B. Davenport Papers, Meyer file; "Report of Committee H of the Institute of Criminal Law and Criminology: Sterilization of Criminals," *Journal of Criminal Law*, 5 (Nov. 1914), 515; Harry H. Laughlin, *The Legal Status of Eugenical Sterilization* (Municipal Court of Chicago, 1930), p. 57.

13. Davenport to Mrs. Harriman, April 28, 1911; Jan. 5, 1911, Davenport Papers, Mrs. E. H. Harriman file; L. F. Whitney to Sidney Cazort, May 19, 1927, American Eugenics Society Papers, box 12. In 1916, Davenport helped inaugurate the *Eugenical News*, a clearinghouse publication for the activities of the Eugenics Record Office and its associates.

14. Peter Tyor, *Segregation or Surgery: The Mentally Retarded in America, 1850–1920* (University Microfilms, 1972), pp. 222–23; Thomas Pogue Weinland, *A History of the I.Q. in America, 1890–1941* (University Microfilms, 1973), p. 113.

15. Rudolph J. Vecoli, "Sterilization: A Progressive Measure?" *Wisconsin Magazine of History*, 43 (Spring 1960), 193–97, 199–200.

16. Arthur Estabrook had drawn upon the data for his follow-up study of the Jukes

to propose to the district attorney of New Haven County, Connecticut, that a man convicted of murder there was mentally deficient by reason of heredity, was thus not responsible for his crime, and should not be sent to the gallows. Davenport rebuked Estabrook for submitting the material on his own, telling him that if others did the same, they might bring Cold Spring Harbor into serious trouble. Davenport added, "A man who kills is just as dangerous whether he is feebleminded or not . . . and society has a perfect right to kill a person who is thus dangerous to society." Estabrook to Harry Laughlin, Nov. 22, 1913; Davenport to Estabrook, Dec. 29, 1913, Davenport Papers, Cold Spring Harbor Series #2.

17. Hassencahl, *Laughlin*, pp. 42–44, 48–50; Laughlin to Davenport, Feb. 25, 1907, Davenport Papers, Cold Spring Harbor Series #2.

18. Laughlin to Davenport, March 30, 1908; Feb. 8, 1916, Davenport Papers, Cold Spring Harbor Series #2; Hassencahl, *Laughlin*, pp. 51–52, 61–62.

19. Hassencahl, *Laughlin*, pp. 45–46, 50–51.

20. Harry H. Laughlin, "Eugenics," *Nature Study Review*, March 1912, pp. 110–11.

21. Before the war, the Eugenics Record Office had established links with the Immigration Restriction League via Prescott Hall, a League officer and a contemporary of Davenport's when he was at Harvard. Laughlin to Davenport, April 14, 1921, Davenport Papers, Cold Spring Harbor Series #2; Haller, *Eugenics*, p. 155; Hassencahl, *Laughlin*, pp. 165–66, 203; John Higham, *Strangers in the Land: Patterns of American Nativism* (2nd ed.; Atheneum, 1963), p. 307.

22. Hassencahl, *Laughlin*, pp. 228–29, 296–97, 250–52, 2, 246–48.

23. When in 1923 Laughlin was to sail for Europe to conduct further studies for the House Committee, Secretary of Labor James J. Davis, as pro-Nordic as they came in the Administration, proposed that the two take the same ship so that they could confer during the voyage. John B. Merriam to Davenport, June 20, 1923; Davenport to Merriam, Aug. 26, 1929, Davenport Papers, Cold Spring Harbor Series #1; "Eugenics Committee of the United States of America," 1924, American Eugenics Society Papers, box 16;

John C. Box to John C. Merriam, Sept. 28, 1929, Records of the Carnegie Institution of Washington, Genetics Department, Immigration Situation file.

24. Shaw to Pearson, June 20, 1893, list 627; Pearson to Ethel Elderton, Aug. 24, 1909, Pearson Papers, list 627; Cabinet VI, D7 (biometric series).

25. Pearson to Galton, Feb. 29, 1908, Pearson Papers, Cabinet VI, D6; Karl Pearson, *The Life, Letters, and Labours of Francis Galton* (3 vols. in 4; Cambridge University Press, 1914–30), IIIA, 362–63, 371–72.

26. Pearson to Major Greenwood, Aug. 2, 1912, Pearson Papers, Cabinet VI, D7; Pearson, *Galton*, IIIA, 312; Pearson, "Eugenics, A Lecture," Dec. 7, 1923, Pearson Papers, list F/9.

27. Pearson to Galton, Oct. 31, 1909; Nov. 8, 1909, Pearson Papers, Cabinet VI, D6.

28. Ludmerer, *Genetics and American Society*, pp. 61–62; Searle, *Eugenics and Politics in Britain*, pp. 16–17; excerpts from Heron's critique, *New York Times*, Nov. 9, 1913, Sec. V, pp. 2–3. See also David Heron, *Mendelism and the Problem of Mental Defect* (Questions of the Day and the Fray; vol. II; Dulau, 1913), especially pp. 2–4, 12, 62. Heron's attack against Davenport's shoddy Mendelism should not be taken to mean that he dissented from Pearson's strongly hereditarian view of the world. He had earlier concluded, for example, that neither nutrition nor any other environmental condition had anything to do with the development of mental ability. Like Pearson, Heron thought that the work of Davenport's group was so egregiously wrong as to jeopardize the eugenic cause. Davenport hotly responded that Pearson's acolytes were after him for his support of Mendelism. David Heron, *The Influence of Defective Physique and Unfavourable Home Environment on the Intelligence of School Children . . .* (Eugenics Laboratory Memoirs; vol. VII; Dulau, 1910), p. 58; Charles B. Davenport and A. J. Rosanoff, *Reply to the Criticisms of Recent American Work by Dr. Heron of the Galton Laboratory* (Eugenics Record Office Bulletin No. 11; Cold Spring Harbor, 1914).

29. Leonard Darwin to Karl Pearson, Jan. 14, 1914; Feb. 15, 1917, Pearson Papers, Cabinet IV, D8. Years later the staunch eugenicist R. Ruggles Gates, professor of

genetics at King's College, London, alluded to the likely consequence of the schism. "I have always thought that the chief strength of the eugenic movement in America depended on the fact that they had a Research Institute [the Eugenics Record Office] connected with it." Ruggles Gates to C. C. Hurst, July 1, 1931, C. C. Hurst Papers, entry 7955/18/62.

30. Edgar Schuster, *Eugenics* (Clear-Type Press, 1912), p. 52; Phyllis Grosskurth, *Havelock Ellis: A Biography* (Alfred A. Knopf, 1980), pp. 409–10; Havelock Ellis, *The Task of Social Hygiene* (Constable, 1912), p. 402, note 1; William Bateson, "Biological Fact and the Structure of Society," The Herbert Spencer Lecture, Feb. 28, 1912, in Beatrice Bateson, *William Bateson, F.R.S.: Naturalist . . .* (Cambridge University Press, 1928), p. 342.

31. Ellis, "Birth Control and Eugenics," *Eugenics Review*, 9 (April 1917), p. 41; Charles A. Boston, "A Protest Against Laws Authorizing the Sterilization of Criminals and Imbeciles," *Journal of Criminal Law*, 4 (Sept. 1913), 332–33.

32. Harry H. Laughlin, *The Legal Status of Eugenic Sterilization*, p. 57; Henry H. Goddard, *Feeble-mindedness: Its Causes and Consequences* (Macmillan, 1914), p. 582.

33. E. S. Gosney and Paul Popenoe, *Sterilization for Human Betterment: A Summary of Results of 6,000 Operations in California* (Macmillan, 1929), pp. 39–40; William J. Robinson, *Eugenics, Marriage, and Birth Control (A Practical Guide)* (The Critic and Guide, 1917), pp. 76–77; Tyor, *Segregation or Surgery*, pp. 169–70; Peter L. Tyor, " 'Denied the Power to Choose the Good': Sexuality and Mental Defect in American Medical Practice, 1850–1920," *Journal of Social History*, 10 (Summer 1977), 473, 480; *Report of the Mental Deficiency Committee, being a Joint Committee of the Board of Education and Board of Control* (3 vols.; His Majesty's Stationer's Office, 1929), III, 138; Simmons, "Explaining Social Policy," pp. 393–94; Goddard, *Feeble-mindedness*, p. 497; Mary Dendy to Galton, Feb. 16, 1909, Francis Galton Papers, file 138/8. Pearson had once told Galton, "I have heard from more than one woman who works among the feeble-minded that at certain ages and certain times they cannot be allowed out for five minutes without offering themselves to the first man they meet. You

have in their cases the imperial passion unrestrained." Pearson, *Galton*, IIIA, 373–74.

34. Simmons, "Explaining Social Policy," p. 392; G. B. Arnold, "A Brief Review of the First Thousand Patients Eugenically Sterilized at the State Colony for Epileptics and Feebleminded," *Proceedings of the American Association for Mental Deficiency*, 43 (1937–38), 59; Robinson, *Eugenics, Marriage, and Birth Control*, pp. 100–1.

35. Lionel S. Penrose, *Mental Defect* (Farrar & Rinehart, 1933), pp. 172–74. The mainline leadership took pains to emphasize that eugenics did not mean free love, feminism, or anything else pertaining to sex. Even in the twenties Charles Davenport, who thought that without sterilization the feebleminded would breed prolifically, continued to condemn birth control. So did Harry Laughlin, a childless epileptic and product, like Davenport, of a repressive late-nineteenth-century family. Laughlin also objected to racial mixing because it would jeopardize racial "purity," explaining that women of lower races were all too inclined to sexual intercourse with men of higher races. Hassencahl, *Laughlin*, pp. 65–66, 263; James Reed, *From Private Vice to Public Virtue* (Basic Books, 1978), pp. 134–35; "Report of Committee H of the Institute of Criminal Law and Criminology: Sterilization of Criminals," p. 526.

36. Marie E. Kopp, "Surgical Treatment as Sex Crime Prevention Measure," *Journal of Criminal Law, Criminology, and Police Science*, 28 (Jan. 1938), 697; Charles V. Carrington, "Asexualization of Hereditary Criminals," *Journal of Criminal Law, Criminology, and Police Science*, 1 (July 1910), 124–25. Carrington was impressed by two cases in particular, one an epileptic masturbator who after vasectomy ceased masturbating altogether and suffered many fewer epileptic attacks. The other was a young black man who, according to Carrington, was the son of a black woman and a "degenerate" white and who was given to masturbation, sodomy, and general deviltry. After sterilization, Carrington reported, he became a "strong well-developed young negro, well behaved, and not a masturbator or sodomist."

37. *Nature*, 128 (July 25, 1931), 131; Paul Popenoe, "Intelligent Eugenics," *Forum*, 94 (July 1935), 28; *Report of the Mental Deficiency Committee*, II, 88; "Second Field Workers

Conference, June 20, 21, 1913," Davenport Papers, Cold Spring Harbor Series #2, Eugenics Record Office file; Haller, *Eugenics*, pp. 133–34; Ludmerer, *Genetics and American Society*, p. 92; Hassencahl, *Laughlin*, pp. 152–55.

38. Gosney and Popenoe, *Sterilization for Human Betterment*, pp. 39–40, 67–68; Jacob H. Landman, *Human Sterilization* (Macmillan 1932), pp. 56–93; "Eugenic Sterilization in the United States: Its Present Status," *Annals of the American Academy*, 149 (May 1930), 28; Arnold, "A Brief Review of the First Thousand Patients . . .," p. 62; Judith Kay Grether, *Sterilization and Eugenics: An Examination of Early Twentieth Century Population Control in the United States* (University Microfilms, 1980), pp. 148–54. Similar psychic fuel seems to have helped drive mainline eugenicists in Britain, where proposals to sexually segregate the mentally deficient obviously implied denying them erotic satisfaction. But in Britain, of course, social radicals like Ellis helped keep the psychic fuel from inflaming the entire movement. The geneticist J. B. S. Haldane, a eugenicist and radical, too, once scoffed away compulsory sterilization as "a piece of crude Americanism like the complete prohibition of alcoholic beverages." Searle, *Eugenics and Politics in Britain*, pp. 94–95.

39. Julius Paul, "State Eugenic Sterilization History," in Jonas Robitscher, ed., *Eugenic Sterilization* (Charles Thomas, 1973), pp. 28–29, 33–34; J. H. Landman, "Race Betterment by Human Sterilization," *Scientific American*, 150 (June 1934), 294.

40. Boston, "A Protest Against Laws Authorizing the Sterilization of Criminals and Imbeciles," pp. 326 note a, 344, 327–28, 339–40.

41. *Ibid.*, pp. 338, 351–52, 332. Justices of various courts were also at times aware of the consequences logically inherent in permitting legislatures to improve society by destroying the ability to procreate in citizens who had not committed any crime. The New Jersey Supreme Court observed: "The feeble-minded and epileptics are not the only persons in the community whose elimination as undesirable citizens would, or might in the judgment of the legislature, be a distinct benefit to society. If the enforced sterility of this class be a legitimate exercise of governmental power, a wide field of legislative activity and duty is thrown open to which it would be difficult to assign a legal limit." Frederick A. Fenning, "Sterilization Laws from a Legal Standpoint," *Journal of Criminal Law*, 4 (March 1914), 809–10.

42. Laughlin, *The Legal Status of Eugenical Sterilization*, pp. 58–59; Donna M. Cone, "The Case of Carrie Buck: Eugenic Sterilization Realized" (unpublished manuscript), p. 6; Robert J. Cynkar, "*Buck v. Bell*: 'Felt Necessities v. Fundamental Values?'" *Columbia Law Review*, 81 (Nov. 1981), 1435.

43. Laughlin, *The Legal Status of Eugenical Sterilization*, pp. 16–17; Cone, "The Case of Carrie Buck," pp. 9–10; Cynkar, "*Buck v. Bell*," pp. 1437–38.

44. Laughlin, *The Legal Status of Eugenical Sterilization*, pp. 16–17, 8; Cynkar, "*Buck v. Bell*," p. 1439.

45. *Buck v. Bell*, 274 U.S. 201–3 (1927); Cynkar, "*Buck v. Bell*," pp. 1439–40, 1446–48, 1450–53.

46. *Buck v. Bell*, 274 U.S. 205, 207 (1927).

47. Hassencahl, *Laughlin*, pp. 98–100, 159; *The American Eugenics Society* (American Eugenics Society, 1927), p. 12, American Eugenics Society Papers, box 9; Gosney and Popenoe, *Sterilization for Human Betterment*, pp. 16–17; Donald K. Pickens, *Eugenics and the Progressives* (Vanderbilt University Press, 1968), p. 91; Arnold, "A Brief Review of the First Thousand Patients . . .," pp. 57, 59; Landman, *Human Sterilization*, pp. 113–14; Ludmerer, *Genetics and American Society*, p. 95.

48. *The Literary Digest*, 93 (May 21, 1927), 11; Cone, "The Case of Carrie Buck," pp. 9–10; Stephen Jay Gould, "Carrie Buck's Daughter," *Natural History*, 93 (July 1984), 18. In 1979, after a long search, K. Ray Nelson, the director of the Lynchburg Training School and Hospital, and a vigorous advocate of the rights of the mentally retarded, located Carrie Buck Detamore in Albermarle County, where she was living in poverty with her sick husband. He also found Carrie's sister Doris. She had been sterilized, too—in the Lynchburg hospital in 1928—but she had been told that the operation she was forced to undergo was an appendectomy. Mrs. Doris Buck Figgins learned the truth during a visit from Nelson. "I broke down and cried," she said later. "My husband and me wanted children desperate—we were crazy about them.

I never knew what they'd done to me." Richmond *Times-Dispatch*, Feb. 24, 1980; Feb. 27,

1980; Los Angeles *Times*, Feb. 24, 1980, part I, p. 4.

Chapter VIII:
A COALITION OF CRITICS

1. *The New York Times*, March 22, 1930, p. 22; Raymond B. Cattell, *The Fight for Our National Intelligence* (P. S. King, 1937), pp. 4–5, 14, 16, 60–61. E. A. Hooton, professor of physical anthropology at Harvard University, declaimed to the St. Louis Harvard Club that modern man was "selling his biological birthright for a mess of morons; . . . the voice may be the voice of democracy, but the hands are the hands of apes." New York *Times*, Feb. 18, 1937, p. 23.

2. *The New York Times*, Aug. 23, 1932, p. 1; *Report of the Mental Deficiency Committee, being a Joint Committee of the Board of Education and Board of Control* (3 vols.; His Majesty's Stationer's Office, 1929), I, 74–75; *Nature*, 128 (July 25, 1931), 129.

3. C. P. Blacker, memorandum, "The Social Problem Group," attached to Blacker to Penrose, Oct. 1, 1931, Lionel Penrose Papers, University College London, file 130/9; Committee for Legalising Eugenic Sterilization, *Eugenic Sterilization* (pamphlet; 2nd ed.; Eugenics Society, undated [1933]), p. 7; *Report of the Mental Deficiency Committee*, II, 79–81; I, 83.

4. Committee for Legalising Eugenic Sterilization, *Eugenic Sterilization*, p. 8; *Report of the Mental Deficiency Committee*, I, 82.

5. In 1932, Henry Fairfield Osborn told the Third International Eugenics Congress in New York City: "While some highly competent people are unemployed, the mass of unemployment is among the less competent." The *New Republic* reported that, to many hard-boiled conservatives, something was "fundamentally wrong with America's 11,500,000 or more of unemployed, that they can never be absorbed into industry and that the best thing for the country would be mass sterilization to prevent their numbers from growing." *The New York Times*, March 9, 1931, p. 15; Aug. 23, 1932, p. 1; *New Republic*, 85 (Feb. 26, 1936), 59; G. R. Searle, "Eugenics and Politics in Britain in the 1930s," *Annals of Science*, 36 (1979), 159–62.

6. Ernest J. Lidbetter, *Heredity and the Social Problem Group* (vol. I; Edward A. Arnold, 1933), pp. 5–6, 11, 17–20. The Mental Deficiency report said that thickly populated areas always contained a large number of mental defectives "who though employable cannot be assimilated by the ordinary industrial system, and these numbers are considerably increased at periods of trade depression." *Report of the Mental Deficiency Committee*, II, 93.

7. Gary Wersky, *The Visible College* (Holt, Rinehart and Winston, 1979), pp. 32–33; *The Literary Digest*, 104 (Jan. 18, 1930), 34; E.W.M., "Cultivation of the Unfit," *Nature*, 137 (Jan. 11, 1936), 45; E. S. Gosney and Paul Popenoe, *Sterilization for Human Betterment: A Summary of Results of 6,000 Operations in California* (Macmillan, 1929), pp. 113–14; Committee for Legalising Eugenic Sterilization, *Eugenic Sterilization*, n.p.; *Fortune*, 16 (July 1937), 106; *The New York Times*, Jan. 3, 1936, VI, 6; Feb. 21, 1937, II, 1, 2; Cattell, *Fight for Our National Intelligence*, p. 69.

8. Gosney and Popenoe, *Sterilization for Human Betterment*, pp. 154–55, 31–33, 35–37, xiii–xiv; Committee for Legalising Eugenic Sterilization, *Eugenic Sterilization*, pp. 16–17.

9. *The New York Times*, May 17, 1936, p. 8; Dec. 29, 1934, p. 7; Jan. 3, 1936, VI, 6; April 25, 1936, p. 19; Marian S. Olden, *Human Betterment Was Our Goal* (Princeton, N.J.: 1970?), pp. 66, 151; Committee for Legalising Eugenic Sterilization, *Better Unborn* (pamphlet; Eugenics Society [undated]), pp. 10 ff; H. L. Mencken, "Utopia by Sterilization," *American Mercury*, 41 (August 1937), 406.

10. Paul Popenoe to Ruggles Gates, Dec. 12, 1933, Ruggles Gates Papers, Kings College London.

11. Cecil Binney, *The Law as to Sterilization* (pamphlet; Committee for Legalising Eugenic Sterilization; Eugenics Society, 193?), p. 3; *Report of the Departmental Committee on Sterilisation (minus appendices)* (pamphlet; Joint Committee on Voluntary

Sterilisation, June 1934), hereafter cited as [*Brock*] *Report . . . on Sterilisation*, p. 5.

12. H. L. Mencken, "Utopia by Sterilization," p. 406.

13. Harry H. Laughlin, "Further Studies on the Historical and Legal Development of Eugenical Sterilization in the United States," *Proceedings of the American Association on Mental Deficiency*, 41 (1936), 99; Frederick Osborn, *Preface to Eugenics* (Harper & Bros., 1940), p. 31. The sterilization figures have been calculated from the data in Jonas Robitscher, ed., *Eugenic Sterilization* (Charles C Thomas, 1973), appendices I and II, and from *U.S. Maps Showing the States Having Sterilization Laws in 1910–1920—1930–1940* (pamphlet; Birthright, Inc., 1942), copy in Penrose Papers, file 77/3.

14. Richmond *Times-Dispatch*, April 6, 1980.

15. Richmond *Times-Dispatch*, Feb. 27, 1980; March 2, 1980.

16. *Newsweek*, 2 (Aug. 1933), 12; *The New York Times*, Jan. 22, 1934, p. 6.

17. Marie E. Kopp, "A Eugenic Program in Operation," Conference on Eugenics in Relation to Nursing, Feb. 24, 1937, American Eugenics Society Papers, box 2; *Newsweek*, 2 (Dec. 30, 1933), 11; Marie E. Kopp, "Legal and Medical Aspects of Eugenic Sterilization in Germany," *American Sociological Review*, 1 (Oct. 1936), 766. By 1939, the total sterilized under the Nazi law was 320,000 people. Gisela Bok, "Racism and Sexism in Nazi Germany: Motherhood, Compulsory Sterilization, and the State," *Signs*, 8 (1983), 413.

18. Kopp, "Legal and Medical Aspects of Eugenic Sterilization in Germany," pp. 761–62; Nazi Minister Paul Joseph Goebbels, who presumably did not apply for the loans, showed the way by fathering four children in five years. Berlin, suffering a low birthrate for twenty years, sweetened the package with the offer of municipal sponsorship for third and fourth children of physically, mentally, and hereditarily sound families (as of 1937 only 311 planned babies out of 2,000 applicants had been admitted to the pantheon of Berlin's most esteemed younger citizens). *The New York Times*, Feb. 20, 1937, p. 19; April 29, 1934, VIII, 6; Kopp, "A Eugenic Program in Operation"; George L. Mosse, *Toward the Final Solution: A History of European Racism* (Howard Fertig, 1978), p. 219.

19. C. P. Blacker, *Eugenics in Prospect and Retrospect* (Hamish Hamilton, 1945), p. 8; J. H. Landman, "Sterilization and Social Betterment," *Survey*, 25 (March 1936), 162; Mosse, *Toward the Final Solution*, pp. 80–81.

20. Kopp, "A Eugenic Program in Operation"; Mosse, *Toward the Final Solution*, pp. 208–9, 217–18, 82.

21. Phyllis Grosskurth, *Havelock Ellis: A Biography* (Alfred A. Knopf, 1980), p. 414; Leon F. Whitney, "Memoirs," Charles B. Davenport Papers, p. 204; Garland E. Allen, "The Founding of the Eugenics Record Office, Cold Spring Harbor, Long Island (New York), 1910–1940; An Essay in Institutional History" (unpublished manuscript), p. 58; Olden, *Human Betterment Was Our Goal*, p. 144; Kopp, "Legal and Medical Aspects of Eugenic Sterilization in Germany," p. 763.

22. The general antipathy of Jews to eugenics puzzled the American zoologist and eugenicist Samuel J. Holmes, since, he wrote, "no one can accuse the Jews as a stock of being deficient in native endowment." Samuel J. Holmes, *The Eugenic Predicament* (Harcourt, Brace, 1933), p. 122.

23. Bertrand L. Conway, "The Church and Eugenics," *Catholic World*, 128 (Nov. 1928), 151; Thomas J. Gerrard, *The Church and Eugenics* (B. Herder, 1912), pp. 19–20, 37; Thomas J. Gerrard, "The Catholic Church and Race Culture," *Dublin Review*, 149 (July 1911), 66–67, 55. Eugenicists back to Galton may have thought it a deplorable eugenic waste for glorious specimens of manhood or womanhood to take religious vows of perpetual virginity, but Gerrard defended clerical celibacy on grounds that it had enlarged man's spirit by fostering art and learning. Besides, the Church hardly encouraged celibacy for all its adherents.

24. Gerrard, *The Church and Eugenics*, 39; *New York Times*, Jan. 9, 1931, pp. 14–15.

25. L. T. Hobhouse, *Social Evolution and Political Theory* (Oxford University Press, 1911), pp. 66–67.

26. James Joyce, *A Portrait of the Artist as a Young Man* (Viking Press, 1968), pp. 208–9.

27. Early in the century Chesterton had attacked Karl Pearson for preaching "the great principle of the survival of the nastiest." G. K. C[hesterton], review of Pearson, *National Life from the Standpoint of Science*, in *The Speaker*, Feb. 2, 1901, p. 488, copy in Karl

Pearson Papers, Cabinet IV, D5; G. K. Chesterton, *Eugenics and Other Evils* (Cassell, 1922), pp. 7, 54. Alexander Graham Bell, his wife deaf since childhood, could appreciate Chesterton's strictures. "The happiness of individuals," Bell insisted, "is often promoted by marriage even in cases where the offspring may not be desirable." Alexander Graham Bell, "A Few Thoughts Concerning Eugenics," *National Geographic*, 19 (Feb. 1908), 122–23.

28. Chesterton, *Eugenics and Other Evils*, pp. 151, 152, ii, 76–77; *Morning Leader*, March 8 [1910], Eugenics Society Press Clippings Scrapbook, Eugenics Society, London.

29. *Illustrated London News*, March 12 [1910]; *The Nation*, 97 (Aug. 14, 1913), 138.

30. J. B. Eggen, "Eugenics Teaching Imperils Civilization," *Current History*, 24 (Sept. 1926), 882; Bertrand Russell, *Icarus, or the Future of Science* (Kegan Paul, Trench, Trubner, 1924), pp. 48–49; Clarence Darrow, "The Eugenics Cult," *American Mercury*, 8 (June 1926), 137.

31. William Davenport to Charles B. Davenport, Oct. 22, 1921, Davenport Papers, William E. Davenport file.

32. John J. Burke to Charles B. Davenport, Jan. 26, 1926, Davenport Papers, Burke file.

33. Sidney Webb had protested to Karl Pearson that he could see no reason for permitting the persistence of social conditions "adverse to the proper development of each generation when born. Even assuming that semi-starvation in childhood, parental neglect, slum life, street trading and existence as a casual laborer, do not in any way affect the germ plasm—well, is that any reason for not remedying such influences? We don't live *merely* for posterity." Sidney Webb to Karl Pearson, Nov. 16, 1909, Pearson Papers, Cabinet IV, D6.

34. In 1916 the anthropologist Alfred L. Kroeber remarked that most anti-eugenic protests had come from "the orthodoxly religious and the professedly skeptical, but rarely from the enlightened camp of science." Among the rare exceptions had been Karl Pearson, who was of course not so much anti-eugenic as opposed to the central dogma of mainline eugenics, the theory of Mendelian inheritance. Pearson wrote a friend in 1913: "I hear we are in great hot water in America for our criticism of the American Eugenists, but I think the day will come when American men of science will themselves protest against the work that has been put forth in the name of Eugenics." Pearson to Raymond Pearl, Dec. 20, 1913, Raymond Pearl Papers, Pearson file; A. L. Kroeber, "Inheritance by Magic," *American Anthropology*, 18 (Jan. 1916), 34–35.

35. William Bateson, as a product of the Oxbridge elite, could sympathize with eugenic notions of a hierarchical social order correlated with hereditary virtue, but he disliked the way Galton and his followers proposed—at least so Bateson thought—to rank the hierarchy in favor of "broadcloth, bank balances and the other appurtenances of the bay-tree type of righteousness" and in disfavor of Beethoven or Keats. Caleb W. Saleeby, *Progress of Eugenics* (Funk & Wagnalls, 1914), pp. 135–36; Donald MacKenzie, "The Development of Statistical Theory in Britain, 1865–1925: A Historical and Sociological Perspective" (doctoral thesis, University of Edinburgh, 1977), pp. 293–94.

36. Predominantly laymen, eugenic activists were usually so much more concerned with propaganda than with knowledge that even pro-eugenic scientists found the situation an embarrassment. Dr. Lewellys F. Barker, professor of medicine at Johns Hopkins and also a eugenicist, remarked that "true eugenics" was more in danger from its friends than from its enemies. "Hasty and ill-advised legislation is preceding not only the cultivation of public opinion, but also that solid foundation of demonstrable fact which alone would justify lawmaking." Kenneth Ludmerer, *Genetics and American Society* (Johns Hopkins University Press, 1972), p. 51; Charles E. Rosenberg, "The Social Environment of Scientific Innovation: Factors in the Development of Genetics in the United States," in Charles Rosenberg, *No Other Gods: On Science and American Social Thought* (Johns Hopkins University Press, 1976), p. 202.

37. Ludmerer, *Genetics and American Society*, pp. 82–83; H. J. Muller, "The Dominance of Economics over Eugenics," *Scientific Monthly*, 37 (July 1933), 40–47; Raymond Pearl, "The Biology of Superiority," *American Mercury*, 12 (Nov. 1927), 260.

38. Ronald W. Clark, *J.B.S.: The Life*

and Work of J. B. S. Haldane (Coward-McCann, 1968), p. 70.

39. J. R. Baker, "Julian Sorrell Huxley," *Biographical Memoirs of Fellows of the Royal Society,* 22 (1976), 213, 210; G. P. Wells, "Lancelot Hogben," *Biographical Memoirs of Fellows of the Royal Society,* 24 (1978), 183–221; Tracy M. Sonneborn, "Herbert Spencer Jennings," *National Academy of Sciences Biographical Memoirs,* 47 (1975), 143–223; N. W. Pirie, "John Burdon Sanderson Haldane," *Biographical Memoirs of Fellows of the Royal Society,* 12 (1966), 219–49.

40. Julian S. Huxley, *Memories* (Harper & Row, 1970), pp. 137, 125, 138; Wells, "Hogben," p. 192; Clark, *Haldane,* pp. 21, 70–71; Sonneborn, "Jennings," 169–70.

41. Wersky, *The Visible College,* pp. 42, 45, 102; Lancelot Hogben, *Genetic Principles in Medicine and Social Science* (Williams and Norgate, 1931), p. 210; Pauline Mazumdar, "Eugenists, Marxists, and the Science of Human Genetics" (unpublished manuscript), p. 133; Jennings to Irving Fisher, May 18, 1922, Jennings Papers, Group J44a, Fisher file; biography of Jennings, early draft, Herbert S. Jennings Papers, American Philosophical Society Library, box 1.

42. Mazumdar, "Eugenists, Marxists, and the Science of Human Genetics," p. 133; Baker, "Huxley," p. 208; Clark, *Haldane,* pp. 24–27, 30–33; Wersky, *The Visible College,* pp. 60–64.

43. Wells, "Hogben," pp. 185–86; Wersky, *The Visible College,* pp. 61–62; Joseph Brenemann to S. W. Geiser, Jan. 5, 1934, Herbert S. Jennings Papers, Section B, J44.a, box 2, Brenemann file; Sonneborn, "Jennings," pp. 143–44.

44. Wells, "Hogben," p. 190; Clark, *Haldane,* p. 65; Sonneborn, "Jennings," pp. 145–46; Baker, "Huxley," p. 231; Maude DeWitt Pearl to Jennings, April 8, 1942, Herbert S. Jennings Papers, Section B, J44, Pearl file.

45. Jennings to Brenemann, May 10, 1896, Herbert S. Jennings Papers, Section B, J44a, box 2, Brenemann file; Clark, *Haldane,* pp. 189, 22, 64, 109; Pirie, "Haldane," p. 235.

46. Wells, "Hogben," pp. 188–89; Julian S. Huxley, "Birth Control," in Huxley, *Essays in Popular Science* (Alfred A. Knopf, 1927), pp. 140–41; Huxley, *Memories,* pp. 114–15, 151, 160; Lancelot Hogben, "Heredity and Human Affairs," in J. G. Crowther, ed., *Sci-ence for a New World* (Harper & Bros., 1934), p. 50; Hogben, *Genetic Principles in Medicine and Social Science,* p. 203.

47. Huxley *Memories,* p. 153, 124; Wersky, *The Visible College,* pp. 66–67, 109.

48. Haldane, chapter 1, in the autobiographical "Why I Am a Cooperator," J. B. S. Haldane Papers, University College London; Clark, *Haldane,* pp. 46–47, 60, 69; Wersky, *The Visible College,* p. 83. When Haldane met his future wife, Charlotte Burghes, she was already married. To obtain a divorce under the laws of England in 1925, Mrs. Burghes had to commit adultery. She and Haldane let it be known that they would accomplish the deed. They even assisted Mr. Burghes's detective in finding their hotel and welcomed him the next day when he brought the morning papers to their bedroom. The authorities of Cambridge University, where Haldane was considered a troublesome member of the faculty, were not so accommodating. A disciplinary board of six, the Sex Viri (in the university's ancient Latin nomenclature), proposed that he be dismissed for gross immorality. The ruling was overturned on Haldane's appeal, in part because some very powerful faculty at Cambridge came to his aid, in part because of the ridicule he managed to heap upon the board by punning the Latin pronunciation of its title into the "Sex Weary." The Haldane case put an end at Cambridge to formal university oversight of the private lives of the faculty. Clark, *Haldane,* pp. 80–83; T. E. Howarth, *Cambridge Between Two Wars* (Collins, 1978), pp. 53–57.

49. Jennings to Brenemann, Aug. 24, 1894; May 10, 1896, Herbert S. Jennings Papers, Section B, J44a, box 2, Brenemann file; Helen Jennings Barrett, "Memories of Herbert Jennings," Herbert S. Jennings Papers, Section B, J44a, box 1, Biographical #2 folder; Sonneborn, "Jennings," p. 168; Herbert Spencer Jennings *The Biological Basis of Human Nature* (W. W. Norton, 1930), pp. 267–68.

50. Wells, "Hogben," pp. 187–89; Wersky, *The Visible College,* pp. 64–65, 103; Clark, *Haldane,* pp. 35–37, 43–45.

51. Clark *Haldane,* pp. 20–21, 23; J. B. S. Haldane, "Eugenics and Social Reforms," *The Nation and the Athenaeum,* 35 (May 21, 1924), p. 291; Wersky, *The Visible College,* pp. 98–99, 94, 55–56, 58–60, 76.

52. Wersky, *The Visible College*, pp. 240–41, 239, 158–60, 162, 164, 64; Wells, "Hogben," p. 200.

53. Sonneborn, "Jennings," pp. 146–47; Jennings to Brenemann, Oct. 20, 1895; July 10, 1892, Herbert S. Jennings Papers, Section B, J44a, box 2, Brenemann file. Jennings reflected, "Isn't it queer how the respectable people can be all on one side . . . and yet that be entirely the wrong side. . . . From my own observation and experience I should say that that class, at least here in the North, would be found *as* a class, in the Republican party. And yet I am just as thoroughly convinced that of *all* the parties at present, the Republican deserves least one's support. . . . I used to think that the older wise people were conservative because they had had much experience and knew that these new ideas were foolish and wouldn't work, but I am coming more and more to the conclusion that it is selfishness, unconscious selfishness." Jennings to Brenemann, Aug. 12, 1894, Herbert S. Jennings Papers, Section B, J44a, box 2, Brenemann file.

54. Wersky, *The Visible College*, pp. 105–6, 241; J. B. S. Haldane, "Towards a Perfected Posterity," *The World Today*, 45 (Dec. 1924), 3.

55. John Dewey considered Jennings, who had once been his student, the authority to turn to on questions in biology, while in England Isaiah Berlin recalled that Haldane was "one of our major intellectual emancipators." Wersky, *The Invisible College*, p. 86; Dewey to S. W. Geiser, Jan. 31, 1934, excerpt in Herbert Spencer Jennings Papers, Section B, J44, box 3, Dewey file.

56. J. B. S. Haldane, *Daedalus, or Science and the Future* (E. P. Dutton, 1924), p. 11; Haldane, "Blood Royal," in J. B. S. Haldane, *Adventures of a Biologist* (Harper & Bros., 1940), p. 137; Jennings, *The Biological Basis of Human Nature*, pp. 203–4, 220; H. G. Wells, Julian S. Huxley, and G. P. Wells, *The Science of Life* (Doubleday, Doran, 1931), p. 1468; Hogben, *The Nature of Living Matter* (Kegan Paul, Trench, Trubner, 1930), p. 208.

Chapter IX:
FALSE BIOLOGY

1. William M. Gemmill, "Genius and Eugenics," *Journal of Criminal Law*, 6 (May 1915), 83–84; "Degeneration and Pessimism," *Edinburgh Review*, 214 (July 1911), 162.

2. Walter Lippmann, "The Mental Age of Americans," *New Republic*, Oct. 25, 1922, pp. 213–15; Walter Lippmann, "Tests of Hereditary Intelligence," *New Republic*, Nov. 22, 1922, pp. 328–30; Walter Lippmann, "A Future for the Tests," *New Republic*, Nov. 29, 1922, pp. 9–11. Just after the First World War, *The New York Times* mocked the notion that the U.S. Army I.Q.-test results revealed that the average Philadelphian possessed a childlike mental capacity: "It looks as if something is the matter with Philadelphia, or the psychologists, or the standard of measurement." New York *Times*, Feb. 25, 1919, p. 10.

3. Carl Campbell Brigham, "Intelligence Tests of Immigrant Groups," *Psychological Review*, 37 (March 1930), pp. 158–60, 164–65; Brigham to C. B. Davenport, Dec. 8, 1929, Charles B. Davenport Papers, Carl Campbell Brigham file; Walter Lippmann, "The Mystery of the 'A' Men," *New Republic*, Nov. 1, 1922, p. 246; Edward L. Thorndike, "Tests of Intelligence: Reliability, Significance, Susceptibility to Special Training and Adaptation to the Nature of the Task," *School and Society*, 9 (Feb. 15, 1919), 189–95; Lancelot Hogben, "Heredity and Human Affairs," in J. G. Crowther, ed., *Science for a New World* (Harper & Bros., 1934), pp. 44–45. In Britain, Cyril Burt thought that the psychophrenology could be avoided through the approach pioneered early in the century by the British psychologist Charles Spearman, who contended that people possessed a general intelligence, an entity that no one mental test could measure but which, in its various factors, individual tests all captured in some proportion. Spearman called the general capability g and developed a complicated mathematical procedure—factor analysis—to determine it. Though a eugenic sym-

pathizer, Spearman had been much less concerned with demonstrating a biological basis of intelligence than with developing a mathematical instrument that would transform theories of mind into constructs as solid as theories of physics. Burt, quite in contrast, in later years built on Spearman's work—he took more credit for the advancement of the formal technique than he deserved—with the aim of exploiting factor analysis to demonstrate that intelligence was inherited. However, the British psychologist Godfrey Thomson, one of the major developers of the mathematical techniques of factor analysis, perceived considerable ambiguity in what the mathematics revealed. To Thomson, the ambiguity did not "prove that we have no such 'factors.' But it does show that perhaps we haven't, that perhaps they are fictions—possibly very useful fictions, but still fictions." L. S. Hearnshaw, *Cyril Burt, Psychologist* (Cornell University Press, 1979), pp. 154–55, 129; Stephen Jay Gould, *The Mismeasure of Man* (W. W. Norton, 1981), pp. 151, 239, 252, 262–63, 272–73, 279; Bernard Norton, "Charles Spearman and the Generation Factor in Intelligence: Genesis and Interpretation in the Light of Sociopersonal Considerations," *Journal of the History of the Behavioral Sciences*, 15 (1979), 150, 147, 148; Godfrey Thomson [autobiography], in *A History of Psychology in Autobiography* (Russell and Russell, 1968), IV, 283.

4. J. L. Gray and Pearl Moshinsky, "Ability and Opportunity in English Education," in Lancelot Hogben, ed., *Political Arithmetic: A Symposium of Population Studies* (George Allen and Unwin, 1938), pp. 368–72, 415.

5. Gillian Sutherland, "Measuring Intelligence: English Local Education Authorities and Mental Testing, 1919–1939," in Charles Webster, ed., *Biology, Medicine and Society, 1840–1940* (Cambridge University Press, 1981), pp. 317, 325–27; Sutherland, "The Magic of Measurement: Mental Testing and English Education, 1900–1940," *Transactions of the Royal Historical Society*, 27 (1977), 146–47, 150–53.

6. "Intelligence and Families," *The Times*, Nov. 17, 1948, clipping in Lionel Penrose Papers, file 65/2.

7. Lancelot Hogben, *Genetic Principles in Medicine and Social Science* (Williams and Norgate, 1931), p. 205; Joseph McCabe, "Is Civilisation in Danger? A Reply," *The Hibbert Journal*, 10 (April 1912), 605–7, 609–10; Mott to Bateson, May 21, 1914, William Bateson Papers, John Innes Horticultural Institution, Norfolk, East Anglia, England, Treasure file; Abraham Myerson et al., *Eugenical Sterilization: A Reorientation of the Problem* (Committee of the American Neurological Association for the Investigation of Eugenical Sterilization; Macmillan, 1936), p. 56; Gemmill, "Genius and Eugenics," p. 85; Hogben, "Heredity and Human Affairs," p. 51; Haldane, "Blood Royal," in J. B. S. Haldane, *Adventures of a Biologist* (Harper & Bros., 1940), pp. 141–42.

8. Jennings to Bruno Lasker, June 20, 1923; Jennings to Irving Fisher, Sept. 27, 1924; Jennings to James McKeen Cattell, Feb. 20, 1924, Herbert S. Jennings Papers, Section B, J44, Lasker file, Fisher file, Cattell file. It was this episode that initiated Jennings's career as a public biologist.

9. Herbert Spencer Jennings, "Undesirable Aliens," *The Survey*, 51 (1923), 309–12, 364; Jennings to Irving Fisher, Nov. 21, 1923, Herbert S. Jennings Papers, Section B, J44, box 3; Herbert Spencer Jennings, *Prometheus, or Biology and the Advancement of Man* (E. P. Dutton, 1925), pp. 57–59; Frances J. Hassencahl, *Harry H. Laughlin, 'Expert Eugenics Agent' for the House Committee on Immigration and Naturalization, 1921 to 1931* (University Microfilms, 1971), p. 268.

10. Thomas Hunt Morgan, *Evolution and Genetics* (2nd ed.; Princeton University Press, 1925), pp. 206–7.

11. Julian S. Huxley and A. C. Haddon, *We Europeans: A Survey of 'Racial' Problems* (Jonathan Cape, 1935), pp. 18, 261, 263, 103, 104, 107.

12. Huxley and Haddon, *We Europeans*, pp. 184, 86, 91, 96–97, 25–26, 267–68.

13. Julian Huxley, "The Concept of Race," *Harper's*, 170 (May 1935), 692; J. B. S. Haldane, *New Paths in Genetics* (Harper & Bros., 1941), pp. 38–39.

14. Russell Marks, *Testers, Trackers and Trustees: The Ideology of the Intelligence Testing Movement in America, 1900–1954* (University Microfilms, 1972), p. 121.

15. Lippmann to Yerkes, Jan. 9, 1923, Robert M. Yerkes Papers, Yale University, box 9, folder 285.

16. Thomas Pogue Weinland, *A History of the I.Q. in America, 1890–1941* (University Microfilms, 1970), pp. 213–16; Franz Boas, *Anthropology and Modern Life* (W. W. Norton, 1928), pp. 102–19; interview with Boas by Mrs. Charles Bosanquet, March 27, 1929, Davenport Papers, Bosanquet file; Edward H. Beardsley, "The American Scientist as Social Activist: Franz Boas, Burt G. Wilder, and the Cause of Racial Justice, 1900–1915," *Isis*, 64 (1973), 59–60; Hamilton Cravens, *The Triumph of Evolution: American Scientists and the Heredity-Environment Controversy, 1900–1941* (University of Pennsylvania Press, 1978), p. 234; Margaret Mead, "Group Intelligence Tests and Linguistic Disability among Italian Children," *School and Society*, 25 (April 16, 1927), 468; Margaret Mead, "The Methodology of Racial Testing: Its Significance for Sociology," *American Journal of Sociology*, 31 (March 1926), 657–58, 665, 667, 668–69.

17. Citation, "American Psychological Association Distinguished Contributions to Psychology in the Public Interest Award, 1979, presented to Otto Klineberg," copy in author's possession.

18. Otto Klineberg, "Reflections of an International Psychologist of Canadian Origin," *International Sociology of Science Journal*, 25 (1973), 39–54; Klineberg [autobiography], in *A History of Psychology in Autobiography*, VI, 163–82. Unless otherwise noted, the rest of this section is based on these two autobiographical pieces plus the author's interview and correspondence with Otto Klineberg.

19. Otto Klineberg, "A Study of Psychological Differences between 'Racial' and National Groups in Europe," *Archives of Psychology*, Sept. 1931, pp. 7–9, 12, 17–23, 25, 27–29, 31, 42, 44.

20. Otto Klineberg, *Negro Intelligence and Selective Migration* (Columbia University Press, 1935; reprint edition, Greenwood Press, 1974), pp. 1, 4, 12–13, 24, 29, 37, 52, 54, 59–60.

21. Otto Klineberg, "Tests of Negro Intelligence," in Otto Klineberg, ed., *Characteristics of the American Negro* (Harper & Bros., 1944), pp. 29, 45–46, 55, 66, 68–69, 85, 95–96.

22. Text of the 1950 "Statement on Race," in United Nations Educational, Scientific, and Cultural Organization, *The Race Concept: Results of an Inquiry* (Greenwood Press, 1970), pp. 98–103.

23. Walter Lippmann, "The Abuse of the Tests," *New Republic*, Nov. 15, 1922, pp. 297–98; Walter Lippmann, "The Great Confusion: A Reply to Mr. Terman," *New Republic*, Jan. 3, 1923, pp. 145–46.

24. Lancelot Hogben, *Nature and Nurture* (W. W. Norton, 1933), pp. 111, 114, 121; Lancelot Hogben, "The Limits of Applicability of Correlation Technique in Human Genetics," *Journal of Genetics*, 27 (Aug. 1933), 379–84, 393–97, 399, 405; Lionel S. Penrose to Edmund O. Lewis, Oct. 10, 1933, Lionel Penrose Papers, University College London, file 147/3; G. P. Wells, "Lancelot Hogben," *Biographical Memoirs of Fellows of the Royal Society*, 24 (1978), 197; Lancelot Hogben, "Some Methodological Aspects of Human Genetics," *American Naturalist*, 68 (May–June 1933), 262.

25. Hogben, *Nature and Nurture*, pp. 108–11, 121; Hogben, *Genetic Principles in Medicine and Social Science*, pp. 30–31; Hogben, "Some Methodological Aspects of Human Genetics," p. 260.

26. Cravens, *The Triumph of Evolution*, pp. 258–59; Hearnshaw, *Cyril Burt*, pp. 60–61, 231–38, 244–53; Barbara Stoddard Burks, "The Relative Influence of Nature and Nurture upon Mental Development: A Comparative Study of Foster Parent–Foster Child Resemblance and True Parent–True Child Resemblance," in Guy Montrose Whipple, ed., *The Twenty-seventh Yearbook of the National Society for the Study of Education: Nature and Nurture, Part I, Their Influence upon Intelligence* (Public School Publishing Co., 1928), pp. 219–25, 240–43, 308–9.

27. Hogben, *Genetic Principles in Medicine and Social Science*, p. 33; Hogben, "Some Methodological Aspects of Human Genetics," pp. 261, 262; interview with Otto Klineberg; Burks, "The Relative Influence of Nature and Nurture . . .," pp. 232, 235.

28. Cravens, *The Triumph of Evolution*, pp. 257–58; Myerson et al., *Eugenical Sterilization*, pp. 133–35.

29. Cravens, *The Triumph of Evolution*, pp. 263–64; Horatio H. Newman, Frank N. Freeman, and Karl J. Holzinger, *Twins: A Study of Heredity and Environment* (University of Chicago Press, 1937), pp. 362, 359. The eminent British psychologist Godfrey

Thomson, who spent a great deal of his life trying to sort out the relative roles of heredity and environment in the shaping of intelligence, wrote in the sixties: ". . . even now all I can venture as a scientist to say is that both are certainly concerned, but in what proportion I do not know." Godfrey Thomson [autobiography], in *A History of Psychology in Autobiography*, IV, 288.

30. Hamilton Cravens, "Inconstancy of the Intelligence Quotient: The Iowa Child Welfare Research Station and the Criticism of Hereditarian Mental Testing" (unpublished manuscript), pp. 7, 9–10; Albert E. Wiggam, "Are Dummies Born or Made?" *Ladies' Home Journal*, 57 (March 1940), 123–24. By the early thirties, the Iowa Station scientists had become strongly sensitive in general to the effects of environmental stimulation on measures of I.Q., and the studies of children in orphanages were paralleled by others on foster children, which yielded similar results. Hamilton Cravens, "The Wandering I.Q.: The Iowa Child Welfare Research Station and Mental Testing, 1925–1940" (unpublished manuscript); *Time*, Nov. 7, 1938, pp. 44–46.

31. Wiggam, "Are Dummies Born or Made?" pp. 123–24.

32. *Ibid.*; Weinland, *A History of the I.Q. in America*, pp. 287–88; Cravens, "Inconstancy of the Intelligence Quotient," note 28; Cravens, *The Triumph of Evolution*, pp. 262–63; Henry L. Minton, "The Iowa Child Welfare Research Station and the 1940 Debate on Intelligence: Carrying on the Legacy of a Concerned Mother," *Journal of the History of the Behavioral Sciences*, 20 (April 1984), 168.

33. Herbert Spencer Jennings, *The Biological Basis of Human Nature* (W. W. Norton, 1930), pp. 95–96.

34. Hogben, *Nurture and Nature*, p. 115; Sonneborn, "Jennings," pp. 182, 185–86; Jennings, *Biological Basis of Human Nature*, 124–25, 128; Julian Huxley, "Eugenics and Society," in Julian S. Huxley, *Man Stands Alone* (Harper & Bros., 1941), p. 44.

35. "Degeneration and Pessimism," *Edinburgh Review*, 214 (July 1911), 162; G. R. Searle, "Eugenics and Politics in Britain in the 1930s," *Annals of Science*, 36 (1979), 165; 9; J. B. S. Haldane, *Human Biology and Politics* (The Science Guild, 1934), p. 16. Major Greenwood, eventually a leading authority

on public health and medical statistics, left Karl Pearson's laboratory in disaffection from its eugenic dogmatism. In 1913 he was outraged by the conclusion of one of Pearson's staff that nutrition had nothing to do with a child's school work. "Simply SHIT from beginning to end," he said. "Give a dog a protein free diet and he will become a corpse after a certain number of days, give him protein but not enough to keep him in nitrogeneous equilibrium and he will equally become a corpse in rather a greater number of days. Now we know that many of the kids are not in nitrogeneous equilibrium." Major Greenwood to G. Udny Yule, June 30, 1913, G. Udny Yule Papers, Royal Statistical Society, London, box 22.

36. Jennings, *Prometheus*, pp. 68–70; J. Arthur Thomson, *Heredity* (2nd ed.; John Murray, 1912), pp. 306–8.

37. H. G. Wells, "Discussion of Galton's Paper, 1904," Francis Galton Papers, file 138/9; H. G. Wells, *Mankind in the Making* (2 vols.; Bernhard Tauchnitz, 1903), I, 75; Arthur E. Fink, *The Causes of Crime: Biological Theories in the United States, 1800–1915* (University of Pennsylvania Press, 1938), pp. 177–78; *The Nation*, 99 (July 1914), 37; Cyril Burt, *Mental and Scholastic Tests* (P. S. King, 1922), p. 208; L. S. Hearnshaw, *Cyril Burt, Psychologist*, pp. 271–72; clipping, meeting of American Association for the Advancement of Science, c. 1927, Davenport Papers, Cold Spring Harbor Series # 2; Carl Murchison, "Criminals and College Students," *School and Society*, 12 (July 3, 1920), 24–30; Cravens, *The Triumph of Evolution*, pp. 246–47. Burt had earned favorable comment from Walter Lippmann for pointing out how environmental deprivation might affect performance on the Binet-Simon test. Burt wrote that "a child's proficiency in the Binet-Simon tests is the complex resultant of a thousand intermingling factors. Besides . . . the intelligence he has inherited, the age he has reached, a host of subsidiary conditions inevitably affects his score." He continued, "Zeal, industry, good will, emotional stability, scholastic information, the accident of social class, the circumstances of sex—each and all of these irrelevant influences, in one case propitious, in another prejudicial, improve or impair the final result. . . . The examiner . . . who notes in the child but the one quality he

means to measure, and ignores the many ac- cidents which embarrass its manifestation, will expose his measurement to the jeopardy of gross distortion. He is like a chemist who weighs salts in a bottle without heeding the weight of the bottle itself." It was obvious to Burt that "in a valid measure of pure intelli- gence accidents of opportunity should have no weight." Given what might influence per- formance in the Binet-Simon test, he urged awarding a "large allowance to the poorer child before we permit ourselves to accept his weak performance as the sign and seal of mental defect." Burt, *Mental and Scholastic Tests*, p. 175, 198–99; Walter Lippmann, "Mr. Burt and the Intelligence Tests," *New Re- public*, May 2, 1923, pp. 263–64; Walter Lipp- mann, "Rich and Poor Girls and Boys," *New Republic*, May 9, 1923, pp. 295–96.

38. Caleb W. Saleeby, *Progress of Eugen- ics* (Funk & Wagnalls, 1914), pp. 145–46; Ig- natius W. Cox, "The Folly of Human Sterili- zation," *Scientific American*, 151 (Oct. 1934), 189. The story of Elizabeth Tuttle and her descendants was, ironically enough, intro- duced into the literature of eugenics by Charles B. Davenport, *Heredity in Relation to Eugenics* (Henry Holt, 1911), pp. 225–28. For comment on the way the story was used by mainliners, see Jacob H. Landman, *Human Sterilization* (Macmillan, 1932), pp. 188–89.

39. H. G. Wells, "Discussion of Gal- ton's Paper, 1904," Galton Papers, file 138/9; Huxley, "Eugenics and Society," p. 60.

40. Raymond Pearl, "The Biology of

Superiority," *American Mercury*, 12 (Nov. 1927), pp. 266, 258–59; Jennings, *Prometheus*, pp. 17–19.

41. Jennings, *Prometheus*, pp. 17–19.

42. Jennings added that so long as gen- erally random sexual reproduction continued in man, there would continue the "surprises, the perplexities, the melodrama, that now present themselves among the fruits of the human vine. . . . Capitalists will continue to produce artists, poets, socialists, and labour- ers; labouring men will give birth to capital- ists, to philosophers, to men of science; fools will produce wise men and wise men will produce fools. Who mounts will fall, who falls will mount; and all the kinds of problems presented to society by the turn of the invisi- ble wheel will remain." Jennings, *Prome- theus*, pp. 75–78, 85–86.

43. *Ibid.*, pp. 79–81; H. L. Mencken, *Prejudices, Sixth Series* (Alfred A. Knopf, 1927), p. 233; Louis Trenchard More, "The Scientific Claims of Eugenics," *The Hibbert Journal*, 13 (Jan. 1915), 358–59. Clarence Dar- row found himself "alarmed at the conceit and sureness of the advocates of this new [eugenic] dream." He added, "I shudder at their ruthlessness in meddling with life. I re- sent their egoistic and stern righteousness. I shrink from their judgment of their fellows." Clarence Darrow, "The Eugenics Cult," *American Mercury*, 8 (June 1926), 135.

44. "A Knockout for the 'Perfect Man,'" *The Literary Digest*, 114 (Oct. 1, 1932), 20.

Chapter X:

LIONEL PENROSE AND THE COLCHESTER SURVEY

1. "Report of the Division of Mental Deficiency, 1919," pp. 70–73, National Com- mittee for Mental Hygiene, Charles B. Dav- enport Papers, Cold Spring Harbor Series #2.

2. J. E. Wallace Wallin, "Who Is Fee- ble-Minded? A Reply to Mr. Kohs," *Journal of Criminal Law*, 6 (March–July 1916), 72–73; Mary Dendy to Karl Pearson, Oct. 26, 1913, Karl Pearson Papers, Cabinet IV, D8; Ste- phen Jay Gould, *The Mismeasure of Man* (W. W. Norton, 1981), pp. 172–74. Edgar A. Doll,

a psychologist who had spent some years with Goddard at Vineland, held in 1917 that an intelligence test did not in itself distinguish between "developmental and degenerative defects, although this distinction is essential to prognosis; . . . between the superficially intelligent feeble-minded and the intellectu- ally stupid normals, although this distinction is essential to the estimation of social compe- tence; . . . among young children . . . between potential defectiveness and potential normal- ity, although this distinction is essential to early recognition; . . . between high-grade borderline defectives and low-grade borderline normals, although this distinction is essential to diagnosis." Necessary to a reliable determination of mental deficiency, Doll insisted, was an assessment of the sum total of mental symptoms. E. A. Doll, "On the Use of the Term 'Feeble-Minded,' " *Journal of Criminal Law*, 8 (July 1917), 216–17.

3. Wallin, "Who Is Feeble-Minded?" p. 77. In the thirties, a committee of the American Neurological Association headed by Abraham Myerson—he was a distinguished Boston psychiatrist who had long held that most mental disabilities were not hereditary —authoritatively concluded that the type of work carried out from the Jukes through the Nams and Kallikaks and on through the subjects of the Eugenics Record Office added up merely to historical rather than scientific interest, not least because the investigative techniques were open to "serious and destructive criticism. The work has largely been done by field workers who have passed judgment rather too glibly and surely on people dead three, four or five generations, concerning whom no real authentic information could be obtained." The report sharply criticized the typical presumption that someone dead for a generation or two could be certainly known to have suffered from syphilis or that the diagnosis of moron could be made by brief inspection. Abraham Myerson et al., *Eugenical Sterilization: A Reorientation of the Problem* (Committee of the American Neurological Association for the Investigation of Eugenical Sterilization; Macmillan, 1936), p. 117.

4. David Heron, *Mendelism and the Problem of Mental Defect* (Questions of the Day and the Fray; vol. VII; Dulau, 1913), p. 50; Bateson to C. C. Hurst, Oct. 19, 1911, March 2, 1912, Dec. 3, 1912, C. C. Hurst Papers, Cambridge University Library, add. 7955/11/7 and 7955/12/1.

5. Herbert Spencer Jennings, *Prometheus, or Biology and the Advancement of Man* (E. P. Dutton, 1925), pp. 24–25, 26–28.

6. Lionel Penrose, "E. O. Lewis," typescript, 1966; "Edmund Oliver Lewis," *The Lancet*, Aug. 21, 1965, p. 395, copies in Lionel S. Penrose Papers, University College London Archives, file 147/3; *Report of the Mental Deficiency Committee, being a Joint*

Committee of the Board of Education and Board of Control (3 vols.; His Majesty's Stationer's Office, 1929), I, 4; III, 44–45, 51, 135–36; C. P. Blacker, memorandum, "The Social Problem Group," attached to Blacker to Penrose, Oct. 1, 1931, Lionel S. Penrose Papers, file 130/9.

7. *Report of the Mental Deficiency Committee*, III, v, 71–73, 77; I, 161, 81.

8. At the time, according to Arthur F. Tredgold, who to the medical and lay public was the leading British authority in the field, primary amentia accounted for about eighty percent of the mentally deficient. No genetic naïf, Tredgold attributed primary mental deficiency to the inheritance not of a single gene but of a *neuropathic diathesis*, or constitution, transmitted, in the explanation of a contemporary review of the field, by "some special pathological germinal material," a "hypothetical substance . . . inherited in some direct, but not very regular manner from parent to offspring, rather like the inheritance of personal property." Lionel S. Penrose, *Mental Defect* (Farrar & Rinehart, 1933), pp. 6–7, 87–88; *Report of the Mental Deficiency Committee*, III, 135–36.

9. Ruth Darwin to Walter Fletcher, June 1, 1929; Hubert Bond, "Application by the Governing Body of the Darwin Trust . . .," March 24, 1930; Ruth Darwin to Walter Fletcher, April 4, 1930, Medical Research Council Records, Medical Research Council, London, England, Darwin Trust file 1588/1; Mental Disorders, Colchester, file 1588, folder I; *Report of the Mental Deficiency Committee*, III, 68, 9.

10. A.L.T., memo, May 12, 1930; Medical Superintendent, Colchester, to Secretary, Medical Research Council, Nov. 7, 1930; Fletcher to Darwin, May 16, 1930; "Appointment of Research Medical Officer, Particulars," June 24, 1930; Ruth Darwin to Walter Fletcher, Oct. 28, 1930, Medical Research Council Records, Mental Disorders, Colchester, file 1588, folder I

11. Penrose, "Personal Records of Fellows of the Royal Society"; "Suggested Amendments to Burke's Landed Gentry," 1952; Lionel S. Penrose Papers, file 20/4, 1/5; Haldane to the Principal, Ruskin College, Oxford, March 12, 1953, J. B. S. Haldane Papers, University College London, file 2. Penrose's wife's father was a professor of physiol-

ogy at Sheffield and a Fellow of the Royal Society. Of the four children produced by the Lionel Penroses, two are professors of mathematics, one a genetic pediatrician, and the other a ten-time British chess champion.

12. "The Peckovers of Wisbech, A Family of Quakers: Short Biographical Notes Compiled Nov. 1972, P.H.C."; photograph book, Lionel S. Penrose Papers, file 1/2; interview with Sir Roland Penrose. Lionel's wife once likened the Penrose family's Quakerism to "the very oxygen they breathed." Margaret Penrose Newman, "The Peckovers of Wisbech," unpublished talk, 1980, pp. 1, 2.

13. Roland Penrose, *Scrapbook, 1900–1981* (Thames & Hudson, 1981), p. 20. The puritanical self-denial of Penrose's Peckover mother was reinforced by the Quaker devotion of the father, who diverged from portraiture long enough to produce two spiritual pictures that now hang in honor in the London Friends headquarters. One, "None Shall Make Them Afraid," depicts a Quaker meeting, serene in the face of an invasion by a band of American Indian braves armed with tomahawks, bows, and arrows. From my firsthand look at the paintings, Friends Institute, Euston Road, London.

14. Interview with Sir Roland Penrose; interviews with Margaret Penrose Newman, Lionel's wife, who married the late Max Newman, a mathematician, after Lionel's death, and with Max Newman. Lionel Penrose, "Personal Records of Fellows of the Royal Society," Lionel S. Penrose Papers, file 20/4.

15. Herbert W. Jones to Mr. Penrose, April 9, 1912; Penrose, "Exercise Book, Scriptures, 1913"; "Report of Studies, 3rd term, 1915"; "Lionel S. Penrose, 1898–1972," *M.A.P.W. Proceedings*, vol. II, part 5; Penrose, "Memoirs: Lunacy," Lionel S. Penrose Papers, file 3/3, 3/2, 20/2, 20/5.

16. Interview with Sir Roland Penrose; interviews with Margaret Penrose Newman and Max Newman; Penrose, "Why Religion Must Die Out," n.d. [1920s], Lionel S. Penrose Papers, 35/6; interview with Shirley Penrose Hodgson.

17. Penrose, "Memoirs: Lunacy." Penrose once described philosophers as belonging "to that class of individuals who spend their time in useless thinking rather than in useless acting." Penrose, "A Train Journey,"

n.d. [1920s], Lionel S. Penrose Papers, file 20/2, 35/5.

18. Penrose, "Memoirs: Lunacy."

19. Interviews with Margaret Penrose Newman and Max Newman; Margaret Gardiner, "Modest Proposals," *Quarto*, Dec. 1980, p. 6; Penrose, "Notes," c. 1923–24, Lionel S. Penrose Papers, file 22; interview with Shirley Penrose Hodgson; interview with Roger Penrose. The discontent with Freudian psychoanalysis was reinforced by Margaret Leathes, his future wife, whom he met while climbing in the Austrian Alps. A medical student at the time, and a woman of a strongly independent and irreverent cast of mind, she perceived a streak of self-importance in the psychoanalytic fraternity. When back in London, accompanying Lionel to meetings of Ernest Jones's Psycho-Analytic Institute, she was struck by Jones's custom of seating himself beneath a wall picture of Freud, giving him the appearance of an anointed monarch on a throne. She thought that Lionel would have made a terrible analyst because he was too absorbed in his own ideas to pay much attention to anyone else's. Interview with Margaret Penrose Newman and Max Newman.

20. Penrose, "Personal Records of Fellows of the Royal Society," Lionel S. Penrose Papers, file 20/4; Penrose, "Thesis for M.D. Degree," University of Cambridge, May 14, 1930, copy in possession of Harry Harris; interview with Roger Penrose; interview with Margaret Penrose Newman. Penrose once mentioned the love affair to one of his sons, leaving the details obscure, and Margaret Penrose Newman recalled that the woman in question had to be committed to a mental hospital.

21. Penrose, "Biographical Résumé" [c. 1944], Lionel S. Penrose Papers, file 49/1; "Abstract of Applications of 10 Candidates Selected for Interview on 28 Oct. 1930," Medical Research Council Records, file 1588, folder I.

22. Interviews with Roger Penrose, Shirley Penrose Hodgson, Margaret Penrose Newman, and Max Newman; Penrose, "Paradoxes and the Theory of Types." In the fifties, with his son Roger, Penrose executed various optical deceptions, one of which became the basis for the Dutch graphic illusionist M. C. Escher's renowned impossible stair-

case. Penrose to M. C. Escher, Dec. 7, 1961, Lionel S. Penrose Papers, file 22, 158/3; Margaret Gardiner, "Modest Proposals," p. 6; Martin Gardner, "Mathematical Games: The Eerie Mathematical Art of Maurits C. Escher," *Scientific American*, 214 (April 1966), 113–14.

23. "Sublimation, sublimation," Margaret recently reflected. "That was the key to Lionel." Interview with Margaret Penrose Newman and Max Newman; Margaret Gardiner, "Modest Proposals," p. 6.

24. Penrose notes, "What Certain People Said When They First Encountered the Self-Reproducing Machine in One Form or Another"; Francis Crick to Penrose, Feb. 10, 1958, Lionel S. Penrose Papers, file 51/16, 51/18.

25. Haldane to the Principal, Ruskin College, Oxford, March 12, 1953, Haldane Papers, University College London, file 2; Penrose, "Exercise Book, Scriptures, 1913," Lionel S. Penrose Papers, file 3/2; interviews with Sir Roland Penrose, Margaret Penrose Newman, and Harry Harris. Among his diversions after that war were mathematical essays in political decision-making—they were aimed at the resolution of conflict among the family of nations—and in the psychology of crowd behavior, which he considered a microcosmic form of national belligerency. Harry Harris, "Lionel Sharples Penrose," *Biographical Memoirs of Fellows of the Royal Society*, 1973, p. 549.

26. Interview with Margaret Penrose Newman and Max Newman; Penrose, *Mental Defect*, p. 6; Penrose, "The Inheritance of Mental Characters" [c. 1933–34], pp. 14–16, Lionel S. Penrose Papers, file 53/4. With typically dry understatement, he observed: "Human society has, from very early times, been built up on the assumption that sons are by nature equipped to occupy the same social positions as their fathers. The belief, though often true, has been shown again and again to be quite false." Lionel S. Penrose, *The Influence of Heredity on Disease* (H. K. Lewis, 1934), p. 4.

27. Penrose, "The Inheritance of Mental Characters," pp. 8–9.

28. Interview with Shirley Penrose Hodgson; Penrose, *Mental Defect*, pp. 141–42, 172–74.

29. Penrose, "Contribution to Eugenics Society Symposium, Jan. 1947," Lionel S. Penrose Papers, file 65/4; "Intelligence: Discussion between Cyril Burt and Lionel Penrose . . .," BBC, Sept. 18, 1948, transcript, Lionel S. Penrose Papers, file 53/12; Penrose, "Memorandum to UNESCO Conference on Differential Fertility and Its Effect on the Intelligence of the Population" [1952?], Penrose Papers, file 65/2. Drawing upon the polygenic theory for graded characters such as intelligence, Penrose explained the seeming constancy of the average intelligence level to Burt and the listeners of the BBC in 1948: "Although the people who are slightly subnormal in intellect may be the most fertile in the community, as you come down the scale, a point is reached where the fertility grows less and less and finally disappears. . . . It seems probable that both extremes of the intelligence scale, high and low, have always been relatively infertile. These extremes, however, are continually being replaced, and the high fertility of a group at one particular level is probably just what is required to keep the equilibrium of the genetical or inborn qualities of the population. And that is why . . . people of high intelligence, who are supposed to be dying out, are all the time having their numbers supplemented by recruits from the children of fertile groups whose average intelligence is much lower on tests." Penrose, it seems, was one of the first to suggest that children in large families tended to score lower on I.Q. tests than those in smaller ones because they tended to talk more to each other than to their parents or other adults. On this model, the negative correlation between I.Q. and family size had nothing to do with genetics. Interview with Cedric Carter.

30. Bespeaking the mysteriousness of mental deficiency, even to people close to it, when Penrose arrived at Colchester, Mrs. Frank Turner, the wife of the superintendent and a hefty woman, bustled into his office. Penrose remembered her abruptly declaring, "So you have come here to find out the causes of mental deficiency? . . . My son was an imbecile. Dr. Turner doesn't like to talk about it so that is why I'm telling you now. There is nothing wrong with my husband's family, or with mine. Perhaps you will find out something about mental deficiency." Mrs. Turner soon hurried off, though not before adding, "Don't let your wife come

near the hospital if she is pregnant." Penrose, "Memoirs—1964: Phenylketonuria," Lionel S. Penrose Papers, file 72/2.

31. *Ibid.*

32. Penrose to Secretary, Pinsent-Darwin Studentship, Dec. 4, 1931, Lionel S. Penrose Papers, file 149/4; Appointment of Research Medical Officer [Colchester], Particulars," Lionel S. Penrose Papers, file 47. Edmund O. Lewis, who took a continuing interest in Penrose's work, was unable to provide detailed guidance. His inquiry for the Mental Deficiency Committee had been confidential in nature, and he was prohibited from sharing his case notes. But from the outset, Lewis was a storehouse of advisory hints. He stressed in particular that it was essential for Penrose to classify the Colchester patients, to differentiate among the various types of deficiency however he could, physically, clinically, or otherwise. Penrose, "E. O. Lewis," 1966, Lionel S. Penrose Papers, file 147/3.

33. Penrose, *Mental Defect*, pp. 85, 93–94, vii; Penrose, "Report of the Research Department of the Royal Eastern Counties' Institution," Sept. 1932, Lionel S. Penrose Papers, file 56/2.

34. Penrose, *Mental Defect*, pp. 91–93; Penrose, "Report of the Research Department of the Royal Eastern Counties' Institution," Sept. 1932.

35. Penrose, *Mental Defect*, pp. 194, 94–95, 90; Penrose, "For Mental Welfare" [1932], Lionel S. Penrose Papers, file 56/1.

36. Penrose, "Report of the Research Department, Royal Eastern Counties' Institution," Sept. 1932; Penrose to Charles Blacker, Oct. 2, 1931; Penrose to Edmund O. Lewis, Jan. 30, 1932; Penrose to Dr. MacCurdy, Dec. 23, 1931, Lionel S. Penrose Papers, files 56/2, 130/9, 147/3, file 149/4; Penrose, *Mental Defect*, pp. 58–60. The magnitude of the job was manifest in the sheer numbers involved—in the end, 1,280 patients, who had 6,629 siblings, not to mention some thousands more parents and collateral relatives. Lionel S. Penrose, *A Clinical and Genetic Study of 1280 Cases of Mental Defect [The Colchester Survey]* (Medical Research Council Special Report Series 229; His Majesty's Stationer's Office, 1938), p. 2.

37. Penrose, *Mental Defect*, pp. 58–59, 180, 194; Penrose, "Report of the Research

Department, Royal Eastern Counties' Institution," Sept. 1932.

38. Haldane to Penrose, Nov. 1934; Penrose to Frederick Gowland Hopkins, Nov. 23, 1934; Penrose to Asbjörn Fölling, June 15, 1936; Penrose to Haldane, April 3, 1935, Lionel S. Penrose Papers, files 136, 139/8, 132/4; Penrose, "Memoirs—1964: Phenylketonuria."

39. Penrose, "Memoirs—1964: Phenylketonuria."

40. Penrose, "Report of the Research Department of the Royal Eastern Counties' Institution," Sept. 1932.

41. J. Langdon Down, "Observations on an Ethnic Classification of Idiots," *London Hospital Reports*, 1866, reprinted in J. Langdon Down, *On Some of the Mental Affections of Childhood and Youth* (J. A. Churchill, 1887), pp. 210–17 (the speculation as to the causes of the syndrome are on pp. 58–63); Beth C. Kevles, "A History of Our Understanding of Down's Syndrome" (unpublished manuscript, 1981).

42. F. G. Crookshank, *The Mongol in Our Midst: A Study of Man and His Three Faces* (E. P. Dutton, 1924), pp. 5–6; Gould, *The Mismeasure of Man*, pp. 134–35.

43. Penrose, "Report of the Research Department of the Royal Eastern Counties' Institution," Sept. 1932; Turner to Penrose, July 3, 1931, Jan. 5, 1932; Penrose to Turner, Jan. 4, 1932; and Essex Voluntary Association for Mental Welfare materials, Lionel S. Penrose Papers, file 130/7. In an early attempt to specify Down's syndrome quantitatively, Penrose took a group of characteristics—very low I.Q., high cephalic index, epicanthic fold on either eye, fissured tongue, and the so-called simian crease, a pronounced transverse palm line—and compared the frequency of their association in fifty patients already classified clinically as mongolian imbeciles with the associational frequency among other victims of mental deficiency. Almost three-quarters of the Down's group had three or more of the characters, while only six out of three hundred and fifty unselected patients had as many. Penrose concluded that "any defective with four or more of these characters is almost certainly a mongol." Penrose, "A Contribution to the Theory of Definition—with special reference to Mongolism," unpublished paper, c. 1932, Lionel S. Penrose Papers, file 61/1.

44. Penrose to Edmund O. Lewis, Feb. 8, 1932, Lionel S. Penrose Papers, file 147/3; Penrose to Peter K. McCowan, Feb. 13, 1932, Lionel S. Penrose Papers, file 149/3; L. S. Penrose, "The Blood Grouping of Mongolian Imbeciles," *The Lancet*, 7 (Feb. 20, 1932), 394–95.

45. Interview with Harry Harris; Penrose, *Mental Defect*, p. 102; interview with Shirley Penrose Hodgson.

46. Penrose, draft Note to *The Lancet*, March 14, 1938, Lionel S. Penrose Papers, file 61/1; Penrose, "The Relative Aetiological Importance of Birth Order and Maternal Age in Mongolism," *Proceedings of the Royal Society*, 115 (1934), 431–32; Kevles, "A History of Our Understanding of Down's Syndrome."

47. Haldane to Egon Pearson, March 27, 1944, Lionel S. Penrose Papers, file 49/1. Penrose's statistics were a bit too lowbrow for R. A. Fisher, who dated Penrose's maternal-age paper as referee for the Royal Society. Fisher wrote to Penrose: "Your family data are much too important for you to be satisfied with an unconvincing statistical analysis. I mean, that no one reading your paper critically will feel sure that a more exact treatment would not have yielded a different result. I may add that I entirely expect your actual conclusions to be the right ones, but that is not sufficient reason why they should not be adequately established." With Fisher's help, Penrose worked out an entirely different, more elaborate approach to the statistical problem. In the end, both methods led to the same conclusions. R. A. Fisher to Penrose, Dec. 8, 1933, Lionel S. Penrose Papers, file 61/2.

48. Penrose, "Report of the Research Department of the Royal Eastern Counties' Institution," Sept. 1932; Penrose to R. A. Fisher, Nov. 2, 1932; Penrose to Frank C. Shrubsall, Aug. 19, 1933; Penrose to R. L. Jenkins, Aug. 14, 1933, Lionel S. Penrose Papers, files 56/2, 61/2, 168/7, 142/5; Penrose, "The Relative Aetiological Importance of Birth Order and Maternal Age in Mongolism,"

Proceedings of the Royal Society, 115 (1934), 431–50.

49. Haldane recognized the flaw in Penrose's argument: "It is admitted that one factor in the causation is abnormal uterine environment. If therefore several sibs are affected an obvious explanation is that the uterine environment, if abnormal during one pregnancy, is more likely 'than would be suspected from chance occurrence' to be abnormal during another pregnancy of the same woman." Haldane thought that the twin evidence might be good, if Penrose had enough of it. But Haldane doubted "'well known facts' in human genetics! Aunt Jobiska's theorem"—relying on what the world in general knows, the apothegm in Edward Lear's verse about the pobble who had no toes—"is of greater value in theology than biology." J. B. S. Haldane to Penrose, Sept. 22, 1934, Lionel S. Penrose Papers, file 136; Penrose, *Mental Defect*, p. 107; Penrose, "Mongolism," May 27, 1960, Lionel S. Penrose Papers, file 62/5.

50. Penrose, *A Clinical and Genetic Study of 1280 Cases of Mental Defect* [*The Colchester Survey*], pp. 69–70, 60–62.

51. Penrose to Ruth Darwin, Aug. 31, 1931, Lionel S. Penrose Papers, file 164/3; E. O. Lewis to David Munro, Sept. 18, 1937, Medical Research Council Records, Mental Disorders, Colchester, file 1588, folder III. Bespeaking the steadily increasing respect in which Penrose's research was held, additional funds for the work came from the Medical Research Council and the Rockefeller Foundation (not to mention his maiden aunt Alexandrina Peckover, who contributed a handsome sum to build a clinical laboratory at the hospital). Memo, Jan. 18, 1937, Medical Research Council Records, L. S. Penrose, file P.F.236; Ruth Darwin to D. O'Brien, Aug. 13, 1936; Norman S. Thompson to Edward Mellanby, April 17, 1937; extract from *East Anglian Times*, Aug. 8, 1936, Medical Research Council Records, Mental Disorders, Colchester, file 1588, folders II and III.

Chapter XI:
A REFORM EUGENICS

1. Hermann J. Muller, *Out of the Night: A Biologist's View of the Future* (Vanguard Press, 1935), pp. ix–x; J. H. Landman, "Race Betterment by Human Sterilization," *Scientific American*, 150 (June 1934), 293. In 1931, *The New Statesman and Nation* editorialized: "Voltaire once remarked about the Jews that it might be true that they were all the unpleasant things they were said to be, but it was not necessary to burn them. We are not sure that all the things that are said about mental defectives are true, and it may not be necessary to sterilize them." "Sterilization of Defectives," *The New Statesman and Nation*, 2 (July 25, 1931), 102.

2. Reginald C. Punnett, "Eliminating Feeblemindedness," *Journal of Heredity*, 8 (1917), 464–65.

3. Using the same unit-character assumption as Punnett, Fisher calculated that the reduction in a single generation would, starting from an initial frequency of 100 per 10,000 of population, come to seventeen percent; or of 30 per 10,000, eleven percent. R. A. Fisher, "Elimination of Mental Defect," *Eugenics Review*, 16 (July 1924), 114–16, reprinted in *Journal of Heredity*, 18 (1927), 529–31.

4. George J. Hecht to Jennings, Feb. 14, 1931, Herbert S. Jennings Papers, Section B, J44, Hecht file; Herbert S. Jennings, *The Biological Basis of Human Nature* (W. W. Norton, 1930), pp. 241–43; Edward M. East, "Hidden Feeblemindedness," *Journal of Heredity*, 8 (1917), 215–17; editorial, New York *Times*, June 22, 1932, quoted in *The Literary Digest*, 114 (Aug. 6, 1932), 16.

5. Paul Popenoe and E. S. Gosney, *Twenty-eight Years of Sterilization in California* (Human Betterment Foundation, 1939), p. 24; J. B. S. Haldane, *Human Biology and Politics* (The Science Guild, 1934), p. 13; Jennings, *Biological Basis of Human Nature*, p. 218. Fisher confused the matter further with the confession to Leonard Darwin that his own estimate in rebuttal to Punnett had been merely an academic calculation. Blacker to Huxley, Oct. 24, 1930, Eugenics Society Records, Wellcome Institute for the History of Medicine, London, file C. 185.

6. Abraham Myerson to *The New York Times*, March 15, 1936, IV, 9; Abraham Myerson et al., *Eugenical Sterilization: A Reorientation of the Problem* (Committee of the American Neurological Association for the Investigation of Eugenical Sterilization; Macmillan, 1936), pp. 56–58; *Report of the Departmental Committee on Sterilisation (minus appendices)* (pamphlet; Joint Committee on Voluntary Sterilisation, June 1934), pp. 17–18 (hereafter cited as [*Brock*] *Report . . . on Sterilisation*).

7. [*Brock*] *Report . . . on Sterilisation*, pp. 37–38; Myerson et al., *Eugenical Sterilization*, pp. 177–83.

8. [*Brock*] *Report . . . on Sterilisation*, pp. 37–40, 59–61; Myerson et al., *Eugenical Sterilization*, pp. 177–83.

9. *Hansard*, vol. 286, Feb. 28, 1934, column 1242; Eugenics Society, *Annual Report*, *1930/31*, p. 7; *1933/34*, p. 3; C. P. Blacker, *Voluntary Sterilization* (pamphlet; reprinted from the *Eugenics Review*, 1961–62), pp. 9–12, 14–15, 19; C. P. Blacker to R. A. Fisher, Dec. 14, 1933, Eugenics Society Records, file C.108; Eugenics Society Press Clippings, Eugenics Society, London, Sterilization, 1931; *Report of the Mental Deficiency Committee, being a Joint Committee of the Board of Education and Board of Control* (3 vols.; His Majesty's Stationer's Office, 1929), II, 88; Lionel S. Penrose, *Mental Defect* (Farrar & Rinehart, 1933), pp. 192–93; [*Brock*] *Report . . . on Sterilisation*, pp. 31–32; Committee for Legalising Voluntary Sterilisation, *Eugenic Sterilization* (2nd ed.), pp. 23, 24.

10. H. G. Wells, Julian S. Huxley, and G. P. Wells, *The Science of Life* (Doubleday, Doran, 1931), pp. 1467, 1468; Committee for Legalising Voluntary Sterilisation, *Better Unborn* (pamphlet; Eugenics Society, 193?), p. 10; Blacker, *Voluntary Sterilization*, pp. 17–19.

11. Blacker, *Voluntary Sterilization*, p. 145; J. B. S. Haldane, *Heredity and Politics* (W. W. Norton, 1938), p. 8; J. B. S. Haldane, *Science and Everyday Life* (Lawrence and Wishart, 1939), p. 261. *Nature* had observed that, under the prevailing circumstances, sterilization was "a luxury for the rich, since State,

municipal and charitable institutions dare not do it," adding, "So it is that the well-to-do, through voluntary sterilisation, are preventing the repetition of hereditary blunders, whilst the poor, who outnumber them, cannot imitate them even though they would." "The Legality of Sterilization," *Nature*, 130 (Dec. 10, 1937), 863.

12. Blacker, *Voluntary Sterilization*, pp. 13–14; Haldane, *Heredity and Politics*, pp. 108–9; Haldane, *Human Biology and Politics*, p. 13; Haldane, "Eugenics and Social Reform," in J. B. S. Haldane, *Possible Worlds* (Harper & Bros., 1928), p. 201. Early on during the legislative initiative, legalization was denounced on the floor of the House of Commons as anti–working class. The Social Problem Group might contain numerous "feebleminded," but, said Hyacinth B. W. Morgan, a physician and Labour Member of Parliament, "there is nothing wrong with the germ plasm itself. At the bottom, mental deficiency"—at least much of the sort that absorbed so many eugenicists—"is an economic problem." *Hansard*, vol. 255, July 21, 1931, columns 1251–56.

13. G. B. Arnold, "A Brief Review of the First Thousand Patients Eugenically Sterilized at the State Colony for Epileptics and Feebleminded," *American Association on Mental Deficiency, Proceedings*, 43 (1937–38), 61; Richmond *Times-Dispatch*, Feb. 23, 1980; Popenoe and Gosney, *Twenty-eight Years of Sterilization in California*, pp. 9–10; Myerson et al., *Eugenic Sterilization*, p. 9; E. S. Gosney and Paul Popenoe, *Sterilization for Human Betterment: A Summary of Results of 6,000 Operations in California* (Macmillan, 1929), pp. 50–51; Judith K. Grether, *Sterilization and Eugenics* (University Microfilms, 1980), pp. 121–22.

14. Thomas J. Gerrard, *The Church and Eugenics* (B. Herder, 1912), p. 36; *National Cyclopedia of American Biography*, vol. A, Butler entry; Caroline Maule to Blacker, Oct. 3, 1935, Eugenics Society Records, file D.150.

15. [*Brock*] *Report . . . on Sterilisation*, p. 33; Benjamin Malzberg, "Notes on Sterilization and Social Control," *Social Forces*, 8 (March 1930), 400–1; John A. Ryan, *Moral Aspects of Sterilization* (Problems of Mental Deficiency No. 3; National Catholic Welfare

Conference, 1930), p. 7; Gosney and Popenoe, *Sterilization for Human Betterment*, pp. 186–87; John A. Ryan, "Futile Immorality," *Forum*, 94 (July 1935), 31; *The New York Times*, Jan. 9, 1931, pp. 14–15; Father Ignatius W. Cox, S.J., Ph.D., explained in 1934: "Sterilization is a mechanism to allow use of the sex faculty, while frustrating positively the primary purpose of finality given to it by Supreme Intelligence. One may legitimately refrain from the use of a faculty, as of speech or of sex. It is quite another thing to use such faculties and in their use to provide positive means for the frustration of their primary purpose, as happens in the case of lying and sterilization." Ignatius W. Cox, "The Folly of Human Sterilization," *Scientific American*, 151 (Oct. 1934), 190.

16. New York *Times*, Oct. 23, 1933, p. 13; Penrose, *Mental Defect*, p. 170; New York *Times*, Dec. 27, 1936, IV, 8; Eugenics Society, *Annual Report, 1936/37*, p. 12.

17. Ryan, "Futile Immorality," p. 31. See also the warnings of Hyacinth Morgan in *Hansard*, vol. 255, July 21, 1931, columns 1251–56.

18. *Review of Reviews*, 94 (Aug. 1936), 48; *The New York Times*, Nov. 17, 1946, IV, 9; Dec. 6, 1942, IV, 11; June 4, 1944, p. 6. In the United States, the nationally publicized case of Ann Cooper Hewitt may have raised the general public's sensitivity to the misuses of sterilization. In 1936, Hewitt, twenty-one years old, filed a civil suit in California against her mother, two physicians, and a California state psychiatrist that demanded damages of five hundred thousand dollars for having had her sterilized. Ann's father was Peter Cooper Hewitt, son of the New York mayor, accomplished inventor, and heir to a fortune. In his will, Peter Cooper Hewitt had provided that two-thirds of the income on a trust fund of one million three hundred thousand dollars should go to his only child, Ann, but if she should die childless, that share of the income would revert to the mother, who had been Mrs. Hewitt in the second of her five marriages. According to the suit, in 1934 Mrs. Hewitt had subjected her daughter to a battery of intelligence tests when she was ill with appendicitis. A psychiatrist in the California State Department of Public Health declared that Ann had a mental age of eleven,

making her a high-grade moron. With this determination in hand, the suit charged, Mrs. Hewitt had Ann sterilized, in August, eleven months before her twenty-first birthday, while the girl underwent an emergency appendectomy. Aaron Shapiro, a New York City attorney representing Mrs. Hewitt, said that the operation had been for "society's sake" and that evidence would be offered to show "erotic tendencies" on Ann's part. However, a New Jersey psychiatrist had tested Ann a few months before she filed the suit and found her free of any mental deficiency. Ann's suit charged that before the operation her mother had already squandered hundreds of thousands of her daughter's expected dollars and had authorized the sterilization out of greed for the rest of the trust fund. "I had no dolls when I was little, and I'll have no children when I'm old," Ann said. "That's my story. That's all there is to it."

On February 4, 1936, Judge Sylvian Lazarus of San Francisco issued criminal charges of mayhem—a felony punishable by one to fourteen years in prison—against Mrs. Hewitt and the two California surgeons who had performed the sterilization operation. Judge Lazarus explained: "The question is whether mothers, fathers or guardians can set themselves up to say whether minors should be sterilized. There are dangers in such a thing." The two physicians were brought to trial, but Judge Raglan Tuttle dismissed the charges against them in August on grounds that they had not committed a criminal act, since Ann had been under age at the time of the operation and her mother had authorized the procedure. Mrs. Hewitt, for her part, had apparently attempted suicide by an overdose of narcotics in mid-February and had been too ill to leave the Jersey City, New Jersey, Medical Center until early July. She faced extradition to California to answer the mayhem charges there when, in mid-December 1936, District Attorney Matthew Brady of San Francisco announced that he was dropping the request for extradition because Ann Cooper Hewitt had said that she would not testify against her mother.

Mrs. Hewitt, though still subject, by order of another San Francisco judge, to the mayhem charges lodged against her, was no longer a fugitive from California criminal justice. And in June 1936 Ann had received an out-of-court settlement of $150,000 against a bond posted by her mother to cover funds that she had allegedly misappropriated from her daughter's income. However, Mrs. Hewitt still faced the half-million-dollar civil-damages suit that Ann had filed, and in June 1937 a San Francisco judge ruled that Mrs. Hewitt had to defend herself against it. In April 1938, Mrs. Hewitt filed for bankruptcy, listing $250 in assets and $693,000 in liabilities, including the $500,000 demanded in Ann's suit. The suit may have been settled out of court by the time Mrs. Hewitt died on April 30, 1939, at the age of 55. *Time,* 27 (Jan. 1936), 42, 44, 46; *Newsweek,* 7 (Jan. 18, 1936), 26; *The New York Times,* Jan. 7, 1936, p. 3; Jan. 8, 1936, p. 3; Jan. 10, 1936, p. 2; Feb. 5, 1936, p. 3; Feb. 29, 1936, p. 1; April 1, 1936, p. 51; June 24, 1936, p. 48; July 3, 1936, p. 23; Aug. 20, 1936, p. 12; Aug. 25, 1936, p. 24; Sept. 1, 1936, p. 44; Oct. 4, 1936, p. 3; Dec. 8, 1936, p. 27; Dec. 16, 1936, p. 2; Dec. 24, 1936, p. 13; Jan. 1, 1937, p. 18; June 6, 1937, II, p. 2; April 8, 1938, p. 19; Dec. 17, 1938, p. 7; May 1, 1939, p. 42.

19. *The New York Times,* Feb. 28, 1946, p. 2; Jan. 29, 1946, p. 4; Feb. 23, 1946, p. 8.

20. Blacker, *Voluntary Sterilization,* pp. 9–10, 22, 146; Marian S. Olden, *Human Betterment Was Our Goal* (Princeton, N.J., 1970?), pp. 39–40, 74–84, 93. The sterilization figures have been calculated from the data in Jonas Robitscher, ed., *Eugenic Sterilization* (Charles C Thomas, 1973), appendices I and II. Catholics found willing political allies. In 1935, in Protestant Alabama, Governor Bibb Graves vetoed a sterilization bill. The measure, said Governor Graves, would result in the death of many women "who have committed no offense against God or man save that, in the opinion of the experts, they should never have been born." *The New York Times,* Sept. 5, 1935, p. 17. Though the lower courts, acting under the authority of *Buck* v. *Bell,* had kept upholding state sterilization statutes as constitutional, the climate of judicial opinion began to show evidence of change, particularly after the 1942 decision of the U.S. Supreme Court in *Skinner* v. *State of Oklahoma.* Jack T. Skinner had been convicted twice for chicken-stealing and a third time for armed robbery. He was to be steril-

ized under a 1935 Oklahoma law that authorized the procedure for anyone convicted of a felony three times. But the law exempted convictions for embezzlement, liquor law violations, and political offenses. The Supreme Court struck down the Oklahoma statute unanimously. The author of the Court's ruling was Justice William O. Douglas, who stressed that the law was unfair because it distinguished between people who violated chicken coops and those who appropriated financial property. In Douglas's opinion, "When the law lays an unequal hand on those who have committed intrinsically the same quality of offense and sterilizes one and not the other, it has made as invidious a discrimination as if it had selected a particular race or nationality for oppressive treatment." *Skinner* v. *State of Oklahoma*, 62 *Supreme Court Reporter*, 316 U.S., pp. 1110–14; "Sterilization of Criminals," *American Law Reports*, 3rd series, 53 (1973), 964, 965.

21. Frederick Osborn, "History of the American Eugenics Society," Jan. 20, 1971, p. 3, copy in American Eugenics Society Papers, American Philosophical Society Library, box 15. Some of the enthusiasts continued to adhere to the racist version of the mainline creed. Typical were George Pitt-Rivers, former secretary of the International Federation of Eugenic Societies, an open anti-Semite who ultimately joined the British Union of Fascists; the geneticist R. Ruggles Gates, who denounced as mere propaganda Julian S. Huxley and A. C. Haddon's *We Europeans;* Harry H. Laughlin, who accepted an award from the Nazis for his contributions to racial hygiene; and Dr. Clarence G. Campbell, Manhattan socialite and president of the American Eugenics Research Association, who, in Berlin in 1935, at the close of the World Population Congress, applauded Nazi racial policies with the toast: "To that great leader, Adolf Hitler!" But among eugenic enthusiasts, such racists constituted a rapidly diminishing minority, most of them isolated on the far political right. G. Pitt-Rivers to Ruggles Gates, April 3, 1936; Huxley to Gates, Feb. 18, 1937, March 18, 1937, Ruggles Gates Papers, Kings College London; Donald MacKenzie, "The Development of Statistical Theory in Britain, 1865–1925: A Historical and Sociological Perspective" (doctoral thesis, University of

Edinburgh, 1977), p. 121; Richmond *Times-Dispatch*, Feb. 24, 1980; *Time*, 26 (Sept. 9, 1935), 20–21.

22. "Eugenics for Democracy," *Time*, 36 (Sept. 9, 1940), 34; Osborn to Patrick J. Kelleher, March 31, 1970; Osborn to Kenneth L. Ludmerer, Nov. 5, 1970, American Eugenics Society Papers, box 3; Charles B. Davenport to John C. Merriam, Oct. 31, 1928, Charles B. Davenport Papers, American Philosophical Society Library, Cold Spring Harbor Series #1.

23. Blacker obituaries in *Munk's Roll*, VI, 49–50 and *The Lancet*, 1 (May 10, 1975), 1096; Blacker to Ursula Grant Duff, July 21, 1936, Eugenics Society Records, file C.133; C. P. Blacker, *Human Values in Psychological Medicine* (Oxford University Press, 1933), pp. v–vii, 4, 39, 132, 175. A cousin of Blacker's was Pedro Beltrán, the fiercely independent publisher of *La Prensa*, who became premier of Peru in 1959. His father finished his life by learning Hebrew so that, when the time came, he could converse with man's Maker in His first language. H. Montgomery Hyde, *Oscar Wilde: A Biography* (Farrar Straus and Giroux, 1975), p. 333 and note 1.

24. Blacker obituary, *Munk's Roll*, VI, 49–50; C. P. Blacker, *Human Values in Psychological Medicine*, pp. v–vii, 4, 39, 132, 175; Blacker to R. H. Felix, Feb. 14, 1948, Eugenics Society Records, file C.21; Blacker, "The Somme: 15 September 1916," fragment of unpublished autobiography, and comments in letter from John Blacker to the author, April 18, 1984; Blacker to R. R. Gates, Nov. 22, 1932, Ruggles Gates Papers. Blacker also told Gates with evident exasperation: "If you collected a sufficient number of pedigrees showing any disease, mode of behavior or eccentricity, from the incidence of cancer to the tendency of male members to go into the Army or female members to get married, or to a preference for Turkish over Virginian cigarettes, or a liking for milk in after-dinner coffee, you would find that a certain proportion of these pedigrees would conform to the hypothesis of dominance. . . . However, it seems to me that we are not entitled to state that the condition in question *is* the result of a dominant gene."

25. Geoffrey R. Searle, "Eugenics and Politics in Britain in the 1930s," *Annals of Science*, 36 (1979), 162–63, 166–68; circular let-

ter on a eugenics conference, Feb. 24, 1937, Raymond Pearl Papers, American Philosophical Society Library; Leon F. Whitney, "Memoirs," p. 219, Charles B. Davenport Papers; Blacker to J. S. Huxley, May 3, 1955; Blacker to Ursula Grant Duff, Oct. 4, 1944, Eugenics Society Records, files C.185, C.134.

26. Frederick Osborn, *Preface to Eugenics* (Harper & Bros., 1940), pp. 291–92; Lord Horder, "Eugenics," *Eugenics Review*, 27 (1936), 277–78; Leonard Darwin, "Henry Twitchin," *Eugenics Review*, 22 (1930), 91–97; "Report of Activities of the American Eugenics Society, 1936–37"; "Program," American Eugenics Society Annual Meeting, Delmonico Hotel, May 7, 1936, American Eugenics Society Papers, box 6, box 1; *Eugenics Society Annual Report, 1935/36*, p. 4; *1936/37*, pp. 10–11; *1937/38*, pp. 10–11.

27. Osborn to Kenneth L. Ludmerer, Nov. 5, 1970, American Eugenics Society Papers, box 3; Joan Fisher Box, *R. A. Fisher: The Life of a Scientist* (John Wiley, 1978), pp. 282–83; A. M. Carr-Saunders to Blacker, June 6, 1932; Blacker to J. S. Huxley, Oct. 24, 1930; Penrose to Blacker, July 14, 1948, Eugenics Society Records, files C.56, C.185, C.271. The thirties, Blacker later wrote, was a "bad time for eugenists to assess their values in terms of class." C. P. Blacker, *Eugenics, Galton and After* (Duckworth, 1952), pp. 140–41.

28. Julian S. Huxley, "Foreword" to Herbert Brewer, *Eugenics and Politics* (pamphlet; Eugenics Society, 1937), copy in Eugenics Society Records, file C.44; Osborn, *Preface to Eugenics*, p. 65. "We must no more forget heredity when we are trying to improve environment than we must forget environment when trying to improve heredity," Haldane declared. From his armchair-socialist through his Communist Party days, Haldane was convinced that psychologists and geneticists might eventually be able to specify the innate inequalities of children by age seven, and sort them out accordingly for education and career. Haldane, *Heredity and Politics*, pp. 189, 193.

29. Julian S. Huxley, "Eugenics and Society," in Julian S. Huxley, *Man Stands Alone* (Harper & Bros., 1941), p. 70; Jennings, *The Biological Basis of Human Nature*, p. 245. Even Lancelot Hogben suspected that genetics might well determine the extremes of intellectual accomplishment and defect, in-

cluding the extreme of the Social Problem Group; "that there does exist a section of genetic types on the borderline of extreme defect not segregated from the rest of the community and more fertile than others of the same social grade." Lancelot Hogben, *The Nature of Living Matter* (Kegan Paul, Trench, Trubner, 1930), p. 213.

30. Frederick Osborn, "Significance of Differential Reproduction for American Educational Policy," *Social Forces*, 14 (Oct. 1935), 23–24, copy in American Eugenics Society Papers, box 15; Frederick Osborn, "Development of a Eugenic Philosophy," *American Sociological Review*, 2 (June 1937), 389; Huxley, "Eugenics and Society," pp. 40–41.

31. Osborn, "Significance of Differential Reproduction for American Educational Policy," pp. 23–24; "Eighth Annual Meeting of the American Eugenics Society," May 12, 1933; Osborn, "Eugenics in the Light of Present Day Scientific Knowledge," Dec. 11, 1937, p. 6, American Eugenics Society Papers, box 2, box 15, box 6; Osborn, *Preface to Eugenics*, pp. 133–34; "Contribution by Dr. C. P. Blacker, at Professor Penrose's Lecture," Jan. 11, 1949; Blacker to the Lord Horder, Jan. 15, 1938, Eugenics Society Records, files C.271, C.172. Osborn remarked that any eugenic program that favored one group against another would "involve many individual injustices and would be the proper target of attack by every minority." Frederick Osborn, "The Comprehensive Program of Eugenics and Its Social Implications," *Living* (Spring–Summer 1939), p. 35, copy in American Eugenics Society Papers, box 15.

32. Osborn, *Preface to Eugenics*, pp. 197–98; American Eugenics Society, "A Eugenics Program for the United States" [1938?], pp. 9–12, American Eugenics Society Papers, box 15; J. S. Huxley to Blacker, May 19, 1935; Huxley to Blacker, Sept. 5, 1933, Eugenics Society Records, file C.185; Huxley, "Eugenics and Society," pp. 71–73; Haldane, "Eugenics and Social Reform," in J. B. S. Haldane, *Possible Worlds*, p. 206; Huxley, "Eugenics and Society," pp. 63, 76, 80–81. To call people on the social bottom genetically unfit, Haldane said, was to make "the appalling assumption that society as at present constituted is perfect." Haldane, "The Possibilities of Human Evolution," in J. B. S.

Haldane, *The Inequality of Man and Other Essays* (Chatto & Windus, 1932), p. 88.

33. Raymond Pearl, *The Natural History of Population* (Oxford University Press, 1939), pp. 244, 246–48; J. B. S. Haldane, "Science and Ethics," *Harper's*, 157 (June 1928), 5. Haldane said elsewhere, "If you desire to check the increase of any population or section of a population, either massacre it or force upon it the greatest practicable amount of liberty, education, and wealth." Haldane, "Eugenics and Social Reform," p. 206. In the twenties, Blacker had disliked Marie Stopes's tying contraception to the cause of women's sexual gratification and had been distressed that some birth-control literature tended, so he thought, to the lascivious promotion of immorality. But in the thirties and forties, Blacker strengthened the Eugenics Society's connections with the British birth-control movement, including members of the Labour Party, who avidly favored the extension of birth control to their constituency. C. P. Blacker, *Birth Control and the State: A Plea and a Forecast* (Kegan Paul, Trench,

Trubner, 1926), pp. 89–95; C. P. Blacker, "Introduction," in C. P. Blacker, ed., *A Social Problem Group?* (Oxford University Press, 1937), pp. 4, 5, 11; C. P. Blacker, "Birth Control and Eugenics," *The Nineteenth Century*, 111 (April 1932), 464; Blacker to Darwin, May 3, 1938; Blacker to J. S. Huxley, Dec. 5, 1934, Eugenics Society Records, files C.172, C.185; *Eugenics Society Annual Report, 1936/1937*, p. 13. See also Osborn, *Preface to Eugenics*, p. 119; "Board of Directors Meeting of the American Eugenics Society," Jan. 9, 1935; "Conference on Eugenics and Birth Control," Jan. 28, 1938; American Eugenics Society, *A Eugenics Program for the United States* [1938?], p. 14, American Eugenics Society Papers, boxes 6, 2, 15.

34. Enid Charles, *The Twilight of Parenthood* (W. W. Norton, 1934), pp. 144–45, 120–21; Blacker, *Eugenics: Galton and After*, p. 148; Osborn, *Preface to Eugenics* (rev. ed.; Harper & Bros., 1950), p. 142.

35. Osborn, *Preface to Eugenics*, pp. 296–97.

Chapter XII:
BRAVE NEW BIOLOGY

1. Hermann J. Muller, *Out of the Night: A Biologist's View of the Future* (Vanguard Press, 1935), p. 44; Lancelot Hogben, *Nature and Nurture* (W. W. Norton, 1933), p. 89; John A. Ryle, "Medicine and Eugenics," *Eugenics Review*, 30 (1938), 14–17; Lord Horder, "Eugenics," *Eugenics Review*, 27 (1936), 278–82.

2. J. B. S. Haldane, *Heredity and Politics* (W. W. Norton, 1938), pp. 106–7; Madge Thurlow Macklin, "A Conference on Heredity as Applied to Man," *Science*, 73 (June 5, 1931), 613–14; Lionel Penrose, "Genetics in Relation to Pediatrics," July 26, 1947, Lionel Penrose Papers, University College London, file 73; F. A. E. Crew to Ruggles Gates, Oct. 2, 1930, Ruggles Gates Papers, Kings College London; Morton D. Schweitzer, "What Use Can the Physician Make of Medical Genetics?" March 31, 1941, American Eugenics Society Papers, American Philosophical Society Library, Philadelphia, Pa., box 10.

3. F. A. E. Crew to C. P. Blacker, May

6, 1930; Nov. 14, 1938; C. P. Blacker to F. A. E. Crew, March 7, 1932, Eugenics Society Records, Wellcome Institute for the History of Medicine, London, file C.79; Margaret Deacon, "The Institute of Animal Genetics at Edinburgh—the First Twenty Years" (unpublished manuscript), p. 33; J. Fraser Roberts, "Human Pathological Conditions Determined by Any One of Several Genes," *Nature*, 130 (Oct. 8, 1932), 543; J. Fraser Roberts to Landsborough Thomson, June 10, 1932; "Report of the Medical Committee on the Correspondence from the Medical Research Council's Committee on Human Genetics," May 3, 1934, Medical Research Council Records, Medical Research Council, London, file 1732/1; James V. Neel, "Human Genetics," in John Z. Bowers and Elizabeth F. Purcell, eds., *Advances in American Medicine: Essays at the Bicentennial* (2 vols.; Josiah Macy, Jr., Foundation, 1976), I, 54; Laurence H. Snyder, *Medical Genetics* (Duke University Press, 1941), pp. 3–4; C. P. Blacker, "Preface," in C. P. Blacker, ed., *The Chances of Morbid Inheritance* (William Wood, 1934), pp. v–vi; J. Fraser Roberts, *An Introduction to Medical Genetics* (Oxford University Press, 1940), p. v; F. A. E. Crew, *Genetics in Relation to Clinical Medicine* (Oliver and Boyd, 1947).

4. Lionel Penrose, "Discussion of 'Clinical and Genetic Study of 1280 Cases of Mental Defect,' " Conference of Medical Officers, July 22, 1938, Lionel Penrose Papers, University College London, file 65/1; Lancelot Hogben, *Genetic Principles in Medicine and Social Science* (Williams and Norgate, 1931), pp. 201–2; Haldane, *Heredity and Politics*, pp. 89–91; Frederick Osborn, *Preface to Eugenics* (Harper & Bros., 1940), p. 34. As for diseases that might respond to postnatal therapy, Herbert Spencer Jennings had no doubts: "far better is the condition of the race in which . . . defective genes have been cancelled as they arise." Herbert Spencer Jennings, *The Biological Basis of Human Nature* (W. W. Norton, 1930), p. 354.

5. Interview with Roger Penrose.

6. Lionel S. Penrose, *The Influence of Heredity on Disease* (H. K. Lewis, 1934), p. 15; Lionel Penrose, "Memoirs—1964: Phenylketonuria," Penrose Papers, file 72/2.

7. Penrose, "Memoirs—1964: Phenylketonuria"; Penrose, "Heredity and Mental Hygiene," 1938, pp. 2–3; Penrose, "Genetics in Relation to Pediatrics," July 26, 1947, Penrose Papers, files 72/2, 53/6, 73.

8. In Britain, the demographer David Glass's *Struggle for Population*, commissioned by the Eugenics Society and published in 1936, stimulated anxious interest, including in the Houses of Parliament, not least because of the attention it gave to the issue of declining fertility. In the United States, eugenicists clutched at straws in the wind suggesting a possible resurgence in upper-class fertility. One was borne down from Poughkeepsie, New York, by President Henry McCracken of Vassar, who reported that, according to a recent questionnaire, college women now considered three or more children their ideal, and that at Vassar a new course of lectures on marriage and the family had attracted over five hundred students. *Eugenics Society Annual Report, 1936/1937*, p. 8; "Summary of Discussion, Conference on Education and Eugenics," March 20, 1937, pp. 5–6, American Eugenics Society Papers, box 2.

9. Herbert Brewer to C. P. Blacker, June 23, 1934, Eugenics Society Records, file C.42; Hogben, *Genetic Principles in Medicine and Social Science*, pp. 188–89.

10. Haldane, "Eugenics and Social Reform," in J. B. S. Haldane, *Possible Worlds* (Harper & Bros., 1928), p. 204. Enid Charles called industrialism "a biological failure because it has lost the capacity to reproduce itself." Enid Charles, *The Twilight of Parenthood* (W. W. Norton, 1934), p. 105.

11. Hermann J. Muller, "The Dominance of Economics over Eugenics," *Scientific Monthly*, 37 (July 1933), 40; Elof Axel Carlson, *Genes, Radiation, and Society: The Life and Work of H. J. Muller* (Cornell University Press, 1981), pp. 133–34; Muller, *Out of the Night*, pp. 103–4. Enid Charles, who bore her husband Lancelot Hogben four children, observed that the modern woman's "intensive culture of personal appearance and bodily fastidiousness" was not readily reconciled with "the corporal realities of reproduction." In Charles's opinion, this "narcissism" was not necessarily regrettable given the "backwardness" of obstetrics and gynecology. Between 1912 and 1928, while the British death rate from respiratory tuberculosis had fallen by twenty-nine percent, the maternal death rate during confinement had shown a fourteen percent increase. Besides, women were limiting their families so that they could work outside the home. Charles, *The Twilight of Parenthood*, pp. 198–200.

12. Joan Fisher Box, *R. A. Fisher: The Life of a Scientist* (John Wiley, 1978), pp. 6, 7, 16, 26–27; Bernard Norton and Egon Pearson, "A Note on the Background to, and Refereeing of, R. A. Fisher's 1918 Paper 'On the Correlation between Relatives on the Supposition of Mendelian Inheritance,' " *Notes and Records of the Royal Society*, 31 (July 1976), pp. 157–58, 161. In 1914, Fisher declared, "The ordinary social reformer sets out with a belief that no environment can be too good for humanity; it is without contradicting this, that the eugenist may add that man can never be too good for his environment." R. A. Fisher, "Some Hopes of a Eugenist" (1914), in *The Collected Papers of R. A. Fisher*, J. H. Bennett, ed. (5 vols.; University of Adelaide, 1971–74), I, 78.

13. Box, *Fisher*, pp. 11, 18–19, 35–37.

14. Early in their marriage, the Fishers

pursued subsistence farming on fields near the school where he was teaching. Amid the food shortages of the war, the venture made for a type of immediate national service, a personal mobilization for England of biology if not of battleships. Fisher expected that raising the offspring on the farm would endow them with all the moral virtue, sturdy character, and healthy bodies of yeoman legend. *Ibid.*, pp. vi, 38–39, 43–44, 96–97, 390–91. Years later a close friend wrote that Fisher was *"the only man I knew* to practice eugenics." *Ibid.*, p. 32.

15. *Ibid.*, pp. 49–51, 60, 441.

16. *Ibid.*, pp. 49–51, 60, 187, 177–78.

17. William B. Provine, *The Origins of Theoretical Population Genetics* (University of Chicago Press, 1971), pp. 121, 129, 152; Box, Fisher, p. 180; Garland E. Allen, *Life Science in the Twentieth Century* (John Wiley, 1975), p. 135.

18. R. A. Fisher, *The Genetical Theory of Natural Selection* (Oxford University Press, 1930), pp. x, 234–36.

19. *Ibid.*, pp. 190–91, 195–98; Fisher, revealing more about himself than about anybody else, suspected that a majority of the group eager to do well by their children regarded contraception "with some degree of reluctance or aversion," and submitted to it under force of economic pressure. "Others, again, would certainly feel themselves disgraced if they were to allow economic motives to curtail in any degree the natural fruit of their marriage." *Ibid.*, pp. 193–94, 221–22.

20. Fisher, "Some Hopes of a Eugenist," pp. 80–81; Fisher, *The Genetical Theory of Natural Selection*, pp. 231, 253.

21. " 'Résumé' " of Address . . . by R. A. Fisher . . . Linnaean Society . . .," April 12, 1932, Eugenics Society Records, file C.107; R. A. Fisher, "Family Allowances in the Contemporary Economic Situation" (1932), in *The Collected Papers of R. A. Fisher*, III, 72, 70.

22. Michael Freeden, "Eugenics and Progressive Thought: A Study in Ideological Affinity," *The Historical Journal*, 3 (1979), 664–66; R. A. Fisher, "The Social Selection of Human Fertility" (1932); R. A. Fisher, "Biological Effects of Family Allowances"; R. A. Fisher, "Modern Eugenics" (1926), in *The Collected Papers of R. A. Fisher*, III, 64–65; II, 554; II, 120; Julian Huxley, "The Vital Importance of Eugenics," *Harper's*, 163 (Aug. 1931),

328–29. Fisher may also have felt justified in helping existing professional families to maintain a hold on the professions. He had written that the large proportion of "new blood" which in every generation came into the professions from lower social groups was "on the whole, inferior to the professional families of long standing." R. A. Fisher, "Positive Eugenics" (1917), in *The Collected Papers of R. A. Fisher*, I, 131–32. Fisher also held that there was no political bias in "finding class differences to be an essential feature of the dysgenic process in civilized life." He elaborated: "Man's only light seems to be his power to recognize human excellence, in some of its various forms. For this it follows that actions, powers, and functions cannot be of equal value. Promotion must be a reality, and the power of promotion a real wealth, whether we call our potentates kings or commissars." Fisher to Leonard Darwin, in J. H. Bennett, ed., *Natural Selection, Heredity, and Eugenics. Including Selected Correspondence of R. A. Fisher with Leonard Darwin and Others* (Clarendon Press, 1983), pp. 137–38.

23. Blacker to J. S. Huxley, Sept. 11, 1933, Eugenics Society Records, file C.185. Enid Charles attacked Fisher for failing to recognize that lower-income women might derive a special sense of self-value from high fertility. "It may be that a woman who produces a large family in conditions which a richer person would describe as squalor, is not very different from the artist or poet doing 'creative' work in a garret. She may not necessarily be of inferior intelligence to her neighbor who prefers the aspidistras and lace curtains of respectability. . . . The fact is that much which has been said and written about the qualitative results of differential fertility is a psychological compensation for the biological inadequacies of the professional type." Charles, *The Twilight of Parenthood*, pp. 140–42. For a discussion by Fisher of claims that psychocultural factors affected fertility, see Fisher to C. S. Myers, Dec. 6, 1932, in Bennett, ed., *Natural Selection, Heredity, and Eugenics*, pp. 240–41.

24. Haldane, "The Possibilities of Human Evolution," in J. B. S. Haldane, *The Inequality of Man and Other Essays* (Chatto & Windus, 1932), p. 87; Haldane, *Heredity and Politics*, p. 132; C. P. Blacker, *Eugenics in Pros-*

pect and Retrospect (Hamish Hamilton, 1945), p. 29; Lionel S. Penrose, "Is Our National Intelligence Declining?" [1938], Eugenics Society Records, file C.271; Blacker to Fisher, Dec. 14, 1933; Huxley to Blacker, May 29, 1933; Sept. 5, 1933, Eugenics Society Records, file C.185. Osborn dissented from direct family allowances: "If money is paid to parents for these purposes there is danger of a dysgenic result. The lower the aspirations of the parents, the larger value such sums will appear to have." Osborn, *Preface to Eugenics*, p. 265; Ellsworth Huntington, *Tomorrow's Children* (John Wiley, 1935), pp. 65, 72–73.

25. Muller, *Out of the Night*, p. 112; "Social Biology and Population Improvement" [the Geneticists' Manifesto], *Nature*, 144 (Sept. 16, 1939), 521–22.

26. J. B. S. Haldane, *Daedalus, or Science and the Future* (E. P. Dutton, 1924), pp. 44–46, 48–50. The book typically mixed Haldane's lack of sexual ease with radical sexual assertiveness. According to Haldane, prehistory had known four fundamental biological innovations, including the domestication of animals, of plants, and of fungi for the production of alcohol. The fourth, he allowed, more ultimate and far-reaching in importance than any of these, was the invention of frontal copulation. That "altered the path of sexual selection, focussed the attention of man as a lover upon woman's face and breasts, and changed our idea of beauty from the steatopygous Hottentot to the modern European, from the Venus of Brassempouy to the Venus of Milo." Haldane, *Daedalus*, pp. 42–43.

27. *Ibid.*, pp. 56–58, 63–64.

28. *Ibid.*, pp. 48–50, 65–69, 92–93.

29. Ronald Clark, *J.B.S.: The Life and Work of J. B. S. Haldane* (Coward-McCann, 1968), pp. 74, 59; Charles, *The Twilight of Parenthood*, pp. 191–92.

30. Carlson, *Genes, Radiation, and Society*, pp. 9–26, 51–52, 55, 57, 89, 150.

31. *Ibid.*, pp. 86, 175; Muller to Huxley, Nov. 22, 1920, Hermann J. Muller Papers, Lilly Library, Indiana University, Huxley file.

32. Carlson, *Genes, Radiation, and Society*, pp. 33, 166, 186–90, 215.

33. *Ibid.*, pp. 231–32.

34. Muller, *Out of the Night*, pp. 113–14, 120.

35. Brewer to Blacker, Jan. 22, 1932, April 28, 1931, August 14, 1934; Brewer, "Memorandum on His Personal Position in regard to Eutelegenesis," 1937; Brewer to Mrs. Collyer, Jan. 9, 1942; Brewer to Blacker, Jan. 11, 1945, Blacker to Colin ———, March 6, 1959, Eugenics Society Records, files C.42, C.43, C.44. Brewer kept up with medical research by slipping new issues of the *British Medical Journal* out of their wrappers at the post office, reading them overnight at home, then returning them the next day for delivery to their Maldon subscribers. Interview with Peggy Brewer Musgrave, April 27, 1984.

36. Herbert Brewer, *Eugenics and Politics* (Eugenics Society, 1937); Brewer to Blacker, Feb. 17, 1936; Brewer, "Eugenics and Socialism" [1932], Eugenics Society Records, files C.42, C.43, C.44.

37. "Eutelegenesis," Summary of Paper read by Mr. Brewer on October 22, 1935, Eugenics Society Records, file C.43.

38. *Ibid.*; Brewer to Blacker, Dec. 7, 1935, Eugenics Society Records, file C.43. Herbert Brewer, "Eutelegenesis," *Eugenics Review*, 27 (1935), 122; John Harvey Caldwell, "Babies by Scientific Selection," *Scientific American*, 150 (March 1934), 124–25.

39. Brewer, "Eutelegenesis," *Eugenics Review*, p. 122; Kenneth Walker to Blacker, April 27, 1934, Eugenics Society Records, file C.42.

40. Gregory Pincus, *The Eggs of Mammals* (Macmillan, 1936), pp. 96–97; Brewer to Blacker, Dec. 27, 1936, Eugenics Society Records, file C.43.

41. Muller, *Out of the Night*, pp. 108–10; Brewer to Blacker, May 19, 1943; Brewer to Blacker, Feb. 8, 1938; Brewer to Mrs. Collyer, Jan. 9, 1942, Eugenics Society Records, files C.43, C.44; Theodore I. Malinin, *Surgery and Life: The Extraordinary Career of Alexis Carrel* (Harcourt Brace Jovanovich, 1979), pp. 124–37.

42. Brewer to Blacker, Sept. 17, 1935, Eugenics Society Records, file C.43; Muller, *Out of the Night*, p. 118; Carlson, *Genes, Radiation, and Society*, pp. 105, 186, 228, 233; Brewer, "Eutelegenesis," *Eugenics Review*, p. 124.

43. Brewer, "Eutelegenesis," *Eugenics Review*, p. 124; Brewer to Blacker, April 25, 1934; "Eutelegenesis," Summary of Paper read by Mr. Brewer on October 22, 1935;

Brewer, *Eugenics and Politics*, 1937, Eugenics Society Records, files C.42, C.43, C.44.

44. Brewer to Blacker, March 17, 1935, Eugenics Society Records, file C.43; Muller, *Out of the Night*, p. 113.

45. "Eutelegenesis," Summary of Paper read by Mr. Brewer on October 22, 1935, Eugenics Society Records, file C.43; Brewer, "Eutelegenesis," *Eugenics Review*, p. 123.

46. Brewer to Blacker, Feb. 8, 1938, Eugenics Society Records, file C.43; Brewer to Ruggles Gates, Feb. 21, 1937, Ruggles Gates Papers.

47. Brewer to Blacker, Feb. 8, 1938; "Eutelegenesis," Summary of Paper read by Mr. Brewer on October 22, 1935; Muller, *Out of the Night*, p. 122.

48. Muller to Allan Bacon, Dec. 4, 1959, Hermann J. Muller Papers, Germinal Choice file, box Ib; Diane B. Paul, "Eugenics and the Left" (unpublished paper), pp. 16–18; Brewer, "Memorandum on His Personal Position in regard to Eutelegenesis," 1937; Brewer, *Eugenics and Politics*; Brewer to Blacker, Jan. 20, 1937; Haldane to Sir [Brewer], Nov. 13, 1935, Eugenics Society Records, files C.43, C.44; Haldane, *Heredity and Politics*, p. 133; Clark, *J.B.S.*, p. 86. Blacker had pronounced Brewer's original drafts on eutelegenesis "brilliant" and had arranged for a small grant from the Eugenics Society so that Brewer could take leave from his postal clerk's duties to develop his idea. R. A. Fisher tended to class enthusiasts of artificial insemination with cranks, and he was joined in his unhappiness about eutelegenesis by Dean William R. Inge, long a stalwart of the mainline wing

in the Eugenics Society, who frankly confessed to Brewer his "natural repugnance in matters of sex," a repugnance so strong as to have made it impossible for the old prelate to have read Havelock Ellis. Lancelot Hogben queried Brewer in 1936: Why worry about breeding humanity with humanity likely to be wiped out by the war sure to come in three years? Blacker to Huxley, May 31, 1934, Eugenics Society Records, file C. 185; Brewer, "Memorandum on Results of Three Months Work," Nov. 14, 1935; Brewer to Chance, March 3, 1935; Brewer, "Memorandum on His Personal Position in regard to Eutelegenesis"; W. R. Inge to Brewer, n.d. [1937]; Blacker to Brewer, Feb. 10, 1937; Brewer to Blacker, Feb. 26, 1936, Eugenics Society Records, file C.43.

49. Julian S. Huxley, "Eugenics and Society," in Julian S. Huxley, *Man Stands Alone* (Harper & Bros., 1941), pp. 78–79; Shaw's letter is reproduced in Brewer to Blacker, Jan. 20, 1937. Huxley neglected to consider the impact of personal taste upon sexually revolutionary principle. When a eugenic enthusiast privately inquired of him whether he might consent to father "my wife's child, possibly by artificial insemination," Huxley had "to confess to feeling, in spite of my intellect, an embarrassment in the matter!" Arthur E. Mason to Huxley, Jan. 3, 1937; Huxley to Blacker, Jan. 11, 1937, Eugenics Society Records, file C. 186.

50. H. G. Wells, Julian S. Huxley, and G. P. Wells, *The Science of Life* (Doubleday, Doran, 1931), pp. 1468–70; Muller, *Out of the Night*, pp. 123–24.

Chapter XIII:
THE ESTABLISHMENT OF HUMAN GENETICS

1. C. C. Hurst, "Proposal to Found a Research Institute in Human Genetics" [1931], C. C. Hurst Papers, Cambridge University Library, Cambridge, England, add. 7955/18/9; J. B. S. Haldane, *Heredity and Politics* (W. W. Norton, 1938), p. 188; Lancelot Hogben, *Genetic Principles in Medicine and Social Science* (Williams and Norgate, 1931), p. 209; Julian S. Huxley, "Eugenics and Society," in Julian S. Huxley, *Man Stands Alone*

(Harper & Bros., 1941), p. 35; Ursula Grant Duff to C. P. Blacker, March 11, 1934, Eugenics Society Records, Wellcome Institute for the History of Medicine, London, file C.133; Ellsworth Huntington, *Tomorrow's Children* (John Wiley, 1935), p. 100; E. P. Lyon, "A Conference on Heredity as Applied to Man," *Science*, 73 (April 17, 1931), 421; Lionel S. Penrose, "Heredity and Mental Hygiene," 1938, Lionel S. Penrose Papers, University College

London, file 53/6. Reform eugenicists also thought it necessary to pursue research on the nature and distribution of intelligence, the physiology of fertility in regard to conception and contraception, and, especially, demographic trends. Frederick Osborn came to figure prominently in the philanthropic sponsorship of population research, and the Eugenics Society encouraged the groundbreaking work of David Glass, beginning with his 1936 *Struggle for Population*. See Frederick Osborn, *Preface to Eugenics* (Harper & Bros., 1940), p. 285; Faith Schenk and A. S. Parkes, *The Activities of the Eugenics Society*, offprint [1969?], pp. 151–52; Glass obituary, *The Times*, Sept. 27, 1978, Eugenics Society Records, file C.125.

2. Ronald A. Fisher, "Eugenics: Academic and Practical" [1935], *The Collected Papers of R. A. Fisher*, J. H. Bennett, ed. (5 vols.; University of Adelaide, 1971–74), III, 45; Lancelot Hogben, "Some Methodological Aspects of Human Genetics," *American Naturalist*, 68 (May–June 1933), 254; Hogben, *Genetic Principles in Medicine and Social Science*, pp. 215–16; Lancelot Hogben, "Heredity and Human Affairs," in J. G. Crowther, ed., *Science for a New World* (Harper & Bros., 1934), p. 27; Lionel S. Penrose, *The Influence of Heredity on Disease* (H. K. Lewis, 1934), pp. 21–22; British Council for Research in Human Genetics, "Scientific Memorandum on the Need of Research in Human Genetics in Great Britain, An Appeal to the Rockefeller Foundation of New York," Sept. 22, 1931, C. C. Hurst Papers, add. 7955/48/19; Penrose to A. L. Thomson, May 19, 1932, Lionel Penrose Papers, file 150/5.

3. Recognizing that much of what was believed about human heredity consisted of extrapolations from the principles of plant and animal genetics, Herbert Jennings had cautioned: "When the biologist, from his knowledge of other organisms, is tempted to dogmatize concerning the possibilities of human development, let him first ask himself: How correctly could I predict the behaviour and social organization of ants from a knowledge of the natural history of the oyster? Man differs from other organisms in these respects certainly as much as the ant does from the oyster; for these distinctive aspects of his biology only the study of man himself is relevant." Herbert Spencer Jennings, *Prometheus, or Biology and the Advancement of Man* (E. F. Dutton, 1925), p. 66; Hogben, *Genetic Principles in Medicine and Social Science*, pp. 11, 167–68; Lancelot Hogben, *The Nature of Living Matter* (Kegan, Paul, Trench, Trubner, 1930), p. 214; Penrose, *The Influence of Heredity on Disease*, p. 71; British Council for Research in Human Genetics, "Scientific Memorandum on the Need of Research in Human Genetics in Great Britain, An Appeal to the Rockefeller Foundation of New York," Sept. 22, 1931.

4. Penrose, *The Influence of Heredity on Disease*, pp. 21–22; Charles B. Davenport to C. C. Hurst, April 16, 1931, C. C. Hurst Papers, add. 7955/18/54.

5. The misleadingly high incidence of schizophrenia was observed by the German scientist Ernst Rüdin, who also made the subsequent correction for ascertainment bias. Abraham Myerson et al., *Eugenical Sterilization: A Reorientation of the Problem* (Committee of the American Neurological Association for the Investigation of Eugenical Sterilization; Macmillan, 1936), pp. 83–84; Penrose, *The Influence of Heredity on Disease*, pp. 21–22, 71; Hogben, *Genetic Principles in Medicine and Social Science*, pp. 214–15; R. A. Fisher, "The Effect of Ascertainment upon the Estimation of Frequencies" [1934], in *The Collected Papers of R. A. Fisher*, III, 169.

6. Myerson et al., *Eugenical Sterilization*, pp. 92–94; Lionel S. Penrose, *Mental Defect* (Farrar & Rinehart, 1934), pp. 76–77; Lionel Penrose, *A Clinical and Genetic Study of 1280 Cases of Mental Defect [The Colchester Survey]* (Medical Research Council Special Report Series 229; His Majesty's Stationer's Office, 1938), pp. 11–12.

7. Curt Stern, "Mendel and Human Genetics," *Proceedings of the American Philosophical Society*, 109 (1965), 218.

8. *Ibid.*, pp. 221–22; Paul Speiser, "Karl Landsteiner," *Dictionary of Scientific Biography*, VII, 622–25; Pauline Mazumdar, "Eugenists, Marxists and the Science of Human Genetics: A History of Human Genetics in Britain, 1900–1940" (unpublished manuscript), pp. 108–9.

9. Hogben, *Genetic Principles in Medicine and Social Science*, p. 64; British Council for Research in Human Genetics, "Scientific Memorandum on the Need of Research in Human Genetics in Great Britain, An Ap-

peal to the Rockefeller Foundation of New York," Sept. 22, 1931.

10. "Historical Note: On the Rôle of Laurence H. Snyder in the Development of Human and Medical Genetics in the United States: An Oral History Tape Interview," *American Journal of Medical Genetics*, 8 (1981), 453; Lancelot Hogben, *Nature and Nurture* (W. W. Norton, 1933), pp. 42–43; Laurence H. Snyder, "Studies in Human Inheritance. IX: The Inheritance of Taste Deficiency in Man," *Ohio Journal of Science*, 32 (1932), 436–40; Joan Fisher Box, *R. A. Fisher: The Life of a Scientist* (John Wiley, 1978), pp. 371–72.

11. Lancelot Hogben, "Comments on Issues Raised at the First Meeting," March 2, 1932, Human Genetics Committee, Medical Research Council Records, London, Human Genetics file 1732/1; Hogben, *Genetic Principles in Medicine and Social Science*, pp. 82, 84; Penrose, *The Influence of Heredity on Disease*, pp. 10, 71–72.

12. British Council for Research in Human Genetics, "Scientific Memorandum on the Need of Research in Human Genetics in Great Britain, An Appeal to the Rockefeller Foundation of New York," Sept. 22, 1931; Mazumdar, "Eugenists, Marxists and the Science of Human Genetics," pp. 109–10; J. B. S. Haldane, "The Investigation of Linkage in Man," Feb. 1, 1934, Medical Research Council Records, Human Genetics file 1732/1, folder II.

13. Laurence H. Snyder, *Medical Genetics* (Duke University Press, 1941), p. 115; Mazumdar, "Eugenists, Marxists and the Science of Human Genetics," pp. 117–18, 140–41; J. B. S. Haldane, *Human Biology and Politics* (The Science Guild, 1934); Box, *Fisher*, pp. 281–82.

14. J. B. S. Haldane, *New Paths in Genetics* (Harper & Bros., 1941), p. 194.

15. Hogben, *Genetic Principles in Medicine and Social Science*, pp. 214, 217–18; *Report of the Departmental Committee on Sterilization (minus appendices)* (pamphlet; Joint Committee on Voluntary Sterilization, June 1934), pp. 51–56; Laurence H. Snyder, "The Study of Human Heredity," *Scientific Monthly*, 51 (Dec. 1940), 541.

16. Hogben to C. C. Hurst, July 7 [1931]; F. A. E. Crew to Hurst, July 14, 1931, C. C. Hurst Papers, add. 7955/18/31, 7955/18/45;

Hogben, *Genetic Principles in Medicine and Social Science*, pp. 214, 216, 219.

17. Charles B. Davenport to John C. Merriam, June 25, 1926, Charles B. Davenport Papers, American Philosophical Society Library, Philadelphia, Cold Spring Harbor Series #1; Harry H. Laughlin, "Eugenics," *Nature Study Review*, March 1912, pp. 111–12.

18. The blue-ribbon committee was headed by L. C. Dunn, a staunch political liberal and a distinguished geneticist, who wrote from abroad that, while recently in Germany, he had been struck "by the consequences of reversing the order as between social program and scientific discovery." Dunn added, "The incomplete knowledge of today, much of it based on a theory of the state which has been influenced by racial, class and religious prejudices of the group in power, has been embalmed in law. . . . The genealogical record offices have become powerful agencies of the state, and medical judgments even when possible, appear to be subservient to political purposes." Dunn to ?, July 3, 1935, Davenport Papers, Dunn file; Davenport to Merriam, Oct. 3, 1925; June 25, 1926, Davenport Papers, Cold Spring Harbor Series #1; Davenport to Charles L. Dana, March 16, 1925; Davenport Papers, Charles L. Dana file; Charles B. Davenport, "History of Eugenics Record Office," 1929; Vannevar Bush to H. H. Laughlin, May 4, 1939; Lewis H. Weed to Vannevar Bush, June 24, 1939, Records of the Carnegie Institution of Washington, Washington, D.C., Harry Laughlin file; Laughlin Retirement file; Frances J. Hassencahl, *Harry H. Laughlin, "Expert Eugenics Agent" for the House Committee on Immigration and Naturalization, 1921 to 1931* (University Microfilms, 1971), pp. 330–35.

19. Karl Pearson, "Prefatory Note," *The Treasury of Human Inheritance. Volume II, Anomalies and Diseases of the Eye: Nettleship Memorial Volume* (Cambridge University Press, 1922), p. v; José Harris, *William Beveridge: A Biography* (Clarendon Press, 1977), pp. 286–87; British Council for Research in Human Genetics, "Scientific Memorandum on the Need of Research in Human Genetics in Great Britain, An Appeal to the Rockefeller Foundation of New York," Sept. 22, 1931; Karl Pearson to C. C. Hurst, July 3, 1931; Julian Huxley to C. C. Hurst, June 22, 1931; "Report of a Meeting at the London School

of Economics," July 1931; F. A. E. Crew to C. C. Hurst, June 23, 1931; C. C. Hurst Papers, add. 7955/48/19, 7955/18/88, 7955/18/213, 7955/18/189, 7955/18/43.

20. Medical Research Council, *Report, 1931/32*, p. 128; Julian Huxley to C. C. Hurst, June 25, 1931; Lancelot Hogben to Hurst, June 24, 1931, C. C. Hurst Papers, add. 7955/18/214, 7955/18/29; Walter Fletcher to Lionel Penrose, Jan. 26, 1932, Lionel Penrose Papers, file 150/5.

21. Medical Research Council, Committee on Human Genetics, Minutes, Feb. 5, 1932; Minute Book, May 20, 1936; Lancelot Hogben, "Comments on Issues Raised at the First Meeting," March 2, 1932; Walter M. Fletcher to Sir, Feb. 24, 1933, Medical Research Council Records, Human Genetics file 1732/1, folder I; Medical Research Council, *Report, 1931/32*, p. 115; *1933/34*, pp. 143–44; *1934/35*, p. 152; A. Landsborough Thomson, *Half a Century of Medical Research* (2 vols; I: *Origins and Policy of the Medical Research Council (UK);* II: *The Programme of the Medical Research Council (UK);* His Majesty's Stationery Office, 1973, 1975), II, 149–50; Manchester *Sunday Chronicle,* June 28, 1936, copy in Ruggles Gates Papers, Kings College London; "A Bureau of Human Heredity," *Nature,* 137 (May 16, 1936), 795–96.

22. Medical Research Council, *Report, 1933/34*, p. 144; *1935/36*, p. 139.

23. Fisher's daughter's rendering of the matter perhaps reveals Fisher's own view of the 1919 job offer from Pearson: "It seemed that the lover was at last to be admitted to his lady's court—on condition that he first submit to castration." Box, *Fisher,* pp. 61, 82, 172–73, 201–2, 273–74; Karl Pearson to Raymond Pearl, Aug. 30, 1935, Raymond Pearl Papers, American Philosophical Society Library, Philadelphia, Karl Pearson file.

24. R. A. Fisher, "Eugenics: Academic and Practical" [1935], *The Collected Papers of R. A. Fisher,* III, 447–48; Rockefeller Foundation, *Annual Report, 1936,* pp. 158–59; Box, *Fisher,* pp. 338–39; R. A. Fisher, "Serological Work and Human Genetics," March 8, 1932; Fisher to C. Todd, Feb. 5, 1932; Human Genetics Committee, Agenda, Oct. 24, 1933; Minute Book, March 22, 1937, Medical Research Council Records, Human Genetics file 1732/1, folder I, folder III; Fisher to Leonard Darwin, Oct. 15, 1930; Nov. 25, 1930, in J. H.

Bennett, ed., *Natural Selection, Heredity, and Eugenics, Including Selected Correspondence of R. A. Fisher with Leonard Darwin and Others* (Clarendon Press, 1983), pp. 129, 134; Mazumdar, "Eugenists, Marxists and the Science of Human Genetics," pp. 139–40; Rockefeller Foundation, *Annual Report, 1935,* pp. 82–83.

25. Medical Research Council, *Report, 1937/38,* pp. 89, 179; *1936/37,* p. 157; Mazumdar, "Eugenists, Marxists and the Science of Human Genetics," p. 142; Box, *Fisher,* p. 347; Lionel Penrose, "1939 Report on Research Department, Royal Eastern Counties' Institution," Nov. 1939, Medical Research Council Records, Mental Disorders, Colchester, file 1588, folder IV; Rockefeller Foundation, *Annual Report, 1937,* pp. 163–64; Box, *Fisher,* p. 349; interview with Ruth Sanger.

26. Haldane, *New Paths in Genetics,* pp. 159–60; Julia Bell to Landsborough Thomson, June 10, 1936, Medical Research Council Records, Human Genetics file 1732/1, folder III; Ronald W. Clark, *J. B. S.: The Life and Work of J. B. S. Haldane* (Coward-McCann, 1968), p. 120; Medical Research Council, *Report, 1936/37,* p. 80.

27. Medical Research Council, *Report, 1935/36,* pp. 139–40; *1936/37,* pp. 157–58; *1938/39,* p. 81; Mazumdar, "Eugenists, Marxists and the Science of Human Genetics," pp. 113–14, 142–43.

28. Haldane, *New Paths in Genetics,* pp. 154–55, 158–59; Penrose, *A Clinical and Genetic Study of 1280 Cases of Mental Defect* [*The Colchester Survey*], p. 58; Medical Research Council, *Report, 1936/37,* p. 80; Lionel Penrose to Munro Fox, Oct. 13, 1952, Lionel Penrose Papers, file 121/3; Lionel Penrose, "The Influence of the English Tradition in Human Genetics," *Proceedings of the Third International Congress of Human Genetics,* James F. Crow and James V. Neel, eds. (Johns Hopkins University Press, 1967), p. 17; Snyder, *Medical Genetics,* p. 114; Mazumdar, "Eugenists, Marxists and the Science of Human Genetics," p. 134.

29. Box, *Fisher,* pp. 350–51, 356–57; Louis K. Diamond, "The Story of Our Blood Groups," in Maxwell M. Wintrobe, ed., *Blood, Pure and Eloquent: A Story of Discovery, of People, and of Ideas* (McGraw-Hill, 1980), pp. 695–96, 701; Ruth Sanger interview.

30. Philip Reilly, *Genetics, Law, and So-*

cial Policy (Harvard University Press, 1977), p. 229; "Historical Note: On the Rôle of Laurence H. Snyder in the Development of Human and Medical Genetics in the United States . . .," *American Journal of Medical Genetics*, 8 (1981), 456; Laurence H. Snyder, "The Study of Human Heredity," *Scientific Monthly*, 51 (Dec. 1940), 536–37; Alexander S. Wiener, "Heredity and the Lawyer," *Scientific Monthly*, 52 (Feb. 1941), 142–46; Snyder, *Medical Genetics*, pp. 17–19.

31. Box, *Fisher*, pp. 421, 423; Laurence H. Snyder, *Blood Grouping in Relation to Clinical and Legal Medicine* (Williams and Wilkins, 1929), pp. 117–18; H. Schadewalt, "Ludwig Hirszfeld," *Dictionary of Scientific Biography*, VI, 432–34.

32. Penrose, "The Influence of the English Tradition in Human Genetics," pp. 17–18; Lionel Penrose and J. B. S. Haldane, "Mutation Rates in Man," *Nature*, 135 (June 1, 1935), 907–8; L. S. Penrose, "The Genetical Analysis of Infectious Disease Susceptibility," June 21, 1965; Penrose, "J. B. S. Haldane," 1966, Lionel Penrose Papers, file 83/3, 52/3; Curt Stern, "Mendel and Human Genetics," *Proceedings of the American Philosophical Society*, 109 (1965), 218. Haldane's line of analysis had actually been pioneered in a paper delivered at the Second International Eugenics Congress in 1921 by Charles H. Danforth, an anatomist at the Washington University, St. Louis, Medical School, who was interested in heredity and variation. Charles H. Danforth, "The Frequency of Mutation and the Incidence of Hereditary Traits in Man," *Eugenics, Genetics, and the Family: Scientific Papers of the Second International Congress of Eugenics . . . 1921* (2 vols.; Williams and Wilkins, 1923), I, 120–28.

33. Penrose to Haldane, Dec. 4, 1952; Penrose, "Heredity and Mental Hygiene," 1938, Lionel Penrose Papers, file 136, 53/6; Penrose, *The Influence of Heredity on Disease*, p. 65. Penrose's estimate proved to be somewhat inaccurate because it was based on a mistaken assessment of the frequency of epiloia in the population. Interview with James V. Neel.

34. Daniel J. Kevles and Stephen Postema, "A Statistical Survey of Human Genetics in the United States and Great Britain, 1930 to 1959" (in preparation); "Historical Note: On the Rôle of Laurence H. Snyder in the Development of Human and Medical Genetics in the United States," p. 457. During the thirties, lectures on genetics began to be offered in British and American medical schools by scientists like F. A. E. Crew. Also, a department of medical genetics was organized in 1936 at the New York State Psychiatric Institute, which was affiliated with Columbia College of Physicians and Surgeons, and in 1941 another was established at the new Bowman Gray School of Medicine at Wake Forest University. C. Nash Herndon, "William Allan: An Appreciation"; "Genetics in the Medical School Curriculum," *American Journal of Human Genetics*, 14 (1962), 97–100; 8 (1956), 2; James V. Neel, "Human Genetics," in John Z. Bowers and Elizabeth F. Purcell, eds., *Advances in American Medicine: Essays at the Bicentennial* (2 vols.; Josiah Macy, Jr., Foundation, 1976), I, 54; John D. Rainer, "Perspectives on the Genetics of Schizophrenia: A Re-evaluation of Kallmann's Contribution—Its Influence and Current Relevance," in Lawrence Kolb, coordinator, *Progress in Psychiatric Research and Education: A Symposium in Honor of the Seventy-fifth Anniversary of the New York State Psychiatric Institute* (Psychiatric Quarterly of New York, 1973), p. 40; Frederick Osborn, "History of the American Eugenics Society," p. 15, American Eugenics Society Papers, American Philosophical Society Library, Philadelphia, box 15.

35. Kevles and Postema, "A Statistical Survey of Human Genetics in the United States and Great Britain, 1930 to 1959"; Kenneth Ludmerer, *Genetics and American Society* (Johns Hopkins University Press, 1972), pp. 47, 155–56; Osborn, *Preface to Eugenics*, pp. 286–87.

36. Charles E. Rosenberg, "The Social Environment of Scientific Innovation: Factors in the Development of Genetics," in Charles E. Rosenberg, *No Other Gods: On Science and American Social Thought* (Johns Hopkins University Press, 1976), pp. 202–03; Ludmerer, *Genetics and American Society*, pp. 148, 156; Sheldon C. Reed, "A Short History of Human Genetics in the U.S.A.," *American Journal of Medical Genetics*, 3 (1979), 287–88; interview with Arthur Steinberg; Snyder, "The Study of Human Heredity," *Scientific Monthly*, 51 (Dec. 1940), 539–40.

37. Not long after the Second World

War, Frederick Osborn suggested that studies in the factors making for differences in traits of mind and personality, defying the geneticist in their complexity, had become "largely the responsibility of psychologists." Osborn, *Preface to Eugenics* (rev. ed.; Harper & Bros., 1952), p. 314; Kevles and Postema, "A Statistical Survey of Human Genetics in the United States and Great Britain, 1930 to 1959"; Abraham Myerson, *The Inheritance of Mental Diseases* (Williams and Wilkins, 1925), pp. 12, 83, 85, 316–20; Albert Deutsch, *The Mentally Ill in America: A History of Their Care and Treatment from Colonial Times* (Doubleday, Doran, 1937), pp. 364–65, 373, 375; Walter E. Fernald, "Feeblemindedness," *Mental Hygiene*, 8 (Oct. 1924), 964–971.

38. Osborn, *Preface to Eugenics*, pp. 22–23; interviews with Barton Childs, Arthur Steinberg, John Fraser Roberts, James V. Neel, and Harry Harris; Richard C. Lewontin, Steven Rose, and Leon Kamin, *Not in Our Genes: Biology, Ideology, and Human Nature* (Pantheon Books, 1984), pp. 206–13; Reed, "A Short History of Human Genetics in the U.S.A.," p. 290; Franz J. Kallmann, *Heredity in Mental Health and Disorder* (W. W. Norton, 1953), pp. 121–23, 142, 179. For an alternative assessment of Kallmann's work, see John D. Rainer, "Perspectives on the Genetics of Schizophrenia: A Re-evaluation of Kallmann's Contribution," pp. 40–46. George A. Jervis, "Metabolic Investigations on a Case of Phenylpyruvic Oligophrenia"; "Studies on Phenylpyruvic Oligophrenia," *Journal of Biological Chemistry*, 126 (1938), 305–13; 169 (1947), 651–56.

39. Interview with John Fraser Roberts; clipping, "Burden Mental Research Trust," *British Medical Journal*, March 4, 1933, pp. 81–82; "The Burden Mental Research Trust: Its Present and Future" [n.d.; 1936?], both in Medical Research Council Records; John Fraser Roberts, *An Introduction to Medical Genetics* (Oxford University Press; Humphrey Milford, 1940); John Fraser Roberts and Marcus E. Pembrey, *An Introduction to Medical Genetics* (7th ed.; Oxford University Press, 1978).

40. Rockefeller Foundation, *Annual Report, 1930*, pp. 156–59; *1931*, p. 174; *1934*, p. 107; *1935*, pp. xiii, 101, 126–27; *1936*, pp. 133, 159–60; Barbara A. Kimmelman, "An Effort in Reductionist Sociobiology: The Rockefeller Foundation and Physiological Genetics, 1930–1942" (unpublished manuscript), pp. 9, 14, 21.

41. John M. Opitz, "Editorial Comment: Biographical Note—Laurence H. Snyder," *American Journal of Medical Genetics*, 8 (1981), 447. "Historical Note: On the Rôle of Laurence H. Snyder in the Development of Human and Medical Genetics in the United States," pp. 449, 451. Snyder remembered that his adviser, William E. Castle, almost rejected his dissertation because Castle, otherwise an able scientist, could not quite comprehend some of Snyder's gene-frequency reasoning. Interview with Laurence H. Snyder.

42. "Historical Note: On the Rôle of Laurence H. Snyder in the Development of Human and Medical Genetics in the United States," pp. 452, 454–55; interview with Laurence Snyder; Kevles and Postema, "A Statistical Survey of Human Genetics in the United States and Great Britain, 1930 to 1959"; Reed, "A Short History of Human Genetics in the U.S.A.," p. 289; interview with James Neel.

43. Interview with Laurence Snyder; Neel, "Human Genetics," in Bowers and Purcell, eds., *Advances in American Medicine*, I, 85.

44. Though the bulk of important work in the development of mathematical methods for human genetics was done in England, some was carried out in the United States. Snyder, "The Study of Human Heredity," *Scientific Monthly*, 51 (Dec. 1940), 540; Kevles and Postema, "A Statistical Survey of Human Genetics in the United States and Great Britain, 1930 to 1959"; Box, *Fisher*, pp. 259–60; interviews with Harry Harris, John Maynard Smith, and Arthur Steinberg; Haldane to Fisher, May 30 [1933], in Bennett, ed., *Natural Selection, Heredity, and Eugenics*, p. 215.

45. Karl Pearson and Ethel M. Elderton, "Foreword," *Annals of Eugenics*, 1 (1925), 1, 4; Bennett, ed., *Natural Selection, Heredity, and Eugenics*, p. 16; Box, *Fisher*, pp. 262, 281; Kevles and Postema, "A Statistical Survey of Human Genetics in the United States and Great Britain, 1930 to 1959." The Eugenics Society granted the *Annals* a subsidy of three hundred pounds annually from 1933, but, much to Fisher's unhappiness, reduced the

amount to one hundred and fifty pounds in 1937. Eugenics Society, *Annual Report, 1937/38,* p. 9; C. P. Blacker to Julian S. Hux-

ley, Nov. 29, 1933; Fisher to Blacker, Nov. 19, 1937; Eugenics Society Records, file C.185; file C.108.

Chapter XIV:
APOGEE OF THE ENGLISH SCHOOL

1. Daniel J. Kevles and Stephen Postema, "A Statistical Survey of Human Genetics in the United States and Great Britain, 1930 to 1959" (in preparation).

2. Interview with Ruth Sanger; Robert R. Race and Ruth Sanger, *The Blood Groups of Man* (Blackwell, 1950).

3. Interviews with Sylvia Lawler and John Fraser Roberts; Lionel Penrose, "J. B. S. Haldane," 1966, Lionel Penrose Papers, file 52/3; N. W. Pirie, "John Burdon Sanderson Haldane," *Biographical Memoirs of Fellows of the Royal Society,* 12 (1966), 232.

4. Haldane to Penrose, Aug. 18, 1943; Haldane to the Provost, University College London, Aug. 9, 1944, Lionel Penrose Papers, files 159, 49/1.

5. Lionel Penrose, "The Galton Laboratory Report, 1949–50"; Penrose to Haldane, June 26, 1943; Nov. 29, 1943; Penrose, "From Eugenics to Human Genetics," lecture 1965; Penrose, "Galton Laboratory," April 27, 1961, Lionel Penrose Papers, files 159, 77/2, 175/5; interview with C. A. B. Smith; Joan Fisher Box, *R. A. Fisher: The Life of a Scientist* (John Wiley, 1978), pp. 400–1; J. B. S. Haldane, "Karl Pearson," in E. S. Pearson and M. G. Kendall, eds., *Studies in the History of Probability and Statistics* (Griffin, 1970), p. 436. Shortly before taking the Galton appointment, Penrose confided to Haldane that the *Annals* certainly had "a reputation for being almost completely unreadable," adding, "If I had a hand in editing it, it would contain more medical and biological material and be less theoretical." Penrose to Haldane, Feb. 25, 1944 [?], Lionel Penrose Papers, file 159.

6. There was constant liaison between the Galton and the departmental group, just downstairs in the Bartlett Building, of Hans Gruneberg, a specialist in mammalian genetics, who carried on the Fisher tradition with mice. Haldane maintained a room for fly genetics in his own area of the department, thinking it important even for theoretical

students to have firsthand experimental experience in classical *Drosophila* genetics. Penrose sent Galton students over to the fly room, where the instruction during part of the postwar years was offered by John Maynard Smith, a Cambridge graduate in aeronautical engineering who had come to University College to pursue a long-standing zoological interest, attracted initially by Haldane's leftist politics and remaining to become his scientific protégé. Interview with John Maynard Smith.

7. Haldane to Hans Kalmus, May 26, 1953; J. B. S. Haldane Papers, University College London, file 2; C. A. B. Smith interview.

8. J. B. S. Haldane, *The Biochemistry of Genetics* (George Allen and Unwin, 1954), pp. 16, 96–99; Pirie, "Haldane," p. 222; Ronald W. Clark, *J. B. S.: The Life and Work of J. B. S. Haldane* (Coward-McCann, 1968), p. 119; conversation with Joseph Scott, Librarian, University College London; interview with Harry Harris.

9. Barton Childs, "Sir Archibald Garrod's Conception of Chemical Individuality: A Modern Appreciation," *New England Journal of Medicine,* 282 (1970), 4–5; Barton Childs, "Garrod, Galton, and Clinical Medicine," *Yale Journal of Biology and Medicine,* 46 (1973), 297; Victor A. McKusick, "The Growth and Development of Human Genetics as a Clinical Discipline," *American Journal of Human Genetics,* 27 (1975), 261; Archibald Garrod to Lionel Penrose, April 5, 1934, Lionel Penrose Papers, file 134/5; Archibald Garrod, *Inborn Errors of Metabolism* (2nd ed.; Henry Frowde and Hodder & Stoughton, 1923), p. 50; Alexander G. Bearn and Elizabeth D. Miller, "Archibald Garrod and the Development of the Concept of Inborn Errors of Metabolism," *Bulletin of the History of Medicine,* 53 (Fall 1979), 315–27.

10. Alexander G. Bearn, "Lettsomian Lectures: Inborn Errors of Metabolism," *Transactions of the Medical Society of London,*

93 (1976), 51, 60; Garrod, *Inborn Errors of Metabolism*, p. 16.

11. Garrod, *Inborn Errors of Metabolism*, pp. 5–6, 10.

12. Lionel S. Penrose, "The Influence of the English Tradition in Human Genetics," in James F. Crow and James V. Neel, eds, *Proceedings of the Third International Congress of Human Genetics* (Johns Hopkins University Press, 1967), p. 18; J. B. S. Haldane, *New Paths in Genetics* (Harper & Bros., 1941), chapter 2; Clark, *J.B.S.*, pp. 113–14; Robert Olby, *The Path to the Double Helix* (Macmillan, 1974), pp. 134–37.

13. Olby, *The Path to the Double Helix*, pp. 131–32; Childs, "Sir Archibald Garrod's Conception of Chemical Individuality," pp. 4–5.

14. Horace Freeland Judson, *The Eighth Day of Creation: Makers of the Biological Revolution* (Simon & Schuster, 1979), pp. 214, 351; Olby, *The Path to the Double Helix*, p. 148; Barbara A. Kimmelman, "An Effort in Reductionist Sociobiology: The Rockefeller Foundation and Physiological Genetics, 1930–1942" (unpublished manuscript), pp. 45–51; George W. Beadle, "Genes and Chemical Reactions in Neurospora," Nobel Lecture, Dec. 11, 1958, *Nobel Lectures, Physiology or Medicine, 1942–1962* (Elsevier, 1964), pp. 591–92, 594, 596.

15. George W. Beadle, "Biochemical Genetics," *Chemical Reviews*, 37 (Aug. 1945), 35; Joseph S. Fruton, *Molecules and Life: Historical Essays on the Interplay of Chemistry and Biology* (Wiley-Interscience, 1972), p. 272; Haldane, *New Paths in Genetics*, p. 82.

16. Unless otherwise indicated, this section from here to the end is based on interviews with Harry Harris. Harry Harris, "The Inheritance of Premature Baldness in Men," *Annals of Eugenics*, 13 (1946), 172–81.

17. L. S. Penrose, "Phenylketonuria: A Problem in Eugenics," *The Lancet*, I (June 29, 1946), 950. While in the Air Force, Harris somehow got the idea that there was a connection between premature baldness and the amount of hair men grew on their chests. Whenever bald patients came in, he would ask whether the baldness was premature, then score the amount of hair elsewhere on their bodies. He continued the research while he helped carry out the medical phase of demobilization at the base

near London. Harris recalled, "There was a series of long Nissen huts. The airmen would come in at one end fully accoutered and, as they went down the huts, hand in their kits, guns, and whatever else. They would eventually arrive stark naked at the center of the hut, where they had to pass by me. I had to certify them as fit, which meant as not having scabies or lice." A waste of time, Harris thought, since if they were not physically fit, they would not have gotten to the Nissen hut. So, he continued, "I conceived the idea that, as they passed me, I'd look at their heads and ask them about their baldness, and then I could make a score—1, 2, 3, 4—for how much hair they had on the chest, back, arms, legs, and scrotum. I kept notebooks with these scores. Hundreds of men passed by me." The research was more valuable to Harris than to the world of science—he found that the greater the degree of body hair, the less the hair on the head—since the data analysis provided an excellent lesson in the correlational statistics he was teaching himself at the time. Harry Harris interview; Harry Harris, "The Relation of Hair Growth on the Body to Baldness," *British Journal of Dermatology and Syphilis*, 59 (Aug.–Sept. 1947), 300–9.

18. L. S. Penrose, "Genetical Analysis in Man," in G. E. W. Wolstenholme and Cecilia M. O'Connor, eds., *CIBA Foundation Symposium, Jointly with the International Union of Biological Sciences, on Biochemistry of Human Genetics* (J. A. Churchill, 1959), p. 11. See, for example, Harry Harris and Hans Kalmus, "The Measurement of Taste Sensitivity to Phenylthiourea (P.T.C.)" and "Chemical Specificity in Genetical Differences of Taste Sensitivity," *Annals of Eugenics*, 15 (1949), 24–31, 32–45.

19. Harry Harris, *Human Biochemical Genetics* (Cambridge University Press, 1959), pp. 83–84; C. E. Dent and H. Harris, "The Genetics of Cystinuria," *Annals of Eugenics*, 16 (1951), 60, 62; H. Harris, Ursula Mittwoch, Elizabeth B. Robson, and F. L. Warren, "Phenotypes and Genotypes in Cystinuria," *Annals of Human Genetics*, 20 (1955), 57; interview with Ursula Mittwoch.

20. Haldane to H. Munro Fox, July 31, 1952, Haldane Papers.

21. Haldane to Penrose, May 17, 1945;

various documents, Lionel Penrose Papers, files 159, 49/6; interviews with Harry Harris, Park Gerald, and Sylvia Lawler.

22. Interviews with Ursula Mittwoch, Harry Harris, Sylvia Lawler.

23. Interviews with Alexander Bearn, Sylvia Lawler.

24. Interviews with Ursula Mittwoch, Park Gerald, C. A. B. Smith, Sylvia Lawler, Barton Childs, Harry Harris, Alexander Bearn. C. A. B. Smith was introduced to the eccentricity of Penrose's Galton when he came to see the director about his letter of appointment. Penrose hurried through the administrative matters in a few sentences, then told Smith that he had there a gramophone record with twenty-five tunes on it, each recorded in a right and a wrong version. Would Smith mind seeing how many he could get right? Smith did so, making only one mistake. Penrose then took Smith to see another staff member, who asked him whether he could distinguish between the smells of lemon and almond. Smith said no, whereupon he was ushered to a specialist in dermatologyphics, who took his fingerprints. Once on the payroll, Smith was pretty much left, like almost everyone else, to do what he pleased. C. A. B. Smith interview.

25. Interviews with Park Gerald, Sylvia Lawler, James V. Neel, Alexander Bearn.

26. Interviews with Harry Harris, C. A. B. Smith, Alexander Bearn, James V. Neel.

27. Interviews with C. A. B. Smith, Ursula Mittwoch; Lionel S. Penrose, *The Biology of Mental Defect* (Sidgwick and Jackson, 1949; 2nd ed., 1953; 3rd ed. 1963); reviews of the book are in Lionel Penrose Papers, file 74; "Lionel S. Penrose—1898–1972," *M.A.P.W. Proceedings*, vol. 2, part 5, copy in Lionel Penrose Papers, file 20/5.

28. James V. Neel interview; list of postgraduate students and workers in the Galton Laboratory, 1945–65, Lionel Penrose Papers, file 49/2.

Chapter XV:

BLOOD, BIG SCIENCE, AND BIOCHEMISTRY

1. Sheldon C. Reed, "A Short History of Human Genetics in the U.S.A.," *American Journal of Medical Genetics*, 3 (1979), 292; interviews with Laurence Snyder, Arthur Steinberg, Park Gerald, Sylvia Lawler, Harry Harris.

2. Unless otherwise noted, the rest of this section is based on an interview and correspondence with James V. Neel. It was from Stern that Neel learned the proverb about the canary and its cage. James V. Neel, "Curt Stern, 1902–1911" (in press).

3. James V. Neel and William J. Schull, *The Effect of Exposure to the Atomic Bombs on Pregnancy Termination in Hiroshima and Nagasaki* (Publication No. 461, National Academy of Sciences–National Research Council; Atomic Bomb Casualty Commission, 1956), pp. 1, 3.

4. *Ibid.*, p. 4.

5. James V. Neel, "The Detection of the Genetic Carriers of Hereditary Disease," *American Journal of Human Genetics*, 26 (March 1949), 19–36; James V. Neel, "The Clinical Detection of the Genetic Carriers of Inherited Disease," *Medicine*, 26 (1947), 115–53.

6. D. J. Weatherall, "Toward an Understanding of the Molecular Biology of Some Common Inherited Anemias: The Story of Thalassemia," in Maxwell M. Wintrobe, ed., *Blood, Pure and Eloquent: A Story of Discovery, of People, and of Ideas* (McGraw-Hill, 1980), pp. 373, 376–80.

7. Confirmation of Neel's conclusion appeared in the same year from the independent effort of E. A. Beet, a physician with the Colonial Medical Service in Northern Rhodesia, whose data, however, consisted of only a single case. C. Lockard Conley, "Sickle-Cell Anemia—the First Molecular Disease," in Wintrobe, ed., *Blood, Pure and Eloquent*, pp. 323–24, 326–27; James V. Neel, "The Inheritance of Sickle Cell Anemia," *Science*, 110 (1949), 64–66, reprinted in Samuel H. Boyer IV, ed., *Papers on Human Genetics* (Prentice-Hall, 1963), pp. 110–14.

8. Harold A. Abramson et al., "Electrophoresis," *Annals of the New York Academy of Sciences*, 29 (Nov. 6, 1939), 105–212; Joseph S.

Fruton, *Molecules and Life: Historical Essays on the Interplay of Chemistry and Biology* (Wiley-Interscience, 1972), p. 144; Kai O. Pedersen, "Arne Wilhelm Kaurin Tiselius," *Dictionary of Scientific Biography*, XIII, 418–22.

9. Conley, "Sickle-Cell Anemia," pp. 338–39; James V. Neel, "Human Genetics," in John Z. Bowers and Elizabeth F. Purcell, eds., *Advances in American Medicine: Essays at the Bicentennial* (2 vols.; Josiah Macy, Jr., Foundation, 1976), I, 56; Linus Pauling, Harvey A. Itano, S. J. Singer, Ibert C. Wells, "Sickle Cell Anemia: A Molecular Disease," *Science*, 110 (1949), 543–48, reprinted in Boyer, ed., *Papers on Human Genetics*, pp. 115–25.

10. Neel and Schull, *The Effect of the Exposure to the Atomic Bombs*, pp. 2, 19, 192, 195–96, 204. In the summer of 1947, Neel went to Washington for a meeting at the National Academy of Sciences because, he recalled, he wanted "formal endorsement from my peers before I started spending those millions of dollars in what would probably be viewed as an inconclusive study." At the meeting, H. J. Muller expressed "deep concern that the absence of a statistically significant effect would be misinterpreted as meaning no effect." Nevertheless, Muller along with everyone else expected the project to prove enormously important for what it would show about the biological impact of radiation in general. Since early in his involvement with the studies in Japan, Neel has repeatedly insisted that the relative lack of genetic consequence must be understood as irrelevant to nuclear weapons policy. The immediate effects of nuclear weapons—blast, fire, radiation—amply suffice to render their use inadmissible. James Neel interview.

11. James Neel interview.

12. Conley, "Sickle-Cell Anemia," p. 331; Harry Harris, *Human Biochemical Genetics* (Cambridge University Press, 1959), p. 167; Anthony C. Allison, "Protection Afforded by Sickle-Cell Trait Against Subtertian Malarial Infection," *British Medical Journal*, 1 (1954), 290–94, reprinted in Boyer, ed., *Papers on Human Genetics*, pp. 153–57; James Neel interview; interview with Anthony C. Allison. To investigate the matter further, Allison also infected with strains of the malarial parasite two comparable groups of fifteen adult Luo male volunteers: one group whose members had sickle-cell trait, the other whose members did

not. During a forty-day period of observation—chemotherapy was administered to halt whatever infections there were at the end of this time—the malarial strains tended to thrive in the Luo without the trait but to fail in those with it. Being adults, both groups doubtless possessed acquired immunity to malaria, but at the time Allison thought the degree of immunity was not sufficient to interfere seriously with the experiment; he later realized, however, that the interference in fact was sufficient to render the experiment scientifically invalid. The volunteer Luo were patients in a mental hospital in Nairobi who were suffering from tertiary neurosyphilis, a disease marked first by physical symptoms, then by mental deterioration ending in insanity. Allison has recalled that the patients were mentally capable enough to give informed consent; that exposure to the malarial parasite was recognized as a form of therapy for their disease; and that the experiment fell within prevailing ethical standards of practice. He adds, nonetheless, that he came to regret having done this experiment on both ethical and scientific grounds. Allison interview; Allison, "Protection Afforded by Sickle-Cell Trait," pp. 157–59. See also Anthony C. Allison, "Polymorphism and Natural Selection in Human Populations," *Cold Spring Harbor Symposia on Quantitative Biology*, 29 (1964), 137–49.

13. James Neel interview.

14. James Neel interview.

15. Interviews with James Neel, Cedric O. Carter, and Arthur Steinberg; Lionel Penrose, "Memorandum on Gene Mutation in Man," Committee on Medical Aspects of Nuclear Radiation, Panel on Genetic Effects, 1955; Penrose to Charlotte Auerbach, April 1, 1955; Guido Pontecorvo to Penrose, April 5, 1955, Lionel Penrose Papers, University College London, file 79; *Sixth Semiannual Report of the United States Atomic Energy Commission*, July 1949, p. 40; *Twenty-first Semiannual Report of the United States Atomic Energy Commission*, January 1957, p. 82; *Twenty-fourth Semiannual Report of the United States Atomic Energy Commission*, July 1958, p. 166; A. Landsborough Thomson, *Half a Century of Medical Research* (2 vols.; I: *Origins and Policy of the Medical Research Council (UK)*; II: *The Programme of the Medical Research Council (UK)*; His Majesty's Stationery Office, 1973, 1975), II, 151.

16. James Neel interview.

17. Elizabeth Robson to Lionel Penrose, Dec. 14, 1954; Lionel Penrose, "Notes of North American Tour" [1958], Lionel Penrose Papers, files 165/9, 20/2.

18. *The New York Times*, Oct. 21, 1960, p. 28; *Time*, 76 (Oct. 31, 1960), 36.

19. Daniel J. Kevles and Stephen Postema, "Human Genetics in the United States and Great Britain, 1930–1959: A Statistical Survey" (in preparation); interviews with Alexander Bearn, Barton Childs, Park Gerald, Arthur Steinberg, James Neel, Victor McKusick, Laurence Snyder; Reed, "A Short History of Human Genetics in the U.S.A.," p. 289; James V. Neel and William J. Schull, *Human Heredity* (University of Chicago Press, 1954).

20. Interviews with Barton Childs, Alexander Bearn, Park Gerald; Lionel Penrose, "Notes of North American Tour."

21. Arthur Steinberg interview.

22. Barton Childs interview.

23. Interviews with Bentley Glass and Victor McKusick.

24. Victor McKusick interview. A. McGhee Harvey, the head of medicine at Hopkins for many years, has written that a major contribution of the Moore program was McKusick's directing a sizable number of "largely undifferentiated young clinicians into medical genetics as an academic clinical discipline. He showed them that hereditary diseases constitute an exciting area for investigation and a satisfying area for clinical service." A. McGhee Harvey, "Clinical Science at Johns Hopkins: Its Successful Pursuit in an Outpatient Setting" (manuscript version), pp. 50–54.

25. Kevles and Postema, "Human Genetics in the United States and Great Britain."

26. James Neel interview.

27. Anthony C. Allison to Lionel Penrose, April 6, 1955, Penrose Papers, file 114/1; Harry Harris interview; Neel, "Human Genetics," p. 66.

28. Harry Harris interview.

29. Letter to the author from Oliver Smithies, Feb. 6, 1984; Oliver Smithies, "Zone Electrophoresis in Starch Gels: Group Variations in the Serum Proteins of Normal Human Adults," *Biochemical Journal*, 61 (1955), 629–41; Harry Harris interview.

30. James Neel interview; Harris, *Human Biochemical Genetics*, pp. 202–3, 207–8; H. Harris, Elisabeth B. Robson, and M. Siniscalco, "Genetics of the Plasma Protein Variants," in G. E. W. Wolstenholme and Cecilia M. O'Connor, eds., *CIBA Foundation Symposium, Jointly with the International Union of Biological Sciences, on Biochemistry of Human Genetics* (J. A. Churchill, 1959), p. 151; Neel, "Human Genetics," pp. 56–57; Conley, "Sickle-Cell Anemia," pp. 344–45.

31. Conley, "Sickle-Cell Anemia," pp. 344–45; Horace Freeland Judson, *The Eighth Day of Creation: Makers of the Biological Revolution* (Simon & Schuster, 1979), pp. 300, 303–8.

32. V. M. Ingram, "A Specific Chemical Difference between the Globins of Normal Human and Sickle-Cell Anaemia Haemoglobin," *Nature*, 178 (1956), 792–94, reprinted in J. Herbert Taylor, ed., *Selected Papers on Molecular Genetics* (Academic Press, 1965), p. 68; Conley, "Sickle-Cell Anemia," pp. 344–45; Judson, *The Eighth Day of Creation*, pp. 303–8.

33. V. M. Ingram, "Gene Mutations in Human Haemoglobins: The Chemical Difference between Normal and Sickle-Cell Haemoglobin," *Nature*, 180 (1957), 326–28, reprinted in Boyer, ed., *Papers on Human Genetics*, pp. 144–45; J. A. Hunt and V. M. Ingram, "Abnormal Human Haemoglobins," in Wolstenholme and O'Connor, eds., *Biochemistry of Human Genetics*, pp. 119–20.

34. Conley, "Sickle-Cell Anemia," p. 343; Harris, *Human Biochemical Genetics*, pp. 6–7, 150, 281, 298; James V. Neel and Oliver Smithies, "General Discussion," in Wolstenholme and O'Connor, eds., *Biochemistry of Human Genetics*, pp. 322, 325.

35. Victor A. McKusick, "The Growth and Development of Human Genetics as a Clinical Discipline," *American Journal of Human Genetics*, 27 (1975), 261–62; Lionel S. Penrose, "Genetical Analysis of Man," in Wolstenholme and O'Connor, eds., *Biochemistry of Human Genetics*, p. 19; Penrose to Haldane, June 4, 1959, Penrose Papers, file 136.

Chapter XVI:
CHROMOSOMES—THE BINDER'S MISTAKES

1. T. C. Hsu, *Human and Mammalian Cytogenetics: An Historical Perspective* (Springer-Verlag, 1979), pp. 27–28.

2. Malcolm Jay Kottler, "From 48 to 46: Cytological Technique, Preconception, and the Counting of Human Chromosomes," *Bulletin of the History of Medicine*, 48 (1974), 467–71; James V. Neel, "Human Genetics," in John Z. Bowers and Elizabeth F. Purcell, eds., *Advances in American Medicine: Essays at the Bicentennial* (2 vols.; Josiah Macy, Jr., Foundation, 1976), I, 66.

3. Kottler, "From 48 to 46," pp. 472–74; Hsu, *Human and Mammalian Cytogenetics*, pp. 9–10; H. J. Muller to Julian Huxley, May 12, 1921, Hermann J. Muller Papers, Lilly Library, University of Indiana, Huxley file; Theophilus S. Painter to Ruggles Gates, July 13, 1921, R. Ruggles Gates Papers, Kings College London.

4. Neel, "Human Genetics," p. 66; Kottler, "From 48 to 46," p. 475; Theophilus S. Painter, "Studies in Mammalian Spermatogenesis," *The Journal of Experimental Zoology*, 37 (Jan.–July 1923), 291–321; Lancelot Hogben, *Genetic Principles in Medicine and Social Science* (Williams and Norgate, 1931), p. 41.

5. Hsu, *Human and Mammalian Cytogenetics*, p. 8; Kottler, "From 48 to 46," pp. 478–79.

6. Curt Stern, "High Points of Human Genetics," *American Biology Teacher*, 37 (March 1975), 144–45.

7. Hsu, *Human and Mammalian Cytogenetics*, p. 10; Kottler, "From 48 to 46," pp. 465, 478, 480, 484–85.

8. Hsu, *Human and Mammalian Cytogenetics*, pp. 16–18.

9. Kottler, "From 48 to 46," pp. 489–90; Hsu, *Human and Mammalian Cytogenetics*, pp. 20–21.

10. Kottler, "From 48 to 46," pp. 465, 489, 490, 493; Hsu, *Human and Mammalian Cytogenetics*, pp. 28–29.

11. Hsu, *Human and Mammalian Cytogenetics*, pp. 5–6; Kottler, "From 48 to 46," pp. 492–93; interview with Charles E. Ford; C. E. Ford and J. L. Hamerton, "The Chromosomes of Man," *Nature*, 178 (Nov. 10, 1956), 1020–23.

12. Interview with Paul E. Polani; Murray L. Barr and Ewart G. Bertram, "A Morphological Distinction Between Neurones of the Male and Female and the Behavior of the Nucleolar Satellite During Accelerated Nucleoprotein Synthesis," *Nature*, 163 (April 30, 1949), 676–77; Paul E. Polani, W. F. Hunter, and Bernard Lennox, "Chromosomal Sex in Turner's Syndrome with Coarctation of the Aorta," *The Lancet*, II (July 17, 1954), 120–21.

13. Hsu, *Human and Mammalian Cytogenetics*, pp. 54–55; Paul E. Polani, M. H. Lessof, and P. M. F. Bishop, "Colour-Blindness in 'Ovarian Agenesis,'" *The Lancet*, II (July 21, 1956), pp. 118–20.

14. P. M. F. Bishop, P. E. Polani, and M. H. Lessof, letter, "Klinefelter's Syndrome," *The Lancet*, II (Oct. 20, 1956), 843; Paul E. Polani, P. M. F. Bishop, B. Lennox, M. A. Ferguson-Smith, and J. S. S. Stewart, "Colour Vision Studies and the X-Chromosome Constitution of Patients with Klinefelter's Syndrome," *Nature*, 182 (Oct. 18, 1958), 1092–93; Paul Polani interview.

15. Paul Polani interview; Paul Polani to the author; November 15, 1984.

16. Interviews with Charles E. Ford, Paul Polani, and Patricia A. Jacobs; C. E. Ford, P. A. Jacobs, and L. G. Lajtha, "Human Somatic Chromosomes," *Nature*, 181 (June 7, 1958), 1565–68.

17. Interviews with Patricia Jacobs and Charles E. Ford; Patricia A. Jacobs and J. A. Strong, "A Case of Human Intersexuality Having a Possible XXY Sex-Determining Mechanism," *Nature*, 183 (Jan. 31, 1959), 302–3; C. E. Ford, P. E. Polani, J. H. Briggs, and P. M. F. Bishop, "A Presumptive Human XXY/XX Mosaic," *Nature*, 183 (April 1959), 1030–32; Lionel Penrose to J. B. S. Haldane, June 4, 1959, Lionel Penrose Papers, University College London, file 136.

18. Lionel S. Penrose, "1939 Report on Research Department, Royal Eastern Counties' Institution," Nov. 1939, Medical Research Council Records, Medical Research Council, London, Mental Disorders, Colchester, file 1588, folder iv; Lionel S. Penrose, "Survey of Recent Literature on Mongolism,

1940–1943"; Lionel Penrose, "Mongolism," May 27, 1960; Penrose to Haldane, June 4, 1959, Lionel Penrose Papers, files 61/1, 62/5, 136; Ursula Mittwoch, "The Chromosome Complement in a Mongolian Imbecile," *Annals of Eugenics*, 17 (1952–53), 37; Hsu, *Human and Mammalian Cytogenetics*, pp. 38–39; interviews with Ursula Mittwoch and Jérôme Lejeune.

19. Penrose to Haldane, June 4, 1959, Lionel Penrose Papers, file 136; Charles Ford interview.

20. Jérôme Lejeune interview.

21. Jérôme Lejeune interview.

22. R. Turpin and J. Lejeune, "Étude Dermatoglyphique des Paumes des Mongoliens et de Leurs Parents et Germains," *La Semaine des Hôpitaux de Paris*, 29 (Dec. 14, 1953), 3955–67; Jérôme Lejeune interview.

23. Raymond Turpin and Jérôme Lejeune, "Analogies entre le type dermatoglyphique palmaire des singes inférieurs et celui des enfants atteints de mongolisme," *Comptes Rendus Académie de Paris*," 238 (Jan. 18, 1954), 395–97.

24. Jérôme Lejeune interview.

25. Jérôme Lejeune interview.

26. Jérôme Lejeune interview.

27. Hsu, *Human and Mammalian Cytogenetics*, pp. 39–40; Jérôme Lejeune interview; Jérôme Lejeune, Marthe Gauthier, and Raymond Turpin, "Les Chromosomes humains en culture de tissus," *Comptes Rendus de l'Académie des Sciences*, 248 (Jan. 26, 1959), 602–3.

28. Jérôme Lejeune interview. Masuo Kodani, "Three Diploid Chromosome Numbers of Man," *Proceedings of the National Academy of Sciences*, 43 (1957), 285–92.

29. Lejeune, Gauthier, and Turpin, "Les Chromosomes humains en culture de tissus," pp. 602–3; Jérôme Lejeune interview; Jérôme Lejeune, Marthe Gauthier, and Raymond Turpin, "Études des chromosomes somatiques de neuf enfants mongoliens," *Comptes Rendus de l'Académie des Sciences*, 248 (March 16, 1959), 1721–22.

30. Letter from O. J. Miller to the author, Dec. 5, 1983, Penrose to Haldane, June 4, 1959, Lionel S. Penrose Papers, file 136; C. E. Ford, K. W. Jones, O. J. Miller, Ursula Mittwoch, L. S. Penrose, M. Ridler, and A. Shapiro, "The Chromosomes in a Patient Showing Both Mongolism and the Kline-

felter Syndrome," *The Lancet*, I (April 4, 1959), 709–10; Patricia A. Jacobs, W. M. Court Brown et al., "The Somatic Chromosomes in Mongolism," *The Lancet*, 1 (April 4, 1959), 710; interviews with Charles Ford and Patricia Jacobs.

31. Sir Ifor Evans to Penrose, May 22, 1959; Lionel Penrose, "Human Chromosomes," Oct. 22, 1959, Lionel S. Penrose Papers, files 175/5, 88/1.

32. Penrose, "Human Chromosomes"; Penrose, "Human Chromosomes for Beginners" [1962], Lionel S. Penrose Papers, file 88/5; Jérôme Lejeune interview; Jacobs, Brown et al., "The Somatic Chromosomes in Mongolism," p. 710; Jérôme Lejeune, Marthe Gauthier, and Raymond Turpin, "The Chromosomes of Man," letter, *The Lancet*, I (April 25, 1959), 885; "A Proposed Standard System of Nomenclature of Human Mitotic Chromosomes," *The Lancet*, I (May 14, 1960), 1063–65.

33. M. Fraccaro, K. Kaijser, and J. Lindsten, "Chromosomal Abnormalities in Father and Mongol Child," *The Lancet*, II (April 2, 1960), 724–27; P. E. Polani, J. H. Briggs, C. E. Ford, C. M. Clarke, and J. M. Berg, "A Mongol Girl with 46 Chromosomes," *The Lancet*, I (April 2, 1960), 721–24; L. S. Penrose, J. R. Ellis, and Joy Delhanty, "Chromosomal Translocations in Mongolism and in Normal Relatives," *The Lancet*, II (Aug. 20, 1960), 409–10; C. O. Carter, J. L. Hamerton, P. E. Polani, A. Gunalp, and S. D. V. Weller, "Chromosome Translocation as a Cause of Familial Mongolism," *The Lancet*, II (Sept. 24, 1960), 678–80; Lionel Penrose, "From Eugenics to Human Genetics," lecture, 1965, Lionel Penrose Papers, file 77/2.

34. Gordon Allen et al., "Mongolism," letter to the editor, *The Lancet*, I (1961), 775.

35. Charles E. Ford, K. W. Jones, Paul E. Polani, J. C. de Almeida, and J. H. Briggs, "A Sex-Chromosome Anomaly in a Case of Gonadal Dysgenesis (Turner's Syndrome)," *The Lancet*, I (April 4, 1959), 711–13; Neel, "Human Genetics," pp. 69, 73; Lionel S. Penrose, *Outline of Human Genetics* (2nd ed.; Heinemann, 1963), p. 136; Penrose, "Human Chromosomes."

36. Daniel J. Kevles and Stephen Postema, "Human Genetics in the United States and Great Britain, 1930–1959: A Statistical Survey" (in preparation); Victor A. McKu-

sick, "The Growth and Development of Human Genetics as a Clinical Discipline." *American Journal of Human Genetics*, 27 (1975), 262; Barton Childs, "Garrod, Galton, and Clinical Medicine," *Yale Journal of Biology and Medicine*, 46 (1973), 298–99.

37. Lionel S. Penrose, "The Influence of the English Tradition in Human Genetics," in James F. Crow and James V. Neel, eds., *Proceedings of the Third International Congress of Human Genetics* (Johns Hopkins University Press, 1967), pp. 19–20.

Chapter XVII:
A NEW EUGENICS

1. Lionel S. Penrose, "The Influence of the English Tradition in Human Genetics," in James F. Crow and James V. Neel, eds., *Proceedings of the Third International Congress of Human Genetics* (Johns Hopkins University Press, 1967), pp. 22–23; Harry Harris, *Human Biochemical Genetics* (Cambridge University Press, 1959), p. v. Harris later reflected: "All the developments have taught us what would seem self-evident: first, that there's no hard-and-fast distinction between what is genetically normal and genetically abnormal. And second, there's no one normality, only different versions of normality." Interview with Harry Harris.

2. James V. Neel and William J. Schull, *Human Heredity* (University of Chicago Press, 1954), pp. 337–38, 341. Sheldon Reed, head of the Dight Institute of Human Heredity in Minnesota, complained that people like Charles Davenport had been made scapegoats for the Nazi murders of Jews. Sheldon C. Reed to Harry L. Shapiro, May 15, 1961, American Eugenics Society Papers, box 4.

3. Harry Harris interview; Lionel S. Penrose, "Phenylketonuria: A Problem in Eugenics," *The Lancet*, I (June 29, 1946), 949; Penrose, "Memorandum to Provost," Dec. 4, 1961; Kenneth Ewart to James Henderson, Aug. 15, 1961, and Provost to Penrose, Dec. 1, 1961, Lionel S. Penrose Papers, 175/5. John Maynard Smith remembered that the Eugenics Society's ongoing interest in sterilization led some people at the Galton to capsule the group as those "off-with-their-cocks boys." Interview with John Maynard Smith.

4. Fisher to Blacker, April 20, 1951; Blacker to Fisher, April 18, 1951; Eugenics Society Records, file C.108; interviews with Harry Harris and John Fraser Roberts; Frederick Osborn, "History of the American Eu-

genics Society," p. 18, American Eugenics Society Papers, box 15.

5. Osborn to Childs Frick, Jan. 12, 1965; Osborn to Bernard Berelson, Jan. 28, 1971; Osborn to Dudley Kirk, Oct. 31, 1972; Frederick Osborn, "History of the American Eugenics Society," p. 4, American Eugenics Society Papers, boxes 2, 3, 4, 15; C. P. Blacker, *Eugenics in Prospect and Retrospect* (Hamish Hamilton, 1945), p. 16. While it lasted, the American society, C. P. Blacker noted, pursued "eugenic ends by less obvious means"—mainly through efforts to encourage the use of genetics for medical purposes and to improve the biological quality of human populations. Blacker called the American program a "policy of crypto-eugenics," which was obviously reform eugenics carried into the postwar period. Faith Schenk and A. S. Parkes, *The Activities of the Eugenics Society* (1969?), p. 154, offprint copy in Eugenics Society Records.

6. Ian H. Porter, "Evolution of Genetic Counseling in America," in Herbert A. Lubs and Felix de la Cruz, eds., *Genetic Counseling: A Monograph of the National Institute of Child Health and Human Development* (Raven Press, 1977), p. 26; Sheldon C. Reed, *Counseling in Medical Genetics* (W. B. Saunders, 1955), pp. 3–4; Sheldon C. Reed, "A Short History of Genetic Counseling," *Social Biology*, 21 (1974), 334–35; Charles B. Davenport to John C. Merriam, May 5, 1922, Charles B. Davenport Papers, Cold Spring Harbor, Long Island, Series #1; John Fraser Roberts interview. Cedric O. Carter, while serving as part-time secretary of the Eugenics Society and working under Fraser Roberts in Great Ormond Street, responded with letters of genetic counseling to the many inquiries concerning a variety of diseases and disabilities that came to the Society. Eugenics Society Records, files D.3 and D.4. Carter also wrote

Human Heredity to inform the lay public about basic and applied human genetics. Published first in 1962 and in a second edition (Penguin Books) in 1977, it had sold some 130,000 copies by 1980. Interview with Cedric O. Carter.

7. James V. Neel, "The Clinical Detection of the Genetic Carriers of Inherited Disease," *Medicine*, 26 (1947), 141–42; James V. Neel, "The Detection of the Genetic Carriers of Hereditary Disease," *American Journal of Human Genetics*, 1 (1949), 26–28; James V. Neel, "The Meaning of Empiric Risk Figures for Disease or Defect," in Helen G. Hammons, ed., *Heredity Counseling: A Symposium Sponsored by the American Eugenics Society* (Harper & Bros., 1959), p. 66; Barton Childs interview; William J. Schull, "The Problem of Inadequate Counseling," in "Discussions," in Hammons, ed., *Heredity Counseling*, p. 106. Arthur Steinberg, who took over responsibility for the Lakeside Heredity Clinic when, in 1957, he joined the faculty of Western Reserve University in Cleveland, recalled the difficulty of dealing with some young couples: "They'd had a baby with some awful business, and all you could say was they had a twenty-five percent risk in its recurring. Couldn't do much of anything for them. I'd come home after clinic and my wife would take a look at me and say, 'You had heredity clinic today.' " Steinberg likened the stress to that on the physicians at the Jimmy Fund in Boston, which specialized in childhood leukemias. Interview with Arthur Steinberg.

8. Sheldon C. Reed, "Types of Advice Given by Heredity Counselors: II," in Hammons, ed., *Heredity Counseling*, p. 87; Reed, "A Short History of Genetic Counseling," p. 335; Arthur Steinberg interview; Ian H. Porter, "Evolution of Genetic Counseling in America," in Lubs and de la Cruz, eds., *Genetic Counseling*, pp. 27–28; Steven M. Spencer, "New Strides in the Battle Against Birth Defects," *Reader's Digest*, 98 (May 1971), 159–64; Osborn, "History of the American Eugenics Society," p. 20; Charles Lindbergh to Osborn, Dec. 9, 1970, American Eugenics Society Papers, box 3.

9. Osborn, "History of the American Eugenics Society," p. 19; Osborn to James R. Sorenson, Aug. 4, 1971, American Eugenics Society Papers, box 4; Barton Childs, "Garrod, Galton, and Clinical Medicine," *Yale Journal of Biology and Medicine*, 46 (1973),

307–8; interview with Victor McKusick.

10. Interview with Victor McKusick; A. McGhee Harvey, "Clinical Science at Johns Hopkins: Its Successful Pursuit in an Outpatient Setting" (manuscript version, pp. 69–70), later published as "Clinical Investigation of Chronic Diseases," *Journal of Chronic Diseases* 33 (1980), 529–66; Marshall D. Levine, J. M. Gursky, and D. L. Rimoin, "Results of a Survey on the Teaching of Medical Genetics to Medical Students, House Officers, and Fellows" (1975), in Lubs and de la Cruz, *Genetic Counseling*, p. 363; Marion Steinman, "Fighting the Genetic Odds," *Life*, 71 (Aug. 6, 1971), 19–25; Childs, "Garrod, Galton, and Clinical Medicine," pp. 307–8.

11. Robert Guthrie, "A Simple Phenylalanine Method for Detecting Phenylketonuria in Large Populations of Newborn Infants," *Pediatrics*, 32 (1963), 338–43; James V. Neel, "Human Genetics," in John Z. Bowers and Elizabeth F. Purcell, eds., *Advances in American Medicine: Essays at the Bicentennial* (2 vols.; Josiah Macy, Jr., Foundation, 1976), I, 86; Committee for the Study of Inborn Errors of Metabolism . . . National Research Council, *Genetic Screening: Programs, Principles, and Research* (National Academy of Sciences, 1975), p. 344; Philip Reilly, *Genetics, Law, and Social Policy* (Harvard University Press, 1977), pp. 27, 37; Gene Bylinsky, "What Science Can Do About Hereditary Diseases," *Fortune*, Sept. 1974, pp. 154–55. By themselves, victims of PKU were estimated to account for one percent of all residents in institutions for the mentally retarded. Charles R. Scriver, "Screening, Counseling, and Treatment for Phenylketonuria: Lessons Learned—A Precis," in Lubs and de la Cruz, eds., *Genetic Counseling*, pp. 255–56.

12. Steinman, "Fighting the Genetic Odds," pp. 19–25; Victor A. McKusick, "The Growth and Development of Human Genetics as a Clinical Discipline," *American Journal of Human Genetics*, 27 (1975), 264; Harris, *Human Biochemical Genetics*, pp. 108–9; Reilly, *Genetics, Law, and Social Policy*, p. 60; telephone conversations with Seymour Packman and George C. Cunningham; George C. Cunningham, "How Cost-Effective Are Our Genetic Screening Methods?" (unpublished manuscript); Committee for the Study of Inborn Errors of Metabolism

. . . National Research Council, *Genetic Screening*, pp. 203–9.

13. Steinman, "Fighting the Genetic Odds," pp. 19–25.

14. Reilly, *Genetics, Law, and Social Policy*, pp. 37, 64–67, 80; C. Lockard Conley, "Sickle-Cell Anemia—the First Molecular Disease," in Maxwell M. Wintrobe, ed., *Blood, Pure and Eloquent: A Story of Discovery, of People, and of Ideas* (McGraw-Hill, 1980), p. 358; Neel, "Human Genetics," p. 65; *The New York Times*, Oct. 27, 1970, p. 51.

15. Tabitha Powledge, "Genetic Screening as a Political and Social Development," in Daniel Bergsma et al., eds., *Ethical, Social, and Legal Dimensions of Screening for Human Genetic Disease* (National Foundation–March of Dimes: Birth Defects, Original Articles Series, vol. X; Stratton Intercontinental Medical Book Group, 1974), p. 34; "Genetics for the Community," *Time*, Sept. 13, 1971, p. 54; Bylinsky, "What Science Can Do About Hereditary Diseases," p. 158.

16. Reilly, *Genetics, Law, and Social Policy*, pp. 30, 81–82, 102–10; this designation was authorized for Title IV of U.S. Public Law 94–278, April 22, 1978, *90 Stat.*, 407 ff.

17. Orlando J. Miller, "An Overview of Problems Arising from Amniocentesis"; Fritz Fuchs, "Amniocentesis: Techniques and Complications," in Maureen H. Harris, ed., *Early Diagnosis of Human Genetic Defects: Scientific and Ethical Considerations* (Fogarty International Center Proceedings No. 6; Government Printing Office, 1972), pp. 23, 11–12; Neel, "Human Genetics," p. 69; Theodore Friedmann, "Prenatal Diagnosis of Genetic Disease," *Scientific American*, 225 (Nov. 1971), 35–42; Harold M. Schmeck, "Fetal Research Reported on Rise," New York *Times*, Feb. 16, 1975, p. 31. President's Commission for the Study of Ethical Problems in Medicine and Biomedical and Behavioral Research, *Screening and Counseling for Genetic Conditions: The Ethical, Social, and Legal Implications of Genetic Screening, Counseling, and Educational 'Programs* (Government Printing Office, 1983), p. 23.

18. Aubrey Milunsky, "Prenatal Genetic Diagnosis: Risks and Needs," in Lubs and de la Cruz, eds., *Genetic Counseling*, pp. 484–85; C. O. Carter, "Practical Aspects of Early Diagnosis," in Harris, ed., *Early Diagnosis of Human Genetic Defects*, p. 18; Jane E.

Brody, "Genetic Defects Sought in Fetus," New York *Times*, May 12, 1976, p. 12.

19. Arthur Salisbury, an officer of the National Foundation–March of Dimes, recalled that at the peak of its program the Foundation was funding eighty-five facilities in thirty-seven states. Interview with Arthur Salisbury. Ian H. Porter, "Evolution of Genetic Counseling in America," in Lubs and de la Cruz, eds., *Genetic Counseling*, p. 28; New York *Times*, May 12, 1976, p. 17; M. A. Ferguson-Smith et al., *The Provision of Services for the Prenatal Diagnosis of Fetal Abnormality in the United Kingdom: Report of the Clinical Genetics Society Working Party on Prenatal Diagnosis in Relation to Genetic Counseling* (Supplements to the Bulletin of the Eugenics Society, No. 3; Eugenics Society, 1978), pp. 4–5.

20. Interviews with John Burn, Cedric O. Carter, and Paul E. Polani.

21. C. O. Carter, "Current Status of Genetic Counseling and Its Assessment," in Arno Motulsky and F. J. G. Egling, eds., *Abstracts of Papers; Fourth International Conference on Birth Defects . . .* (Excerpta Medica, 1974), p. 277; Barton Childs, "Genetic Counseling: A Critical Review of the Published Literature," in Bernice H. Cohen, Abraham M. Lilienfeld, and P. C. Huang, eds., *Genetic Issues in Public Health and Medicine* (Charles C Thomas, 1978), p. 337; Arthur Steinberg interview; Reed, *Counseling in Medical Genetics*, p. 14; James R. Sorenson, "Some Social and Psychologic Issues in Genetic Screening: Public and Professional Adaptation to Biomedical Innovation," in Bergsma et al., eds., *Ethical, Social, and Legal Dimensions of Screening for Human Genetic Disease*, p. 169. In 1946, Penrose had told C. P. Blacker: "My own experience is that eugenic prognosis is much more often a guess about the social value of the offspring rather than a mathematical calculation of the chances of Mendelian inheritance." Sheldon Reed recalled that he had hit upon the term "genetic counseling" because he wanted to identify it as "a kind of genetic social work without eugenic connotations." As John Fraser Roberts typically reflected, "It seemed to me that you wanted to give potential parents all the information there was, and give it to them in detail, but the decision was entirely theirs." Penrose to Blacker, Jan. 25, 1946, Eugenics

Society Records, file C.271; Sheldon C. Reed, "A Short History of Genetic Counseling," p. 338; John Fraser Roberts interview.

22. Frederick Osborn, *The Future of Human Heredity* (Weybright and Talley, 1968), pp. 29–30; Lionel S. Penrose, "Genetics and Society," 1969, manuscript in Lionel S. Penrose Papers, file 103; James V. Neel, "Social and Scientific Priorities in the Use of Genetic Knowledge," in Bruce Hilton, Daniel Callahan et al., eds., *Ethical Issues in Human Genetics: Genetic Counseling and the Uses of Genetic Knowledge* (Plenum Press, 1973), p. 361.

23. J. B. S. Haldane, "Biological Possibilities in the Next Ten Thousand Years," in Gordon Wolstenholme, ed., *Man and His Future* (A CIBA Foundation volume; Little, Brown, 1963), p. 341; Osborn to William Shockley, March 8, 1966, American Eugenics Society Papers, box 2; Osborn, *The Future of Human Heredity*, pp. 29–30; P. B. Medawar, *The Future of Man* (Methuen, 1959), p. 84; Frederick Osborn and Carl Jay Bajema, "The Eugenic Hypothesis," *Social Biology*, 19 (1972), reprinted in Carl Jay Bajema, ed., *Eugenics Then and Now* (Benchmark Papers in Genetics, No. 5; Dowden Hutchinson and Ross, 1976), pp. 288–90. In Michigan and Minnesota, and likely everywhere else in the nation, Sheldon Reed reported, the net birthrate of the mentally deficient was only two-thirds that of people with high I.Q.s and did not threaten the overall national intelligence. Sheldon C. Reed, "The Normal Process of Genetic Change in a Stable Physical Environment," in John D. Roslansky, ed., *Genetics and the Future of Man* (Appleton-Century-Crofts, 1966), pp. 19–20.

24. Julian S. Huxley, "Eugenics in Evolutionary Perspective" [Galton lecture, 1962], in Julian S. Huxley, *Essays of a Humanist* (Harper & Row, 1964), pp. 266–67. Osborn argued that whether the social qualities of children from lower-income homes were in origin genetic, environmental, or some combination of the two, their "relative inadequacy" tended to be handed on to the next generation. Frederick Osborn, "Qualitative Aspects of Population Control: Eugenics and Euthenics," *Law and Contemporary Problems*, Summer 1960, p. 416, copy in American Eugenics Society Papers, box 15.

25. Osborn expected that the racially differential birthrate would diminish, as it had done among other minority groups, with the still wider spread of birth control. Huxley, less sanguine about the voluntary responsiveness of the "social problem group," suggested, given how "grave" was the threat, the imposition of birth control upon its members by "compulsory or semi-compulsory" means. Huxley, "Eugenics in Evolutionary Perspective," p. 270; Osborn to Dwight J. Ingle, May 14, 1969, Osborn to Joseph C. Wilson, Sept. 14, 1967; Osborn to James W. Reed, May 2, 1973, American Eugenics Society Papers, box 3, box 4; Linda Gordon, *Woman's Body, Woman's Right: A Social History of Birth Control in America* (Grossman, 1976), pp. 395–97.

26. Elof Axel Carlson, *Genes, Radiation, and Society: The Life and Work of H. J. Muller* (Cornell University Press, 1981), pp. 341–43.

27. H. J. Muller, "Our Load of Mutations," *American Journal of Human Genetics*, 2 (June 1950), 145, 142.

28. *Ibid.*, p. 142.

29. *Ibid.*, pp. 145–46; Hermann J. Muller, "The Guidance of Human Evolution," *Perspectives in Biology and Medicine*, 3 (Autumn 1959), 9.

30. Stalin's brutal repression of Lysenko's opponents had, it appears, shifted the balance for Muller somewhat toward heredity in the nature-nurture equation. He did not seem to consider genetic load when in 1959 he identified dangerously large retinues of children, "legitimate or otherwise," with people who were "dominated by superstitious taboos," or were "unduly egotistical," or "shiftless or bungling." Privately, Muller allowed that alcoholics and criminals "must contain a much higher than average proportion of genetically defective people." Muller to P. S. Barrows, Oct. 7, 1965, Hermann J. Muller Papers, Lilly Library, University of Indiana, Germinal Choice file, box IV; H. J. Muller, "Progress and Prospects in Human Genetics," *American Journal of Human Genetics*, 1 (Sept. 1949), 2–3; Muller, "The Guidance of Human Evolution," p. 13.

31. Neel, "Social and Scientific Priorities in the Use of Genetic Knowledge," pp. 357–58; Julian Huxley, "The Future of Man," in Wolstenholme, ed., *Man and His Future*, p. 17. In his last book, Frederick Osborn was occupied with genetic load and asserted that

at least two percent of the American population, or about four million people, suffered some sort of chronic weakness because of a "poor heredity." Osborn, *The Future of Human Heredity*, pp. 96–97.

32. Muller, "Our Load of Mutations," pp. 150, 171; Carlson, *Genes, Radiation, and Society*, p. 397; Muller to P. S. Barrows, Oct. 7, 1965, Hermann J. Muller Papers, Germinal Choice file, box IV; Hermann J. Muller, "The Guidance of Human Evolution," short version in *Studies in Genetics: The Selected Papers of H. J. Muller* (Indiana University Press, 1962), pp. 589–90.

33. Muller, "The Guidance of Human Evolution," pp. 31–32; Carlson, *Genes, Radiation, and Society*, pp. 398–99; Reilly, *Genetics, Law, and Social Policy*, p. 193.

34. Hermann J. Muller, "Human Evolution by Voluntary Choice of Germ Plasm," *Science*, 134 (Sept. 8, 1961), 648; Muller, "The Guidance of Human Evolution," p. 25; Muller to P. S. Barrows, Oct. 7, 1965, Hermann J. Muller Papers, Germinal Choice file, box IV; Carlson, *Genes, Radiation, and Society*, pp. 400, 404; Hermann J. Muller, "What Genetic Course Will Man Steer?" in Crow and Neel, eds., *Proceedings of the Third International Congress of Human Genetics*, pp. 532–33.

35. Muller, "What Genetic Course Will Man Steer?" p. 532; Carlson, *Genes, Radiation, and Society*, p. 400; Muller to Govind Dev, Dec. 7, 1959, Hermann J. Muller Papers, Germinal Choice file, box Ia.

36. Blacker to H. J. Muller, Oct. 9, 1964, Eugenics Society Records, file C.24; letters to James Crow, George Beadle, Fred Osborn, J. B. S. Haldane, Julian Huxley et al., March 20, 1961; C. W. Kline to Muller, March 19, 1958; agreement for Germinal Choice, June 5, 1963, signed by Hermann J. Muller and Robert K. Graham, all in Hermann J. Muller Papers, Germinal Choice file, box Ic.

37. Noting that Graham had backed Barry Goldwater in the 1964 elections, Muller wrote with irritation that Graham would "doubtless consider *him* a good donor." Muller to Raymond B. Cattell, Nov. 18, 1965; see also Muller to Cattell, May 8, 1964, Hermann J. Muller Papers, Germinal Choice file, box Ib; Muller to Robert Graham, May 29, 1963; Robert Graham to Dorothea Muller, March 10, 1980; Thea Muller,

"Statement," March 1, 1980, Hermann J. Muller Papers, Germinal Choice file, box IX; Elof Axel Carlson, "Eugenics Revisited: The Case for Germinal Choice," *Stadler Symposium*, 5 (1973), 22; Lori B. Andrews, "Brave New Baby," *The Student Lawyer*, Dec. 1983, pp. 26–27; Marjorie Wallace, "Waiting for a Sperm-Bank Genius," *The Sunday Times* (London), July 4, 1982, pp. 13–14; Los Angeles *Times*, March 2, 1980, section II, pp. 1, 5; Feb. 29, 1980, p. 1; July 31, 1981, p. 3; brochure, "Repository for Germinal Choice," enclosed with letter from Warren A. Greaves to Fred E. C. Culick, March 22, 1984, copies supplied me by Mr. Culick.

38. Paul E. Smith to Muller, April 9, 1962, Hermann J. Muller Papers, Germinal Choice file, box Id; Philip H. Abelson, "Who Shall Live?" mimeographed text of address given at the Johns Hopkins School of Hygiene and Public Health, 50th Anniversary Celebration, Oct. 5, 1966, Hermann J. Muller Papers, Germinal Choice file, box Ib; H. J. Muller to Blacker, Nov. 28, 1960, Eugenics Society Records, file C.24; "Eugenics and Genetics: Discussion," in Wolstenholme, ed., *Man and His Future*, pp. 296–97. "Just the idea of a sperm bank!" someone exclaimed to a Louis Harris pollster. "It would be like a blood bank—they'd end up having a sperm drive every year to replenish the supply. Spermmobiles would be around like bloodmobiles." Louis Harris, "The Life Poll," *Life*, 23 (June 13, 1969), 52.

39. Ernst Mayr, *Animal Species and Evolution* (Harvard University Press, 1963), pp. 661–62; James F. Crow, "Mechanisms and Trends in Human Evolution," *Daedalus*, 90 (Summer 1961), 430; Francis Crick, in "Eugenics and Genetics: Discussion," in Wolstenholme, ed., *Man and His Future*, pp. 274–75.

40. Albert Rosenfeld, "The Second Genesis," *Life*, 23 (June 13, 1969), 40.

41. President's Commission for the Study of Ethical Problems in Medicine and Biomedical and Behavioral Research, *Splicing Life: A Report on the Social and Ethical Issues of Genetic Engineering with Human Beings* (Government Printing Office, 1982); p. 8; Paul Ramsey, "Shall We Clone a Man?" in Kenneth Vaux, ed., *Who Shall Live? Medicine, Technology, and Ethics* (Fortress Press, 1970), pp. 81–82.

42. With cloning, one could "order up carbon copies of people," the Caltech biologist James F. Bonner predicted. Ramsey, "Shall We Clone a Man?" p. 81.

43. *Ibid.*, p. 82; Anne McLaren, "Methods and Success of Nuclear Transplantation in Mammals," *Nature*, 309 (June 21, 1984), 671–72.

44. R. G. Edwards, "Aspects of Human Reproduction," in Watson Fuller, ed., *The Social Impact of Modern Biology* (Routledge and Kegan Paul, 1970), p. 110.

45. *Ibid.*, pp. 110, 115–16; Frederick Ausubel, Jonathan R. Beckwith, and Kaaren Janssen, "The Politics of Genetic Engineering: Who Decides Who's Defective?" *Psychology Today*, June 1974, p. 32; Glass, "Science: Endless Horizons or Golden Age?" p. 28; Muller, "The Guidance of Human Evolution," p. 33; R. G. Edwards, "Advances in Reproductive Biology and Their Implications for Studies on Human Congenital Defects," in Motulsky and Ebling, eds., *Birth Defects*, pp. 100–1; Rosenfeld, "The Second Genesis," p. 44.

46. W. French Anderson, "Genetic Therapy," in Michael P. Hamilton, ed., *The New Genetics and the Future of Man* (William B. Eerdmans, 1972), pp. 112–13.

47. Rosenfeld, "The Second Genesis," pp. 44–45; Glass, "Science: Endless Horizons or Golden Age?" p. 28; John Maynard Smith, "Eugenics and Utopia," *Daedalus*, 94 (1965), 487–88; Theodore Friedmann, "The Future for Gene Therapy—Reevaluation," in Marc Lappé and Robert S. Morison, eds., *Ethical and Scientific Issues Posed by Human Uses of Molecular Genetics* (Annals of the New York Academy of Sciences, vol. 265; New York Academy of Sciences, 1976), pp. 147–48; Robert L. Sinsheimer, "The Prospect of Designed Genetic Change," *Engineering and Science*, 32 (April 1969), 11–12; Theodore Friedmann and Richard Roblin, "Gene Therapy for Human Genetic Disease?" *Science*, 175 (March 3, 1972), 951. At the time, the method most commonly suggested by which the preferred DNA might be incorporated in the human cell nucleus drew upon the fact that certain viruses were known on occasion to take part of the DNA from one bacterial host, then introduce the foreign fragment into the DNA strand of another. It was thus speculated that existing viruses might be controlled—or perhaps even new ones designed —to take a snip of healthy human DNA from a test tube and insert it into a human being's genetically diseased or deficient cell. Even if this could be accomplished, germinal choice remained far more attractive than molecular genetic engineering to H. J. Muller, who thought the latter highly problematic. Muller pointed out that it would require an immense storehouse of as yet unknown data—minute knowledge of the role played by individual genes in the organism; the locations of the genes on the chromosomes; the constitution of the genes down to every base pair; the effect on the phenotype of changing one of the three billion nucleotide pairs that the total human genome was estimated to contain. Muller called it a "monstrous" task simply to identify the nucleotides, and "titanic" to determine how all the genetic parts, "through devious successions of intricately interwoven processes, finally make the man as we find him." Muller, "The Guidance of Human Evolution," p. 7.

48. T. C. Hsu, *Human and Mammalian Cytogenetics: An Historical Perspective* (Springer-Verlag, 1979), pp. 99–101; Neel, "Human Genetics," p. 74; interview with O. J. Miller.

49. Hsu, *Human and Mammalian Cytogenetics*, pp. 134–35.

50. *Ibid.*; Neel, "Human Genetics," pp. 76–77; James V. Neel, "Our Twenty-fifth," *American Journal of Human Genetics*, 26 (March 1974), 141.

51. Richard Roblin, "Reflections on Issues Posed by Recombinant DNA Molecule Technology, I"; Theodore Friedmann, "The Future for Gene Therapy—a Reevaluation," in Lappé and Morison, eds., *Ethical and Scientific Issues Posed by Human Uses of Molecular Genetics*, pp. 59–60, 147–48; Anderson, "Genetic Therapy," pp. 112–13; Glass, "Science: Endless Horizons or Golden Age?" p. 28; Sinsheimer, "The Prospect of Designed Genetic Change," pp. 11–12; Friedmann and Roblin, "Gene Therapy for Human Genetic Disease?" p. 951.

52. Sinsheimer, "The Prospect of Designed Genetic Change," pp. 8, 13.

53. *Ibid.*, p. 8.

Chapter XVIII:
VARIETIES OF PRESUMPTUOUSNESS

1. *The Observer* (London), Oct. 20, 1974, p. 1.

2. Arthur S. Jensen, *Genetics and Education* (Harper & Row, 1972), pp. 51, 65, 157–58; Arthur S. Jensen, "How Much Can We Boost IQ and Scholastic Achievement?" *Harvard Educational Review*, 33 (1969), 1–123, reprinted in *Genetics and Education*, pp. 69–203 (citations here and henceforth are to this reprint of the article); Jensen, "The Ethical Issues," *Genetics and Education*, pp. 328–29.

3. Jensen, "How Much Can We Boost IQ and Scholastic Achievement?" pp. 162–63, 178–79.

4. *Ibid.*, pp. 159–60, 162–63; Jensen, *Genetics and Education*, pp. 8–9.

5. Richard Herrnstein, "I.Q.," *The Atlantic*, 228 (Sept. 1971), 63–64; H. B. Gibson, *Hans Eysenck: The Man and His Work* (Peter Owen, 1981), pp. 30–34; Hans J. Eysenck, *The I.Q. Argument: Race, Intelligence and Education* (The Library Press, 1971), p. i; Hans J. Eysenck, *The Inequality of Man* (EdITS, 1975). R. A. Fisher had adumbrated Herrnstein's argument: "If desirable characters, intelligence, enterprise, understanding of our fellow men, capacity to arouse their admiration or confidence, exert any net average social advantage, then it follows that they will become correlated with social class. The more thoroughly we carry out the democratic programme of giving equal opportunities to talent wherever it is found, the more thoroughly we insure that genetic class differences of eugenic value shall be built up." Fisher to E. B. Wilson, Aug. 2, 1930, in J. H. Bennett, ed., *Natural Selection, Heredity and Eugenics, Including Selected Correspondence of R. A. Fisher with Leonard Darwin and Others* (Clarendon Press, 1983), p. 272.

6. Eysenck, *The Inequality of Man*, pp. 144–45, 219–20, 224–27, 250–51, 255–56; Eysenck, *The I.Q. Argument*, pp. ii–iv.

7. Jensen, "The Ethical Issues," p. 331; Jensen, "How Much Can We Boost IQ and Scholastic Achievement?" pp. 178–79; *Time*, 89 (Feb. 3, 1967), 65; *Science*, 168 (May 8, 1970), 685; 172 (May 7, 1971), 539–41; *Newsweek*, 77 (May 10, 1971), 69–70; *The New York Times*, May 3, 1970, p. 58; April 29, 1971, p. 24.

8. Edward O. Wilson, "Human Decency Is Animal," *The New York Times Magazine*, Oct. 12, 1975, p. 39.

9. Edward O. Wilson, *Sociobiology: The New Synthesis* (Harvard University Press, 1975); Edward O. Wilson, "Human Decency Is Animal"; Edward O. Wilson, *On Human Nature* (Harvard University Press, 1978); *Time*, 110 (August 1, 1977); Stephen Jay Gould, "Biological Potential vs. Biological Determinism," *Natural History Magazine*, 85 (May 1976), reprinted in Arthur L. Caplan, ed., *The Sociobiology Debate: Readings on Ethical and Scientific Issues* (Harper and Row, 1978), p. 344; Albert Rosenfeld, "Sociobiology Stirs a Controversy over Limits of Science," *Smithsonian*, 11 (Sept. 1980), p. 73.

10. Wilson, *On Human Nature*, pp. 32, 171; Edward O. Wilson, *Sociobiology: The Abridged Edition* (Harvard University Press, 1980), p. 275. All references here are to this edition of *Sociobiology*, which includes the complete Chapter 27 of the original.

11. Wilson, "Human Decency Is Animal," pp. 39–42; Loren R. Graham, *Between Science and Values* (Columbia University Press, 1981), pp. 203–04.

12. Wilson, *Sociobiology*, pp. 275, 279, 287; Wilson, "Human Decency Is Animal," pp. 45, 48; Wilson, *On Human Nature*, pp. 4–5, 124–25, 143–45, 153–54.

13. Wilson, "Human Decency Is Animal," pp. 48, 50; Wilson, *On Human Nature*, pp. 6–7, 96–97, 147, 208; Graham, *Between Science and Values*, p. 211; interview with Edward O. Wilson.

14. Rosenfeld, "Sociobiology Stirs a Controversy," pp. 73, 79; Wilson, *On Human Nature*, p. 50; *Time*, 110 (Aug. 1, 1977), 55; Edward O. Wilson, "Academic Vigilantism and the Political Significance of Sociobiology," *BioScience*, 26 (March 1976), reprinted in Caplan, ed., *The Sociobiology Debate*, pp. 291–93.

15. Wilson, "Human Decency Is Animal," pp. 48, 50.

16. *Time*, 110 (Aug. 1, 1977), 54–55; Paul Samuelson, "Social Darwinism," *Newsweek*, 86 (July 7, 1975), 55; Graham, *Between Science and Values*, pp. 209–12. Wilson gave the im-

pression of contradictoriness by, for example, writing in one place: "*What the genes prescribe is not necessarily a particular behavior but the capacity to develop certain behaviors* . . . ," and saying in another: "The question of interest is no longer whether human social behavior is genetically determined; it is to what extent." Wilson, "Human Decency Is Animal," pp. 46, 48; Wilson, *On Human Nature*, pp. 18–19.

17. Jonas Robitscher, "Eugenic Sterilization: A Biomedical Intervention," pp. 5, 8, 12; Julius Paul, "State Eugenic Sterilization . . . ," pp. 25–26, 34–35; Donald Giannella, "Eugenic Sterilization and the Law," p. 76, all in Jonas Robitscher, ed., *Eugenic Sterilization* (Charles C Thomas, 1973); C. P. Blacker, *Voluntary Sterilization* (reprints from the *Eugenics Review*, 1961, 1962), p. 147.

18. Richmond *Times-Dispatch*, Feb. 28, 1980, March 16, 1980; Monroe E. Price and Robert A. Burt, "Sterilization, State Action, and the Concept of Consent," in *The Mentally Retarded Citizen and the Law* (Free Press, 1976), p. 74. Julius Paul, a professor of political science at the State University of New York in Fredonia, noted in 1970 that the trend in sterilization was not overtly eugenic. Nevertheless, it marked a return to the early twentieth century's punitive attack against illegitimacy, public welfare, and the "under-privileged," and to "the same economic, racial and *moral* overtones that were advanced in the days of the extreme hereditarians." Paul, "State Eugenic Sterilization," pp. 34–35.

19. James V. Neel, "Human Genetics," in John Z. Bowers and Elizabeth F. Purcell, eds., *Advances in American Medicine: Essays at the Bicentennial* (2 vols.; Josiah Macy, Jr., Foundation, 1976), I, 70–71; Geoffrey Beale, "Social Effects of Research in Human Genetics," in Watson Fuller, ed., *The Social Impact of Modern Biology* (Routledge and Kegan Paul, 1970), pp. 88–89; Frederick Ausubel, Jonathan R. Beckwith, and Kaaren Janssen, "The Politics of Genetic Engineering: Who Decides Who's Defective?" *Psychology Today*, June 1974, p. 39; Barton Childs et al., "Human Behavior Genetics," in Harry Harris and Kurt Hirschhorn, eds., *Advances in Human Genetics* (vol. 7; Plenum Press, 1976), p. 88; Orlando J. Miller, "An Overview of Problems Arising from Amniocentesis,"

in Maureen H. Harris, ed., *Early Diagnosis of Human Genetic Defects: Scientific and Ethical Considerations* (Fogarty International Center Proceedings No. 6; Government Printing Office, 1972), p. 26; Stephen Jay Gould, *The Mismeasure of Man* (W. W. Norton, 1981), pp. 144–45; Patricia A. Jacobs, "The William Allan Award Address: Human Population Cytogenetics: the First Twenty-five Years," *American Journal of Human Genetics*, 34 (1982), 693–94; Digamber S. Borgaonkar and Saleem A. Shah, "The *XYY* Chromosome Male—Or Syndrome," in A. G. Steinberg and A. G. Bearn, eds., *Progress in Medical Genetics* (vol. 10; New York: Grune & Stratton, 1974), pp. 138–39, 177, 191–92, 197–98; Patricia A. Jacobs et al., "Aggressive Behavior, Mental Sub-normality, and the *XYY* Male," *Nature*, 208 (1965), 1351–52; Herman A. Witkin et al., "Criminality in *XYY* and *XXY* Men," *Science*, 193 (Aug. 13, 1976), 547–55.

20. David M. Rorvik, "The Brave New World of the Unborn," *Look*, 33 (Nov. 1969), 82; Miller, "An Overview of Problems Arising from Amniocentesis," p. 28; J. S. Fitzsimmons et al., *The Provision of Regional Genetic Services in the United Kingdom: Report of the Clinical Genetics Society Working Party on Regional Genetic Services* (Supplements to the Bulletin of the Eugenics Society, No. 4; Eugenics Society, 1982), p. 5.

21. Jane E. Brody, "Genetics Clinics Predict Defects," New York *Times*, Feb. 2, 1969, p. 76; Theodore Friedmann, "Prenatal Diagnosis of Genetic Disease," *Scientific American*, 225 (Nov. 1971), 34–42; Philip Reilly, *Genetics, Law, and Social Policy* (Harvard University Press, 1977), pp. 92, 146.

22. Gene Bylinsky, "What Science Can Do About Hereditary Disease," *Fortune*, Sept. 1974, p. 152; Paul Ramsey, *Fabricated Man: The Ethics of Genetic Control* (Yale University Press, 1970), pp. 97–98; David M. Rorvik, "The Brave New World of the Unborn," *Look*, 33 (Nov. 1969), 82; Ausubel, Beckwith, and Janssen, "The Politics of Genetic Engineering: Who Decides Who's Defective?" p. 40; Miller, "An Overview of Problems Arising from Amniocentesis," p. 28. See also Bentley Glass, "Science: Endless Horizons or Golden Age?" *Science*, 171 (Jan. 8, 1971), 28.

23. Beale, "Social Effects of Research in Human Genetics," pp. 89–90; "The Right to

Bad Genes," *Time*, June 26, 1972, pp. 46, 51; series of articles on industrial genetic screening, *The New York Times*, Feb. 3, 1980, p. 1; Feb. 4, 1980, p. 1; Feb. 5, 1980, p. 1; Feb. 6, 1980, p. 1.

24. Reilly, *Genetics, Law, and Social Policy*, pp. 32–33; Neel, "Human Genetics," p. 65; Marc Lappé, *Genetic Politics: The Limits of Biological Control* (Simon & Schuster, 1979), p. 110; Committee for the Study of Inborn Errors of Metabolism . . . National Research Council, *Genetic Screening: Programs, Principles, and Research* (National Academy of Sciences, 1975), p. 119.

25. Reilly, *Genetics, Law, and Social Policy*, pp. 67, 74; Tabitha Powledge, "Genetic Screening as a Political and Social Development," in Daniel Bergsma et al., eds., *Ethical, Social, and Legal Dimensions of Screening for Human Genetic Disease* (National Foundation–March of Dimes: Birth Defects, Original Articles Series, vol. X; Stratton Intercontinental Medical Book Group, 1974), pp. 36–37; Robert F. Murray, Jr., "Discussion," in Marc Lappé and Robert S. Morison, eds., *Ethical and Scientific Issues Posed by Human Uses of Molecular Genetics* (Annals of the New York Academy of Sciences, vol. 265; New York Academy of Sciences, 1976), p. 165; New York *Times*, Feb. 7, 1970, p. 22; Feb. 4, 1981, p. 1; *Science*, 211 (Jan. 16, 1981), 257. There were two hundred fifty sickle-cell programs in operation in 1972, many of them funded with federal assistance.

26. Lappé, *Genetic Politics*, pp. 70, 90–93; Samuel P. Bessman and Judith P. Swazey, "Phenylketonuria: A Study of Biochemical Legislation," in E. Mendelsohn, J. P. Swazey, and I. Taviss, eds., *Human Aspects of Biomedical Innovation* (Harvard University Press, 1971), pp. 50–51; President's Commission for the Study of Ethical Problems in Medicine and Biomedical and Behavioral Research, *Screening and Counseling for Genetic Conditions: The Ethical, Social, and Legal Implications of Genetic Screening, Counseling, and Education Programs* (Government Printing Office, 1983), pp. 13–14; Committee for the Study of Inborn Errors of Metabolism . . . National Research Council, *Genetic Screening: Programs, Principles, and Research*, pp. 24–25, 28, 29, 51, 92–93. Ausubel, Beckwith, and Janssen, "The Politics of Genetic Engineering: Who Decides Who's Defective?" p.

45. Tracy Sonneborn, a colleague of H. J. Muller, though sympathetic to germinal choice, privately challenged its control by any centralized group of sages. "Even if the sages were all truly Sages, there is a Hitlerian overtone that is repugnant." Sonneborn to Muller, Sept. 28, 1964, Hermann J. Muller Papers, Germinal Choice file, box IV.

27. Barton Childs, "Garrod, Galton, and Clinical Medicine," *Yale Journal of Biology and Medicine*, 46 (1973), 305; "The Hastings Center: A Short and Long 15 Years," Special Supplement, in *The Hastings Center Report*, 14 (April 1984); "Doctor's Dilemmas," *The Wall Street Journal*, Nov. 23, 1983. In 1968, the American Society of Human Genetics created an Advisory Committee on Public Issues in Genetics. Records of the American Society of Human Genetics, American Philosophical Society Library, Philadelphia, box 11, Board file.

28. Interviews with John Burn, Malcolm Ferguson-Smith, Martin Bobrow, Harry Harris, Aubrey Milunsky.

29. Richard C. Lewontin, "Race and Intelligence," in N. J. Block and Gerald Dworkin, eds., *The IQ Controversy: Critical Readings* (Pantheon Books, 1976), p. 78; Michael Schudson, "A History of the *Harvard Educational Review*," in John Snarey et al., eds., *Conflict and Continuity: A History of Ideas on Social Equality and Human Development* (Harvard Educational Review Reprint Series #15, 1981), pp. 15–17.

30. Rosenfeld, "Sociobiology Stirs a Controversy," pp. 73–74; Nicholas Wade, "Sociobiology: Troubled Birth for New Discipline," *Science*, 191 (March 19, 1976), 1151–55.

31. Jensen, "How Much Can We Boost IQ and Scholastic Achievement?" pp. 88–89, 162.

32. L. S. Hearnshaw, *Cyril Burt, Psychologist* (Cornell University Press, 1979), pp. 233–34; Gould, *The Mismeasure of Man*, pp. 232–33, 235; Jensen, "How Much Can We Boost IQ and Scholastic Achievement?" p. 122; Lewontin, "Race and Intelligence," pp. 86–87; Leon J. Kamin, *The Science and Politics of I.Q.* (Lawrence Erlbaum Associates, 1974), pp. 35–47.

33. Brian Evans and Bernard Waites, *IQ and Mental Testing* (Macmillan, 1980), chapters 5 and 6; Lewontin, "Race and Intelligence," p. 89; Richard C. Lewontin, "The

Analysis of Variance and the Analysis of Causes," in Block and Dworkin, eds., *The IQ Controversy*, pp. 184, 192, 189.

34. Lewontin, "Race and Intelligence," p. 89. Similar critiques of Jensen's claim were advanced, though with less force, by Eysenck, *The I.Q. Argument*, pp. 66–67, 113–114; and Herrnstein, "I.Q.," p. 57.

35. Lewontin, "Race and Intelligence," p. 89.

36. Walter F. Bodmer and L. L. Cavalli-Sforza, "Intelligence and Race," *Scientific American*, 223 (Oct. 1970), pp. 28–29; interviews with Bodmer and with Cavalli-Sforza.

37. Bodmer and Cavalli-Sforza, "Intelligence and Race," pp. 27, 29. See also Eysenck, *The I.Q. Argument*, pp. 129–30; Frederick Osborn to Paul J. Kern, Dec. 21, 1966; Osborn to C. C. Aronsfeld, May 28, 1969, American Eugenics Society Papers, box 2.

38. Oliver Smithies to Members, Genetics Society of America, July 15, 1975, and attached "Revised Statement of GSA Members on Genetics, Race, and Intelligence (July 14, 1975)"; Norman H. Horowitz to Elizabeth S. Russell, Jan. 29, 1975, in Genetics Society of America Records, American Philosophical Society Library, Philadelphia, Pa., box 37; Norman H. Horowitz to the author, Oct. 16, 1984. Horowitz also thought the draft statement "weak morally": "Racists assert that blacks are genetically inferior in IQ and therefore need not be treated as equals. The proposed statement disputes the premise of this assertion, but not the logic of the conclusion. . . . Even if the premise were correct, the conclusion would not be justified. Yet the proposed statement directs its main fire at the premise, and by so doing seems to accept the racist logic. It places itself in a morally vulnerable position, for if, at some future time, it is found that the premise is correct, then the whole G[enetics] S[ociety] of A[merica] case collapses, together with its justification for equal opportunity." Horowitz to Russell, Jan. 29, 1975. For a different emphasis in re the differentiation of groups in the context of human evolution and genetics, see Stephen Jay Gould, "Human Equality Is a Contingent Fact of History," *Natural History*, 93 (Nov. 1984), 26–28, 30 ff.

39. "Report of the Ad Hoc Committee," *Genetics*, 83 (1976), s99–101. Some members of the Society were distressed that the

statement was not issued in its original form. See David Suzuki to Oliver Smithies, Aug. 1, 1975, and the Jonathan King file in Genetics Society of America Records, Committee on Genetics, Race, and Intelligence, box 37.

40. Wade, "Sociobiology: Troubled Birth for New Discipline," pp. 1151–55; Elizabeth Allen et al., "Against 'Sociobiology,' " *New York Review of Books*, 22 (Nov. 13, 1975), 43–44; Sociobiology Study Group of Science for the People, "Sociobiology—Another Biological Determinism," *BioScience*, 26 (March 1976), reprinted in Caplan, ed., *The Sociobiology Debate*, 280–90.

41. *Time*, 110 (Aug. 1, 1977), 58; interview with Edward O. Wilson; Stephen Jay Gould, "Sociobiology and Human Nature: A Postpanglossian Vision," *Human Nature*, 1 (Oct. 1978), reprinted in Ashley Montagu, ed., *Sociobiology Examined* (Oxford University Press, 1980), pp. 285–86; Gould, "Biological Potential vs. Biological Determinism," p. 344.

42. Gould, "Sociobiology and Human Nature," pp. 283–84, 287–88; Gould, "Biological Potential vs. Biological Determinism," pp. 345–46, 348–49.

43. Gould, "Biological Potential vs. Biological Determinism," p. 345.

44. Penrose to S. P. H. Mandel, June 28, 1961, Lionel S. Penrose Papers, University College London, file 165/4; Theodosius Dobzhansky, "Comments on Genetic Evolution," *Daedalus*, 90 (Summer 1961), 461–62.

45. Lionel S. Penrose, *Outline of Human Genetics*, pp. 143–44, 122; Anthony C. Allison, "Protection Afforded by Sickle-Cell Trait Against Subtertian Malarial Infection," in Samuel H. Boyer IV, ed., *Papers on Human Genetics* (Prentice-Hall, 1963), pp. 153–63. As early as 1966, Joshua Lederberg remarked: "Is there much point in setting eugenic standards relevant only to a small minority of the world's population even as we watch the unprecedented breakdown of intercultural barriers? The jet airplane has already had an incalculably greater effect on human population genetics than any conceivable program of calculated eugenics." Joshua Lederberg, "Experimental Genetics and Human Evolution," *American Naturalist*, 100 (Sept.–Oct. 1966), 523.

46. Harry Harris, *Prenatal Diagnosis and Selective Abortion* (Harvard University Press, 1975), pp. 50–51; Friedmann, "Prenatal

Diagnosis of Genetic Disease," pp. 34–42; James Neel commented: "Abortion based on prenatal diagnosis, coupled with reproductive compensation, will probably slow the rate of elimination of undesirable genes." James V. Neel, "Ethical Issues Resulting from Prenatal Diagnosis," in Harris, ed., *Early Diagnosis of Human Genetic Defects*, p. 223.

47. Paul Ramsey, "Genetic Therapy: A Theologian's Response," in Michael P. Hamilton, ed., *The New Genetics and the Future of Man* (William B. Eerdmans, 1972), pp. 161–62; Jérôme Lejeune interview; Neel, "Ethical Issues Resulting from Prenatal Diagnosis," p. 221; Bernard Nathanson, with Richard N. Ostling, *Aborting America* (Doubleday, 1979), pp. 172–76; Gillian Peele, *Revivalism and Reaction: The Right in Contemporary America* (The Clarendon Press, 1984), pp. 94–95.

48. Interviews with Harry Harris and James Neel.

49. Robert H. Blank, *The Political Implications of Human Genetic Technology* (Westview Press, 1981), p. 156; Nathanson, *Aborting America*, pp. 233–34; Mitchell S. Golbus et al., "Prenatal Diagnosis in 3000 Amniocenteses," *New England Journal of Medicine*, 300 (Jan. 25, 1979), 157–63; M. A. Ferguson-Smith et al., *The Provision of Services for the Prenatal Diagnosis of Fetal Abnormality in the United Kingdom: Report of the Clinical Genetics Society Working Party on Prenatal Diagnosis in Relation to Genetic Counseling* (Supplements to the Bulletin of the Eugenics Society, No. 3; Eugenics Society, 1978), pp. 6–7; President's Commission, *Screening and Counseling for Genetic Conditions*, pp. 33–34, 54–55; Charles J. Epstein and Mitchell S. Golbus, "Prenatal Diagnosis of Genetic Diseases," *American Scientist*, 65 (1977), 710; interview with Kurt Hirschhorn. Diseased fetuses are, of course, found on the average in about twenty-five percent of women who undergo amniocentesis for reasons of risk for recessive biochemical abnormalities, but only in one to two percent of those at risk for chromosomal disorders. Only a few percent of the total are found to be abnormal because suspicions of chromosomal problems lie behind the vast majority of amniocenteses. In recent years it has become known that at least half of all human preg-

nancies end in spontaneous abortions and that perhaps twenty-five percent of these conceptuses suffer from chromosomal anomalies. Harry Harris has written, "The normal chromosomal constitution of the species, which tended in the past to be thought of as a rather stable affair prone only to occasional aberrations, can now be seen as being maintained by an intense pressure of natural selection at an early stage of foetal life, which culls out the aberrations by spontaneous abortion. There is indeed some justification for the statement that 'nature is the greatest' abortionist." James Neel to the author, Jan. 20, 1984; Harris, *Prenatal Diagnosis and Selective Abortion*, pp. 6, 11, 65; C. J. Roberts and C. R. Lowe, "Where Have All the Conceptions Gone?" *The Lancet*, I (1975), 498–99.

50. Arthur Salisbury, vice-president for medical services of the National Foundation, later pointed out that its policy was to fund service activities for only five years. He added, however, that the right-to-life people "say we're on a search-and-destroy mission to eliminate all but the perfect and [they] compare us with the Third Reich." Interview with Arthur Salisbury. New York *Times*, Sept. 16, 1976, p. 1; "March of Dimes Ends Its Support of Genetic Services," New York *Times*, March 9, 1978, p. 18; March 10, 1978, section IV, p. 14; March 12, 1978, section IV, p. 7; Jérôme Lejeune interview; Linda Greenhouse, "Abortion Goes before the Supreme Court Again," New York *Times*, April 20, 1980, p. E8; U.S. Congress, House Committee on Appropriations, *Departments of Labor and Health, Education, and Welfare Appropriations*, 1980, part 2, p. 495; 1983, part 2, p. 448; Maya Pines, "Heredity Insurance," *The New York Times Magazine*, April 30, 1978, pp. 88–89; President's Commission, *Screening and Counseling for Genetic Conditions*, p. 34.

51. Interview with Jérôme Lejeune; Jérôme Lejeune, "On the Nature of Men," *American Journal of Human Genetics*, 22 (1970), 124–25.

52. Jérôme Lejeune interview. Arthur Salisbury doubts that there has been any diminution of research aimed at helping people with Down's syndrome as a result of the ability to diagnose and abort trisomy-21 fetuses. Arthur Salisbury interview.

53. Daniel Callahan, "New Beginnings

in Life: A Philosopher's Response," in Hamilton, ed., *The New Genetics and the Future of Man,* p. 95; Lionel Penrose, "Notebooks," c. 1950–62, Lionel S. Penrose Papers, file 77/1. Lederberg, "Experimental Genetics and Human Evolution," p. 524; Kurt Hirschhorn, "On Re-Doing Man," *Annals of the New York Academy of Science,* 184 (June 7, 1971), 103–12. Peter Medawar had told a BBC audience, "It is simply not true to say that advances in medicine and hygiene must cause a genetical deterioration of mankind." Richard Lewontin emphasized that it was arbitrary to talk even theoretically about genetic load without adequate consideration of the real processes of human birth and mortality, which was missing from the subject at the opening of the seventies. P. B. Medawar, *The Future of Man* (Methuen, 1959), pp. 32–33; Richard C. Lewontin, "A Proposal for a Training Program in Population Genetics and Demography" [June 1965], American Eugenics Society Papers, box 13.

54. Medawar, *The Future of Man,* pp. 99–100; Reilly, *Genetics, Law, and Social Policy,* pp. 141–42; "Discussion," in Harris, ed., *Early Diagnosis of Genetic Defects,* p. 99; Arthur Steinberg interview; Arno G. Motulsky, "Genetic Therapy: A Clinical Geneticist's Response," in Hamilton, ed., *The New Genetics and the Future of Man,* p. 128; Joshua Lederberg, "Biological Innovation and Genetic Intervention," in John A. Behnke, ed., *Challenging Biological Problems* (Oxford University Press, 1972), p. 18. The University of Chicago biochemist Leon R. Kass gibed that perhaps the "crusaders against genetic deterioration are worried about the wrong genes," and elaborated: "After all, how many architects of the Vietnam war have suffered from Down's syndrome?" Leon R. Kass, "New Beginnings in Life," in Hamilton, ed., *The New Genetics and the Future of Man,* pp. 8–10.

55. Penrose, "Genetics and Society," 1969, Penrose Papers, file 103; "Eugenics and Genetics: Discussion," in Gordon Wolstenholme, ed., *Man and His Future* (A CIBA Foundation volume; Little, Brown, 1963), pp. 280, 295–96; Brooks Atkinson, Miami *Herald,* July 22, 1961, clipping in Hermann J. Muller Papers, Germinal Choice file, box Ib; Dobzhansky, "Comments on Genetic Evolution," p. 463. Haldane, sympathetic as he was to the principle of germinal choice, had to conclude: "I do not think we know much more about how to bring it about than Galileo or Newton knew about how to fly." J. B. S. Haldane, "The Implications of Genetics for Human Society," in S. J. Geerts, ed., *Genetics Today: Proceedings of the Eleventh International Congress of Genetics . . . 1963* (Pergamon Press, 1964), II, xcvi. Little attention was given to the effects genetic intervention might have on the female genome. See Lederberg, "Experimental Genetics and Human Evolution," p. 525.

56. Kass, "New Beginnings in Life," pp. 34–35, 48.

57. Paul Ramsey, "Moral and Religious Implications of Genetic Control," in John D. Roslansky, ed., *Genetics and the Future of Man* (Appleton-Century-Crofts, 1966), p. 156; Rorvik, "The Brave New World of the Unborn," p. 83; A. D. Holcombe to H. J. Muller, Sept. 8, 1960, Hermann J. Muller Papers, Germinal Choice file, box Ia; George F. Will, "Our New Baby Technology May Grow into a Monster," Los Angeles *Times,* July 30, 1978, section VI, p. 5. Perhaps many people agreed with what Marjorie Foster of Toledo, Ohio, mother of three, told a Louis Harris pollster about artificial insemination: "It would take manhood from the father. We could forget about the woman's role. I just wouldn't feel the child was mine. It might sprout horns or wings or something." Louis Harris, "The Life Poll," *Life,* 23 (June 13, 1969), 53–54.

58. Penrose, "Genetics and Society," 1969, Lionel Penrose Papers, file 103.

59. President's Commission for the Study of Ethical Problems in Medicine and Biomedical and Behavioral Research, *Splicing Life: A Report on the Social and Ethical Issues of Genetic Engineering with Human Beings* (Government Printing Office, 1982), pp. 2–3.

Chapter XIX:
SONGS OF DEICIDE

1. "Gene Therapy," *Nature*, 306 (Dec. 1, 1983), 414; Lionel S. Penrose, *Outline of Human Genetics* (2nd ed.; Heinemann, 1963), p. 127; Harry Harris, *Prenatal Diagnosis and Selective Abortion* (Harvard University Press, 1975), p. 84; "Discussion," Richard Roblin, Stanfield Rogers, and Alexander Morgan Capron, "Reflections on Issues Posed by Recombinant DNA Molecule Technology, I, II, and III," in Marc Lappé and Robert S. Morison, eds., *Ethical and Scientific Issues Posed by Human Uses of Molecular Genetics* (Annals of the New York Academy of Sciences, vol. 265; New York Academy of Sciences, 1976), pp. 80–81. In the United States in 1979–80, screening programs detected 195 children with PKU, 536 with hypothyroidism, 25 with galactosemia, 8 with maple syrup urine disease, and 8 with homocystinuria. In North America prior to 1970, fifty to one hundred children were born each year with Tay-Sachs disease; in 1980, presumably because of the use of prenatal diagnosis and abortion, the number was only thirteen. President's Commission for the Study of Ethical Problems in Medicine and Biomedical and Behavioral Research, *Screening and Counseling for Genetic Conditions: The Ethical, Social, and Legal Implications of Genetic Screening, Counseling, and Educational Programs* (Government Printing Office, 1983), pp. 19, 33–34. R. G. Edwards averred that the primary motive for his work with Steptoe on in-vitro fertilization was the relief of infertility among couples who desired children, and the acquisition of knowledge about the human species and its environment —objectives that Edwards called "worthy and humane." R. G. Edwards, "Aspects of Human Reproduction," in Watson Fuller, ed., *The Social Impact of Modern Biology* (Routledge and Kegan Paul, 1970), p. 110. People who write off gene therapy as repugnant, Robert Sinsheimer had remarked at the end of the sixties, are not "among the losers in that chromosomal lottery that so firmly channels our human destinies." Robert L. Sinsheimer, "The Prospect of Designed Genetic Change," *Engineering and Science*, 32 (April 1969), 13.

2. Herbert A. Lubs, "Frequency of Genetic Disease," in Herbert A. Lubs and Felix de la Cruz, eds., *Genetic Counseling: A Monograph of the National Institute of Child Health and Human Development* (Raven Press, 1977), pp. 3, 5, 9; Marc Lappé, *Genetic Politics: The Limits of Biological Control* (Simon & Schuster, 1979), p. 26; Harris, *Prenatal Diagnosis and Selective Abortion*, p. 38; Gilbert S. Omenn, "Prenatal Diagnosis of Genetic Disorders," *Science*, 200 (May 26, 1978), 952–58; Charles R. Scriver, Claude Laberge, Caroline L. Clow, and F. Clarke Fraser, "Genetics and Medicine: An Evolving Relationship," *Science*, 200 (May 26, 1978), 946–52; President's Commission, *Screening and Counseling for Genetic Conditions*, pp. 5, 64, 87–88; Charles J. Epstein and Mitchell S. Golbus, "Prenatal Diagnosis of Genetic Diseases," *American Scientist*, 65 (1977), 703–4, 710; J. S. Fitzsimmons et al., *The Provision of Regional Genetic Services in the United Kingdom: Report of the Clinical Genetics Society Working Party on Regional Genetic Services* (Supplements to the Bulletin of the Eugenics Society, No. 4; Eugenics Society, 1982), p. 1; John Burn, "Clinical Genetics," *British Medical Journal*, 283 (Oct. 8, 1983), 999.

3. Victor A. McKusick, "The Growth and Development of Human Genetics as a Clinical Discipline," *American Journal of Human Genetics*, 27 (1975), 270; Marshall D. Levine, J. M. Gursky, and D. L. Rimoin, "Results of a Survey on the Teaching of Medical Genetics to Medical Students, House Officers, and Fellows (1975)," in Lubs and de la Cruz, eds., *Genetic Counseling*, pp. 359–61; Barton Childs, Carl A. Huether, and Edmond A. Murphy, "Human Genetics Teaching in U.S. Medical Schools," *American Journal of Human Genetics*, 33 (1981), 1–2, 8–9; President's Commission, *Screening and Counseling for Genetic Conditions*, pp. 11, 36–37; Kurt Hirschhorn interview; John Burn interview; American Board of Medical Genetics, Inc., *Membership Directory, Spring 1982*; letter to the author from Arthur Salisbury, March 13, 1984; Fitzsimmons et al., *The Provision of Regional Genetic Services in the United Kingdom*, pp. 1–2, 9; Burn, "Clinical

Genetics," p. 999; M. A. Ferguson-Smith et al., *The Provision of Services for the Prenatal Diagnosis of Fetal Abnormality in the United Kingdom: Report of the Clinical Genetics Society Working Party on Prenatal Diagnosis in Relation to Genetic Counseling* (Supplements to the Bulletin of the Eugenics Society, No. 3; Eugenics Society, 1978), pp. 8–9, 16–17, 25; R. Harris et al., *Clinical Genetics Society: Report of the Working Party on the Role and Training of Clinical Geneticists* (Supplements to the Bulletin of the Eugenics Society, No. 5; Eugenics Society, 1983), pp. 9–11.

4. *The New York Times*, Dec. 28, 1978, p. 1; Jan. 30, 1979, p. C1; *Los Angeles Times*, July 23, 1979, part IV, p. 21; Barbara J. Culliton, "Physicians Sued for Failing to Give Genetic Counseling," *Science*, 203 (Jan. 19, 1979), 251; President's Commission . . . , *Screening and Counseling for Genetic Conditions*, p. 76; Aubrey Milunsky, "Prenatal Genetic Diagnosis and the Law"; and Alexander Morgan Capron, "The Continuing Wrong of 'Wrongful Life,' " both in Aubrey Milunsky and George J. Annas, eds., *Genetics and the Law II* (Plenum Press, 1980), pp. 63, 84–85. The cases were *Becker* v. *Schwartz* and *Park* v. *Chessin*. Full citations are given in Margery W. Shaw, "The Potential Plaintiff: Preconception and Prenatal Torts," *Genetics and the Law II*, pp. 230, 232.

5. Capron, "The Continuing Wrong . . . ," p. 90; Culliton, "Physicians Sued for Failing to Give Genetic Counseling," p. 251; Maya Pines, "Heredity Insurance," *The New York Times Magazine*, April 30, 1978, p. 85; *Los Angeles Times*, June 12, 1980, p. 3. See also *"Turpin* v. *Sortini,"* Sup., 182, *California Reporter* 337, 115–27.

6. Ruth Halcomb, "Winning the War Against Birth Defects," *Parents' Magazine*, 52 (May 1977), 54; Aubrey Milunsky, *Know Your Genes* (Avon Books, 1979), pp. 8–9.

7. "Test to Check on Fetuses Held Safe," *The New York Times*, Jan. 26, 1979, p. 8; Arthur Steinberg interview; Harris, *Prenatal Diagnosis and Selective Abortion*, pp. 33–34; D. J. H. Brock, "Antenatal Diagnosis of Spina Bifida and Anencephaly," in Lubs and de la Cruz, eds., *Genetic Counseling*, pp. 225–26; Philip Reilly, *Genetics, Law, and Social Policy* (Harvard University Press, 1977), pp. 13, 26; President's Commission, *Screening and Counseling for Genetic Conditions*, pp. 27–28;

Arthur Salisbury interview; Ferguson-Smith et al., *The Provision of Services for the Prenatal Diagnosis of Fetal Abnormality in the United Kingdom*, p. 20.

8. President's Commission for the Study of Ethical Problems in Medicine and Biomedical and Behavioral Research, *Splicing Life: A Report on the Social and Ethical Issues of Genetic Engineering with Human Beings* (Government Printing Office, 1982); pp. 38–40; Arthur Steinberg interview; Kathleen McAuliffe and Sharon McAuliffe, "Keeping Up with the Genetic Revolution," *The New York Times Magazine*, Nov. 6, 1983, pp. 93–94; interview with Charles J. Epstein.

9. Arthur Steinberg interview; President's Commission, *Splicing Life*, p. 40; Harry Harris interview; Frank H. Ruddle, "Cell Fusion as a Tool in the Study of Cellular Biology," in Arno Motulsky and F. J. G. Ebling, eds., *Abstracts of Papers; Fourth International Conference on Birth Defects . . .* (Excerpta Medica, 1974), p. 57.

10. *Science*, 222 (Nov. 25, 1983), 913–15; *Nature*, 306 (Nov. 17, 1983), 222; Maya Pines, "In the Shadow of Huntington's," *Science '84*, 5 (May 1984), 32–39.

11. Harry Harris interview; Maya Pines, "Genetic Profiles Will Put Our Health in Our Hands," *Smithsonian*, 7 (July 1976), 86–91; Arno G. Motulsky, "Doomsaying Genetically: The Fear Is Groundless," *Science Digest*, 86 (Aug. 1979), 20–23.

12. Barton Childs, "Garrod, Galton, and Clinical Medicine," *Yale Journal of Biology and Medicine*, 46 (1973), 307; Arno G. Motulsky, George R. Fraser, and Joseph Felsenstein, "Public Health and Long-Term Genetic Implications of Intrauterine Diagnosis and Selective Abortion," *Birth Defects: Original Article Series*, 7, No. 5 (April 1971), 31.

13. Reilly, *Genetics, Law, and Social Policy*, pp. 77–78, 82, 105, 131–32, 242–43; Lappé, *Genetic Politics*, p. 109; conversation with Dr. Charles Mahan, University of Florida, April 16, 1980; Richmond *Times-Dispatch*, March 16, 1980; President's Commission, *Screening and Counseling for Genetic Conditions*, p. 32.

14. A. G. Motulsky, "Brave New World? Ethical Issues in Prevention, Treatment and Research of Human Birth Defects"; P. A. Marks et al., "Isolation and Synthesis of Human Genes," in Motulsky and Ebling, eds., *Birth Defects*, 1974, pp. 326–27, 77;

Theodore Friedmann and Richard Roblin, "Gene Therapy for Human Genetic Disease?" *Science*, 175 (March 3, 1972), 952, 954; Arno G. Motulsky, "Genetic Therapy: A Clinical Geneticist's Response"; W. French Anderson, "Genetic Therapy," in Michael P. Hamilton, ed., *The New Genetics and the Future of Man* (William B. Eerdmans, 1972), pp. 126, 117; James v. Neel, "Social and Scientific Priorities in the Use of Genetic Knowledge," in Bruce Hilton, Daniel Callahan et al., eds., *Ethical Issues in Human Genetics: Genetic Counseling and the Uses of Genetic Knowledge* (Plenum Press, 1973), p. 364; Frederick Osborn to Raymond B. Cattell, Jan. 4, 1968, American Eugenics Society Papers, box 3; President's Commission, *Splicing Life*, pp. 42–43, 46–47; McAuliffe and McAuliffe, "Keeping Up with the Genetic Revolution," pp. 96–97; conversation with Lee Hood. Among the exceptional possibilities for gene therapy are the Mendelian blood diseases, which may lend themselves to genetic therapy in that the bone marrow, where blood is manufactured, is a readily accessible tissue that can be removed, perhaps genetically repaired, then replaced. Still, even in the blood diseases, as D. J. Weatherall, the leading British authority on hemoglobin disorders, noted, many difficulties remain to be crossed "in the 'no man's land' between the patient and his DNA." D. J. Weatherall, "Toward an Understanding of the Molecular Biology of Some Common Inherited Anemias: The Story of Thalassemia," in Maxwell M. Wintrobe, ed., *Blood, Pure and Eloquent: A Story of Discovery, of People, and of Ideas* (McGraw-Hill, 1980), p. 408.

15. McAuliffe and McAuliffe, "Keeping Up with the Genetic Revolution," p. 48; conversation with Lee Hood; Anne McLaren, "Methods and Success of Nuclear Transplantation in Mammals," *Nature*, 309 (June 21, 1984), 671–72. Barton Childs, like many physicians, has said that the speculative genetic prognostications—not to mention the fears of them—reveal simply a profound ignorance of the complexities of the human organism. Such people ought "to know how hard it is to muck up the homeostasis of an organism which is based on eons of evolution." Barton Childs interview.

16. McAuliffe and McAuliffe, "Keeping Up with the Genetic Revolution," pp. 42, 19,

30, 32. Lee Hood, the chairman of the Division of Biology at the California Institute of Technology, reflected in 1984: "If you had asked me a few years ago: When would we be able to introduce genes into mammals and get them expressed? I would have predicted, in ten to fifteen years. Yet it was accomplished in just three. The field is moving so fast that the future is unpredictable." Conversation with Lee Hood.

17. President's Commission, *Splicing Life*, pp. 33–35; Robert Sinsheimer, "Life by Design," unpublished talk, M.I.T., Oct. 18, 1983; conversation with Robert Sinsheimer.

18. A study of 2,300 people tested and counseled for sickle-cell trait in Orchemenos, Greece, in the early seventies found that about a quarter of those discovered to be carriers concealed the fact from their spouses. "When Genetic Counseling Backfires," *Psychology Today*, Sept. 1975, pp. 20, 80; Lappé, *Genetic Politics*, pp. 51, 67–68, 84; McAuliffe and McAuliffe, "Keeping Up with the Genetic Revolution," p. 96; President's Commission, *Screening and Counseling for Genetic Conditions*, p. 16. James Sorenson, "Some Social and Psychologic Issues in Genetic Screening: Public and Professional Adaptation to Biomedical Innovation," in Daniel Bergsma et al., eds., *Ethical, Social, and Legal Dimensions of Screening for Human Genetic Disease* (National Foundation–March of Dimes: Birth Defects, Original Articles Series, vol. X; Stratton Intercontinental Medical Book Group, 1974), pp. 179–80; President's Commission, *Splicing Life*, p. 62; Earl Lane, "Genetic Counseling: Is It Working?" *Los Angeles Times*, Sept. 14, 1980, section V, p. 25; Park Gerald interview; Michael M. Kaback, "Detection of Tay-Sachs Disease Carriers: Lessons and Ramifications," in Lubs and de la Cruz, eds., *Genetic Counseling*, p. 221; Gene Bylinsky, "What Science Can Do About Hereditary Diseases," *Fortune*, Sept. 1974, p. 156; Richard Restak, "The Danger of Knowing Too Much," *Psychology Today*, Sept. 1975, p. 23.

19. Barton Childs, "Genetic Counseling: A Critical Review of the Published Literature," in Bernice H. Cohen, Abraham M. Lilienfeld, and P. C. Huang, eds., *Genetic Issues in Public Health and Medicine* (Charles C. Thomas, 1978), pp. 340–41; Lappé, *Genetic Politics*, p. 32; Restak, "The Danger of

Knowing Too Much," pp. 22, 88; Jane E. Brody, "Genetic Defects Sought in Fetus," New York *Times*, May 12, 1976, p. 17; interviews with Park Gerald and Charles Epstein; Epstein and Golbus, "Prenatal Diagnosis of Genetic Diseases," p. 709.

20. C. P. Blacker to Muller, Nov. 15, 1961, Hermann J. Muller Papers, Germinal Choice file, box Ia.; Louis Harris, "The Life Poll," *Life*, 23 (June 13, 1969), 53; the figures on abortion are from *Statistical Abstract of the United States, 1984*, pp. 64, 71, and New York *Times*, July 10, 1984, p. 7; Kurt Hirschhorn interview.

21. Reilly, *Genetics, Law, and Social Policy*, pp. 190–91; Los Angeles *Times*, Feb. 4, 1984, p. 1; Sept. 17, 1982, part V, p. 26; April 19, 1983, part V, p. 1; Ellen Goodman, "Wombs for Rent . . . ," Los Angeles *Times*, Feb. 8, 1983, part II, p. 5; Afton Blake, "First Word," *Omni*, 5 (Aug. 1983), 6.

22. "Discussion," Roblin, Rogers, and Capron, "Reflections on Issues Posed by Recombinant DNA Molecule Technology, I, II, and III," in Lappé and Morison, eds., *Ethical and Scientific Issues Posed by Human Uses of Molecular Genetics*, pp. 80–81.

23. John Neary, "A Scientist's Disturbing Variations on a Racial Theme," *Life*, 68 (1970), 58D; McAuliffe and McAuliffe, "Keeping Up with the Genetic Revolution," pp. 95–96; Bernard D. Davis, "Pythagoras, Genetics, and Workers' Rights," New York *Times*, Aug. 14, 1980, p. 23; New York *Times*, Feb. 4, 1981, p. 1; June 23, 1982, p. 12.

24. Reilly, *Genetics, Law, and Social Policy*, pp. 147–48; C. O. Carter, *Human Heredity* (2nd ed.; Penguin Books, 1977), pp. 259–60.

25. Joshua Lederberg put it succinctly: "Like many other messianic visions, eugenics is faulted by a confusion between the needs of an abstract *mankind* and those of individual men and women," adding, "The principal task of genetics is scientific understanding; the principal target for its applications to man is the alleviation of individual distress— which the physician cannot repudiate no matter what the general state of the world." Lederberg, "Biological Innovation and Genetic Intervention," in John A. Behnke, ed., *Challenging Biological Problems* (Oxford University Press, 1972), pp. 13, 14.

Essay on Sources

THE PUBLISHED primary and secondary literature of eugenics is enormous, and so is that concerning its related and descendant subjects—for example, genetics, medical genetics, and intelligence testing. Comprehensive access to the primary literature in the United States and Britain may be obtained from Samuel J. Holmes, *A Bibliography of Eugenics* (Berkeley, Calif., 1924); from successive series of the *Index-Catalogue of the Library of the Surgeon-General's Office, United States Army: Authors and Subjects* (1st through 5th Series; Washington, D. C., 1880–1961); and by searching out the listings under appropriate headings in the *Cumulative Book Index*. A valuable introduction to the secondary literature of eugenics, which has been growing rapidly in recent years, is Lyndsay Farrall, "The History of Eugenics: A Bibliographical Review," *Annals of Science*, 36 (March 1979), 111–23. The sources consulted for this book include a large sample of the primary literature—a representation of the advocates as well as the critics of eugenics and of its lay as well as scientific figures—in addition to biographies and autobiographies, historical treatments, manuscript collections, articles in popular and scientific periodicals, and interviews. Most of the interviews were tape-recorded and copies of them have been deposited in the Archives of the California Institute of Technology. The bibliographical notes that follow are selective, especially with respect to the periodical literature used. References to pertinent scientific articles can be found in the note citations in those sections of the book where particular subjects of interest are treated. Entry to the popular periodical literature in both the United States and Britain can be obtained by consulting such subject headings as "heredity," "eugenics," "sterilization," "mental testing," "genetic counseling," "genetic research," etc., and the cross-references given to other subjects in *Nineteenth Century Guide to Periodical Literature, 1890–1899* (2 vols.; New York, 1944) and the *Reader's Guide to Periodical Literature* (1900–).

The starting point for Francis Galton is Karl Pearson, *The Life, Letters, and Labours of Francis Galton* (3 vols. in 4; Cambridge, 1914–30). Among Galton's large body of writings, particularly important for my purposes were: *Memories of My Life* (London, 1908); "Hereditary Talent and Character," *Macmillan's Magazine*, 12 (1865), 157–66, 318–27; *Hereditary Genius: An Inquiry into Its Laws and Consequences* (London, 1869), and the second edition (London, 1892; reprinted, Cleveland, 1962); *English Men of Science: Their Nature and Nurture* (London, 1874; reprinted, London, 1970); *Natural Inheritance* (London, 1889); and *Essays in Eu-*

genics (London, 1909), which collects Galton's post-1900 writings on the subject. Key articles for Galton's work in heredity, regression, and correlation are his "Hereditary Improvement," *Fraser's Magazine*, 87 (1873), 116–30; "A Theory of Heredity," *Contemporary Review*, 27 (1875), 80–95; "Typical Laws of Heredity," *Proceedings of the Royal Institution*, 8 (Feb. 9, 1877), 282–301; "Opening Address . . . President of the Section [II, Anthropology]," *Nature*, 32 (Sept. 24, 1885), 507–10; "Regression Towards Mediocrity in Hereditary Stature," *Journal of the Royal Anthropological Institute of Great Britain and Ireland*, 15 (1886), 246–63; "Family Likeness in Stature," *Proceedings of the Royal Society*, 40 (Jan. 21, 1886), 42–73; "President's Address," *Journal of the Royal Anthropological Institute*, 15 (1886), 489–99; and "Co-relations and Their Measurement, Chiefly from Anthropometric Data," *Proceedings of the Royal Society*, 45 (1888), 135–45. A sizable collection of Galton's correspondence is in the Francis Galton Papers in the University College London Archives.

In recent years Galton scholarship has benefited greatly from the efforts of a number of people who have used the Galton materials with far more critical detachment than did Pearson. Insightful on Galton's personal life are Derek W. Forrest, *Francis Galton: The Life and Work of a Victorian Genius* (New York, 1974), which contains a bibliography of Galton's published writings, and Raymond E. Fancher, "Biographical Sources of Francis Galton's Psychology" (unpublished manuscript, 1980), but Eliot Slater's psychologically oriented "Galton's Heritage," *Eugenics Review*, 52 (July 1960), 91–103, is disappointing. Exceptionally important analyses of the way that Galton's science was interwoven with his social circumstances and eugenic convictions are Ruth Schwartz Cowan's dissertation, *Sir Francis Galton and the Study of Heredity in the Nineteenth Century* (Ann Arbor, 1969), and her masterful series of articles: "Francis Galton's Statistical Ideas: The Influence of Eugenics," *Isis*, 63 (1972), 509–28; "Francis Galton's Contribution to Genetics," *Journal of the History of Biology*, 5 (Fall 1972), 389–412; and "Nature and Nurture: The Interplay of Biology and Politics in the Work of Francis Galton," *Studies in the History of Biology*, 1 (1977), 133–207. A notable additional study in a similar vein is Donald MacKenzie, "The Development of Statistical Theory in Britain, 1865–1925: A Historical and Sociological Perspective (doctoral dissertation, University of Edinburgh, 1977), the uncut version of his compactly provocative *Statistics in Britain, 1865–1900: The Social Construction of Scientific Knowledge* (Edinburgh, 1981). Galton is authoritatively set in the context of the history of statistics in Theodore M. Porter, *The Calculus of Liberalism: The Development of Statistical Thinking in the Social and Natural Sciences in the Nineteenth Century* (Ann Arbor, 1981), and Victor L. Hilts, *Statist and Statistician: Three Studies in the History of Nineteenth Century English Statistical Thought* (New York, 1981), as well as in Hilts's "Statistics and Social Science," in R. N. Giere and R. S. Westfall, eds., *Foundations of Statistical Method: The Nineteenth Century* (Bloomington, Ind., 1973). Useful for another aspect of Galton's work are Raymond E. Fancher, "Francis Galton's African Ethnography and Its Role in the Development of His Psychology," *British Journal for the History of Science*, 16 (March 1983), 67–79; Allan R. Buss, "Galton and the Birth of Differential Psychology and Eugenics: Social, Political, and Economic Forces," *Journal of the History of the Behavioral Sciences*, 12 (1976), 47–58.

A guide to the immense corpus of Pearson's published writings is included in Churchill Eisenhart's straightforward "Karl Pearson," *Dictionary of Scientific Biography* (16 vols.; New York, 1970–80), X, 447–73. Essential to understanding the interplay of the man's social views and his scientific work are Karl Pearson's *The Ethic of Freethought* (London, 1888); *The Chances of Death and Other Studies in Evolution* (2 vols.; London, 1897); *The Grammar of Science* (London, 1892; 2nd ed., London, 1900); *National Life from the Standpoint of Science* (London, 1901); his successive *Eugenics Laboratory Lectures*, notably *The Scope and Importance to the State of the Science of National Eugenics* (London, 1909), *The Groundwork of Eugenics* (London, 1909), *Nature and Nurture: The Problem of the Future* (London, 1910), *Tuberculosis, Heredity, and Environment* (London, 1912), and *The Problem of Practical Eugenics* (London, 1912); and the various publications of the Galton Laboratory which appeared in the series *Questions of the Day and of the Fray*. Important among Pearson's scientific writings are his Huxley Lecture, "On the Inheritance of the Mental and Moral Characters in Man, and Its Comparison with the Inheritance of the Physical Characters," *Journal of the Anthropological Institute of Great Britain and Ireland*, 33 (1903), 179–237, and the papers he published in the eighteen-nineties as "Mathematical Contributions to the Theory of Evolution," especially "Regression, Heredity and Panmixia," *Philosophical Transactions*, A, 187 (1896), 253–318, and "On the Law of Ancestral Heredity," *Proceedings of the Royal Society*, 62 (1898), 386–412. Indispensable for Pearson's life and work and for the development of the Galton and Biometric laboratories are the Karl Pearson Papers in the Archives at University College London, an enormous collection rich in correspondence among Pearson, Galton, and Weldon as well as between Pearson and his numerous collaborators, friends, and foes. The locations given in this book's note-citations of documents in the Pearson Papers are out of date, since the papers have recently been reorganized, but the documents can be found by using the splendidly detailed catalogue by M. Merrington et al., *A List of the Papers and Correspondence of Karl Pearson (1857–1936) Held in the Manuscripts Room, University College London Library* (London, 1983). Useful supplements to the Pearson Papers are the small Karl Pearson and W. F. R. Weldon collections at the Archives of the Royal Society of London; and the College Records and the Sharpe Family Papers, the latter providing information on Maria Sharpe's background, in the University College London Archives.

Dutiful glimpses of Pearson the man are provided by his son Egon Pearson in *Karl Pearson: An Appreciation of Some Aspects of His Life and Work* (Cambridge, 1938), and tart ones by his early colleague G. Udny Yule in "Karl Pearson," *Obituary Notices of Fellows of the Royal Society*, 2 (1936–38), 73–104. Information on the Men and Women's Club is to be found in Ruth First and Ann Scott, *Olive Schreiner* (New York, 1980), and in Phyllis Grosskurth, *Havelock Ellis: A Biography* (New York, 1980). The shape of the statistical school that Pearson fostered is outlined in E. S. Pearson and M. G. Kendall, eds., *Studies in the History of Statistics and Probability* (London, 1970), and in Hilts, *Statist and Statistician*. An early treatment of Pearson's socioeconomic views is in Bernard Semmell, *Imperialism and Social Reform: English Social Thought, 1895–1914* (London, 1960). In recent years, Pearson's scientific efforts have been set in social and political context by a number of striking studies, including MacKenzie's dissertation and his *Statistics in Britain;* Bernard Norton, "Karl Pearson and the Galtonian Tradition: Studies in the Rise of Quantitative Social Biology" (doctoral dissertation, History of Science, University College London, 1978), as well as Norton's "Biology and Philosophy: The Methodological Foundations of Biometry," *Journal of the History of Biology,* 8 (Spring 1975), 85–93; and his "Karl Pearson and Statistics: The Social Origins of Scientific Innovation," *Social Studies of Science,* 8 (Feb. 1978), 3–34. Essential for Weldon, his relationship to Pearson, and Pearson's institutionalization of their research program is Lyndsay Farrall, *The Origins and Growth of the English Eugenics Movement, 1865–1912* (Ann Arbor, 1970), which can be profitably supplemented by Lyndsay Farrall, "W. F. R. Weldon, Biometry, and Population Biology" (unpublished manuscript); by Karl Pearson, "Walter Frank Raphael Weldon," *Biometrika,* 5 (1906), 1–52; and by Ruth Schwartz Cowan, "Walter Frank Raphael Weldon," *Dictionary of Scientific Biography,* XIV, 251–52. The careers of two of the women in the Galton Laboratory are explored in Rosaleen Love, " 'Alice in Eugenics-Land': Feminism and Eugenics in the Scientific Careers of Alice Lee and Ethel Elderton," *Annals of Science,* 36 (1979), 145–58. Provocative perspectives on Pearson are advanced in the correspondence between Major Greenwood and G. Udny Yule in the Yule Papers at the Royal Statistical Society, which can be used with F. Yates, "George Udny Yule," and Lancelot Hogben, "Major Greenwood, 1880–1949," *Obituary Notices of Fellows of the Royal Society,* 8 (1952–53), 309–23, and 7 (1950), 139–54.

Increasing scholarly attention has been given in recent years to the response of late-nineteenth-century scientists to the substantive problems in Darwin's theory of evolution, including the conundrum of heredity and natural selection, and to the growing call for the use of experimental and statistical methods. Reliable introductions to the issues—and to their resolution—are Garland E. Allen, *Life Science in the Twentieth Century* (New York, 1975); William B. Provine, *The Origins of Theoretical Population Genetics* (Chicago, 1971); and Ernst Mayr and William B. Provine, eds., *The Evolutionary Synthesis: Perspectives on the Unification of Biology* (Cambridge, Mass., 1980). These may be supplemented by Bernard Norton, "Metaphysics and Population Genetics: Karl Pearson and the Background to Fisher's Multifactorial Theory of Inheritance," *Annals of Science,* 32 (1975), 537–53; and Bernard Norton and E. S. Pearson, "A Note on the Background to, and Refereeing of, R. A. Fisher's 1918 Paper 'On the Correlation Between Relatives on the Supposition of Mendelian Inheritance,' " *Notes and Records of the Royal Society of London,* 31 (July 1976), 151–62. Though not available in time for my work on this book, a fundamentally important biography of a principal figure in the mathematical making of the modern evolutionary synthesis is William B. Provine, *Sewall Wright: Geneticist and Evolutionist* (Chicago, forthcoming, 1986). Convenient access to Haldane's ideas on genetics and evolution may be gained through his *The Causes of Evolution* (London, 1932). Particularly valuable works in their special subjects are Cowan's on Galton, Farrall's on Weldon, and Norton's on Pearson, in addition to Peter J. Vorzimmer, *Charles Darwin: The Years of Controversy* (Philadelphia, 1970); Peter J. Bowler, *The Eclipse of Darwinism: Anti-Darwinian Evolution Theories in the Decades around 1900* (Baltimore, 1983); Robert C. Olby, "Charles Darwin's Manuscript of *Pangenesis,*" *British Journal for the History of Science,* 1 (1963), 251–63; Peter Vorzimmer, "Charles Darwin and Blending Inheritance," *Isis,* 54 (1963), 371–90; Gerald L. Geison, "Darwin and Heredity: The Evolution of His Hypothesis of Pangenesis," *Journal of the History of Medicine and Allied Sciences,* 24 (1969), 375–411. An introduction to Mendel and the literature concerning his life and work is V. Kruta and V. Orel, "Johann Gregor Mendel," *Dictionary of Scientific Biography,* IX, 277–83. My own assessment of why Mendel went so long unappreciated owes a great deal to Elizabeth Gasking, "Why Was Mendel's Work Ignored?" *Journal of the History of Ideas,* 20 (1959), 60–84. For the early development of Mendelian genetics, important treatments include J. S. Wilkie, "Some Reasons for the Rediscovery and Appreciation of Mendel's Work in the First Years of the Present Century," *British Journal for the History of Science,* 1 (June 1962), 5–17; Beatrice Bateson, *William Bateson, F.R.S., Naturalist: His Essays and Addresses, together with a Short Account of His Life* (Cambridge, 1928); R. C. Punnett, "Early Days of Genetics," *Heredity,* 4 (April 1950), 1–10; Alfred H. Sturtevant, "The Early Mendelians," *Proceedings of the American Philosophical Society,* 109 (Aug. 1965), 199–204; William E. Castle, "The Beginnings of Mendelism in America," in L. C. Dunn, ed., *Genetics in the Twentieth Century: Essays on the Progress of Genetics during Its First Fifty Years* (New York, 1951), which contains a number of other essays that illuminate the title topic of the book. A. H. Sturtevant, *A History of Genetics* (New York, 1965), provides a useful overview, while indispensable for its subject is Garland E. Allen, *Thomas Hunt Morgan: The Man and His Science* (Princeton, 1978). A critical introduction to the historiography of the debate between the so-called biometricians and Mendelians in the early years of the century is Daniel J. Kevles, "Genetics in the United States and Great Britain, 1890–1930: A Review with Speculations," *Isis,* 71 (Sept. 1980), 441–55, reprinted in Charles Webster, ed., *Biology, Medicine and Society, 1840–1940* (Cambridge, 1981), pp. 193–215.

An arresting treatment of Charles B. Davenport the man is E. Carleton MacDowell, "Charles Benedict Davenport, 1866–1944: A Study of Conflicting Influences," *Bios,* 17 (1946), 3–50, which includes a bibliography of Davenport's published writings, among them his most important book, *Heredity in Relation to Eugenics* (New York, 1911). Critical insight into his work in human heredity can be gained from Charles E. Rosenberg, "Charles B. Davenport and the Irony of American Eugenics," in Charles E. Rosenberg, *No Other Gods: On Science and American Social Thought* (Baltimore, 1976), pp. 89–97. A study of Davenport and the institutionalization of eugenics is Garland E. Allen, "The Eugenics Record Office, Cold Spring Harbor, 1910–1940," forthcoming in *Osiris.* Information concerning the development of the Station for Experimental Evolution and of the Eugenics Record Office, including lists of their publications, can be gleaned from the annual *Yearbooks* of the Carnegie Institution of Washington. The major source for Davenport's life and work as well as the history of the Eugenics Record Office is the Charles B. Davenport Papers, an extensive collection of memoranda, correspondence, and other unpublished materials housed at the American Philosophical Society Library. Important additional materials exist in the Records of the Carnegie Institution of Washington, particularly those for the Department of Genetics, located at the headquarters of the Institution in Washington, D.C.

Further information on Davenport and his activities may be obtained from the general and specialized studies of the eugenics movement in the United States, starting with Mark Haller's pioneering *Eugenics: Hereditarian Attitudes in American Thought* (New Brunswick, N.J., 1963). Important also, not least for their use of manuscript sources, are Kenneth L. Ludmerer, *Genetics and American Society: A Historical Appraisal* (Baltimore, 1972), which ably assesses the role of geneticists in the eugenics movement; Barbara Kimmelman, "The American Breeders' Association: Genetics and Eugenics in an Agricultural Context, 1903–1913," *Social Studies of Science*, 13 (May 1983), 163–204; and Hamilton Cravens, *Triumph of Evolution: American Scientists and the Heredity-Environment Controversy, 1900–1914* (Philadelphia, 1978), which explores the wide range of ideas upon which eugenicists drew. Of the immense literature on social Darwinism and hereditarianism, particularly valuable for the background to eugenics are Arthur E. Fink, *The Causes of Crime: Biological Theories in the United States, 1800–1915* (Philadelphia, 1938); Charles E. Rosenberg, "The Bitter Fruit: Heredity, Disease, and Social Thought," in his *No Other Gods*, pp. 25–53; Robert C. Bannister, *Social Darwinism: Science and Myth in British-American Social Thought* (Philadelphia, 1979); Gareth Steadman Jones, *Outcast London* (Oxford, 1971); Greta Jones, *Social Darwinism and English Thought: The Interaction between Biological and Social Theory* (Atlantic Highlands, N.J., 1980). Allan Chase's angry *The Legacy of Malthus: The Social Costs of the New Scientific Racism* (New York, 1977) is a mine of information, and Nancy Stepan, *The Idea of Race in Science: Great Britain, 1800–1960* (London, 1982) is insightful on the relatively low degree of racism in British eugenics. A Marxist interpretation of eugenics is proposed in Garland E. Allen, "Genetics, Eugenics, and Class Struggle," *Genetics*, 79 (1975), suppl., 29–45, and "Genetics, Eugenics, and Society: Internalists and Externalists in Contemporary History of Science," *Social Studies of Science*, 6 (1976), 105–22. Robert V. Bruce, *Bell: Alexander Graham Bell and the Conquest of Solitude* (Boston, 1973), discusses its subject's interest in eugenics. The social composition of the British wing is analyzed in Farrall's *Origin and Growth of the English Eugenics Movement;* in Donald MacKenzie, "Eugenics in Britain," *Social Studies of Science*, 6 (Sept. 1976), 499–532; and in Geoffrey R. Searle, "Eugenics and Class," in Charles Webster, ed., *Biology, Medicine and Society*, pp. 217–42. For its public program, see Geoffrey R. Searle, *Eugenics in Britain, 1900–1914* (Leyden, 1976).

Donald K. Pickens, *Eugenics and the Progressives* (Nashville, Tenn., 1968), though skimpy, spotlights the involvement in eugenics of American social reformers, as does Bartlett C. Jones, "Prohibition and Eugenics," *Journal of the History of Medicine and Allied Sciences*, 18 (1963), 158–72. The attraction to eugenics of parts of the British left is adumbrated in Michael Freeden, *The New Liberalism:*

An Ideology of Social Reform (Oxford, 1978), and explicitly argued in his "Eugenics and Progressive Thought: A Study in Ideological Affinity," *The Historical Journal,* 22 (1979), 645–71, which drew a rebuttal from Greta Jones, "Eugenics and Social Policy between the Wars," *The Historical Journal,* 25 (1982), 717–28. The appeal of eugenics to parts of the European left is made clear in the rewarding comparative study by Loren R. Graham, "The Eugenics Movement in Germany and Russia in the 1920s," *American Historical Review,* 82 (1977), 1133–64. An introduction to a key issue for the eugenic left is Jeffrey Weeks, *Sex, Politics, and Society: The Regulation of Sexuality since 1800* (New York, 1981), and eugenics figures in part of Hal D. Sears, *The Sex Radicals: Free Love in High Victorian America* (Lawrence, Kan., 1977). On John Humphrey Noyes's experiment, see Maren Lockwood Carden, *Oneida: Utopian Community to Modern Corporation* (New York, 1977), and Raymond Lee Muncy's more general *Sex and Marriage in Utopian Communities* (Baltimore, 1974). The linkages among eugenics, sexuality, and the "woman issue" are suggested in Jane Hume Clapperton, *Scientific Meliorism and the Evolution of Happiness* (London, 1885) and *A Vision of the Future Based on the Application of Ethical Principles* (London, 1904); Victoria C. Woodhull, *The Scientific Propagation of the Human Race* (n.p., 1893) and *The Rapid Multiplication of the Unfit* (n.p., 1891), copies of which are in the London School of Economics Library; Havelock Ellis, *The Problem of Race-Regeneration* (New Tracts for the Times; London, 1911), *The Task of Social Hygiene* (London, 1912), and *The Philosophy of Conflict and Others Essays in Wartime* (2nd Series; London, 1919), which should be supplemented by Phyllis Grosskurth's splendid *Havelock Ellis;* Scott Nearing, *The Super Race: An American Problem* (New York, 1912); Charles A. L. Reed, *Marriage and Genetics: Laws of Human Breeding and Applied Eugenics* (Cincinnati, 1913); William J. Robinson, *Practical Eugenics: Four Means of Improving the Human Race* (New York, 1912); T. W. Shannon et al., *Scientific Knowledge of the Laws of Sex Life and Heredity or Eugenics* (Marietta, Ohio, 1917; Replica Edition, 1970); Mary Ries Melendy, *Sex-Life, Love, Marriage, Maternity* (Philadelphia, 1914), in Robert K. Leslie, ed., *The Science of Eugenics and Sex Life, Love, Marriage, Maternity: The Regeneration of the Human Race . . . from the Notes of Walter J. Hadden . . . Charles H. Robinson . . . Mary R. Melendy . . .* (New York, 1927). The rapidly growing corpus of scholarship on the history of women and/or contraception contains a number of excellent studies pertinent to the subject of this book, among them Linda Gordon, *Woman's Body, Woman's Right: A Social History of Birth Control in America* (New York, 1976); Carl N. Degler, *At Odds: Women and the Family in America from the Revolution to the Present* (New York, 1980); James Reed, *From Private Vice to Public Virtue: The Birth Control Movement and American Society since 1830* (New York, 1978); and Richard Allen Soloway, *Birth Control and the Population Question in England, 1877–1930* (Chapel Hill, N.C., 1982). Also helpful are Angus McLaren, *Birth Control in Nineteenth-Century England* (New York, 1978) and Ruth Hall, *Passionate Crusader: The Life of Marie Stopes* (New York, 1977).

Major sources for the activities of organized eugenics in both its mainline and reform phases are the American Eugenics Society Papers at the American

Philosophical Society Library and the Eugenics Society Records at the Wellcome Institute for the History of Medicine in London. Essential supplements are the *Annual Reports* of the Eugenics Education Society, renamed the Eugenics Society in 1926, which were separately published from 1908–9 to 1938–39 and appeared thereafter in the *Eugenics Review;* and the scrapbooks of press cuttings concerning eugenics, all of which are at the Eugenics Society in London. Richly important for many aspects of the evolving relationship to eugenics of geneticists and other biologists are the Charles C. Hurst Papers at the Cambridge University Library; the Julian Huxley Papers at Rice University; and the Davenport Papers, the Raymond Pearl Papers, and the Herbert Spencer Jennings Papers at the American Philosophical Society Library. Somewhat useful for the same purpose are the Reginald Ruggles Gates Papers at Kings College, London; the William Bateson Papers at the John Innes Institute, Norwich, East Anglia; the small J. B. S. Haldane collection at University College London; and the Samuel J. Holmes Papers, in the Bancroft Library, University of California, Berkeley.

The primary literature of mainline eugenics and its variants produced by geneticists and other biologists is exemplified in the writings of, among others, G. Archdall Reid, *The Laws of Heredity* (London, 1910); Robert Heath Lock, *Recent Progress in the Study of Variation, Heredity, and Evolution* (3rd ed.; London, 1911); Michael F. Guyer, *Being Well-Born: An Introduction to Eugenics* (Indianapolis, 1916); Samuel J. Holmes, *The Eugenic Predicament* (New York, 1933); Edward M. East, *Heredity and Human Affairs* (New York, 1929); Horatio Hackett Newman, *Evolution, Genetics, and Eugenics* (Chicago, 1925); *Problems in Eugenics: Papers Communicated to the First International Eugenics Congress . . .* (2 vols.; London, 1912–13); *Eugenics, Genetics, and the Family: Scientific Papers of the Second International Congress of Eugenics, 1921* (2 vols.; Baltimore, 1923). Typical of general mainline writings are the articles in the *Eugenics Review* in England and in the *Journal of Heredity* in the United States; Henry Smith, *A Plea for the Unborn* (London, 1897); Albert E. Wiggam, *The New Decalogue of Science* (Indianapolis, 1923) and *The Fruit of the Family Tree* (Indianapolis, 1924); Edgar Schuster, *Eugenics* (London, 1912); Caleb W. Saleeby, *The Progress of Eugenics* (London, 1914); William C. D. Whetham and Catherine D. Whetham, *The Family and the Nation: A Study in Natural Inheritance and Social Responsibility* (London, 1909); Blanche Eames, *Principles of Eugenics: A Practical Treatise* (New York, 1914); William Ralph Inge, *Lay Thoughts of a Dean* (New York, 1926) and *Outspoken Essays (Second Series)* (New York, 1927); Lothrop Stoddard, *The Revolt Against Civilization: The Menace of the Underman* (New York, 1922); Madison Grant, *The Passing of the Great Race* (New York, 1916); Paul Popenoe and Roswell Hill Johnson, *Applied Eugenics* (New York, 1926); and Leonard Darwin, *What Is Eugenics?* (New York, 1929).

The questions increasingly raised about mainline eugenics are evident in numerous articles in popular periodicals; in such books as Franz Boas, *Anthropology and Modern Life* (New York, 1928); Leonard T. Hobhouse, *Social Evolution and Political Theory* (Oxford, 1911); Bertrand Russell, *Icarus, or the Future of Science* (London, 1924); and G. K. Chesterton's biting *Eugenics and Other Evils* (London, 1922), which may be better understood with the helpful study of Margaret Canovan,

G. K. Chesterton: Radical Populist (New York, 1977). The Catholic position on eugenics is spelled out in Thomas J. Gerrard, *The Church and Eugenics* (St. Louis, 1912), which appeared in a third edition (St. Louis, 1921); Charles P. Bruehl, *Birth Control and Eugenics in the Light of Fundamental Ethical Principles* (New York, 1928). Fundamentally important to the emergence of the dissent from all or parts of mainline eugenics were the writings of a number of biologists, including J. Arthur Thomson, *Heredity* (London, 1908); Thomas Hunt Morgan, *Evolution and Genetics* (Princeton, 1925) and *The Scientific Basis of Evolution* (New York, 1932); the successive editions of Edwin Grant Conklin, *Heredity and Environment in the Development of Men* (1st ed.; Princeton, 1915) and Conklin's *The Direction of Human Evolution* (New York, 1922).

The anti-mainline leadership may be approached through a number of biographical treatments that include bibliographies of their subjects' writings: J. R. Baker, "Julian Sorell Huxley"; G. P. Wells, "Lancelot Hogben"; N. W. Pirie, "John Burdon Sanderson Haldane," all in *Biographical Memoirs of Fellows of the Royal Society*, respectively, 22 (1976), 207–39; 24 (1978), 183–221; 12 (1966), 219–49; Ronald W. Clark, *J.B.S.: The Life and Work of J. B. S. Haldane* (New York, 1968); Tracy M. Sonneborn, "Herbert Spencer Jennings, 1868–1947," *National Academy of Sciences Biographical Memoirs*, 47 (1975), 143–223. Gary Wersky's insightful *The Visible College* (New York, 1979) is required reading for anyone interested in the British group. Julian Huxley's autobiographical *Memories* (New York, 1970) is helpful, and so are Robert E. Filner, "The Social Relations of Science Movement (SRS) and J. B. S. Haldane," *Science and Society*, 41 (Fall 1977), 303–16; Diane B. Paul, "Eugenics and the Left," *Journal of the History of Ideas*, 45 (Oct.–Dec. 1984), 567–90, and "A War on Two Fronts: J. B. S. Haldane and the Response to Lysenkoism in Britain," *Journal of the History of Biology*, 16 (Spring 1983), 1–37; and T. E. B. Howarth, *Cambridge between Two Wars* (London, 1978). Of the considerable body of writings produced by the anti-mainline leadership, I found especially important the following: Herbert S. Jennings, *Prometheus, or Biology and the Advancement of Man* (New York, 1925), *The Biological Basis of Human Nature* (New York, 1930), " 'Undesirable Aliens,' " *The Survey*, 51 (Dec. 15, 1923), 309–12, 364, and "The Laws of Heredity and Our Present Knowledge of Human Genetics on the Material Side," in Herbert S. Jennings et al., *Scientific Aspects of the Race Problem* (Washington, D.C., 1941); Julian S. Huxley, *Essays in Popular Science* (New York, 1927), *Science and Social Needs* (New York, 1935), and *Man Stands Alone* (New York, 1941); H. G. Wells, Julian S. Huxley, and G. P. Wells, *The Science of Life* (2 vols.; New York, 1931); Julian S. Huxley and A. C. Haddon, *We Europeans: A Survey of "Racial" Problems* (London, 1935); Julian S. Huxley et al., *Reshaping Man's Future: Biology in the Service of Man* (London, 1944); J. B. S. Haldane, *Daedalus, or Science and the Future* (New York, 1924), *Possible Worlds* (New York, 1928), *The Inequality of Man* (London, 1932), *Human Biology and Politics* (London, 1934), *Heredity and Politics* (New York, 1938), *Science and Everyday Life* (London, 1939), *Adventures of a Biologist* (New York, 1940), and *Science Advances* (New York, 1947); Lancelot Hogben, *The Nature of Living Matter* (London, 1930), *Genetic Principles in Medicine and Social Science* (London, 1931),

Nature and Nurture (New York, 1933), "Heredity and Human Affairs," in J. G. Crowther, ed., *Science for a New World* (New York, 1934), and his edited *Political Arithmetic: A Symposium of Population Studies* (London, 1938).

The ideas of reform eugenics are contained in many of the writings of the anti-mainline leadership mentioned above. Necessary additions are Hermann J. Muller, *Out of the Night: A Biologist's View of the Future* (New York, 1935), and the informative biography by Elof Axel Carlson, *Genes, Radiation, and Society: The Life and Work of H. J. Muller* (Ithaca, N.Y., 1981). Herbert Brewer's concern with eutelegenesis is revealed in various files in the Eugenics Society Records and in Herbert Brewer, "Eutelegenesis," *Eugenics Review*, 27 (1935), 121–26. Brewer the man was fleshed out in an interview with his daughter, Peggy Brewer Musgrave, in San Francisco, April 1984. Part of the work that excited Brewer and Muller is explored in Theodore L. Malinin, *Surgery and Life: The Extraordinary Career of Alexis Carrel* (New York, 1979), and reported in Gregory Pincus, *The Eggs of Mammals* (New York, 1936). John Blacker provided information concerning his father, C. P. Blacker, in an interview in London in March 1984, and Harry Shapiro did the same for Frederick Osborn and the activities of the American Eugenics Society in an interview in New York City the same month. The basic published writings of organized reform eugenics are: C. P. Blacker, *Birth Control and the State: A Plea and a Forecast* (London, 1926), *Human Values in Psychological Medicine* (London, 1933), *A Social Problem Group* (London, 1937), *Eugenics in Prospect and Retrospect* (London, 1945), and *Eugenics: Galton and After* (London, 1952), also Frederick Osborn's edition of Gladys C. Schwesinger, *Heredity and Environment: Studies in the Genesis of Psychological Characteristics* (New York, 1933), his *Preface to Eugenics* (New York, 1940; 2nd ed., New York, 1951), *The Future of Human Heredity* (New York, 1968), and Frederick Osborn and Carl Jay Bajema, "The Eugenic Hypothesis," in Carl Jay Bajema, ed., *Eugenics Then and Now* (Stroudsburg, Pa., 1976), pp. 283–91. See also Geoffrey R. Searle, "Eugenics and Politics in the 1930s," *Annals of Science*, 36 (1979), 159–79. An important history remains to be written of the general relationship among eugenics, demography, and population control. In the meantime, reform-eugenic attitudes concerning the fertility issue in the nineteen-thirties may be gleaned from Ellsworth Huntington, *Tomorrow's Children: The Goal of Eugenics* (New York, 1935); Enid Charles, *The Twilight of Parenthood* (New York, 1934); Raymond Pearl, *The Natural History of Population* (London, 1939); David V. Glass, *The Struggle for Population* (Oxford, 1936); and C. P. Blacker and David V. Glass, *The Future of Our Population?* (pamphlet; London: Population Investigation Committee, n.d.). Ronald Fisher's special theory of differential human fertility may be studied in Ronald A. Fisher, *The Genetical Theory of Natural Selection* (Oxford, 1930); *The Collected Papers of R. A. Fisher*, ed. J. H. Bennett (5 vols.; Adelaide, 1971–74); and J. H. Bennett, ed., *Natural Selection, Heredity, and Eugenics, Including Selected Correspondence of R. A. Fisher with Leonard Darwin and Others* (Oxford, 1983). See also Joan Fisher Box's biography of her father, *R. A. Fisher: The Life of a Scientist* (New York, 1978).

The development of the 1913 Mental Deficiency Act is treated in Kathleen

Jones, *A History of the Mental Health Services* (London, 1972), and in Harvey G. Simmons, "Explaining Social Policy: The English Mental Deficiency Act of 1913," *Journal of Social History*, 11 (1978), 387–403. Lloyd P. Gartner, *The Jewish Immigrant in England, 1870–1914* (Detroit, 1960), is helpful for its subject. The standard work on immigration restriction in the United States is John Higham, *Strangers in the Land: Patterns of American Nativism, 1860–1925* (2nd ed.; New York, 1963), and a convenient summary of the history of immigration policy is in Richard A. Easterlin et al., *Immigration* (Cambridge, Mass., 1982). The role of eugenicists in the restrictionist movement, including its legislative culmination, is well illuminated in Ludmerer, *Genetics and American Society*, and Frances J. Hassencahl, *Harry H. Laughlin, "Expert Eugenics Agent" for the House Committee on Immigration and Naturalization, 1921 to 1931* (Ann Arbor, 1971), which includes a biographical treatment of Laughlin. For its special subject, see Charles B. Davenport, *State Laws Limiting Marriage Selection Examined in the Light of Eugenics* (Eugenics Record Office Bulletin No. 9; Cold Spring Harbor, N.Y., 1913). The paramount programmatic interest of mainline eugenics is treated in Rudolph J. Vecoli, "Sterilization: A Progressive Measure," *Wisconsin Magazine of History*, 43 (Spring 1960), 190–202, and Jonas Robitscher, ed., *Eugenic Sterilization* (Springfield, Ill., 1973). The legislative and administrative history of sterilization in California awaits its historian, but access to the subject may be obtained from Ezra S. Gosney and Paul Popenoe, *Sterilization for Human Betterment: A Summary of Results of 6,000 Operations in California, 1909–1929* (New York, 1930); *Twenty-eight Years of Sterilization in California* (Pasadena, 1939); and Richard W. Fox, *So Far Disordered in Mind: Insanity in California, 1870–1930* (Berkeley, 1978). Very useful for its attempt to analyze who was sterilized in California is Judith K. Grether, *Sterilization and Eugenics: An Examination of Early Twentieth Century Population Control in the United States* (Ann Arbor, 1980). Similarly helpful for Virginia is G. B. Arnold, "A Brief Review of the First Thousand Patients Eugenically Sterilized at the State Colony for Epileptics and Feebleminded," *American Association on Mental Deficiency Proceedings*, 43 (1938), 56–63, and for North Carolina, Moya Woodside, *Sterilization in North Carolina: A Sociological and Psychological Study* (Chapel Hill, N.C., 1950). The state-by-state legal history of sterilization is reviewed in Jacob H. Landman, *Human Sterilization: The History of the Sexual Sterilization Movement* (New York, 1932), which may be supplemented by Harry Laughlin's successive compilations: *The Legal, Legislative and Administrative Aspects of Sterilization* (Cold Spring Harbor, N.Y., 1914), *Eugenical Sterilization in the United States* (Chicago, 1922), and *The Legal Status of Eugenical Sterilization* (Chicago, 1930), which contains the briefs, court rulings, etc., of *Buck v. Bell*. Valuable treatments of the case are Donna M. Cone's unpublished article, "The Case of Carrie Buck: Eugenic Sterilization Realized"; the authoritative legal analysis of R. J. Cynkar, *"Buck v. Bell:* 'Felt Necessities' v. Fundamental Values?" *Columbia Law Review*, 81 (1981), 1418–61; and Stephen Jay Gould, "The Case of Carrie Buck's Daughter," *Natural History*, 93 (July 1984), 14–18. Sterilization in the thirties and forties may be approached through Harry H. Laughlin, "Further Studies on the Historical and Legal Development of Eugenical Sterilization in the United States," *Proceedings of the American Association on Mental*

Deficiency, 41 (1936), 96–110; Leon F. Whitney, *The Case for Sterilization* (New York, 1934); Marian S. Olden, *Human Betterment Was Our Goal* ([Princeton], n.d.); The Committee of the American Neurological Association for the Investigation of Eugenical Sterilization, *Eugenical Sterilization: A Reorientation of the Problem* (New York, 1936). Also important are the books advancing the Catholic position on eugenics as well as John A. Ryan, *Moral Aspects of Sterilization* (Washington, D.C., 1930); a series of retrospective articles on sterilization published in the Richmond *Times-Dispatch* in February, March, and April 1980; and the many articles on the subject in the New York *Times*, 1930–50, which is a rewarding source for developments in the Nazi eugenic program. There is no overall history of eugenics under the Nazis, but indicative of the valuable work underway is Gisela Bok, "Racism and Sexism in Nazi Germany: Motherhood, Compulsory Sterilization, and the State," *Signs*, 8 (1983), 400–21. Also useful is the brief treatment in George L. Mosse, *Toward the Final Solution: A History of European Racism* (New York, 1978), as well as Marie E. Kopp, "Eugenic Sterilization Laws in Europe," *American Journal of Obstetrics and Gynecology*, 34 (Sept. 1937), 499–504, and her "Legal and Medical Aspects of Eugenic Sterilization in Germany," *American Sociological Review*, 1 (Oct. 1936), 761–70. On aspects of Nazi positive eugenics, see Marc Hillel and Clarissa Henry, *Of Pure Blood* (New York, 1976). Essential for the debates and activities concerning sterilization in Britain during the thirties are the Eugenics Society Papers, including relevant correspondence files, the pamphlets issued by the Committee for Legalising Eugenic Sterilization—notably *Better Unborn, The Law as to Sterilization*, and *Eugenic Sterilization*—and *Report of The Departmental Committee on Sterilisation (minus appendices), June 1934* (pamphlet; London, 1934); Geoffrey R. Searle's "Eugenics and Politics in Britain"; the press cuttings scrapbooks at the Eugenics Society; C. P. Blacker, *Voluntary Sterilization* (London, 1934), and his *Voluntary Sterilization: Introduction and Summary, the Last Sixty Years . . .* (pamphlet; London, 1962).

The large literature on intelligence testing includes a growing number of general historical treatments: Thomas P. Weinland, *A History of the I.Q. in America, 1890–1941* (Ann Arbor, 1970), and Russell Marks, *Testers, Trackers and Trustees: The Ideology of the Intelligence Testing Movement in America, 1900–1954* (Ann Arbor, 1972). A major episode in the history of I.Q. testing in the United States is treated in Daniel J. Kevles, "Testing the Army's Intelligence: Psychologists and the Military in World War I," *Journal of American History*, 55 (Dec. 1968), 565–81; Franz Samelson, "World War I Intelligence Testing and the Development of Psychology," *Journal of the History of the Behavioral Sciences*, 13 (1977), 274–282; and Joel Spring, "Psychologists and the War: The Meaning of Intelligence in the Alpha and Beta Tests," *History of Education Quarterly*, 12 (Spring 1972), 3–15. Correspondence concerning intelligence testing in relation to eugenics is in the Robert M. Yerkes Papers at Yale University. The major source for the social interpretation of the wartime test results is Robert M. Yerkes, ed., *Psychological Examining in the United States Army* (Washington, D.C., 1921), which led to Carl Campbell Brigham's *A Study of American Intelligence* (Princeton, 1923). Additional insight into the history of mental testing in Britain may be gained from Cyril Burt, *Mental and Scholastic Tests* (London, 1922); Leslie S. Hearnshaw's sympathetic yet critical biography, *Cyril Burt, Psychologist* (Ithaca, N.Y., 1979); Gillian Sutherland and Stephen Sharp, " 'The Fust Official Psychologist in the Wurrld': Aspects of the Professionalization of Psychology in Early Twentieth Century Britain," *History of Science*, 18 (1980), 181–208; Gillian Sutherland, "Measuring Intelligence: English Local Education Authorities and Mental Testing, 1919–1939," in J. V. Smith and D. Hamilton, eds., *The Meritocratic Intellect: Studies in the History of Education Research* (Aberdeen, 1980), pp. 79–95; and Bernard Norton, "Charles Spearman and the General Factor in Intelligence: Genesis and Interpretation in the Light of Sociopersonal Considerations," *Journal of the History of the Behavioral Sciences*, 15 (1979), 142–54. In *The Mismeasure of Man* (New York, 1981), Stephen Jay Gould neatly exposes the flaws in the foundational theories, methods, and uses of intelligence testing in both the United States and Britain. The methodological ricketiness in hereditarian theories of intelligence are stressed in Brian Evans and Bernard Waites, *IQ and Mental Testing: An Unnatural Science and Its Social History* (London, 1980), and the changing views of American psychologists on the issue are recounted in Cravens's *Triumph of Evolution* and in his unpublished papers: "The Wandering I.Q.: The Iowa Child Welfare Research Station and Mental Testing, 1925–1940" and "Inconstancy of the Intelligence Quotient: The Iowa Child Welfare Research Station and the Criticism of Hereditarian Mental Testing, 1917–1939." The response to the Iowa station's results is explored in Henry L. Minton, "The Iowa Child Welfare Research Station and the 1940 Debate on Intelligence: Carrying on the Legacy of a Concerned Mother," *Journal of the History of the Behavioral Sciences*, 20 (April 1984), 160–76. The anti-hereditarian position is framed in Carl Brigham's mea culpa, "Intelligence Tests of Immigrant Groups," *Psychological Review*, 37 (March 1930), 158–65; in Horatio Hackett Newman, Frank N. Freeman, and Karl J. Holzinger, *Twins: A Study of Heredity and Environment* (Chicago, 1937); and in Walter Lippmann's arresting series of articles in the *New Republic*, vol. 32: "The Mental Age of Americans," Oct. 25, 1922, pp. 213–15; "The Mystery of the 'A' Men," Nov. 1, 1922, pp. 246–48; "The Reliability of Intelligence Tests," Nov. 8, 1922, pp. 275–77; "The Abuse of the Tests," Nov. 15, 1922, pp. 297–98; "Tests of Hereditary Intelligence," Nov. 22, 1922, pp. 328–30; and vol. 33: "A Future for the Tests," Nov. 29, 1922, pp. 9–11, which should be supplemented by the short series in vol. 34 of the magazine: "Mr. Burt and the Intelligence Tests," May 2, 1923, pp. 263–64; "Rich and Poor, Girls and Boys," May 9, 1923, pp. 295–96; and "A Judgment of the Tests," May 16, 1923, pp. 322–23. Lancelot Hogben spotlighted the methodological pitfalls in hereditarian interpretations of intelligence in his *Genetic Principles in Medicine and Social Science, Nurture and Nature*, and the technical "The Limits of the Applicability of Correlation Techniques in Human Genetics," *Journal of Genetics*, 27 (Aug. 1933), 379–406. Otto Klineberg has recalled his career in "Otto Klineberg," in *A History of Psychology in Autobiography*, vol. VI (Englewood Cliffs, N.J., 1974), 163–82, and "Reflections of an International Psychologist of Canadian Origin," *International Social Science Journal*, 25 (1973), 39–54. My assessment of the man and his work benefited greatly from an interview with him in New York City in May 1982. Klineberg's key articles on race and intelligence include "An Experimental Study of Speed and Other Factors in 'Racial' Differences," *Archives of Psychology*, No. 93 (Jan. 1928); "A Study of Psychological Differences Between 'Racial' and National Groups in Europe," *Archives of Psychology*, No. 132 (Sept. 1931); *Negro Intelligence and Selective Migration* (New York, 1935); "Mental Testing of Racial and National Groups," in Herbert Spencer Jennings et al., *Scientific Aspects of the Race Problem* (Washington, D.C., 1941), pp. 253–94; "Tests of Negro Intelligence," in Otto Klineberg, ed., *Characteristics of the American Negro* (New York, 1944), pp. 23–96; and *Race and Psychology* (Paris, 1951). The increasing pervasiveness of ideas like Klineberg's concerning genetics, race, and intelligence is clear in *The Race Concept* (Paris, 1952) and Margaret Mead, Theodosius Dobzhansky, Ethel Tobach, and Robert E. Light, eds., *Science and the Concept of Race* (New York, 1968).

Arthur R. Jensen's 1969 dissent from those ideas, "How Much Can We Boost IQ and Scholastic Achievement?" *Harvard Educational Review*, 39 (1969), 1–123, is reprinted in his *Genetics and Education* (New York, 1972), which includes an extended preface recounting how he came to write the article and his problems once it was published as well as a lengthy bibliography of the pro and con writings it provoked. Useful also is Michael Schudson, "A History of the *Harvard Educational Review*," in John R. Snarey et al., *Conflict and Continuity: A History of Ideas on Social Equality and Human Development* (Cambridge, Mass., 1981). Richard J. Herrnstein expanded his "I.Q.," *The Atlantic*, 228 (Sept. 1971), 43–58, 63–64, into *I.Q. in the Meritocracy* (Boston, 1973). Hans J. Eysenck's views are vigorously argued in his *The IQ Argument: Race, Intelligence, and Education* (New York, 1971) as well as in *The Inequality of Man* (London, 1973). A profile of Jensen is Lee Edson, "*Jensenism*, n.: the Theory That I.Q. Is Largely Determined by the Genes," *The New York Times Magazine*, Aug. 31, 1969, pp. 10–11, 40–47, and a more extensive one of Eysenck is H. B. Gibson, *Hans Eysenck: The Man and His Work* (London, 1981). Important articles forming part of the initial scholarly response to Jensen appeared in the *Harvard Educational Review*, 39 (Summer 1969), and a selection of many others, including Richard C. Lewontin's critiques, are collected in N. J. Block and Gerald Dworkin, eds., *The I.Q. Controversy: Critical Readings* (New York, 1976), and also in Ken Richardson, David Spears, and Martin Richards, *Race and Intelligence: The Fallacies Behind the Race-IQ Controversy* (Baltimore, 1972), which includes a version by W. F. Bodmer of his article with L. L. Cavalli-Sforza, "Intelligence and Race," *Scientific American*, 223 (1970), 19–29. Indispensable for the views of numerous American geneticists on the race-intelligence controversy, and for the development of the position eventually taken on it by their professional society, are the Genetics Society of America Papers, particularly the files of the Committee on Genetics, Race, and Intelligence, at the American Philosophical Society Library. See also Theodosius Dobzhansky, *Genetic Diversity and Human Equality* (New York, 1973), and James C. King, *The Biology of Race* (rev. ed.; Berkeley, 1981). William Shockley's encounters with the National Academy of Sciences are pursued in Jerry Hirsch, "To 'Unfrock the Charlatans,'" *Sage Race Relations Abstracts*, 6 (May 1981), 1–65. Brian

Evans and Bernard Waites ably set the controversy in Anglo-American historical perspective in their *IQ and Mental Testing,* and a rewarding sociohistorical treatment is Jonathan Harwood, "The Race-Intelligence Controversy: A Sociological Approach, I—Professional Factors and II—'External Factors,' " in *Social Studies of Science,* 6 (Sept. 1976), 369–94; 7 (Feb. 1977), 1–30. The fraudulence of Cyril Burt's twin studies is fully exposed in Leon J. Kamin, *The Science and Politics of I.Q.* (Potomac, Md., 1974), and Hearnshaw attempts to account for why Burt did it in his biography of the man.

Many of the works about intelligence testing may be consulted regarding the special subject of mental deficiency. See also Peter L. Tyor, *Segregation or Surgery: The Mentally Retarded in America, 1850–1920* (Ann Arbor, 1972), and his excellent " 'Denied the Power to Choose the Good': Sexuality and Mental Defect in American Medical Practice, 1850–1920," *Journal of Social History,* 10 (1977), 272–89. Materials for Henry H. Goddard and his work include Tyor's unpublished paper, "Henry H. Goddard: Morons, Mental Defect, and the Origins of Intelligence"; John McPhee, *The Pine Barrens* (New York, 1968), a treatment of Goddard and his field workers; the Goddard files in the Charles B. Davenport Papers; and Goddard's own writings, notably "Four Hundred Feeble-Minded Children Classified by the Binet Method," *Journal of Psycho-Asthenics,* 15 (Sept. and Dec. 1910), 17–30; *Feeble-mindedness: Its Causes and Consequences* (New York, 1914); and "Mental Tests and the Immigrant," *Journal of Delinquency,* 2 (Sept. 1917), 243–77. Exemplary of ideas on its subject in Britain is A. F. Treadgold, *Mental Deficiency (Amentia)* (5th ed.; Baltimore, 1929), and the ongoing worry about British intellectual decline is expressed in Raymond B. Cattell, *The Fight for Our National Intelligence* (London, 1937); Cyril Burt, *Intelligence and Fertility* (London, 1946); and Godfrey Thomson, *The Trend of National Intelligence: The Galton Lecture, 1946, with a Symposium in 1947 by Sir Alexander Carr-Saunders, Sir Cyril Burt, Professor Lionel Penrose, Professor Godfrey Thomson* (London, 1947). To my knowledge, there is no general history of post-Goddard theories of mental deficiency. A useful collection of original papers is Marvin Rosen, Gerald R. Clark, and Marvin S. Kivitz, eds., *The History of Mental Retardation* (2 vols.; Baltimore, 1976), and the subject is dealt with to some extent in Albert Deutsch, *The Mentally Ill in America: A History of Their Care and Treatment from Colonial Times* (2nd ed.; New York, 1949). A special critique of the Goddard-Davenport school is David Heron, *Mendelism and the Problem of Mental Defect: A Criticism of Recent American Work* (Questions of the Day and of the Fray, VII; London, 1913), and the general breakaway from its simplicities is manifest in J. E. Wallace Wallin, *Problems of Subnormality* (Yonkers-on-Hudson, N.Y., 1917); Abraham Myerson, *The Inheritance of Mental Diseases* (Baltimore, 1925); and Walter E. Fernald, "Feeblemindedness," *Mental Hygiene,* 8 (Oct. 1924), 964–971. E. O. Lewis's skepticism of prevailing theories runs through the influential *Report of the Mental Deficiency Committee, being a Joint Committee of the Board of Education and Board of Control* (3 vols.; London, 1929).

A fine introduction to its subject is Harry Harris, "Lionel Sharples Penrose," *Biographical Memoirs of Fellows of the Royal Society,* 19 (1973), 521–61, which contains a bibliography of Penrose's published work. Helpful on the Penrose family background is Roland Penrose, *Scrapbook, 1900–1981* (London, 1981), and Frances Partridge, *Love in Bloomsbury* (Boston, 1981) includes glimpses of Penrose. My understanding of Penrose was greatly aided by interviews conducted in London, Oxford, and Cambridge in June–July 1982 and March 1984 with his widow, Margaret Penrose Newman; her husband, Max Newman, who was Lionel's lifelong friend; his brother Roland Penrose; his daughter, Shirley Penrose Hodgson; his son Roger Penrose; his friend Cyril Clark; and a number of the geneticists mentioned elsewhere in this essay. Essential for Penrose's life and work is the large collection of Lionel S. Penrose Papers in the University College London Archives, which may be used with the excellent guide compiled by M. Merrington et al., *A List of the Papers and Correspondence of Lionel Sharples Penrose (1898–1972)* . . . (London, 1979). Penrose's views on genetics, disease, and mental deficiency are advanced in his *Influence of Heredity on Disease* (London, 1934) and *Mental Defect* (New York, 1934). Significant material for the origins and development of Penrose's position at the Royal Eastern Counties' Institution is in the Records of the Medical Research Council at the Council's headquarters in London, and the results of Penrose's work are summarized in Lionel S. Penrose, *A Clinical and Genetic Study of 1280 Cases of Mental Defect (The "Colchester Survey")* (Medical Research Council Special Report 229; London, 1938; reissued, London: Institute for Research into Mental and Multiple Handicap, 1975). J. Langdon Haydon Down summarized his views on the syndrome eventually named after him in *On Some of the Mental Affections of Childhood and Youth* (London, 1887). Rudolf F. Vollman, *Down's Syndrome (Mongolism): A Reference Bibliography* (Washington, D.C., 1969), reprints Down's original paper of 1866 and provides a guide to the literature on the subject since. Studies correlative to Penrose's are R. L. Jenkins, "Etiology of Mongolism," and Adrien Bleyer, "Role of Advanced Maternal Age in Causing Mongolism," both in *American Journal of Diseases of Children*, respectively, 45 (1933), 506–19; 55 (1938), 79–92. Penrose's notable early works on Down's syndrome include "On the Interaction of Heredity and Environment in the Study of Human Genetics (with Special Reference to Mongolian Imbecility)," *Journal of Genetics*, 25 (1932), 407–22; "The Relative Aetiological Importance of Birth Order and Maternal Age in Mongolism," *Proceedings of the Royal Society*, B, 115 (1934), 431–50. On another subject of major importance to Penrose, see his "Phenylketonuria—A Problem in Eugenics," *The Lancet*, June 29, 1946, 949–51.

Exemplary of the state of its field at the opening of the nineteen-thirties is Erwin Baur, Eugen Fischer, and Fritz Lenz, *Human Heredity*, Eden and Cedar Paul, trans. (New York, 1931), which advances both racist eugenic theories and the new mathematical methods of human genetics. Lancelot Hogben made the new methodological tools conveniently available to Anglo-American scientists in "The Genetic Analysis of Familial Traits: I. Single Gene Substitutions; II. Double Gene Substitutions, with Special Reference to Hereditary Dwarfism; III. Matings Involving One Parent Exhibiting a Trait Determined by a Single Recessive Gene Substitution with Special Reference to Sex-Linked Conditions," *Journal of Genetics*, 25 (1932), 97–112, 211–40, 293–314. See also Lancelot Hogben, "Some Methodological Aspects of Human Genetics," *American Naturalist*, 57 (May–June 1933), 254–63, and

J. B. S. Haldane, "A Method for Investigating Recessive Characters in Man," *Journal of Genetics*, 25 (1932), 251–55. Haldane called attention to the tools for studying human genetics in his influential *New Paths in Genetics* (New York, 1941). The overall development of human genetics from 1930 to the early sixties may be followed in such scientific texts as Curt Stern, *Principles of Human Genetics* (London, 1949; 2nd ed., 1960); James V. Neel and William J. Schull, *Human Heredity* (Chicago, 1954); Lionel S. Penrose, ed., *Recent Advances in Human Genetics* (London, 1961), as well as his *Outline of Human Genetics* (2nd ed.; London, 1963). F. Vogel and A. G. Motulsky, *Human Genetics* (New York, 1979), a valuable scientific reference, is somewhat attentive to the history of its subject. More directly so are Curt Stern, "High Points in Human Genetics," *American Biology Teacher*, March 1975, pp. 144–49, and his "Mendel and Human Genetics," *Proceedings of the American Philosophical Society*, 109 (1965), 216–26; James V. Neel, "Human Genetics," in John Z. Bowers and Elizabeth F. Purcell, eds., *Advances in American Medicine: Essays at the Bicentennial* (2 vols.; New York, 1976), I, 39–99. Key papers in human blood-group genetics, biochemical genetics, and cytogenetics are reprinted in Samuel H. Boyer, ed., *Papers on Human Genetics* (Englewood Cliffs, N.J., 1963), and important papers on human aspects of its subject are in J. Herbert Taylor, ed., *Selected Papers on Molecular Genetics* (New York, 1965).

There is no comprehensive historical study of human genetics, and nothing more than a few fragmentary autobiographical reminiscences by its practitioners. The starting point for my analysis of the relative national strengths within the Anglo-American human-genetics community was Daniel J. Kevles and Stephen Postema, "A Statistical Survey of Human Genetics in the United States and Great Britain, 1930 to 1959," in preparation, which reports the results of a survey of a number of scientific journals done to determine who was publishing in the field and also analyzes the results in terms of its leadership, their type of training and research, and their institutional locations. Crucially important for my understanding of developments in the two countries were interviews with the following human, medical, and molecular geneticists: Anthony C. Allison, Palo Alto, Dec. 1984; Alexander Bearn, Rahway, N.J., May 1982; Martin Bobrow, London, March 1984; Walter Bodmer, London, July 1982; John Burn, London, March 1984; Cedric Carter, London, June 1982; L. L. Cavalli-Sforza, via telephone from Palo Alto, April 1984; James Crow, Pasadena, Feb. 1982; Barton Childs, Baltimore, May 1982; Bernard Davis, Boston, March 1984; John Edwards, Oxford, March 1984; Charles J. Epstein, San Francisco, April 1984; Malcolm Ferguson-Smith, via telephone from Glasgow, March 1984; Charles E. Ford, Oxford, June 1982; Park Gerald, Boston, June 1982; Harry Harris, Philadelphia, Oct. 1982 and May 1983; Kurt Hirschhorn, New York, March 1984; Patricia A. Jacobs, via telephone from Hawaii, Oct. 1983; Hans Kalmus, London, July 1982; Sylvia Lawler, London, June 1982; Joshua Lederberg, New York, March 1984; Jérôme Lejeune, Paris, Sept. 1980 and July 1982; Richard Lewontin, Cambridge, Mass., June 1982; Victor McKusick, Baltimore, May 1982; Matthew Meselson, Cambridge, Mass., July 1982; O. J. Miller, New York, March 1984; Aubrey Milunsky, Boston, March 1984; Ursula Mittwoch, London, June 1982; James V. Neel, Ann Arbor, Mich., May 1982; Paul E. Polani, London, June 1982 and

March 1984; James Renwick, London, March 1984; John A. Fraser Roberts, London, June 1982; Leon Rosenberg, New Haven, Conn., March 1984; Frank Ruddle, Boston, March 1984; Ruth Sanger, London, June 1982; C. A. B. Smith, London, June 1982; John Maynard Smith, Kingston, Sussex, U.K., June 1982; Laurence Snyder, via telephone from Hawaii, April 1983; Arthur Steinberg, Cleveland, May 1982.

For the background of the English school of human genetics as it emerged in the thirties, see P. Froggart and N. C. Nevin, "The 'Law of Ancestral Heredity': Its Influence on the Early Development of Human Genetics," *History of Science*, 10 (1971), 1–27. Also useful are the successive volumes of *The Treasury of Human Inheritance* (London, 1909–   ) and of the *Annals of Eugenics* (1925–   ). Helpful perspective is provided by Lionel Penrose, "The Influence of the English Tradition in Human Genetics," *Proceedings of the Third International Congress of Human Genetics*, James F. Crow and James V. Neel, eds. (Baltimore, 1967), pp. 13–25. Aspects of the development of the British school during the thirties may be gleaned from the *Annual Reports* of the Medical Research Council; the Council's archival records at its London headquarters; and A. Landsborough Thomson's history of the Council, *Half a Century of Medical Research* (2 vols.; London, 1973, 1975). Insightful on the relationship to eugenics of part of the British human-genetics research program is Pauline M. H. Mazumdar, "Eugenists, Marxists and the Science of Human Genetics: A History of Human Genetics in Britain, 1900 to 1940" (unpublished manuscript). The creation of the social biology group at the London School of Economics is discussed in José Harris, *William Beveridge: A Biography* (Oxford, 1977), and the work at the Galton Laboratory is recounted in Joan Fisher Box's *R. A. Fisher*. Aspects of Haldane's role in the British school are examined in K. R. Dronamraju, ed., *Haldane and Modern Biology* (Baltimore, 1968). For the blood-group work and its implications, see Robert R. Race and Ruth Sanger, *The Blood Groups of Man* (Oxford, 1950); William C. Boyd, *Genetics and the Races of Man* (Boston, 1950); and Pauline M. Mazumdar, *Karl Landsteiner and the Problem of Species, 1838–1968* (Ann Arbor, 1976). Essential for the development of the Galton Laboratory under Penrose are the Penrose Papers at University College London, and much of the research of the Laboratory was published in *Annals of Eugenics* and its renamed successor, *Annals of Human Genetics* (1954–   ). Useful for their subjects are Sheldon C. Reed, "A Short History of Human Genetics in the USA," *American Journal of Medical Genetics*, 3 (1979), 282–95, and James V. Neel's review of the first quarter-century of the American Society of Human Genetics, "Our Twenty-fifth," *American Journal of Human Genetics*, 26 (March 1974), 136–44. The Papers of the American Society of Human Genetics at the American Philosophical Society Library contain little of historical usefulness, but the development of the American branch of the field may be traced through the successive volumes of the *American Journal of Human Genetics* (1949–   ). The preference of the Rockefeller Foundation for plant and animal over human genetics is evident from its *Annual Reports* for the nineteen-thirties and is made explicitly clear in Barbara Kimmelman, "An Effort in Reductionist Sociobiology: The Rockefeller Foundation and Physiological Genetics, 1930–1942" (unpublished manuscript, 1981). For Franz Kallmann's approach to the genetics of mental disorder, see his *Heredity in Health and Mental Disorder* (New York, 1953) and the sympathetic appreciation of it in Lawrence Kolb, coordinator, *Progress in Psychiatric Research and Education: A Symposium in Honor of the Seventy-fifth Anniversary of the New York State Psychiatric Institute* (New York, 1973). A powerful critique of Kallmann's work on the genetics of schizophrenia is in Richard C. Lewontin, Steven Rose, and Leon Kamin, *Not in Our Genes: Biology, Ideology, and Human Nature* (New York, 1984).

For the general history of biochemical genetics, see Robert Olby's informatively detailed *The Path to the Double Helix* (New York, 1974), and for molecular genetics, Horace Freeland Judson's compellingly vital *The Eighth Day of Creation: Makers of the Revolution in Biology* (New York, 1979). George W. Beadle, "Biochemical Genetics," *Chemical Reviews*, 37 (Aug. 1945), 15–96, is an excellent review of the field at its time of publication, and Beadle made clear the pioneering importance of Archibald E. Garrod in his "Genes and Chemical Reactions in Neurospora," Nobel Lecture, Dec. 11, 1958, *Nobel Lectures, Physiology or Medicine, 1942–1962* (Amsterdam, 1964), pp. 587–99. See also Garrod's *Inborn Errors of Metabolism* (2nd ed.; London, 1923) and the discerning studies: Alexander G. Bearn and Elizabeth D. Miller, "Archibald Garrod and the Development of the Concept of Inborn Errors of Metabolism," *Bulletin of the History of Medicine*, 53 (Fall 1979), 315–27; Barton Childs, "Garrod, Galton, and Clinical Medicine," *Yale Journal of Biology and Medicine*, 46 (1973), 297–313; and Barton Childs, "Sir Archibald Garrod's Conception of Chemical Individuality: A Modern Appreciation," *New England Journal of Medicine*, 28 (1970), 71–77. Haldane awarded recognition to research in human biochemical genetics in *The Biochemistry of Genetics* (London, 1954). The rapid development in the human area during the nineteen-fifties is evident from Harry Harris, *An Introduction to Human Biochemical Genetics* (London, 1953); Harris's *Human Biochemical Genetics* (London, 1959); and G. E. W. Wolstenholme and Cecilia M. O'Connor, eds., *CIBA Foundation Symposium, Jointly with the International Union of Biological Sciences, on Biochemistry of Human Genetics* (London, 1959). Excellent accounts of the blood diseases and disorders and their relationship to genetics are in Maxwell M. Wintrobe, ed., *Blood, Pure and Eloquent: A Story of Discovery, of People, and of Ideas* (New York, 1980). My discussion of the problem of the human chromosome number relies heavily on Malcolm Jay Kottler's outstanding study, "From 48 to 46: Cytological Technique, Preconception, and the Counting of Human Chromosomes," *Bulletin of the History of Medicine*, 48 (1974), 465–502. Of great value for post-1956 developments is T. C. Hsu, *Human and Mammalian Cytogenetics: An Historical Perspective* (New York, 1979). The revolutionary progress in human cytogenetics from 1956 to 1959 is exemplified in two volumes edited by D. Robertson Smith and William M. Davidson, *Symposium on Nuclear Sex* (London, 1958) and *Human Chromosomal Abnormalities* (London, 1961). A bibliography of Jérôme Lejeune's scientific works is available as *Titres et Travaux Scientifiques de Jérôme Lejeune* (Paris, 1972). The *XYY* controversy was set off by Patricia A. Jacobs et al., "Aggressive Behavior, Mental Sub-normality, and the *XYY* Male," *Nature*, 208 (1965), 1351–52, and the issues are reviewed in D. S. Borgaonkar and S. A. Shah, "The *XYY* Chromosome Male—Or Syndrome?" in *Progress in*

Medical Genetics, vol. 10, A. G. Steinberg and A. G. Bearn, eds. (New York, 1974), 135–222. See also the study by Herman Witkin et al., "Criminality in *XYY* and *XXY* Men," *Science*, 193 (Aug. 13, 1976), 547–55, and Patricia A. Jacobs, "The William Allan Memorial Award Address: Human Population Cytogenetics: The First Twenty-five Years," *American Journal of Human Genetics*, 34 (1982), 689–98.

Like human genetics in general, the special subject of medical genetics has yet to find a historian. The reform-eugenic concern with the uses of genetics in medicine is manifest in a number of writings: John A. Ryle, "Medicine and Eugenics," *Eugenics Review*, 30 (1938), 9–20; The Lord Horder, "Eugenics" and "Eugenics As I See It," *Eugenics Review*, 27 (1936), 277–84; 28 (1937), 265–72; C. P. Blacker, ed., *The Chances of Morbid Inheritance* (Baltimore, 1934); Charles B. Davenport, Madge Thurlow Macklin, et al., *Medical Genetics and Eugenics* (2 vols.; Philadelphia, 1940, 1943); H. J. Muller, C. C. Little, and Laurence H. Snyder, *Genetics, Medicine, and Man* (Ithaca, N.Y., 1947); F. A. E. Crew, *Genetics in Relation to Clinical Medicine* (Edinburgh, 1947); Laurence Snyder, "Genetics and Medicine," *Ohio State Medical Journal*, 29 (1933), 705–8, and his *Medical Genetics* (Durham, N.C., 1941). Useful on aspects of the history of medical genetics are C. Nash Herndon, "William Allan: An Appreciation," *American Journal of Human Genetics*, 14 (1962), 97–101; Laurence H. Snyder, *Blood Grouping in Relation to Clinical and Legal Medicine* (Baltimore, 1929); and John M. Opitz, "Historical Note: On the Role of Laurence H. Snyder in the Development of Human and Medical Genetics in the United States: An Oral History Interview," *American Journal of Medical Genetics*, 8 (1981), 447–68; Victor McKusick, "The Growth of Human Genetics as a Clinical Discipline"; Barton Childs, Carl A. Huether, and Edmond Murphy, "Human Genetics Teaching in U.S. Medical Schools," both in *American Journal of Human Genetics*, respectively, 27 (1975), 261–73; 33 (1981), 1–10. The evolution of basic medical genetics as such may be traced through the successive editions of John A. Fraser Roberts, *An Introduction to Medical Genetics* (1st ed.; Oxford, 1940).

The expansion in the possibilities of genetic screening is suggested by James V. Neel, "The Detection of the Genetic Carriers of Hereditary Disease," *American Journal of Human Genetics*, 1 (1949), 19–36, and Barton Childs, "The Prospects for Genetic Screening," *The Journal of Pediatrics*, 87 (Dec. 1975), 1125–32. For the screening issues that erupted in the seventies, see Daniel Bergsma et al., eds., *Ethical, Social, and Legal Dimensions of Screening for Human Genetic Disease* (New York, 1974); Philip Reilly, *Genetics, Law, and Social Policy* (Cambridge, Mass., 1977); Samuel P. Bessman and Judith P. Swazey, "PKU: A Study of Biomedical Legislation," in E. Mendelsohn, J. P. Swazey, and I. Taviss, eds., *Human Aspects of Biomedical Innovation* (Cambridge, Mass., 1971), pp. 49–76; Committee for the Study of Inborn Errors of Metabolism . . . National Research Council, *Genetic Screening: Programs, Principles, and Research* (Washington, D.C., 1975); and Marc Lappé, *Genetic Politics: The Limits of Biological Control* (New York, 1979).

The literature of genetic counseling and prenatal diagnosis is vast. Basic for the early history of these subjects are Sheldon C. Reed, "A Short History of Genetic Counseling," *Social Biology*, 21 (1974), 332–39; Sheldon C. Reed, *Counseling in Medical Genetics* (Philadelphia, 1955); and Helen G. Hammons, ed., *Heredity Counseling:*

A Symposium Sponsored by the American Eugenics Society . . . (New York, 1959). Valuable for developments since the nineteen-sixties are the many volumes published by the National Foundation–March of Dimes in its *Birth Defects: Original Article Series*—for example, Daniel Bergsma and Harold Abramson, eds., *Genetic Counseling* (Baltimore, 1970), and Daniel Bergsma and James V. Neel, eds., *Contemporary Genetic Counseling* (New York, 1973). Very helpful to me in understanding the role of the Foundation in the development of prenatal counseling and diagnosis was an interview with Arthur Salisbury in New York, May 1982. Useful perspectives on the field may also be gained from Herbert A. Lubs and Felix de la Cruz, eds., *Genetic Counseling: A Monograph of the National Institute of Child Health and Human Development* (New York, 1977); Harry Harris, *Prenatal Diagnosis and Selective Abortion* (Cambridge, Mass., 1975); Barton Childs, "Genetic Counseling: A Critical Review of the Published Literature," in Bernice H. Cohen, Abraham M. Lilienfeld, and P. C. Huang, eds., *Genetic Issues in Public Health and Medicine* (Springfield, Ill., 1978), pp. 329–57; Mitchell S. Golbus et al., "Prenatal Diagnosis in 3000 Amniocenteses," *New England Journal of Medicine*, 300 (Jan. 25, 1979), 157–63; Charles J. Epstein and Mitchell S. Golbus, "Prenatal Diagnosis of Genetic Diseases," *American Scientist*, 65 (1977), 703–11; Aubrey Milunsky, ed., *Genetic Disorders and the Fetus: Diagnosis, Prevention, Treatment* (New York, 1979); Ian H. Porter and Richard Skalko, eds., *Heredity and Society* (New York, 1973); Y. Edward Hsia, Kurt Hirschhorn, Ruth L. Silverberg, and Lynn Godmilow, *Counseling in Genetics* (New York, 1979); and President's Commission for the Study of Ethical Problems in Medicine and Biomedical and Behavioral Research, *Screening and Counseling for Genetic Conditions: The Ethical, Social, and Legal Implications of Genetic Screening, Counseling, and Educational Programs* (Washington, D.C., 1983). Special aspects of genetic counseling are dealt with in Seymour Kessler, ed., *Genetic Counseling: Psychological Dimensions* (New York, 1979); James R. Sorenson, Judith P. Swazey, and Norman A. Scotch, *Reproductive Pasts, Reproductive Futures: Genetic Counseling and Its Effectiveness* (New York, 1981); John C. Fletcher, *Coping with Genetic Disorders: A Guide for Clergy and Parents* (New York, 1981); and two volumes edited by Aubrey Milunsky and George J. Annas, *Genetics and the Law* (New York, 1976) and *Genetics and the Law II* (New York, 1980). Early popular books about genetics, medicine, and reproduction include Amram Scheinfeld, *You and Heredity* (New York, 1939) and *Your Heredity and Environment* (Philadelphia, 1965); more recent versions are C. O. Carter, *Human Heredity* (2nd ed.; London, 1977); Aubrey Milunsky, *Know Your Genes* (New York, 1977); and David Hendin and Joan Marks, *The Genetic Connection* (New York, 1979).

For information pertaining to the development and contemporary state of genetic counseling and prenatal diagnosis in Great Britain, see M. A. Ferguson-Smith et al., *The Provision of Services for the Prenatal Diagnosis of Fetal Abnormality in the United Kingdom: Report of the Clinical Genetics Society Working Party on Prenatal Diagnosis in Relation to Genetic Counseling* (London, 1978); J. S. Fitzsimmons et al., *The Provision of Regional Genetic Services in the United Kingdom: Report of the Clinical Genetics Society Working Party on Regional Genetic Services* (London, 1982); Paul E. Polani et al., "Sixteen Years' Experience of Counselling, Diagnosis,

and Prenatal Detection in One Genetic Centre: Progress, Results, and Problems,"
Journal of Medical Genetics, 16 (1979), 166–75; R. Harris et al., *Clinical Genetics
Society: Report of the Working Party on the Role and Training of Clinical Geneticists*
(London, 1983); John Burn, "Clinical Genetics," *British Medical Journal*, 283 (Oct.
8, 1983), 999–1000; Paul E. Polani et al., *Paediatric Research Unit, Guy's Hospital
Medical School: An Abridged Record of Research and Service* (London, 1981); C. H.
Rodeck and K. H. Nicolaides, eds., *Prenatal Diagnosis* (London, 1984); M. A.
Ferguson-Smith, ed., *Early Prenatal Diagnosis* (London, 1983); Eva Alberman and
K. J. Dennis, eds., *Late Abortions in England and Wales: Report of a National
Confidential Study, Royal College of Obstetricians and Gynaecologists* (London, 1984).

   Condensed versions of some of Hermann J. Muller's papers on genetic load
and germinal choice are conveniently collected in *Studies in Genetics: The Selected
Papers of H. J. Muller* (Bloomington, Ind., 1962). Papers of special importance are
"Our Load of Mutation," *American Journal of Human Genetics*, 2 (June 1950), 111–76;
"The Guidance of Human Evolution," *Perspectives in Biology and Medicine*, 3 (1959),
1–43; "Human Evolution by Voluntary Choice of Germ Plasm," *Science*, 134 (1961),
643–49; and "What Genetic Course Will Man Steer?" in *Proceedings of the Third
International Congress of Human Genetics*, James F. Crow and James V. Neel, eds.
(Baltimore, 1967), pp. 521–43. Essential for the story of germinal choice in the sixties
are the Hermann J. Muller Papers at the Lilly Library, University of Indiana. See
also Julian Huxley, *Essays of a Humanist* (New York, 1964), and Elof Axel Carlson,
"Eugenics Revisited: The Case for Germinal Choice," *Stadler Symposium*, 5 (1973),
13–34. Among the important responses to the eugenic implications that Muller drew
from genetic load are Bruce Wallace and Theodosius Dobzhansky, *Radiation,
Genes, and Man* (New York, 1959), and Theodosius Dobzhansky, *Heredity and the
Nature of Man* (New York, 1964), both of which stress the desirability of genetic
variation in human populations—a theme echoed in W. F. Bodmer and L. L.
Cavalli-Sforza, *Genetics, Evolution, and Man* (San Francisco, 1976), and in Richard
Lewontin, *Human Diversity* (New York, 1982).

   Among the numerous writings stimulated by Muller's proposal as well as by
the general discussions of genetic engineering and the possibilities of gene therapy
are the following published symposia: H. Hoagland and R. W. Burhoe, eds.,
Evolution and Man's Progress (New York, 1962); Gordon Wolstenholme, ed., *Man
and His Future* (Boston, 1963); Tracy M. Sonneborn, ed., *The Control of Human
Heredity and Evolution* (New York, 1965); John D. Roslansky, ed., *Genetics and the
Future of Man* (New York, 1966); Maureen H. Harris, ed., *Early Diagnosis of
Human Genetic Defects: Scientific and Ethical Considerations* (Washington, D.C.,
1972); Kenneth Vaux, ed., *Who Shall Live? Medicine, Technology, and Ethics* (Phila-
delphia, 1970); Watson Fuller, ed., *The Social Impact of Modern Biology* (London,
1970); Michael P. Hamilton, ed., *The New Genetics and the Future of Man*, (Grand
Rapids, Mich., 1972); Marc Lappé and Robert S. Morison, eds., *Ethical and Scientific
Issues Posed by Human Uses of Molecular Genetics* (New York, 1976). Additional
contributions to the mounting debate on human genetics and society are Peter B.
Medawar, *The Future of Man* (London, 1959); John Maynard Smith, "Eugenics and
Utopia," *Daedalus*, 94 (1965), 487–505; Joshua Lederberg's "Experimental Genetics

and Human Evolution," *American Naturalist*, 100 (Sept.–Oct. 1966), 519–31, and his later "Biological Innovation and Genetic Intervention," in John A. Behnke, ed., *Challenging Biological Problems: Directions Towards Their Solution* (New York, 1972), 7–27. The diverse perspectives of the nineteen seventies on human genetic engineering may be gained from H. Bentley Glass, "Science: Endless Horizons or Golden Age?" *Science*, 171 (1971), 23–29; Daniel Bergsma, ed., *Advances in Human Genetics and Their Impact on Society . . .* (Birth Defects: Original Article Series, vol. 8, No. 4, July 1972); Paul Ramsey, *Fabricated Man: The Ethics of Genetic Control* (New Haven, Conn., 1970); Amitai Etzioni, *Genetic Fix* (New York, 1973); Macfarlane Burnet, *Genes, Dreams, and Realities* (New York, 1971); Joseph Fletcher, *The Ethics of Genetic Control: Ending Reproductive Roulette* (New York, 1974); Laurence E. Karp, *Genetic Engineering: Threat or Promise?* (Chicago, 1976); Bruce Hilton, Daniel Callahan, et al., eds., *Ethical Issues in Human Genetics: Genetic Counseling and the Uses of Genetic Knowledge* (New York, 1973); Kurt Hirschhorn, "On Redoing Man," *Annals of the New York Academy of Sciences*, 184 (June 7, 1971), 103–12; Bernard D. Davis, "Ethical and Technical Aspects of Genetic Intervention," *New England Journal of Medicine*, 285 (Sept. 30, 1971), 799–801; A. G. Motulsky and W. Lenz, *Birth Defects* (Amsterdam, 1974); and T. Friedmann and R. Roblin, "Gene Therapy for Human Genetic Disease?" *Science*, 175 (1972), 949–55.

In *Between Science and Values* (New York, 1981), Loren R. Graham discerningly sets the issues of human genetic engineering and sociobiology in the context of his main subject. The extension of sociobiology to man began with the last chapter in Edward O. Wilson, *Sociobiology: The New Synthesis* (Cambridge, Mass., 1975), which he expanded into *On Human Nature* (Cambridge, Mass., 1978). Access to the large critical literature on human sociobiology may be obtained through a number of collections, notably Arthur L. Caplan, ed., *The Sociobiology Debate: Readings on Ethical and Scientific Issues* (New York, 1978), and Ashley Montagu, ed., *Sociobiology Examined* (Oxford, 1980). See also the sharp critique in Lewontin, Rose, and Kamin, *Not in Our Genes*. For the impact of recombinant DNA techniques upon the prospects of human genetic engineering, medical and otherwise, since the end of the seventies, see *Human Genetics: Possibilities and Realities* (Amsterdam, 1979); Kathleen McAuliffe and Sharon McAuliffe, "Keeping Up with the Genetic Revolution," *The New York Times Magazine*, Nov. 6, 1983, pp. 41–44, 92 ff.; President's Commission for the Study of Ethical Problems in Medicine and Biomedical and Behavioral Research, *Splicing Life: A Report on the Social and Ethical Issues of Genetic Engineering with Human Beings* (Washington, D.C., 1982); C. O. Carter, ed., *Developments in Human Reproduction and Their Eugenic, Ethical Implications* (London, 1983); Zsolt Harsanyi and Richard Hutton, *Genetic Prophecy: Beyond the Double Helix* (New York, 1981); Theodore Friedmann, *Gene Therapy: Fact and Fiction in Biology's New Approach to Disease* (Cold Spring Harbor, N.Y., 1983); Yvonne Baskin, *The Gene Doctors: Medical Genetics at the Frontier* (New York, 1984); and W. French Anderson, "Prospects for Human Gene Therapy," *Science*, 226 (Oct. 26, 1984), 401–9.

Acknowledgments

M ANY PEOPLE HELPED in a variety of ways to advance the writing of this book, and I take pleasure in thanking them here. For consenting to interviews, providing materials relevant to their careers, and, in some cases, commenting on draft chapters concerning them, I am grateful to Anthony C. Allison, Alexander Bearn, Martin Bobrow, Walter Bodmer, John Burn, Cedric O. Carter, L. L. Cavalli-Sforza, Barton Childs, Cyril Clark, James Crow, Bernard Davis, John Edwards, Charles J. Epstein, Malcolm Ferguson-Smith, Charles E. Ford, Park Gerald, Harry Harris, C. Nash Herndon, Kurt Hirschhorn, Shirley Penrose Hodgson, Patricia A. Jacobs, Hans Kalmus, Otto Klineberg, Sylvia Lawler, Joshua Lederberg, Jérôme Lejeune, Richard Lewontin, Victor McKusick, O. J. Miller, Aubrey Milunsky, Ursula Mittwoch, Richard and Peggy Brewer Musgrave, James V. Neel, Margaret Penrose Newman, Roger Penrose, Roland Penrose, Paul E. Polani, James Renwick, John A. Fraser Roberts, Leon Rosenberg, Frank Ruddle, Arthur Salisbury, Ruth Sanger, Harry Shapiro, Cedric A. B. Smith, John Maynard Smith, Laurence Snyder, Arthur Steinberg, and Edward O. Wilson.

For facilitating my consultation of the published and, especially, unpublished sources on which this book rests heavily, I greatly appreciate the help of S. E. Walters at the Eugenics Society; J. Percival and her assistants at the University College London Archives; William Prowting at the Medical Research Council; Leslie Hall and Julia Sheppard at the Wellcome Institute for the History of Medicine; Elizabeth Atchison at the John Innes Institute; Stephen Catlett at the American Philosophical Society Library; and the staffs at the Bancroft Library of the University of California in Berkeley, the British Library, the Archives of the Carnegie Institution of Washington, the Cambridge University Library, the Worshipful Company of Drapers, the Harvard University Library, the Huntington Library, the Kings College London Library, the Lilly Library at the University of Indiana, the Rice University Archives, the Library of the Royal College of Physicians, the Royal Statistical Society Library, and the Library of Yale University.

I am grateful for the advice, information, and, in a number of cases, copies of published or unpublished work kindly provided by Nancy Bekavac, Daniel Bell, Betty Booker, Richard Burian, William Bynum, Klara Carmely, Donna M. Cone, Hamilton Cravens, George C. Cunningham, David Edge, Raymond E. Fancher, Donald Fleming, Margaret Gowing, David Grether, Susan Grether, Jonathan Harwood, Howard Hiatt, Jerry Hirsch, Michael Holroyd, Barbara Kimmelman,

Clayton Koppes, Edward B. Lewis, Kenneth L. Ludmerer, Donald MacKenzie, Pauline Mazumdar, Seymour Packman, Diane Paul, George W. Pigman III, John D. Roberts, Robert Scott Root-Bernstein, Michael Schudson, Richard Allen Soloway, Oliver Smithies, Gillian Sutherland, Peter L. Tyor, and Leila Zenderland. For assistance in research on particular topics, I wish to thank Beth C. Kevles, Susan Kolden-Dimotakis, Janet Krober, Stephen Postema, Kenneth Schneyer, Robert Sorscher, and George Tolomiczenko; and for helpful comments on individual chapters in the book, Garland E. Allen, Brian Barry, John F. Benton, Louis Breger, Bruce E. Cain, Ruth Schwartz Cowan, Alan Donagan, W. T. Jones, Bernard Norton, Talbot Page, Theodore M. Porter, Robert A. Rosenstone, and James Woodward. I owe a considerable debt to Lyndsay Farrall, Harry Harris, Leroy Hood, Norman Horowitz, James V. Neel, Marilyn Nissenson, Robert Olby, and Martin Ridge for their valuable criticism of the complete manuscript or major parts of it. An important theme in this study derives from an insight of Hugh Nissenson's. Bettyann Kevles gave the manuscript discerning multiple readings, and the index of the book was prepared with characteristically exceptional care by Carol Pearson.

My work benefited immensely from the support of the California Institute of Technology, especially the Division of Humanities and Social Sciences. I was fortunate to be guided into the world of computer-assisted word-processing by Hardy C. Martel, and to be aided in the overall project by the able short-term secretarial support of Terry Atwell and Irene Baldon and by the excellent long-term services of Susan M. Cave and Lynne Schlinger. Susan Davis untangled a variety of administrative knots and Joy Hansen supplied important omnibus assistance. I owe special thanks to the staff in the Millikan Library at the Institute, including Janet Casebier, Judy Nollar, Kathleen Potter, and Belen Gelle, as well as Roderick J. Casper, Dana Roth, and Jeanne F. Tatro.

I am very glad to thank the public and private agencies that generously assisted my work. The beginning of the research was aided by a grant from the National Science Foundation, the completion of it by one from the Alfred P. Sloan Foundation, and a significant part of the writing was accomplished during fellowship years made possible by the National Endowment for the Humanities and the John Simon Guggenheim Memorial Foundation. I am grateful for the hospitality provided me during a brief, final stint of research by the Ciba Foundation in London, and during two earlier, extended ones by the History and Social Studies of Science Unit of the University of Sussex and the History and Sociology of Science Department at the University of Pennsylvania. I also wish to thank the Charles Warren Center of Harvard University for the privileges of a residential fellowship year and for making available while I was there the cheerful help of Patricia Denault and Barbara DeWolfe.

I am grateful to the following individuals and institutions for permission to quote from unpublished materials authored by the individuals or contained in the manuscript collections listed in parentheses after their names: Anthony Allison (himself); the American Philosophical Society Library (American Eugenics Society Papers, Charles B. Davenport Papers, Herbert Spencer Jennings Papers, Raymond

Pearl Papers); Grett Archer (Cyril Burt); Mary Bateson (William Bateson); John Blacker (C. P. Blacker); G. C. E. Crew (F. A. E. Crew); the Eugenics Society (Eugenics Society Papers); Laura Ruggles Gates (Reginald Ruggles Gates); Carel Goldschmidt (Smith Ely Jelliffe); Roger M. Greenwood (Major Greenwood); Norman Horowitz (himself); Lady [M. J.] Huxley (Julian Huxley); Naomi Mitchison (J. B. S. Haldane); Thea Muller (Hermann J. Muller); Peggy Brewer Musgrave (Herbert Brewer); Sara Pearson (Karl Pearson); Oliver and Margaret Penrose (Lionel S. Penrose); The Society of Authors on Behalf of the Estate of Bernard Shaw (Bernard Shaw); University College London (Francis Galton Papers and College Records); University of Adelaide (Ronald A. Fisher); Yale University Library (Walter Lippmann).

A substantial part of this book first appeared as a series in *The New Yorker*. I am deeply grateful to William Shawn, the editor of the magazine, for his sustained encouragement and to Sara Lippincott for her sharp and caring editorial judgment. Additional thanks go to Hal Espen, Nancy Franklin, and Lex Kaplen of the Checking Department at *The New Yorker* for saving me from a number of errors. I am greatly indebted to Ashbel Green, my editor at Alfred A. Knopf, for his friendship and patience over the years and for his wise advice concerning the shape and content of the book.

I wish to extend special thanks for a variety of favors and kindnesses to Daniel Aaron, Bruce Cain, David Herbert Donald, John A. Ferejohn, Eric F. Goldman, John Heilbron, Mark Kac, Jane Kramer, Roy and Kay MacLeod, Dorothy Nelkin, Hugh and Marilyn Nissenson, Roger Noll, Bernard Norton, Barbara Rosenkrantz, Martin and Susan Sherwin, John Sutherland, Stephan and Abigail Thernstrom, and Arnold Thackray. My debt is immeasurable to Bettyann, Beth, and Jonathan Kevles, who gamely provided good company—and much more—throughout this long adventure.

D.J.K.

Index

Index 417

upper class, 173; *see also* genes; genes, human
Gerrard, Thomas J., 119 and *n.*, 168
Gesell, Arnold E., 324
Glass, Bentley, 232
Glass, David, 350, 354
Goddard, Henry H., 80; and heritability of intelligence, 77, 84; and mental deficiency, 77–9, 80, 82, 92–3 and *n.*, 107, 148–9, 206, 325
goiter, heritability of, 56, 203
Goldman, Emma, 64, 90, 106
Goldstein, Sidney, 172
Gosney, Ezra, 114–15, 118
Gotto, Sybil, 316
Gould, Stephen Jay, 284–5
government: and human breeding, 12, 13, 20–1, 34, 47, 70, 90–5; and funds for scientific research, 54, 69, 220; *see also* civil liberties; eugenics: and role of government; federal government, U.S.; House of Representatives, U.S.; Parliament, Great Britain; state governments, U.S.
Graham, Robert K., 262–3 and *n.*, 299
Grant, Madison, 75 and *n.*, 133, 134–5, 252
Green, Howard, 266–7
Greenwood, Major, 337
Gruneberg, Hans, 359
Guthrie, Robert, 255, 278
Guyer, Michael F., 66–7, 75, 79, 94, 101

Haddon, A. C., 133, 347
hair color, genetics of, 44, 51, 196
Haldane, Charlotte Burghes, 333
Haldane, J. B. S., 176, 231, 334, 348, 352; and genetics, ix, 37, 69, 149, 177, 181–2, 195, 198–220 *passim*, 258; and eugenics, 122, 123, 127–8, 170, 172, 184–6; background, training, and career of, 122–3, 124, 126, 127, 202, 210–11, 214–15 and *n.*, 249, 333, 359; and his colleagues, 123, 125, 210; temperament of, 125, 214–15, 352; sociopolitics of, 126, 127, 215; publications of, 127–8, 198, 217; and mental deficiency, 131, 166; on racial differences, 133–4; and Penrose, 152, 155, 161, 162 *n.*, 213–15, 221; and artificial insemination, 167, 191, 262, 329, 377; and sterilization, 167, 329; and differential birth rate, 174 and *n.*, 179, 183; and biological innovation, 190, 298
Hale, Howard, 116
Hall, G. Stanley, 73, 77, 79

Hall, Prescott, 327
Hamerton, John, 242
Hardy, G. H., 195, 209
harelip, heritability of, 39, 200, 203
Harriman, Mrs. E. H., 54–5, 56
Harriman, Mary, 54
Harris, Harry, 295, 366; background, training, and research of, 217–18 and *n.*, 219–20, 221, 233, 236; and abortion, 286, 376
Harvard Educational Review, 269, 280
Hastings Center, 279
hemophilia, 202, 300; genetics of, 39, 46, 48, 127–8, 200, 204–5
Herbert, Solomon, 76
hereditarianism, pre-Mendelian, 71
heredity, statistical theory of, *see* Fisher, Ronald A.; Galton, Francis; Haldane, J. B. S.; Pearson, Karl
heredity, human, 90 and *n.;* and Galton, 3–5, 13–19, 16 *n.*, 18 *n.*, 30; and Pearson, 30–1, 36, 39; studies of, 38, 39–40; *see also* genetics, human; pedigrees, family
Heredity Clinic, University of Michigan, 225, 253
heritability, as technical term, 280–1
Heron, David, 74, 105 and *n.*, 149, 327
Herrnstein, Richard J., 270, 271 and *n.*
Hewitt, Ann Cooper, 345–6
Hiroshima, Japan, 224–5, 227–8 and *n.*
Hirschhorn, Kurt, 288
Hirszfeld, Ludwig, 204
Hitler, Adolf, 117–18, 164
Hobhouse, L. T., 119
Hogben, Enid Charles, 124, 125
Hogben, Lancelot, 167, 179, 197, 223, 231, 353; and eugenics, 122, 123, 127–8, 170, 172, 176; background and career of, 122–4, 200; sociopolitics of, 124–5, 126–7; temperament of, 125, 126; publications of, 127–8, 197; and hereditarian intelligence, 130, 131, 139, 140, 199, 281, 282, 348; and human genetic research, 194 and *n.*, 198, 199, 200–3, 212
Holmes, Justice Oliver Wendell, 111, 276
Holmes, Samuel J., 75, 101, 331
Holocaust, the, x, 251 and *n.*
Holzinger, Karl J., 140–1, 142
homosexuality, 66, 273–4
Hood, Lee, 380
Hooton, E. A., 114, 330
Hoover, Herbert, 114
Hopkins, Sir Frederick Gowland, 178, 216
Horder, Thomas, Lord, 115, 176
Horowitz, Norman H., 283 and *n.*